DELMAR PUBLISHERS INC. • ALBANY, NEW YORK 12205

Preface

Mathematics for Machine Technology is written to overcome the often mechanical "plug in" approach found in many trade-related mathematics textbooks. An understanding of mathematical concepts is stressed in all topics – which range from general arithmetic processes to oblique trigonometry and numerical control.

Both content and method are those used by the author in teaching applied machine technology mathematics classes for apprentices in the machine, tool and die, and tool design trades. Each unit is developed as a learning experience based on preceding units – making prerequisites unnecessary.

Presentation of basic concepts is accompanied by realistic industry-related examples and actual industrial applications. These applications progress from the simple to those with solutions which are relatively complex. Many problems require the student to work with illustrations such as are found in machine trade handbooks and engineering drawings.

An analytical approach to problem solving is emphasized in the geometry, trigonometry, and numerical control sections. This approach is necessary in actual practice in translating engineering drawing dimensions to machine working dimensions. Integration of algebraic and geometric principles with trigonometry by careful sequence and treatment of material also helps the student in solving industrial applications. The *Instructor's Guide* provides answers and solutions for all problems.

Robert D. Smith has experience in both the manufacturing industry and in education. He held positions as tool designer, quality control engineer, and chief manufacturing engineer prior to teaching. Mr. Smith taught mathematics, physics, and industrial materials and processes on the secondary school level before becoming Senior Instructor of Related Subjects at the A.I. Prince Vocational-Technical School in Hartford, Connecticut. where he taught applied mathematics. Mr. Smith is presently Assistant Professor of Vocational-Technical Education at Central Connecticut State College, New Britain, Connecticut. He has membership in several professional organizations in his field of interest: the American Technical Education Association and the Society of Manufacturing Engineers.

Elinor Gunnerson
Editor

The author and editorial staff at Delmar Publishers are interested in continually improving the quality of this instructional material. The reader is invited to submit constructive criticism and questions. Responses will be reviewed jointly by the author and source editor. Send comments to:

Director of Publications
Delmar Publishers Inc.
50 Wolf Road
Albany, New York 12205

Contents

SECTION I FRACTIONS AND DECIMALS

Unit		Page
1	Introduction to Fractions and Mixed Numbers	7
2	Addition of Fractions and Mixed Numbers	12
3	Subtraction of Fractions and Mixed Numbers	16
4	Multiplication of Fractions and Mixed Numbers	20
5	Division of Fractions and Mixed Numbers	24
6	Combined Operations of Fractions and Mixed Numbers	28
7	Introduction to Decimals	33
8	Changing Decimals and Fractions; Rounding-off Decimals	37
9	Addition and Subtraction of Decimal Fractions	42
10	Multiplication of Decimal Fractions	46
11	Division of Decimal Fractions	49
12	Powers	53
13	Roots	58
14	Table of Decimal Equivalents and Combined Operations of Decimals	64

SECTION II MEASUREMENT

Unit		Page
15	Degree of Precision, Tolerance and Clearance	69
16	Steel Rules and Gage Blocks	76
17	Vernier Instruments: Caliper and Height Gage	82
18	Micrometers	89
19	Metric System	95

SECTION III FUNDAMENTALS OF ALGEBRA

Unit		Page
20	Introduction to Symbolism	99
21	Signed Numbers	105
22	Algebraic Operations of Addition, Subtraction, and Multiplication	112
23	Algebraic Operations of Division, Powers, and Roots	117
24	Introduction to Equations	121
25	Solution of Equations by Subtraction, Addition, and Division	128

Unit		Page
26	Solution of Equations by Multiplication, Roots, and Powers	135
27	Solution of Equations Consisting of Combined Operations and Rearrangement of Formulas	141
28	Ratio and Proportion	146
29	Direct and Inverse Proportions	152
30	Applications of Formulas to Cutting Speeds, Rpm, and Cutting Time	157
31	Applications of Formulas to Spur Gears	163

SECTION IV FUNDAMENTALS OF PLANE GEOMETRY

32	Introduction to Geometric Figures	169
33	Protractors – Simple and Caliper	176
34	Angles	180
35	Triangles	185
36	Triangles and Other Common Polygons	190
37	Circles: Part I	198
38	Circles: Part II	204
39	Fundamental Geometric Constructions	210

SECTION V TRIGONOMETRY

40	Introduction to Right Angle Trigonometry	216
41	Analysis of Trigonometric Functions and the Cartesian Coordinate System	223
42	Basic Calculation of Angles and Sides of Right Triangles	228
43	Simple Practical Machine Applications	233
44	More Complex Practical Machine Applications	240
45	Oblique Triangles: Law of Sines and Law of Cosines	248

SECTION VI NUMERICAL CONTROL

46	Introduction to Numerical Control	254
47	Point-to-point Programming on Two-axis Machines	257
48	Binary Numeration System and Binary Coded Tape	264
	Appendix	270
	Index	277

SECTION I

Fractions and Decimals

Unit 1 Introduction to Fractions and Mixed Numbers

OBJECTIVES

After studying this unit the student should be able to

- Reduce fractions to lowest terms.
- Convert fractions to equivalent fractions.
- Change mixed numbers to improper fractions.
- Change improper fractions to mixed numbers.

Most measurements and calculations made by a machinist are not limited to whole numbers. Blueprint dimensions are often given as fractions and certain measuring tools are graduated in fractional units. The machinist must be able to make calculations using fractions and to measure fractional values.

FRACTIONAL PARTS

A fraction is a value which shows the number of equal parts taken of a whole quantity or unit. Figure 1-1 shows a line divided into 4 equal parts.

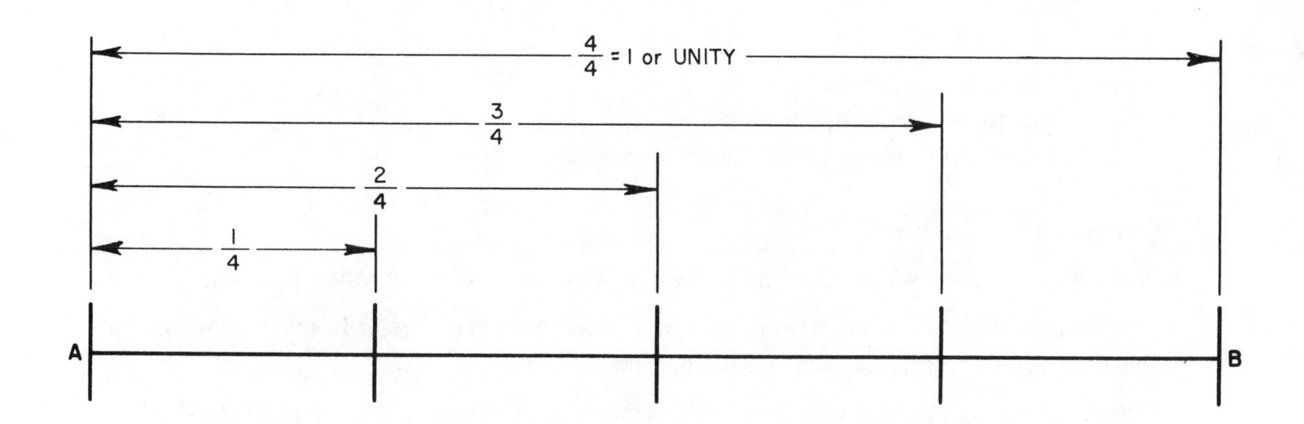

Fig. 1-1

1 part $= \frac{\text{1 part}}{\text{total parts}}$ or 1/4 of the length of the line

2 parts $= \frac{\text{2 parts}}{\text{total parts}}$ = 2/4; 3 parts $= \frac{\text{3 parts}}{\text{total parts}}$ = 3/4

4 parts $= \frac{\text{4 parts}}{\text{total parts}}$ = 4/4 Note: 4 parts make up the whole.
4/4 = 1, or unity

Figure 1-2 shows a line of the same length as that in figure 1-1. Each of the 4 equal parts is divided into 8 equal parts.

There is a total of 4 x 8 or 32 parts.

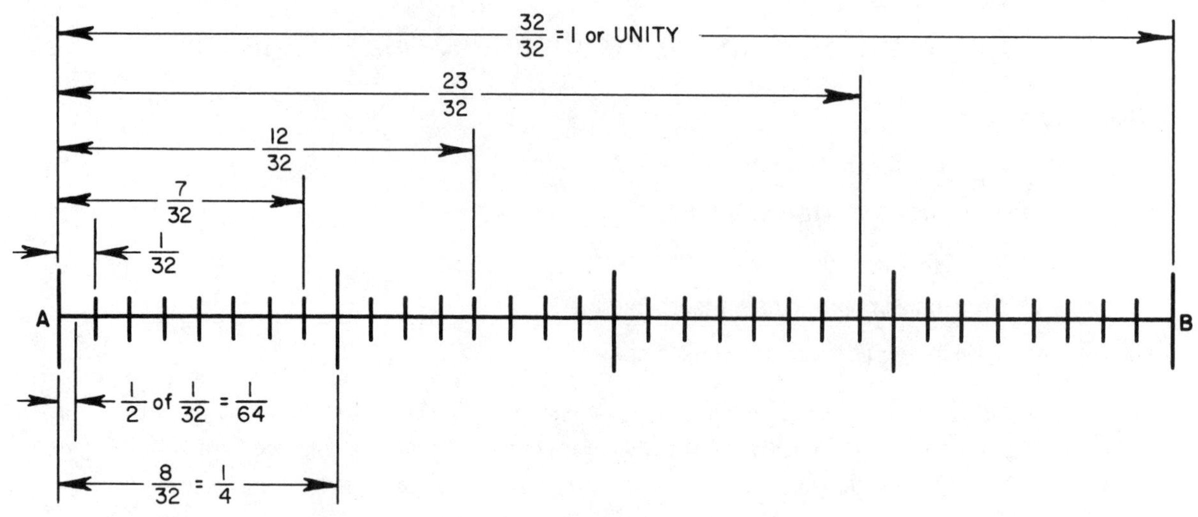

Fig. 1-2

1 part = 1/32 of the total length; 7 parts = 7/32.

12 parts = 12/32; 23 parts = 23/32; 32 parts = 32/32 = 1, or unity.

1/2 of 1 part = 1/2 of 1/32 = 1/64 of the total length.

1/4 of 1 part = 1/4 of 1/32 = 1/128.

Note: 8 parts = 8/32 of the total length.
8 parts = 1 of the 4 large parts.
1 large part = 1/4 of the total length. Therefore, 8/32 = 1/4.

DEFINITIONS OF FRACTIONS

- A *fraction* is a value which shows the number of equal parts taken of a whole quantity or unit.
- The *denominator* of a fraction is the number that shows into how many equal parts the whole or unit has been divided. The denominator is written below the line.
- The *numerator* of a fraction is the number that shows how many equal parts of the whole have been taken. The numerator is written above the line.
- An *improper* fraction is a fraction in which the numerator is larger than the denominator, as 3/2, 5/4, 15/8.
- A *mixed number* is a number composed of a whole number and a fraction, as 3 7/8, 7 1/2.

  Note: 3 7/8 means 3 + 7/8. It is read "three and seven-eighths." 7 1/2 means 7 + 1/2. It is read "seven and one-half."
- A *complex fraction* is a fraction in which one or both of the terms are fractions or mixed numbers as

$$\frac{3/4}{6}, \quad \frac{32}{15/4}, \quad \frac{8\ 3/4}{3}, \quad \frac{7/16}{2\ 2/5}, \quad \frac{4\ 1/4}{7\ 5/8}.$$

REDUCTION OF FRACTIONS

The value of a fraction is not changed by multiplying the numerator and denominator by the same quantity. The value of a fraction is not changed by dividing the numerator and denominator by the same number.

For example, $1/2 = \frac{1 \times 4}{2 \times 4} = 4/8$. Both the numerator and denominator are multiplied by 4. Because 1/2 and 4/8 have the same value, they are called *equivalent*. Also, $8/12 = \frac{8 \div 4}{12 \div 4} = 2/3$. Both numerator and denominator are divided by 4. Since 8/12 and 2/3 have the same value, they are *equivalent*.

A fraction is in its *lowest terms* when both the numerator and denominator cannot be divided by the same number without getting a remainder, as 5/9, 7/8, 3/4, 11/12, 15/32, 9/11.

Rule: To reduce a fraction to lowest terms

- Divide both numerator and denominator by the greatest possible value.

Example 1: Reduce 9/12 to lowest terms.

▲ Both terms can be divided by 3. $\frac{9 \div 3}{12 \div 3} = \frac{3}{4}$

Example 2: Reduce 12/42 to lowest terms.

▲ Both terms can be divided by 2. $\frac{12 \div 2}{42 \div 2} = \frac{6}{21}$

Note: The fraction is reduced, but not to lowest terms.

▲ Further reduce 6/21.

▲ Both terms can be divided by 3. $\frac{6 \div 3}{21 \div 3} = \frac{2}{7}$

Note: The value 2/7 may be obtained in one step if 12/42 is divided by 2 x 3 or 6. $\frac{12 \div 6}{42 \div 6} = \frac{2}{7}$

CHANGING MIXED NUMBERS AND IMPROPER FRACTIONS

Rule: To change a mixed number to an improper fraction

- Multiply the whole number by the denominator.
- Add the numerator to obtain the numerator of the improper fraction.
- The denominator is the same as that of the original fraction.

Example 1: Change 4 1/2 to an improper fraction.

▲ Multiply the whole number by the denominator.

▲ Add the numerator to obtain numerator for the improper fraction.

▲ The denominator is the same as that of the original fraction.

$\frac{4 \times 2 + 1}{2} = \frac{9}{2}$

Example 2: Change 12 3/16 to an improper fraction.

▲ $\frac{12 \times 16 + 3}{16} = \frac{195}{16}$

Rule: To change an improper fraction to a mixed number

- Divide the numerator by the denominator.

Examples: Change the improper fractions to mixed numbers.

▲ 11/4 = 11 ÷ 4 = 2 3/4 ▲ 43/3 = 43 ÷ 3 = 14 1/3 ▲ 931/8 = 931 ÷ 8 = 116 3/8

APPLICATION

A. FRACTIONAL PARTS

1. Write the fractional part which each length, A through F, represents of the total shown on the scale.

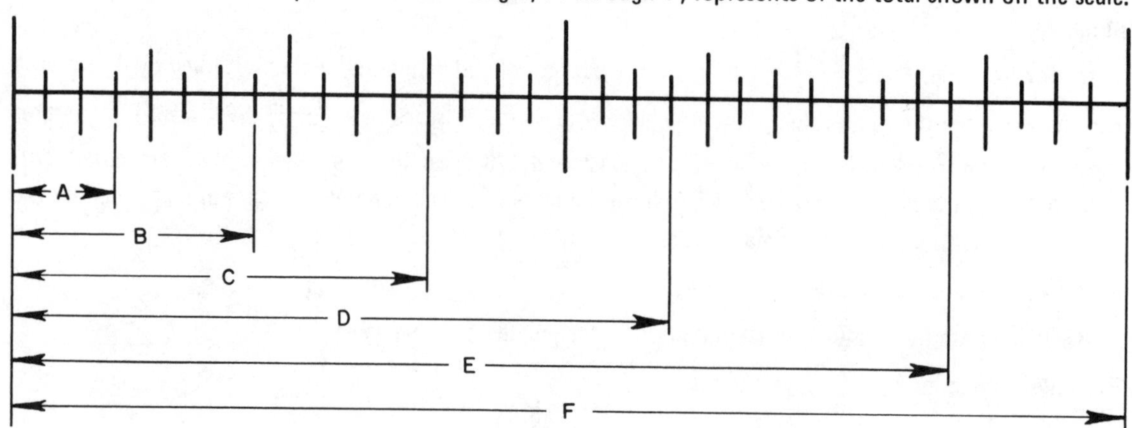

Fig. 1-3

2. The circle is divided into equal parts. Write the fractional part each of the following represents.

a. 1 part ____ f. 1/2 of 1 part ____

b. 3 parts ____ g. 1/3 of 1 part ____

c. 9 parts ____ h. 1/4 of 1 part ____

d. 5 parts ____ i. 1/10 of 1 part ____

e. 16 parts ____ j. 1/16 of 1 part ____

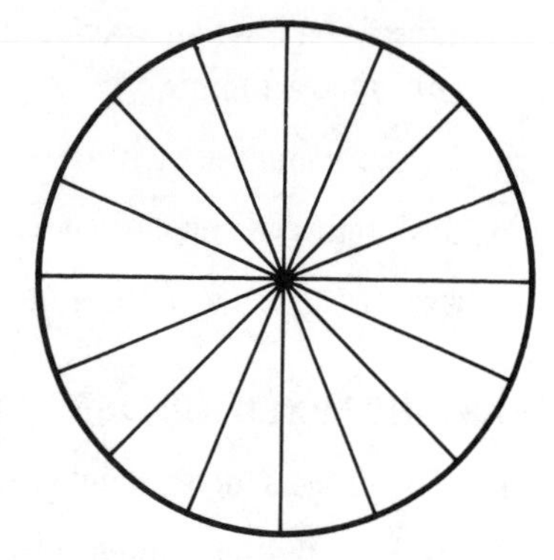

Fig. 1-4

B. REDUCTION OF FRACTIONS

1. Reduce to halves.

a. 4/8 ____ c. 100/200 ____ e. 15/10 ____ g. 54/36 ____

b. 9/18 ____ d. 121/242 ____ f. 18/12 ____ h. 125/50 ____

2. Reduce to lowest terms.

a. 6/8 ____ c. 6/10 ____ e. 11/44 ____ g. 27/6 ____ i. 25/150 ____

b. 12/4 ____ d. 35/5 ____ f. 14/6 ____ h. 65/15 ____ j. 8/128 ____

3. Change to thirty-seconds.

a. 1/4 ____ c. 9/8 ____ e. 21/16 ____ g. 197/16 ____

b. 3/4 ____ d. 7/16 ____ f. 17/2 ____ h. 33/8 ____

4. Change to equivalent fractions as indicated.

a. 3/4 = ?/8 ____ d. 17/14 = ?/42 ____ g. 7/16 = ?/128 ____

b. 5/12 = ?/36 ____ e. 22/9 = ?/45 ____ h. 13/8 = ?/48 ____

c. 6/15 = ?/60 ____ f. 14/3 = ?/18 ____ i. 19/16 = ?/160 ____

C. MIXED NUMBERS AND IMPROPER FRACTIONS

1. Change these mixed numbers to improper fractions.

a. 2 2/3 ____ d. 3 3/8 ____ g. 10 1/3 ____ j. 4 63/64 ____

b. 1 7/8 ____ e. 5 9/32 ____ h. 8 2/5 ____ k. 51 3/4 ____

c. 6 4/5 ____ f. 7 4/7 ____ i. 100 1/2 ____ l. 408 13/16 ____

2. Change these improper fractions to mixed numbers.

a. 10/3 ____ d. 87/4 ____ g. 127/32 ____ j. 235/16 ____

b. 21/2 ____ e. 63/18 ____ h. 43/13 ____ k. 514/4 ____

c. 9/8 ____ f. 127/124 ____ i. 150/9 ____ l. 337/64 ____

3. Change these mixed numbers to improper fractions. Then change the improper fractions to the equivalent fractions indicated.

a. 2 1/2 = ?/8 ____ d. 12 2/3 = ?/18 ____

b. 1 3/4 = ?/16 ____ e. 9 7/8 = ?/64 ____

c. 6 4/5 = ?/15 ____ f. 15 1/2 = ?/128 ____

4. Sketch and redimension this plate. Reduce all proper fractions to lowest terms. Reduce all improper fractions to lowest terms and change to mixed numbers.

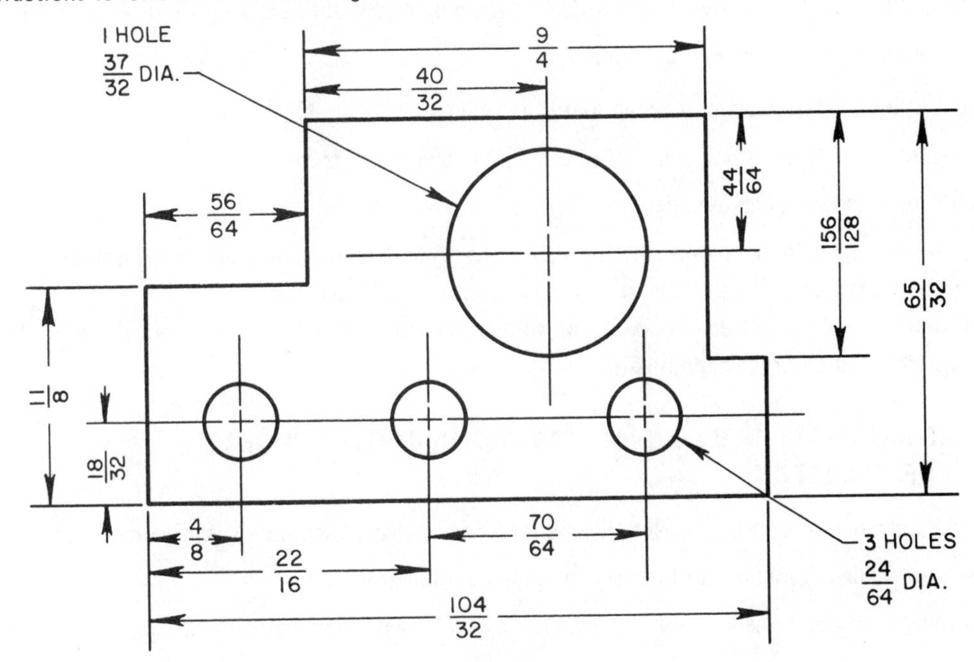

Fig. 1-5

Unit 2 Addition of Fractions and Mixed Numbers

OBJECTIVES

After studying this unit the student should be able to

- Determine least common denominators.
- Convert fractions to equivalent fractions having least common denominators.
- Add fractions and mixed numbers.

A machinist must be able to add fractions and mixed numbers in order to determine the length of stock required for a job, the distances between various parts of a machined piece, and the depth of holes and cutouts in a workpiece.

CHANGING DENOMINATORS TO LEAST COMMON DENOMINATORS

Fractions cannot be added unless they have a common denominator. *Common denominator* means that the denominators of each of the fractions are the same, as 5/8, 7/8, 15/8.

In order to add fractions which do not have common denominators, such as 3/8 + 1/4 + 7/16, it is necessary to determine the least common denominator.

The *least common denominator* is the smallest denominator into which each denominator can be divided without leaving a remainder.

Rule: To find the least common denominator

- Determine the smallest number into which all denominators can be divided without leaving a remainder.
- Use this number as a common denominator.

Example 1: Find the least common denominator of 3/8, 1/4, and 7/16.

- ▲ The smallest number into which 8, 4, and 16 can be divided without leaving a remainder is 16.
- ▲ Write 16 as the least common denominator.

Example 2: Find the least common denominator of 3/4, 1/3, 7/8, and 5/12.

- ▲ The smallest number into which 4, 3, 8, and 12 can be divided is 24.
- ▲ Write 24 as the least common denominator.

Note: In this example, denominators such as 48, 72, and 96 are common denominators because 4, 3, 8, and 12 divide into these numbers, but they are not the least common denominators.

Although any common denominator can be used when adding fractions, it is generally easier and faster to use the least common denominator.

CHANGING FRACTIONS INTO EQUIVALENT FRACTIONS WITH THE LEAST COMMON DENOMINATOR

Rule: To change fractions into equivalent fractions having the least common denominator

- Divide each denominator into the least common denominator.
- Multiply both the numerator and denominator of each fraction by the value obtained.

Example 1: Change 2/3, 7/15, and 1/2 to equivalent fractions having a least common denominator.

- ▲ The least common denominator is 30.
- ▲ Divide each denominator into 30. $30 \div 3 = 10; \quad 30 \div 15 = 2; \quad 30 \div 2 = 15$
- ▲ Multiply each term of the fraction by the value obtained. $\frac{2 \times 10}{3 \times 10} = \frac{20}{30}; \frac{7 \times 2}{15 \times 2} = \frac{14}{30}; \frac{1 \times 15}{2 \times 15} = \frac{15}{30}$

Example 2: Change 5/8, 15/32, 3/4, and 9/16 to equivalent fractions having a least common denominator.

- ▲ The least common denominator is 32.
- ▲ $32 \div 8 = 4; \ 32 \div 32 = 1; \quad 32 \div 4 = 8; \ 32 \div 16 = 2$
- ▲ $\frac{5 \times 4}{8 \times 4} = \frac{20}{32}; \quad \frac{15 \times 1}{32 \times 1} = \frac{15}{32}; \quad \frac{3 \times 8}{4 \times 8} = \frac{24}{32}; \quad \frac{9 \times 2}{16 \times 2} = \frac{18}{32}$

ADDING FRACTIONS

Rule: To add fractions

- ■ Change the fractions into equivalent fractions having the least common denominator.
- ■ Add the numerators and write their sum over the least common denominator.
- ■ Reduce the fraction to lowest terms.
- ■ Change an improper fraction to a mixed number and reduce the fractional part to lowest terms.

Example 1: Add 1/2 + 3/5 + 7/10 + 5/6

- ▲ Change 1/2, 3/5, 7/10, and 5/6 to equivalent fractions with 30 as the least common denominator.

  1/2 = 15/30; 3/5 = 18/30; 7/10 = 21/30; 5/6 = 25/30
- ▲ Add the numerators and write their sum over the least common denominator.

  15/30 + 18/30 + 21/30 + 25/30 = 79/30
- ▲ Change 79/30 to a mixed number.

  79/30 = 79 ÷ 30 = 2 19/30

Example 2: Determine the total length of the shaft in figure 2-1.

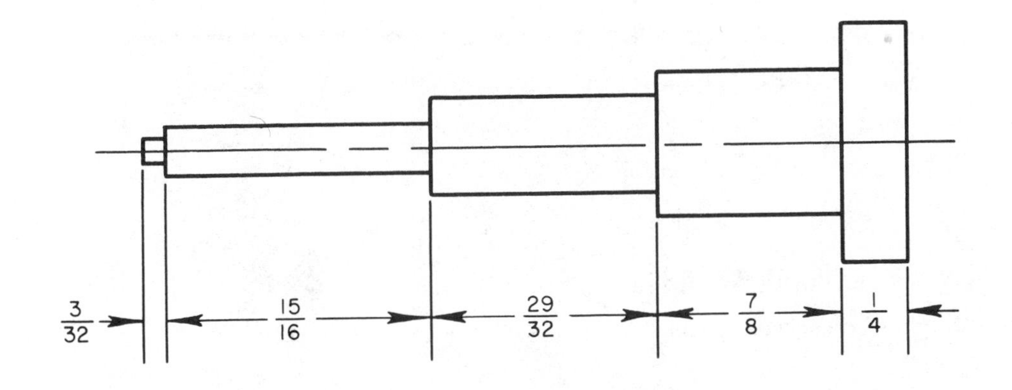

Fig. 2-1

- ▲ Change 3/32, 15/16, 29/32, 7/8, and 1/4 to equivalent fractions with 32 as the least common denominator.

  3/32 = 3/32, 15/16 = 30/32, 29/32 = 29/32, 7/8 = 28/32, 1/4 = 8/32

▲ Add the numerators and write their sum over the least common denominator.

3/32 + 30/32 + 29/32 + 28/32 + 8/32 = 98/32

▲ Change 98/32 to a mixed number and reduce.

98/32 = 3 2/32 3 2/32 = 3 1/16 total length

ADDING FRACTIONS, MIXED NUMBERS, AND WHOLE NUMBERS

Rule: To add fractions, mixed numbers, and whole numbers

- Add the whole numbers.
- Add the fractions separately.
- Combine.

Example 1: Add 1/3 + 7 + 3 1/2 + 5/12 + 2 19/24.

▲ Add the whole numbers. 7 + 3 + 2 = 12

▲ Add the fractions. 1/3 + 1/2 + 5/12 + 19/24 =

8/24 + 12/24 + 10/24 + 19/24 = 49/24 = 2 1/24

▲ Combine: 12 + 2 1/24 = 14 1/24

Example 2: Find the distance between the two 1/2-inch diameter holes in the plate shown in figure 2-2.

▲ 1 + 1 = 2

▲ 13/32 + 47/64 + 3/16 = 26/64 + 47/64 + 12/64 = 85/64

▲ 85/64 = 1 21/64

▲ 2 + 1 21/64 = 3 21/64

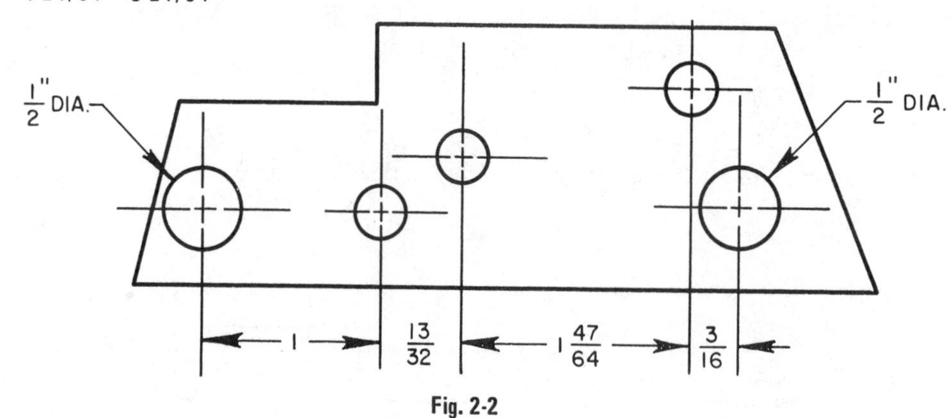

Fig. 2-2

APPLICATION

A. LEAST COMMON DENOMINATORS

Determine the least common denominators of the following sets of fractions.

1. 2/3, 1/6, 5/12 ____
2. 3/5, 9/10, 1/3 ____
3. 5/6, 7/12, 3/16, 19/24 ____
4. 4/5, 3/4, 7/10, 1/2 ____

B. EQUIVALENT FRACTIONS WITH LEAST COMMON DENOMINATORS

Change these fractions to equivalent fractions having the least common denominator.

1. 1/2, 2/3, 5/12 ____
2. 7/16, 3/8, 1/2 ____
3. 9/10, 1/4, 3/5, 1/2 ____
4. 3/16, 9/32, 17/64, 3/4 ____

C. ADDING FRACTIONS

1. Determine the dimensions A, B, C, D, E, and F of this profile gage.

   A = ____ C = ____ E = ____

   B = ____ D = ____ F = ____

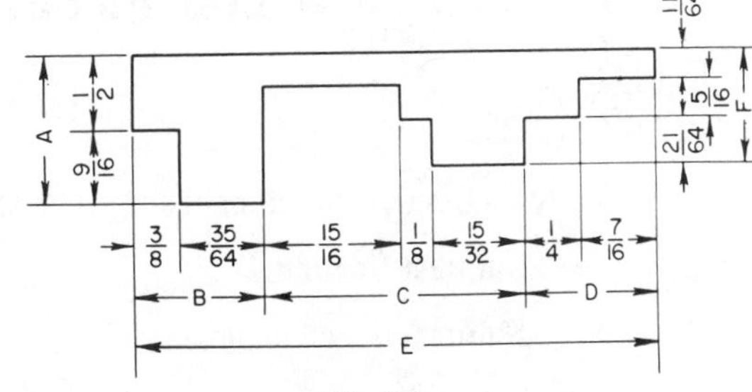

Fig. 2-3

2. Determine the length, width, and height of this casting.

   length = ____

   width = ____

   height = ____

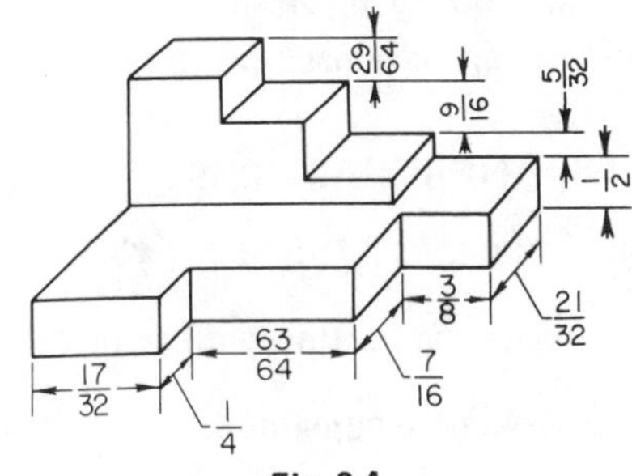

Fig. 2-4

D. ADDING FRACTIONS AND MIXED NUMBERS

1. Determine dimensions A, B, C, D, E, F, and G of this plate. Reduce where necessary.

   A = ____
   B = ____
   C = ____
   D = ____
   E = ____
   F = ____
   G = ____

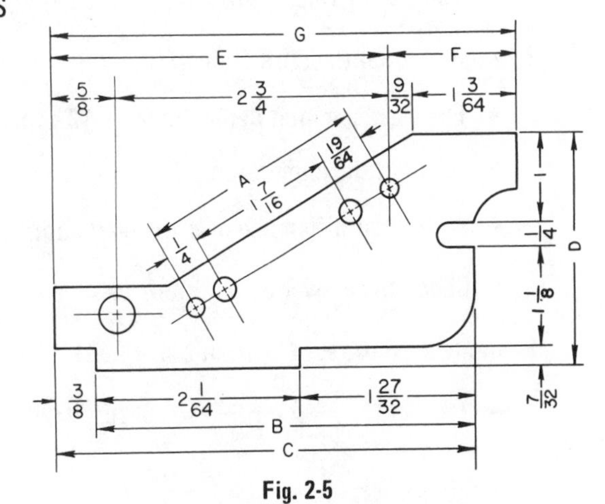

Fig. 2-5

2. Determine dimensions A, B, C, and D of this pin.

   A = ____
   B = ____
   C = ____
   D = ____

Fig. 2-6

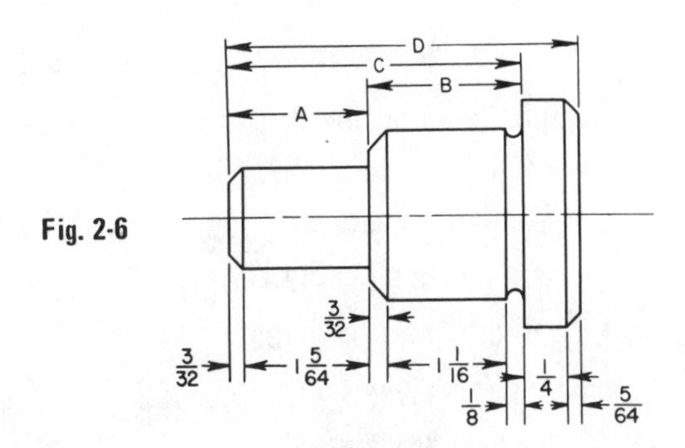

3. The operation sheet for machining an aluminum housing specifies 1 hour for facing, 2 3/4 hours for milling, 5/6 hour for drilling, 3/10 hour for tapping, and 2/5 hour for setting up. What is the total time allotted for this job?

Unit 3 Subtraction of Fractions and Mixed Numbers

OBJECTIVES

After studying this unit the student should be able to

- Subtract fractions.
- Subtract mixed numbers.

While making a part from a blueprint, a machinist often finds it necessary to convert blueprint dimensions into working dimensions. Subtraction of fractions and mixed numbers is required in order to properly position a part on a machine, to establish hole locations, and to determine depths of cut.

SUBTRACTING FRACTIONS

Rule: To subtract fractions

- Change the fractions into equivalent fractions having the least common denominator.
- Subtract the numerators.
- Write their difference over the least common denominator.
- Reduce the fraction to lowest terms.

Example 1: Subtract 3/8 from 9/16.

- ▲ The least common denominator is 16. Change 3/8 to 16ths. $\frac{3 \times 2}{8 \times 2} = 6/16$
- ▲ Subtract the numerators. $9/16 - 6/16 = 3/16$
- ▲ Write their difference over the least common denominator.
- ▲ Check the answer by addition. $3/16 + 6/16 = \frac{3 + 6}{16} = 9/16$

Example 2: Subtract 2/5 from 3/4

- ▲ $3/4 - 2/5 = \frac{3 \times 5}{4 \times 5} - \frac{2 \times 4}{5 \times 4} = 15/20 - 8/20$
- ▲ $\frac{15}{20} - \frac{8}{20} = \frac{7}{20}$
- ▲ Check the answer. $7/20 + 8/20 = 15/20$, $15/20 = 3/4$

Example 3: Find the distances x and y between the centers of the pairs of holes in the strap shown in figure 3-1.

- ▲ To find distance x:

  $7/8 - 11/32 = 28/32 - 11/32$

  $28/32 - 11/32 = 17/32$

- ▲ Check: $17/32 + 11/32 = 28/32$

  $28/32 = 7/8$

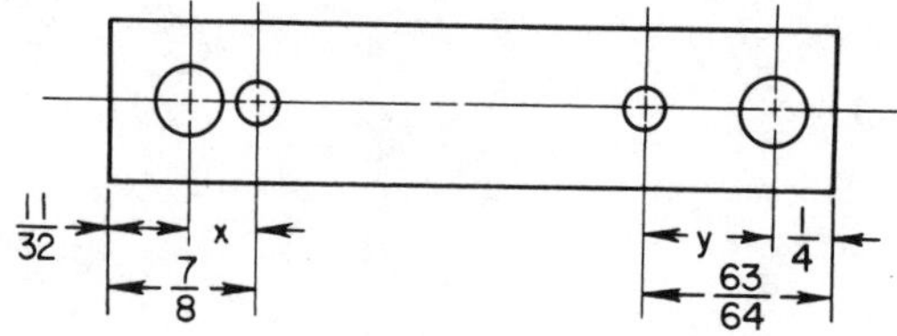

Fig. 3-1

- ▲ To find distance y: $63/64 - 1/4 = 63/64 - 16/64$; $63/64 - 16/64 = 47/64$
- ▲ Check: $47/64 + 16/64 = 63/64$

SUBTRACTING MIXED NUMBERS

Rule: To subtract mixed numbers,

- subtract the whole number
- subtract the fractions separately
- combine.

Example 1: Subtract 2 1/4 from 9 3/8.

- ▲ Subtract the whole numbers 9 - 2 = 7
- ▲ Subtract the fractions 3/8 - 1/4 = 3/8 - 2/8 = 1/8
- ▲ Combine: 7 + 1/8 = 7 1/8
- ▲ Check: 7 1/8 + 2 1/4 = 7 1/8 + 2 2/8 = 9 3/8

Example 2: Find the length of thread of the bolt shown in figure 3-2.

- ▲ 2 7/8 - 1 3/32 = (2 - 1) + (7/8 - 3/32)
- ▲ 2 - 1 = 1
- ▲ 7/8 - 3/32 = 28/32 - 3/32 = 25/32
- ▲ 1 + 25/32 = 1 25/32
- ▲ Check: 1 25/32 + 1 3/32 = 2 28/32 = 2 7/8

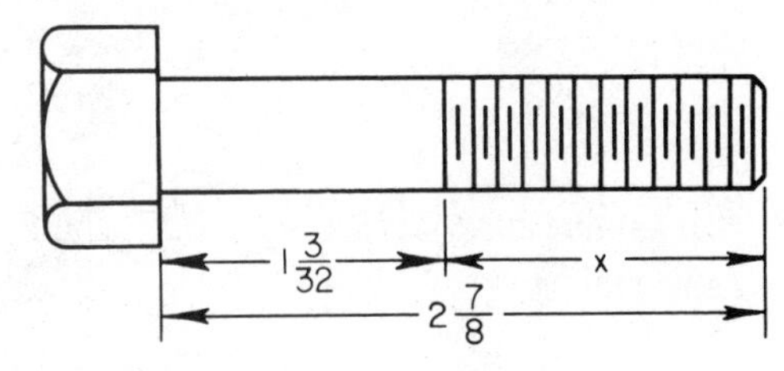

Fig. 3-2

Example 3: Subtract 7 15/16 from 12 5/8.

- ▲ 12 5/8 - 7 15/16 = 12 10/16 - 7 15/16 = 11 26/16 - 7 15/16 = 4 11/16

Note: Since 15/16 cannot be subtracted from 10/16, it is necessary to borrow one unit from the whole number 12. Change the borrowed unit to 16/16; add 16/16 + 10/16; (11 + 16/16 + 10/16) - 7 15/16.

- ▲ Check: 4 11/16 + 7 15/16 = 11 26/16 = 12 10/16 = 12 5/8

Example 4: Subtract 52 31/64 from 75.

- ▲ 75 - 52 31/64 = 74 64/64 - 52 31/64 = 22 33/64
- ▲ Check: 22 33/64 + 52 31/64 = 74 64/64 = 75

Example 5: Find dimension y of the counterbored block shown in figure 3-3.

- ▲ 2 3/8 - 29/32 = 2 12/32 - 29/32 =

  (1 + 32/32 + 12/32) - 29/32 =
  1 44/32 - 29/32 = 1 15/32

- ▲ Check: 1 15/32 + 29/32 =

  1 44/32 = 2 12/32 =
  2 3/8

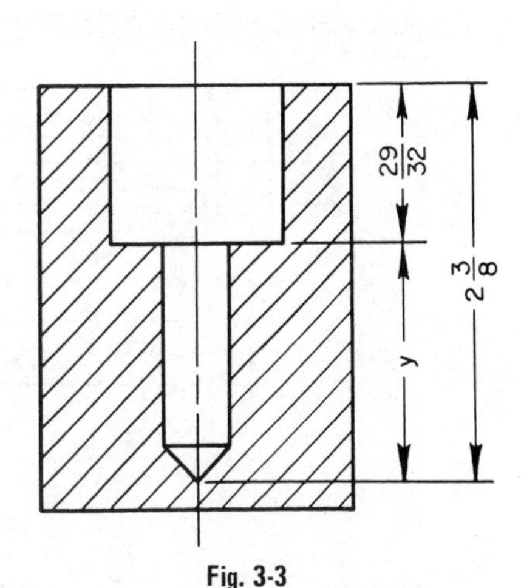

Fig. 3-3

APPLICATION

A. SUBTRACTING FRACTIONS

1. Subtract these fractions. Reduce to lowest terms where necessary and check.

a. 5/8 - 11/32 ____	c. 9/10 - 21/50 ____	e. 9/16 - 13/64 ____
b. 7/8 - 5/8 ____	d. 5/8 - 9/64 ____	f. 19/24 - 1/16 ____

2. Determine dimensions A, B, C, and D of this casting.

A = ____

B = ____

C = ____

D = ____

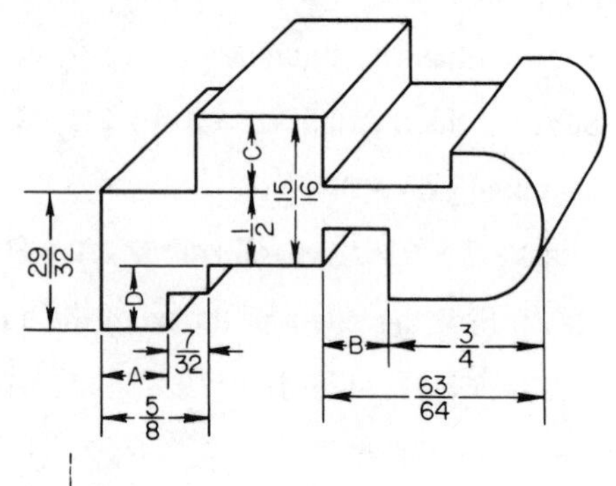

Fig. 3-4

3. Determine dimensions A, B, C, D, E, and F of this drill jig.

A = ____ D = ____

B = ____ E = ____

C = ____ F = ____

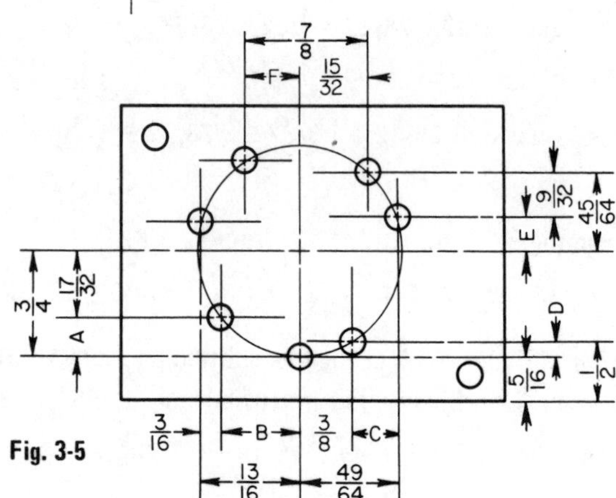

Fig. 3-5

B. SUBTRACTING MIXED NUMBERS

1. Determine dimensions A, B, C, D, E, F, and G of this tapered pin.

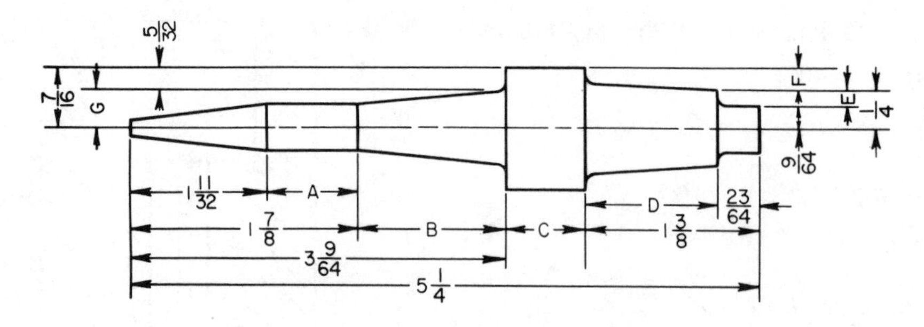

Fig. 3-6

A = ____, B = ____, C = ____, D = ____, E = ____, F = ____, G = ____

2. Determine dimensions A, B, C, D, E, F, G, H, and J of this plate.

A = ____ B = ____ C = ____ D = ____ E = ____ F = ____

G = ____ H = ____ J = ____

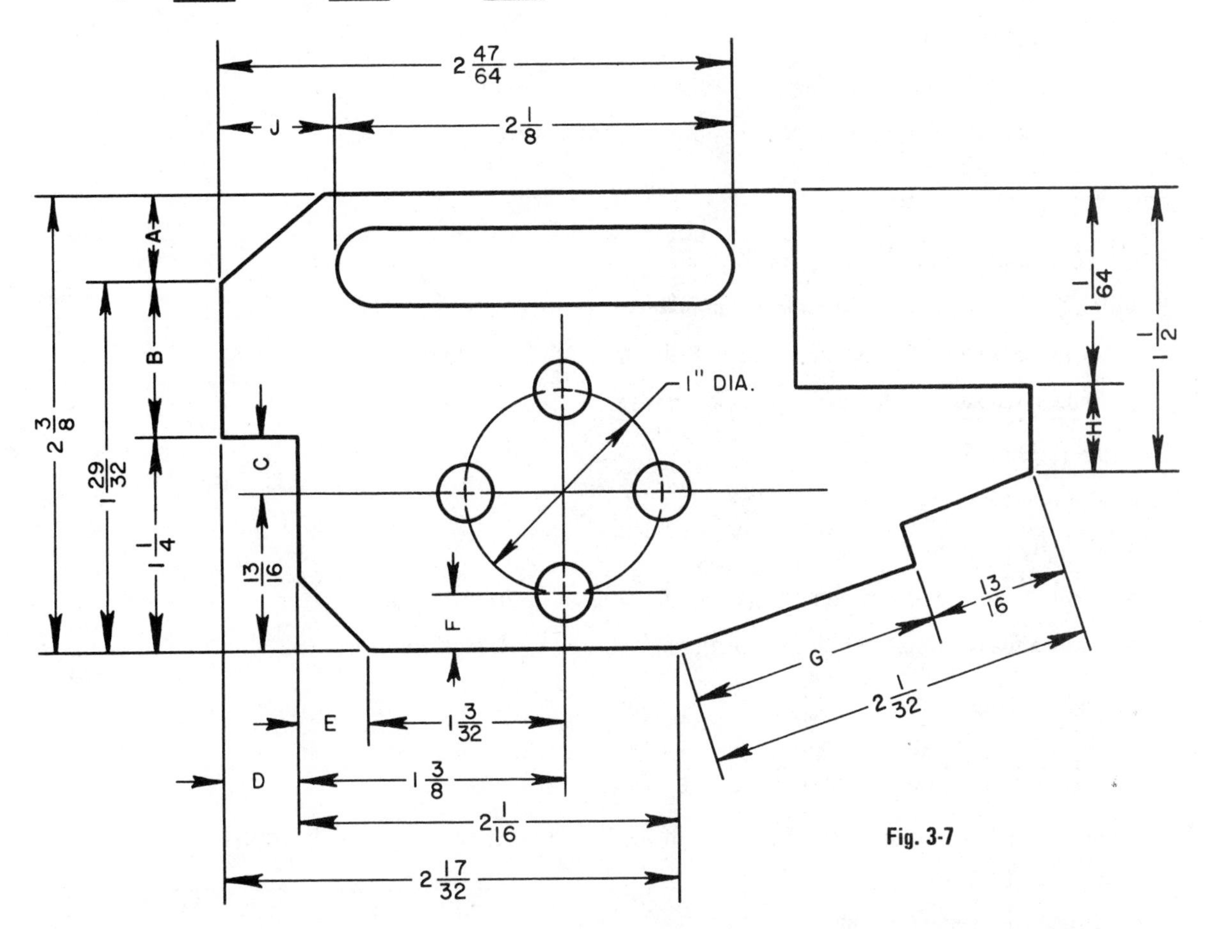

Fig. 3-7

3. Three holes are bored in a checking gage. The lower left edge of the gage is the reference point for the hole locations. Sketch the hole locations and determine the missing distances. From the reference point:

Hole #1 is 1 3/32" to the right, and 1 5/8" up.
Hole #2 is 2 1/64" to the right, and 2 3/16" up.
Hole #3 is 3 1/4" to the right, and 3 1/2" up.

Determine:

a. The horizontal distance between hole #1 and hole #2. ____
b. The horizontal distance between hole #2 and hole #3. ____
c. The horizontal distance between hole #1 and hole #3. ____
d. The vertical distance between hole #1 and hole #2. ____
e. The vertical distance between hole #2 and hole #3. ____
f. The vertical distance between hole #1 and hole #3. ____

Unit 4 Multiplication of Fractions and Mixed Numbers

OBJECTIVES

After studying this unit the student should be able to

- Multiply fractions.
- Multiply mixed numbers.
- Reduce fractions and mixed numbers by cancellation.

MULTIPLYING FRACTIONS

Rule: To multiply two or more fractions

- Multiply the numerators and the denominators separately.
- Write the product of the numerators over the product of the denominators.
- Reduce the resulting fraction to lowest terms.

Example 1: Multiply 3/4 by 8/9.

- ▲ Multiply the numerators. $\frac{3 \times 8}{4 \times 9} = 24/36$
- ▲ Multiply the denominators.
- ▲ Write the product of the numerators over the product of the denominators.
- ▲ Reduce the resulting fraction to lowest terms. $24/36 = \frac{24 \div 12}{36 \div 12} = 2/3$

Example 2: Multiply $\frac{2}{3} \times \frac{5}{6} \times \frac{3}{10}$

- ▲ $\frac{2 \times 5 \times 3}{3 \times 6 \times 10} = \frac{30}{180}$; $\frac{30 \div 30}{180 \div 30} = \frac{1}{6}$

Example 3: Find the distance between centers of the first and last holes shown in figure 4-1.

- ▲ Multiply 6 x 7/16.

  Note: The value of a number remains unchanged when the number is placed over a denominator of 1.

- ▲ Multiply $6 \times \frac{7}{16} = \frac{6}{1} \times \frac{7}{16} =$

  $\frac{6 \times 7}{1 \times 16} = \frac{42}{16}$

- ▲ Reduce $\frac{42}{16} = 2\frac{10}{16}$; $2\frac{10}{16} = 2\frac{5}{8}$

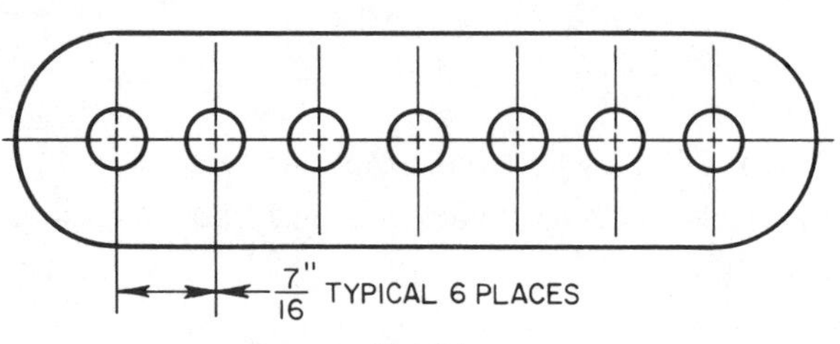

Fig. 4-1

CANCELLATION

Cancellation is a method of reducing combinations of fractions to lowest terms before the fractions are multiplied. Generally, problems involving multiplication of fractions are solved more quickly and easily by applying the principle of cancellation.

Cancellation consists of dividing fractions by a factor which is common both numerator and denominator. Factors are whole numbers which, when multiplied, produce a given number. For example, 2 and 3 are factors of 6.

Rule: To reduce fractions by cancellation

- Divide fractions by a factor common to both a numerator and a denominator.

Example 1: Multiply by cancellation method $\frac{3}{4} \times \frac{8}{9}$

▲ Divide by 3 which is the factor common to both the numerator 3 and the denominator 9.
$3 \div 3 = 1; \quad 9 \div 3 = 3$

▲ Divide by 4 which is the factor common to both the denominator 4 and the numerator 8.
$4 \div 4 = 1; \quad 8 \div 4 = 2.$

▲ Write in the form $\frac{3}{4} \times \frac{8}{9} = \frac{\overset{1}{\cancel{3}}}{\underset{1}{\cancel{4}}} \times \frac{\overset{2}{\cancel{8}}}{\underset{3}{\cancel{9}}}$

▲ Multiply reduced fractions. $\frac{1 \times 2}{1 \times 3} = \frac{2}{3}$

Example 2: Multiply $\frac{4}{7} \times \frac{5}{18} \times \frac{14}{15}$

▲ Cancel 2 into 4 and 18

▲ Cancel 7 into 7 and 14

▲ Cancel 5 into 5 and 15

$\frac{\overset{2}{\cancel{4}}}{\underset{1}{\cancel{7}}} \times \frac{\overset{1}{\cancel{5}}}{\underset{9}{\cancel{18}}} \times \frac{\overset{2}{\cancel{14}}}{\underset{3}{\cancel{15}}} =$

▲ Multiply: $\frac{2 \times 1 \times 2}{1 \times 9 \times 3} = \frac{4}{27}$

Example 3: Multiply $\frac{5}{14} \times \frac{8}{9} \times \frac{7}{10}$

▲ Cancel 5 into 5 and 10.

▲ Cancel 2 into 14 and 8.

▲ Cancellation is continued by cancelling 7 into 7 and 7, and cancelling 2 into 4 and 2.

$\frac{\overset{1}{\cancel{5}}}{\underset{7}{\cancel{14}}} \times \frac{\overset{4}{\cancel{8}}}{9} \times \frac{7}{\underset{2}{\cancel{10}}} = \frac{\overset{1}{\cancel{5}}}{\underset{\underset{1}{\cancel{7}}}{\cancel{14}}} \times \frac{\overset{\overset{2}{\cancel{4}}}{\cancel{8}}}{9} \times \frac{\overset{1}{\cancel{7}}}{\underset{\underset{1}{\cancel{2}}}{\cancel{10}}}$

▲ Multiply: $\frac{1 \times 2 \times 1}{1 \times 9 \times 1} = \frac{2}{9}$

Rule: To multiply mixed numbers

- Change the mixed numbers to improper fractions.
- Follow the procedure for multiplying proper fractions.

Example 1: Multiply: $2\frac{2}{5} \times 6\frac{7}{8}$

▲ Change $2\frac{2}{5}$ and $6\frac{7}{8}$ to improper fractions. $2\frac{2}{5} \times 6\frac{7}{8} = \frac{12}{5} \times \frac{55}{8}$

▲ Cancel 5 into 5 and 55.

▲ Cancel 4 into 12 and 8.

$\frac{\overset{3}{\cancel{12}}}{\underset{1}{\cancel{5}}} \times \frac{\overset{11}{\cancel{55}}}{\underset{2}{\cancel{8}}}$

▲ Multiply and change the product to a mixed number. $\frac{3 \times 11}{1 \times 2} = \frac{33}{2} = 33 \div 2 = 16\frac{1}{2}$

Example 2: The block of steel shown in figure 4-2 is to be machined. The block measures 8 3/4 inches long, 4 9/16 inches wide, and 7/8 inch thick. Find the volume of the block. (Volume = length x width x thickness.)

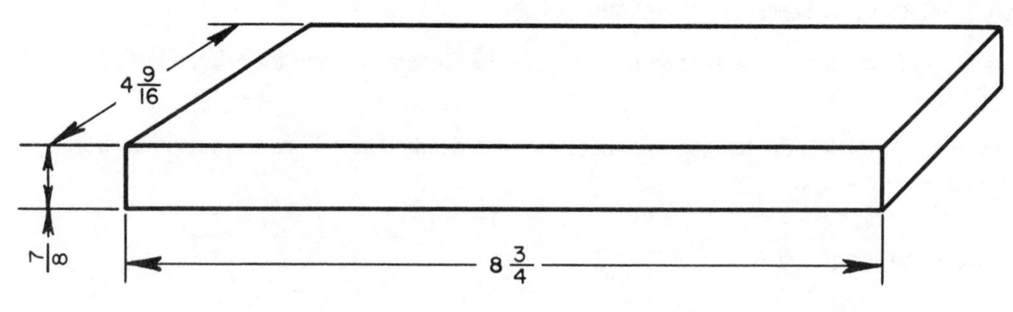

Fig. 4-2

▲ Volume $= 8\frac{3}{4} \times 4\frac{9}{16} \times \frac{7}{8} =$

▲ $\frac{35}{4} \times \frac{73}{16} \times \frac{7}{8} = \frac{35 \times 73 \times 7}{4 \times 16 \times 8} = \frac{17885}{512} = 17885 \div 512 = 34\frac{477}{512}$ cubic inches

APPLICATION

A. MULTIPLYING FRACTIONS

1. Multiply these fractions. Reduce to lowest terms where necessary.

a. $\frac{2}{3} \times \frac{1}{6}$ ____

b. $\frac{1}{2} \times \frac{1}{8}$ ____

c. $\frac{5}{8} \times \frac{13}{64}$ ____

d. $\frac{3}{4} \times \frac{3}{5} \times \frac{2}{3}$ ____

e. $7 \times \frac{9}{14} \times 3$ ____

f. $\frac{8}{25} \times \frac{5}{8} \times \frac{3}{7}$ ____

2. Determine dimensions A, B, C, D, and E of the template in figure 4-3.

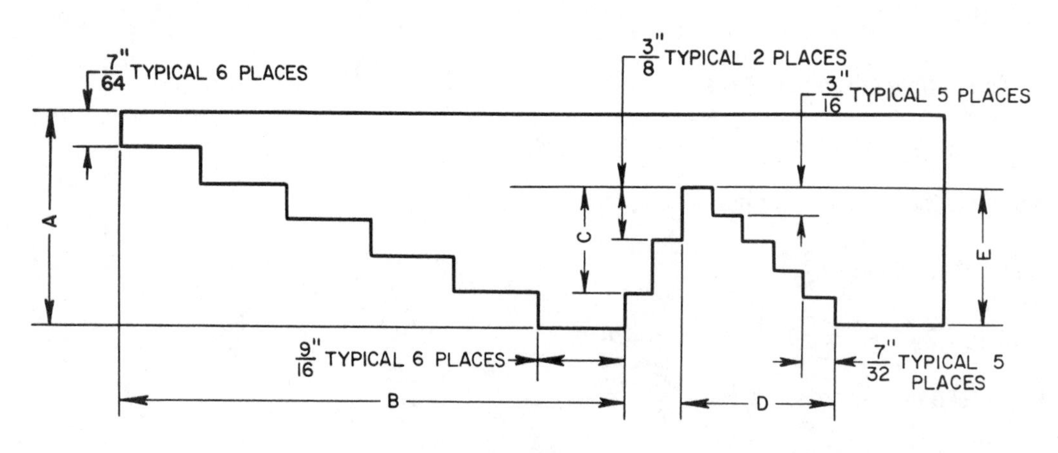

Fig. 4-3

A = ____ B = ____ C = ____ D = ____ E = ____

3. Determine the distance across flats and the washer thickness of this special washer-faced nut.

   a. Distance across flats = ____________

   b. Washer thickness = ____________

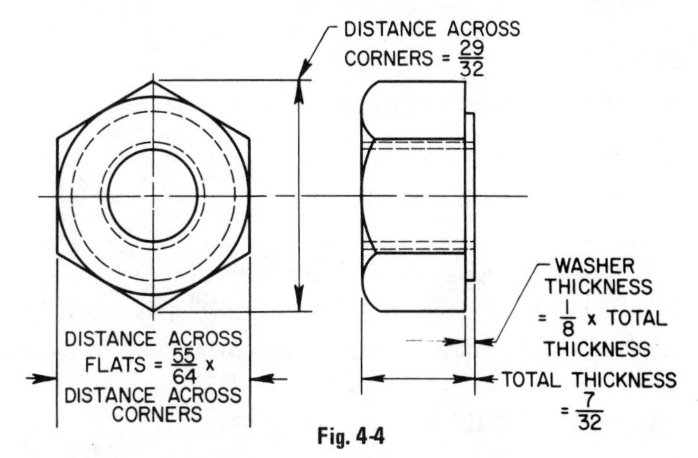

Fig. 4-4

4. An American Standard thread is a sharp V-thread with flat tops and bottoms. Pitch is the distance between two threads. The length of the flat A equals 1/8 of the pitch. The distance B between the top and bottom flats is equal to 3/4 of the depth. Determine the following dimensions.

   a. Depth = $\frac{5''}{16}$

      Distance B = ____

   b. Depth = $\frac{3''}{8}$

      Distance B = ____

   c. Depth = $\frac{15''}{64}$

      Distance B = ____

   d. Depth = $\frac{1''}{2}$

      Distance B = ____

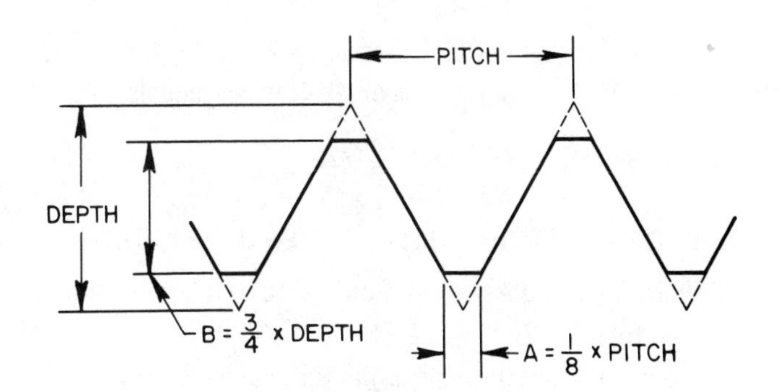

Fig. 4-5

   e. Pitch = $\frac{3''}{4}$, Flat A = ____

   f. Pitch = $\frac{17''}{32}$, Flat A = ____

   g. Pitch = $\frac{3''}{16}$, Flat A = ____

   h. Pitch = $\frac{1''}{4}$, Flat A = ____

B. MULTIPLYING MIXED NUMBERS

1. Multiply these mixed numbers. Reduce to lowest terms where necessary.

   a. 1 2/3 x 6 3/10 ____

   b. 3 5/16 x 7 3/4 ____

   c. 3 7/8 x 2 1/2 ____

   d. 1 2/3 x 10 1/4 x 3/8 ____

   e. 1 7/32 x 2 x 1/10 ____

   f. 2 2/3 x 2 2/3 x 5 1/4 ____

2. How many inches of drill rod are required in order to make 20 drills each 3 3/16" long? Allow 3/32 waste for each drill. ____________

Unit 5 Division of Fractions and Mixed Numbers

OBJECTIVES

After studying this unit the student should be able to

- Divide fractions
- Divide mixed numbers

In machine technology, division of fractions and mixed numbers is used in determining production times and costs per machined unit, in calculating the pitch of screw threads, and in computing the number of parts that can be manufactured from a given amount of raw material.

DIVIDING FRACTIONS

A fraction expresses the process of division. For example, the fraction 5/2 means that 5 is divided by 2. The fraction 5/2 is also written as 5 ÷ 2. Both forms ask the question, "How many 2s are there in 5?" There are 2 1/2. Therefore, 5/2 = 5 ÷ 2 = 2 1/2.

The fraction $\frac{3/4}{1/2}$ means that 3/4 is divided by 1/2. The fraction $\frac{3/4}{1/2}$ is also written 3/4 ÷ 1/2. It asks how many 1/2s are there in 3/4? Because 3/4 = 1/2 + 1/4 and 1/4 = 1/2 of 1/2, 3/4 equals one 1/2 plus one half of a second 1/2. One 1/2 plus one-half of 1/2 equals one and one-half 1/2s.

Therefore, $\frac{3/4}{1/2}$ = 3/4 ÷ 1/2 = 1 1/2.

DIVISION OF FRACTIONS AS THE INVERSE OF MULTIPLICATION OF FRACTIONS

Dividing by 2 is the same as multiplying by 1/2. For example, 5/2 = 5 ÷ 2 = 2 1/2 and 5 x 1/2 = 5/2 = 2 1/2. Therefore, 5 ÷ 2 = 5 x 1/2.

Dividing by 1/2 is the same as multiplying by 2. For example, $\frac{3/4}{1/2}$ = 3/4 ÷ 1/2 = 1 1/2 and 3/4 x 2/1 = 6/4 = 1 1/2. Therefore, 3/4 ÷ 1/2 = 3/4 x 2/1.

Two is the *inverse* of 1/2, and 1/2 is the *inverse* of 2.

Inverting a fraction means turning the fraction "upside down." For example, 1/3 inverted is 3/1, 8/7 inverted is 7/8, 63/64 inverted is 64/63; 9/16 inverted is 16/9.

Rule: To divide fractions

- Invert the divisor.
- Follow the procedure for multiplying fractions.

Example 1: Divide 5/8 by 3/4

- ▲ Invert the divisor.
- ▲ Change the division operation to a multiplication operation. 5/8 ÷ 3/4 = 5/8 x 4/3
- ▲ Follow the procedure for multiplication. $5/8 \times 4/3 = 5/\underset{2}{\cancel{8}} \times \overset{1}{\cancel{4}}/3 = 5/6$
- ▲ Check $5/6 \times 3/4 = 5/\overset{2}{\cancel{6}} \times \overset{1}{\cancel{3}}/4 = 15/24 = 5/8$

Example 2: The machine bolt shown in figure 5-1 has a pitch of 1/16 inch. Find the number of threads in 7/8 inch.

▲ Divide $7/8 \div 1/16$

▲ Invert the divisor:

$7/8 \div 1/16 = 7/8 \times 16/1$

▲ Multiply: $7/\not{8}_1 \times \not{16}^2/1 = 14$

▲ Check: $14 \times 1/16 = 14/16 = 7/8$

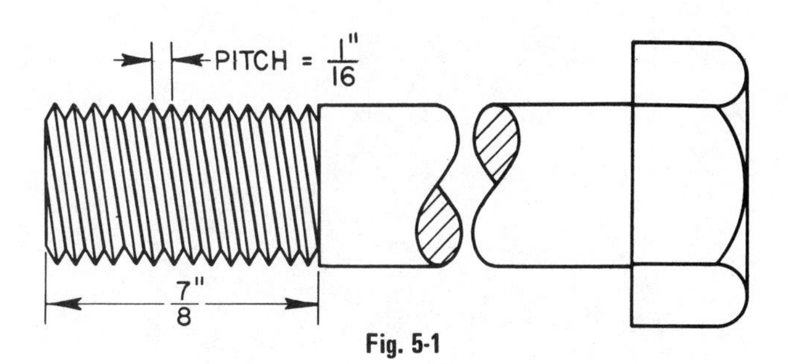

Fig. 5-1

DIVISION OF MIXED NUMBERS

Rule: To divide mixed numbers

- Change the mixed numbers to improper fractions.
- Follow the procedure for dividing fractions.

Example 1: Divide 7 1/2 by 2 3/8

▲ Change 7 1/2 and 2 3/8 to improper fractions: $7\ 1/2 \div 2\ 3/8 = 15/2 \div 19/8$

▲ Invert the divisor.

▲ Change the division operation to a multiplication operation: $15/2 \div 19/8 = 15/2 \times 8/19$

▲ Multiply: $\frac{15}{\not{2}_1} \times \frac{\not{8}^4}{19} = \frac{60}{19}; \frac{60}{19} = 3\frac{3}{19}$

▲ Check: $3\frac{3}{19} \times 2\frac{3}{8} = \frac{\not{60}^{15}}{\not{19}_1} \times \frac{\not{19}^1}{\not{8}_2} = \frac{15}{2} = 7\ 1/2$

Example 2: Figure 5-2 shows a section of strip stock with 5 equally spaced holes. Determine the distance between two consecutive holes.

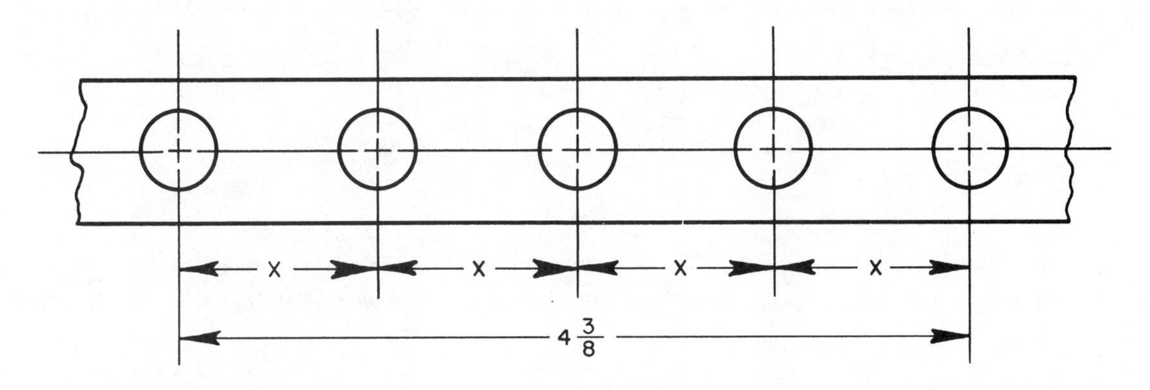

Fig. 5-2

Note: The number of spaces between the holes is one less than the number of holes.

▲ Change to improper fractions: $4\ 3/8 \div 4 = 35/8 \div 4/1$

▲ Invert the divisor and multiply: $35/8 \times 1/4 = 35/32$; $35/32 = 1\ 3/32$

▲ Check: $1\ 3/32 \times 4 = 35/32 \times 4/1 = 140/32 = 35/8 = 4\ 3/8$

APPLICATION

A. INVERTING FRACTIONS

Invert these fractions.

1. 7/8 ____ 2. 1/4 ____ 3. 97/8 ____ 4. 2 ____

B. DIVIDING FRACTIONS

1. The casting in figure 5-3 shows seven tapped holes. The number of threads is determined by dividing the depth of thread by the thread pitch. Find the number of threads in

   a. Hole A ____ b. Hole B ____ c. Hole C ____ d. Hole D ____ e. Hole E ____
   f. Hole F ____ g. Hole G ____

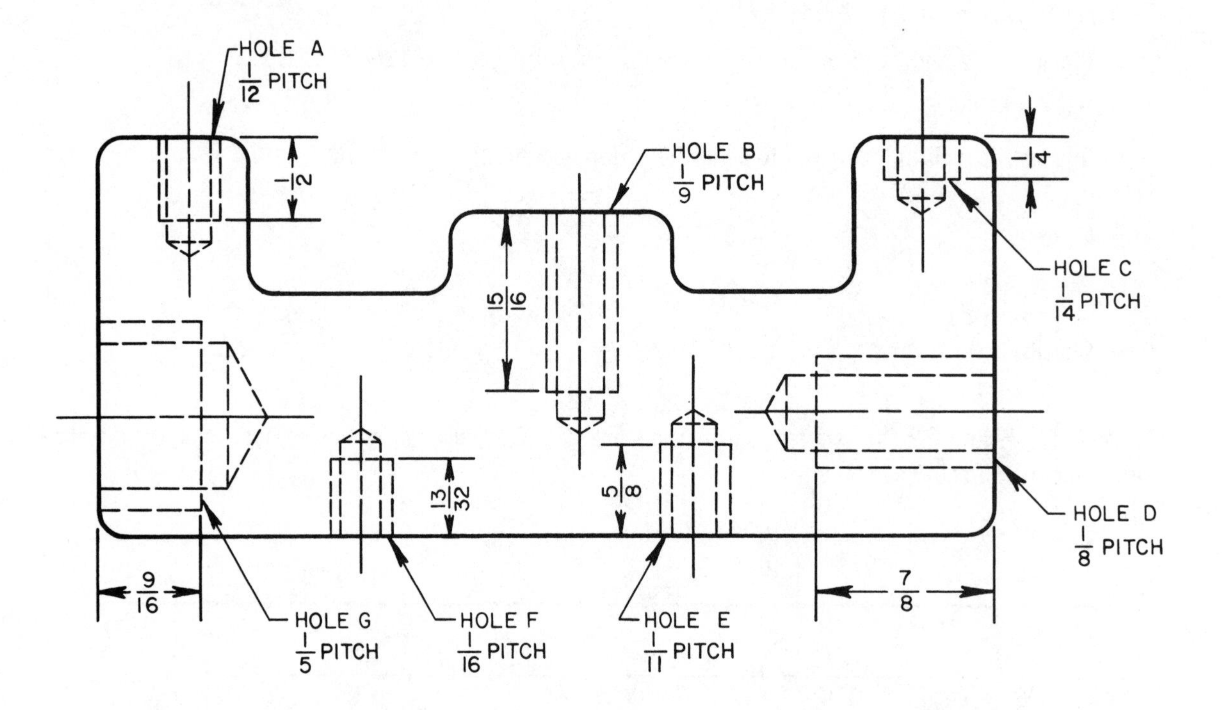

Fig. 5-3

2. Bar stock is being cut on a lathe. The tool feeds (advances) 1/32 inch each time the stock turns once (1 revolution). How many revolutions will the stock make when the tool advances 7/8 inch? ____________

3. A groove 15/16 inch deep is to be milled in a steel plate. How many cuts are required if each cut is 3/16 inch deep? ____________

C. DIVIDING MIXED NUMBERS

1. The sheet metal section in figure 5-4 has 5 sets of drilled holes: A, B, C, D, and E. The holes within a set are equally spaced in the horizontal direction. Compute the horizontal distance between 2 consecutive holes for each set.

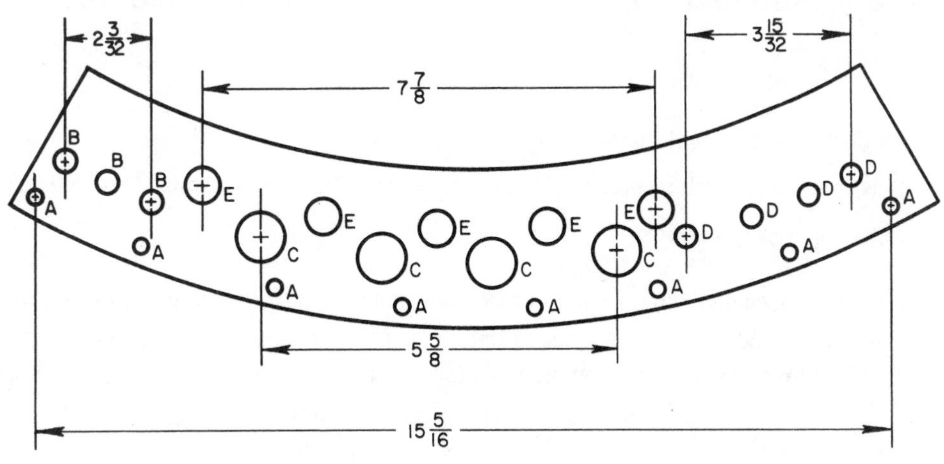

Fig. 5-4

a. Distance between 2 A holes = ____

b. Distance between 2 B holes = ____

c. Distance between 2 C holes = ____

d. Distance between 2 D holes = ____

e. Distance between 2 E holes = ____

2. The feed on a boring mill is set for 1/64 inch. How many revolutions does the work make when the tool advances 4 1/2 inches?

3. How many pieces can be blanked from a strip of steel 27 1/4 feet long if each stamping requires 2 3/16 inches of material plus an allowance of 5/16 inch at one end of the strip? ________

4. A groove is milled the full length of a steel plate which is 3 1/4 feet long. This operation takes a total of 4 1/16 minutes. How many feet of steel are cut in one minute? ________

5. How many binding posts can be cut from a brass rod 37 1/2 inches long if each post is 1 7/8 inches long? Allow 3/32 inch waste for each cut. ________

6. A bar of steel 22 3/4 feet long weighs 107 11/16 pounds. How much does a one-foot length of bar weigh?

7. Shown in figure 5-5 is a single-threaded square thread screw. The lead of a screw is the distance that the screw advances in one turn (revolution). The lead is equal to the pitch in a single-threaded screw. Given the number of turns and the amount of screw advance, determine the leads.

	SCREW ADVANCE	NUMBER OF TURNS	LEAD
a.	2 1/4″	10	
b.	7 37/64″	24 1/4	
c.	2 7/16″	6 1/2	
d.	1 1/2″	15	
e.	6 3/10″	12 3/5	

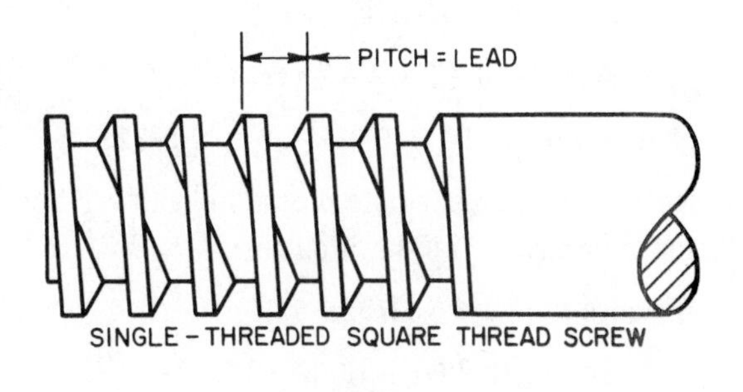

Fig. 5-5

Unit 6 Combined Operations of Fractions and Mixed Numbers

OBJECTIVES

After studying the unit the student should be able to

- Solve problems which involve combined operations of fractions and mixed numbers.
- Solve complex fractions.

Before a part is machined, the sequence of machining operations, the machine setup, and the working dimensions needed to produce the part must be determined. In actual practice, calculations of machine setup and working dimensions require not only the individual operations of addition, subtraction, multiplication, and division, but a combination of two or more of these operations.

ORDER OF OPERATIONS FOR COMBINED OPERATIONS

Rule:

- *Do all the work in the parentheses first.* Parentheses are used to group numbers.
- *Do multiplication and division next.* Perform multiplication and division operations in the order in which they occur.
- *Do addition and subtraction last.* These may be done in any order provided the number subtracted is smaller than the number from which it is subtracted.

COMBINING ADDITION AND SUBTRACTION

Example 1: Find the value of 3 1/2 - 3/8 + 5/16

▲ Subtract 3/8 from 3 1/2: 3 1/2 - 3/8 = 3 1/8

▲ Add 3 1/8 to 5/16: 3 1/8 + 5/16 = 3 7/16

Solving the Same Example Using a Different Order of Operation

▲ Add 3 1/2 to 5/16: 3 1/2 + 5/16 = 3 13/16

▲ Subtract 3/8 from 3 13/16: 3 13/16 - 3/8 = 3 7/16

Note: Although the order of operations is different, the same answer, 3 7/16 is obtained.

Example 2: In figure 6-1 find the distance from the base of the plate to the center of hole #2.

Distance x = 9/16 + 2 1/8 - 13/32

▲ Add: 9/16 + 2 1/8 = 2 11/16

▲ Subtract: 2 11/16 - 13/32 = 2 9/32

Note: Using an alternate procedure with a different order of operations, the same answer, 2 9/32, is obtained.

Distance x = 2 1/8 - 13/32 + 9/16

▲ Subtract: 2 1/8 - 13/32 = 1 23/32

▲ Add: 1 23/32 + 9/16 = 2 9/32

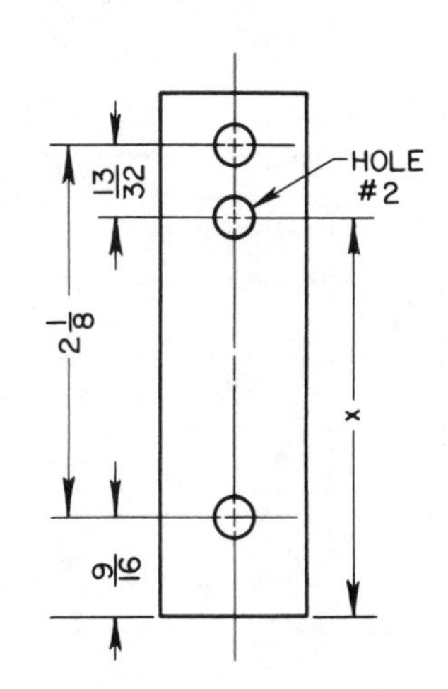

Fig. 6-1

COMBINING MULTIPLICATION AND DIVISION

Example 1: Find the value of 2/3 x 8 ÷ 2 1/2

▲ Multiply: $2/3 \times 8 = \frac{2 \times 8}{3 \times 1} = 16/3$

▲ Divide: 16/3 ÷ 2 1/2 = 16/3 x 2/5 = 32/15 = 2 2/15

Example 2: The stainless steel plate shown in figure 6-2 has grooves which are of uniform length and equally spaced within a distance of 33 1/2 inches. The time required to rough and finish mill a one-inch length of groove is 7/10 minute. How many minutes are required to cut all the grooves? Disregard the time required to reposition the part.

The number of grooves in 33 1/2″ = 33 1/2 ÷ 4 3/16.

The time required to cut 1 groove = 7/10 x 11 5/8.

Total time equals the number of grooves multiplied by the time for each groove:

33 1/2 ÷ 4 3/16 x 7/10 x 11 5/8

Divide: $33\ 1/2 \div 4\ 3/16 = \frac{\overset{1}{\cancel{67}}}{2} \times \frac{\overset{8}{\cancel{16}}}{\underset{1}{\cancel{67}}} = 8$ grooves

Multiply: $8 \times 7/10 \times 11\ 5/8 = \frac{\cancel{8} \times 7 \times 93}{1 \times 10 \times \underset{1}{\cancel{8}}} = \frac{651}{10} = 65\ 1/10$ Min.

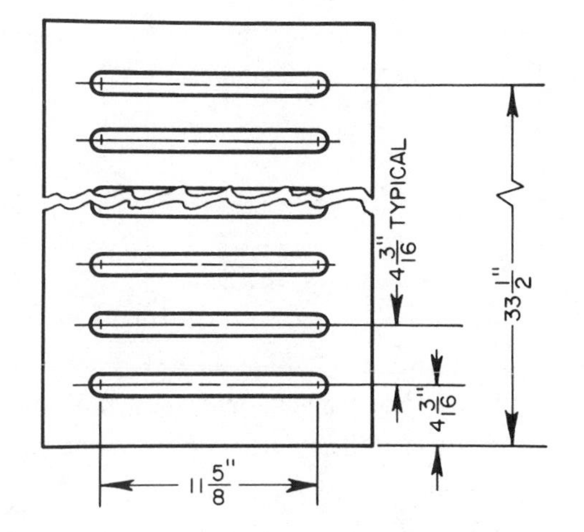

Fig. 6-2

COMBINING ADDITION, SUBTRACTION, MULTIPLICATION, AND DIVISION

Example 1: Find the value of 7 5/6 + 5 1/2 ÷ 3/4 – 10 x 7/16

▲ First divide and multiply: $5\ 1/2 \div 3/4 = \frac{11}{\underset{1}{\cancel{2}}} \times \frac{\overset{2}{\cancel{4}}}{3} = \frac{22}{3}$; $10 \times 7/16 = \frac{\overset{5}{\cancel{10}} \times 7}{1 \times \underset{8}{\cancel{16}}} = 4\ 3/8$

▲ Next add and subtract: 7 5/6 + 22/3 = 47/6 + 44/6 = 91/6 = 15 1/6
15 1/6 – 4 3/8 = 15 4/24 – 4 9/24 = 14 28/24 – 4 9/24 = 10 19/24

Example 2: Find the value of (7 5/6 + 5 1/2) ÷ 3/4 – 10 x 7/16

Note: This example is the same as the preceding example except for the parenthesis.

The expression (7 5/6 + 5 1/2) ÷ 3/4 states that the sum of 7 5/6 + 5 1/2 is divided by 3/4. Another way of writing (7 5/6 + 5 1/2) ÷ 3/4 is $\frac{7\ 5/6 + 5\ 1/2}{3/4}$.

▲ First do the work in parenthesis: (7 5/6 + 5 1/2) = 7 5/6 + 5 3/6 = 13 1/3

▲ Next divide and multiply: 13 1/3 ÷ 3/4 = 40/3 x 4/3 = 160/9 = 17 7/9

$$10 \times 7/16 = \frac{\overset{5}{\cancel{10}}}{1} \times \frac{7}{\underset{8}{\cancel{16}}} = 35/8 = 4\ 3/8$$

▲ Then add and subtract: 17 7/9 – 4 3/8 = 17 56/72 – 4 27/72 = 13 29/72

COMPLEX FRACTIONS

A *complex fraction* is an expression in which either the numerator or denominator or both are fractions or mixed numbers.

$$\frac{\frac{5}{9}}{\frac{1}{3}} \qquad \frac{17}{3\ 1/2} \qquad \frac{4\ 9/16}{1\ 3/8} \qquad \frac{5\ 7/8\ + 2\ 3/4}{3\ 15/16 - 1\ 1/8}$$

A fraction indicates a division operation. Therefore, complex fractions can be solved by dividing the denominator into the numerator.

$$\frac{\frac{5}{9}}{\frac{1}{3}} = \frac{5}{9} \div \frac{1}{3} \qquad \frac{4\ 9/16}{1\ 3/8} = 4\ 9/16 \div 1\ 3/8$$

$$\frac{5\ 7/8 + 2\ 3/4}{3\ 15/16 - 1\ 1/8} = (5\ 7/8 + 2\ 3/4) \div (3\ 15/16 - 1\ 1/8)$$

Note: The complete numerator is divided by the complete denominator. Therefore, parentheses are used to indicate that addition in the numerator and subtraction in the denominator must be performed before division.

$$\frac{5\ 7/8 + 2\ 3/4}{3\ 15/16 - 1\ 1/8} = (5\ 7/8 + 2\ 3/4) \div (3\ 15/16 - 1\ 1/8) = (5\ 7/8 + 2\ 6/8) \div (3\ 15/16 - 1\ 2/16) =$$

$$8\ 5/8 \div 2\ 13/16 = 69/8 \div 45/16 = \frac{69}{\cancel{8}_1} \times \frac{\cancel{16}^2}{45} = \frac{138}{45} = 3\ 1/15$$

APPLICATION

1. Solve the following examples of combined operations.

   a. 1/2 + 3/16 - 1/8 ____

   b. 3 7/8 - 2 3/16 + 3/8 ____

   c. 3/10 + 8 2/5 - 3 1/25 ____

   d. 18 - 2 2/3 + 4 1/6 ____

   e. 32 1/8 + 2 3/16 x 3/4 ____

   f. 7/9 x (2/3 + 3 5/6) ____

   g. 16 - 4 1/2 ÷ 1/2 + 2 1/8 ____

   h. (16 - 4 1/2) ÷ 1/2 + 2 1/8 ____

   i. (16 - 4 1/2) ÷ (1/2 + 2 1/8) ____

   j. 15 1/4 x 1 1/3 + 2 2/3 ÷ 4 5/6 ____

2. Solve the following complex fractions.

   a. $\frac{\frac{1}{3}}{\frac{1}{2}}$ ____

   b. $\frac{3\frac{7}{8}}{5}$ ____

   c. $\frac{\frac{15}{16}}{2\frac{1}{8}}$ ____

   d. $\frac{1/3 + 5/6}{7\ 1/2}$ ____

   e. $\frac{6\frac{3}{4} - 2\frac{7}{8}}{3\frac{1}{2} + 1\frac{1}{16}}$ ____

   f. $\frac{10\frac{1}{2} \times \frac{1}{2}}{4 \div 2\ 1/4}$ ____

3. Refer to the shaft shown in figure 6-3. Determine the missing dimensions in a-f using the dimensions given in the table.

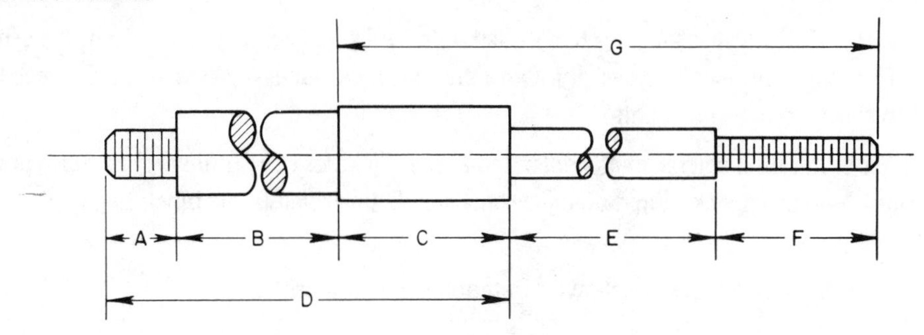

Fig. 6-3

	A	B	C	D	E	F	G
a.	1/2		1 3/8	6 3/4		15/16	8 1/8
b.		3 13/16	1 5/8	5 37/64	4 3/8	3/4	
c.	9/16	4 3/32			5 1/8	27/32	7 1/32
d.	5/8		1 7/16	5 31/32		7/8	7 15/16
e.		3 3/4	1 11/16	6 1/32	4 61/64	25/32	
f.	11/16	4 1/8			5 3/16	7/8	7 3/64

4. The outside diameter of an aluminum tube is 3 1/16 inches. The wall thickness is 5/32 inch. What is the inside diameter? _____

5. Four studs of the following lengths in inches are to be machined from bar stock: 2 3/4, 1 7/8, 2 5/16, and 1 13/32. Allow 1/8 inch waste for each cut and 1/32 inch on each end of each stud for facing. What is the total length of bar stock required? _____

6. Find dimensions A, B, C, and D of the idler bracket in figure 6-4.

A = _____

B = _____

C = _____

D = _____

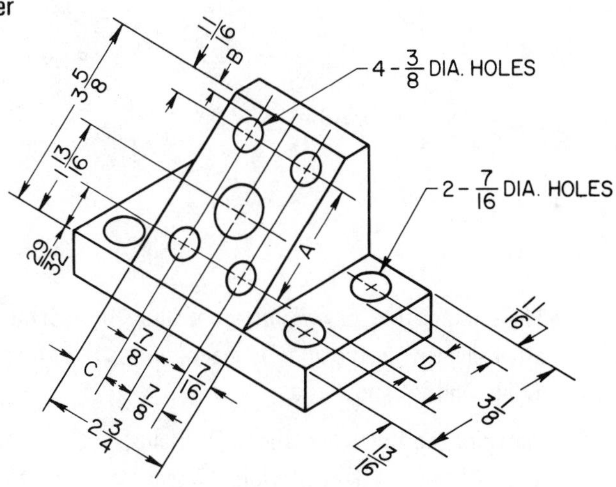

Fig. 6-4

7. How long does it take to cut a distance of 2 1/2 feet along a shaft that turns 15 revolutions per minute with a tool feed of 1/32 inch per revolution? _____

8. An angle iron 47 1/2 inches long has two drilled holes which are equally spaced from the center of the piece. The center distance between the two holes is 19 7/8 inches. What is the distance from each end of the piece to the closest hole? _____

9. A tube has an inside diameter of 3/4 inch and a wall thickness of 1/16 inch. The tube is to be fitted in a drilled hole in a block. What diameter hole should be drilled in the block to give 1/64 inch total clearance? _____

10. Two views of mounting block are shown. Determine dimensions A-G.

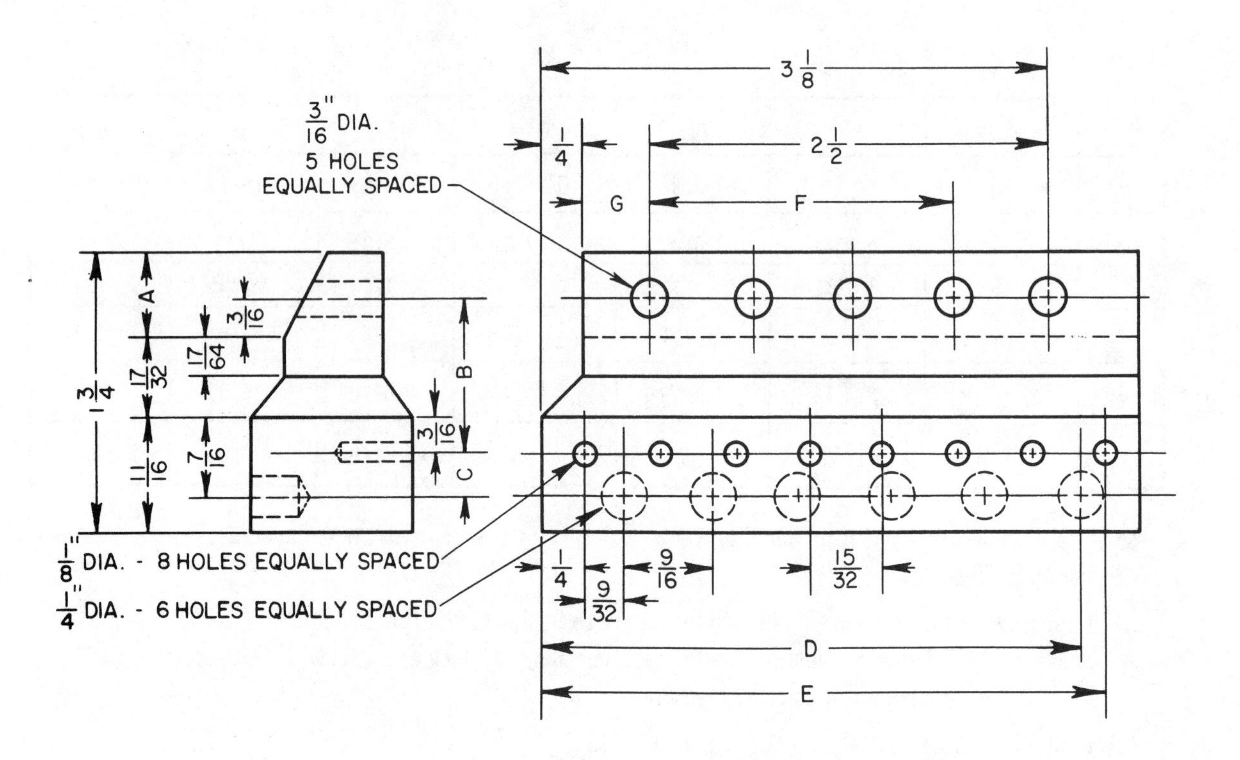

Fig. 6-5

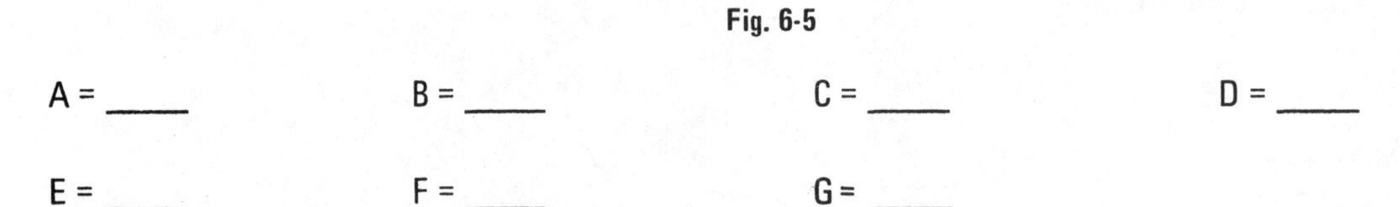

A = _____ B = _____ C = _____ D = _____

E = _____ F = _____ G = _____

11. The composition of an aluminum alloy by weight is 24/25 aluminum and 1/40 copper. The only other element in the alloy is magnesium. How many pounds of magnesium are required for casting 125 pounds of alloy? _____

12. Pieces of the following lengths in inches, 2 1/2, 1 3/4, 1 7/8, and 5/16, are cut from an 18-inch steel bar. Allowing 1/8-inch waste for each cut, what is the length of bar left after the pieces are cut? _____

Unit 7 Introduction to Decimals

OBJECTIVES

After studying this unit the student should be able to

- Locate decimal fractions on a number line.
- Convert a common fraction that has a denominator which is a power of ten to a decimal fraction.
- Change decimal fractions from word form to number form.
- Change decimal fractions from number form to word form.

Most blueprints are dimensioned with decimal fractions rather than common fractions. The dials which are used in establishing machine settings and movement, in determining tool speeds and travel, and in measuring dimensions of parts are usually graduated in decimal units.

EXPLANATION OF DECIMAL FRACTIONS

Decimal fractions are fractions whose denominators are powers of 10, such as 10; 100; 1,000; 10,000; 100,000; and 1,000,000. A decimal fraction is not written as a common fraction. The denominator is omitted and is replaced by a decimal point placed to the left of the numerator.

The line in figure 7-1 is one unit long. It is divided into ten equal smaller parts. The locations of common fractions and their decimal fraction equivalents are shown on the line.

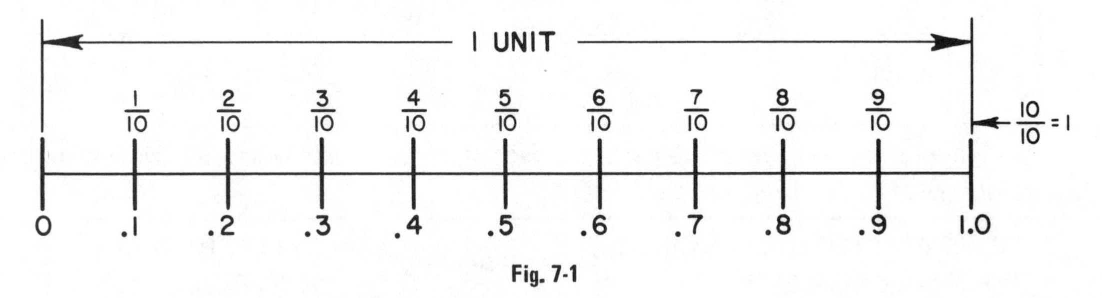

Fig. 7-1

One of the ten equal small parts, 1/10 (.1) of the line; is shown in figure 7-2 in magnified form. Here the 1/10 or .1 unit is divided into 10 equal smaller parts. The location of common fractions and their decimal fraction equivalents are shown on this line.

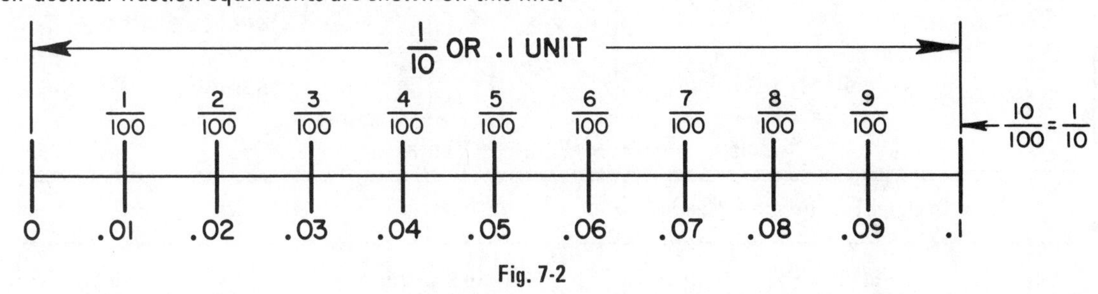

Fig. 7-2

If the 1/100 (.01) division is divided into 10 equal smaller parts, the resulting parts are 1/1000 (.001); 2/1000 (.002); 3/1000 = (.003); ... 9/1000 (.009); 10/1000 = 1/100 (.01).

Each time a decimal point is moved one place to the right, a value 10 times greater than the previous value is obtained.

- Each time the decimal point is moved one place to the left a value 1/10 of the previous value is obtained.
- Each time a value is multiplied by 10 the decimal point is moved one place to the right. Each step in the following example shows both the decimal fraction and its equivalent common fraction.

DECIMAL FRACTION	COMMON FRACTION
.000003	$\frac{3}{1,000,000}$
.000003 x 10 = 0.00003	$\frac{3}{1,000,000} \times 10 = \frac{3}{100,000}$
.00003 x 10 = 0.0003	$\frac{3}{100,000} \times 10 = \frac{3}{10,000}$
.0003 x 10 = 0.003	$\frac{3}{10,000} \times 10 = \frac{3}{1,000}$
.003 x 10 = 0.03	$\frac{3}{1,000} \times 10 = \frac{3}{100}$
.03 x 10 = 0.3	$\frac{3}{100} \times 10 = \frac{3}{10}$
.3 x 10 = 3.	$\frac{3}{10} \times 10 = 3$

The following example gives the names of the parts of a number with respect to their position from the decimal point.

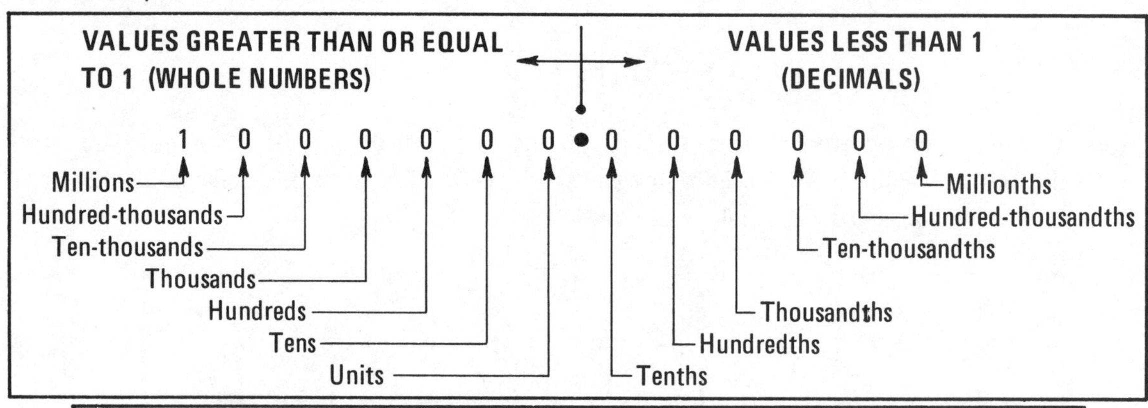

How Decimal Fractions are Read		
5/10	.5	five tenths
7/100	.07	seven hundredths
11/1,000	.011	eleven thousandths
219/10,000	.0219	two hundred nineteen ten-thousandths
43/100,000	.00043	forty-three hundred-thousandths
817/1,000,000	.000817	eight hundred seventeen millionths

Mixed Decimals

A number that consists of a whole number and a decimal fraction is called a *mixed decimal.*

How Mixed Decimals are Read

3.4	three and four tenths
1.002	one and two thousandths
16.0793	sixteen and seven hundred ninety-three ten-thousandths
8.00032	eight and thirty-two hundred-thousandths

APPLICATION

A. Find the decimal value of each of the distances shown in the lines 1, 2, and 3 below. Note the total unit value of each line.

1.

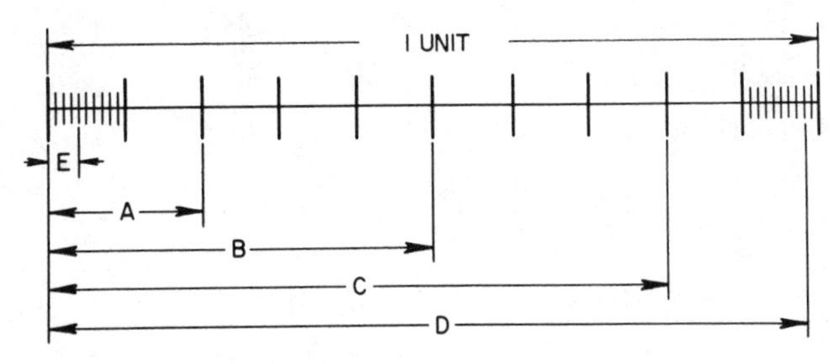

Fig. 7-3

A = ____ B = ____ C = ____ D = ____ E = ____

2.

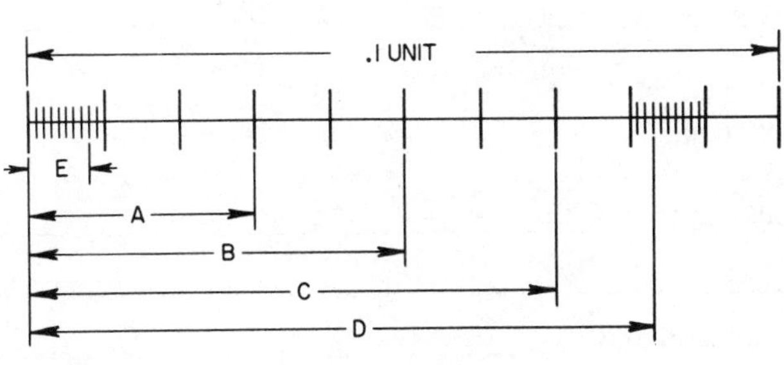

Fig. 7-4

A = ____ B = ____ C = ____ D = ____ E = ____

3.

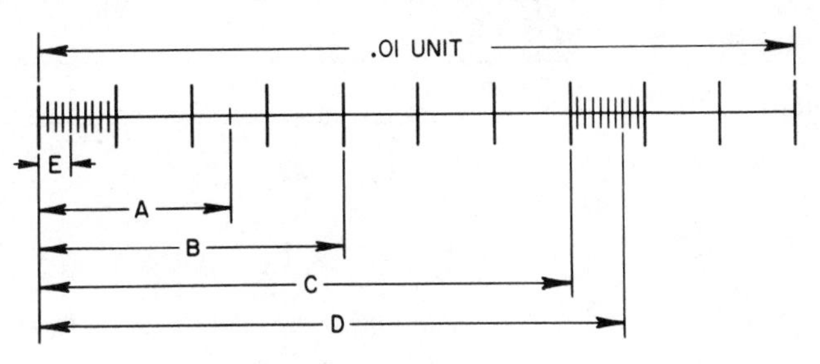

Fig. 7-5

A = ____ B = ____ C = ____ D = ____ E = ____

B. Write the equivalent decimal fraction for each of the following common fractions.

1. 9/10 = ____
2. 3/10,000 = ____
3. 17/100 = ____
4. 43/100 = ____
5. 99/1,000 = ____
6. 999/10,000 = ____
7. 73/1,000 = ____
8. 1973/100,000 = ____
9. 63,917/100,000 = ____

C. In each of the following problems, the value on the left must be multiplied by one of the following numbers: 1/10,000; 1/1,000; 1/100; 1/10; 10; 100; 1,000; or 10,000 in order to obtain the value on the right of the equal sign.

Determine the proper number.

1. .9 x ____ = .0009
2. .7 x ____ = .007
3. .03 x ____ = .3
4. .0003 x ____ = .003
5. .135 x ____ = .00135
6. 4 x ____ = .4
7. .0643 x ____ = .000643
8. .0643 x ____ = 6.43
9. .00643 x ____ = 64.3
10. 643 x ____ = .643

D. Write these numbers as words.

1. .032 ____________
2. .007 ____________
3. .132 ____________
4. .0075 ____________
5. .108 ____________
6. 1.5 ____________
7. 10.37 ____________
8. 17.0009 ____________
9. 4.0012 ____________
10. 13.103 ____________

E. Write these words as numbers.

1. eighty-four ten thousandths ____________
2. seven tenths ____________
3. forty-three and eight hundredths ____________
4. two and seven hundred thousandths ____________
5. thirty-five ten thousandths ____________
6. ten and two tenths ____________
7. five and one ten-thousandths ____________
8. twenty and seventy-one hundredths ____________

Unit 8 Changing Decimals and Fractions; Rounding-off Decimals

OBJECTIVES

After studying this unit the student should be able to

- Round-off decimals to any required number of places.
- Change common fractions to decimal fractions.
- Change decimal fractions to common fractions.

When blueprint dimensions of a part are given in fractional units, a machinist is usually required to convert these fractional values to decimal working dimensions. In computing material requirements and in determining stock waste and scrap allowances, it is sometimes more convenient to change decimal values to approximate fractional equivalents.

ROUNDING-OFF DECIMALS

After solving a problem using decimal fractions, the answer often has more decimal places than are required. For example, a length of .875376 inch can not be cut on a milling machine. In cutting to the closest thousandths of an inch, the machinist would consider .875376 inch as .875 inch. *Rounding-off a decimal* means expressing the decimal with a fewer number of decimal places.

Rule: To round-off a decimal

- Determine the number of decimal places required in an answer.
- If the digit directly following the last decimal place required is less than 5, drop all digits which follow the required number of decimal places.
- If the digit directly following the last decimal place required is 5 or larger, add one to the last required digit and drop all digits which follow the required number of decimal places.

Example 1: Round off .873429 to three decimal places.

- ▲ The digit following the third decimal place is 4. — .873 (4) 29
- ▲ Because 4 is less than 5, drop all digits after the third decimal place. — .873

Example 2: Round off .36845 to two decimal places.

- ▲ The digit following the second decimal place is 8. — .36 (8) 45
- ▲ Because 8 is greater than 5, add 1 to the 6. — .37

Example 3: Round off 18.738257 to four decimal places.

- ▲ The digit following the fourth decimal place is 5. — 18.7382 (5) 7
- ▲ Add 1 to the 2. — 18.7383

CHANGING COMMON FRACTIONS TO DECIMAL FRACTIONS

A common fraction is an indicated division. For example, 3/4 is the same as $3 \div 4$; 5/16 is the same as $5 \div 16$; 99/171 is the same as $99 \div 171$.

Because both the numerator and the denominator of a common fraction are whole numbers, changing a common fraction to a decimal fraction requires division with whole numbers.

Rule: To change a common fraction to a decimal fraction

- Divide the denominator into the numerator.

Note: Decimals called *terminating decimals* end without having a remainder, although the division may have to be carried to a great number of decimal places.

The conversion of some common fractions to decimal form results in decimals that continue without end, called *nonterminating decimals.*

- Either of the following two methods can be used in carrying out a decimal to a desired number of places.
    1. The division should be carried out to one more place than the number of places required in the answer, then rounded-off one place.

       or

    2. The division should be carried out to the required number of places. If there is a remainder equal to or greater than the fraction 1/2, add 1 to the last digit. If the remainder is less than 1/2, leave the last digit unchanged.

Example 1: Change 2/3 to a 4-place decimal.

Method 1:

▲ Divide the numerator by the denominator.

▲ Add one more zero after the 2 than the required number of decimal places (Add 5 zeros.)

$$3\overline{)2.00000} = .66666$$

▲ Round-off .66666 to 4 places. .6667

Method 2:

▲ Add the same number of zeros after the 2 as the required number of decimal places. (Add 4 zeros).

$$3\overline{)2.0000} = .6666\ 2/3$$

▲ The remainder 2/3 is greater than 1/2; therefore, add 1 to the fourth place digit 6. .6667

Example 2: Change 5/7 to a 2 place decimal.

Method 1:

▲ Add 3 zeros after the 5.

$$7\overline{)5.000} = .714$$

▲ Round-off to 2 places. .71

Method 2:

▲ Add 2 zeros after the 5.

$$7\overline{)5.00} = .71\ 3/7$$

▲ The remainder 3/7 is less than 1/2. Leave the second place digit 1 unchanged. .71

CHANGING DECIMAL FRACTIONS TO COMMON FRACTIONS

A decimal is the numberator of a common fraction with a denominator of 1.

Rule: To change a decimal fraction to a common fraction

- Make the numerator of the common fraction the decimal with the decimal point omitted.
- The denominator is 1 followed by the same number of zeros as there are decimal places in the decimal fraction.
- Reduce to lowest terms.

Example 1: Change .375 to a common fraction.

- ▲ The decimal .375 without the decimal point is the numerator. The numerator is 375.
- ▲ The denominator is 1 with the same number of zeros as there are decimal places in the decimal fraction. There are three zeros. The denominator is 1000.
- ▲ Therefore, $.375 = \frac{375}{1000}$.
- ▲ Reduce $\frac{375}{1000}$ to lowest terms: $\frac{375}{1000} = \frac{3}{8}$

Note: In the preceding example, by following the rule the same result was obtained as by multiplying both the numerator and denominator of $\frac{.375}{1.000}$ by 1000.

$$\frac{.375}{1.000} = \frac{.375}{1.000} \times \frac{1000}{1000} = \frac{375}{1000} = \frac{3}{8}$$

Example 2: Change .27 to a common fraction.

- ▲ The numerator is 27.
- ▲ The denominator is 100.
- ▲ Therefore, $.27 = \frac{27}{100}$.

Example 3: Change .03125 to a common fraction.

- ▲ The numerator is 3125.
- ▲ The denominator is 100,000.
- ▲ Therefore, $.03125 = \frac{3125}{100{,}000}$

Reduce $\frac{3125}{100{,}000}$ to lowest terms. $\frac{3125}{100{,}000} = \frac{1}{32}$

APPLICATION

A. ROUNDING-OFF DECIMALS

Round-off these decimals to the indicated number of decimal places.

1. .78273 (3 places) ____
2. .1247 (2 places) ____
3. .23975 (3 places) ____
4. .01723 (3 places) ____
5. .01723 (2 places) ____
6. .90039 (2 places) ____
7. .72008 (4 places) ____
8. .0001 (3 places) ____
9. .0005 (3 places) ____
10. .099 (2 places) ____

B. CHANGING COMMON FRACTIONS TO DECIMAL FRACTIONS

Change these common fractions to decimal fractions. Give answers to 4 decimal places.

1.	13/32 ____	5.	2/3 ____	9.	7/32 ____
2.	7/8 ____	6.	10/11 ____	10.	1/2 ____
3.	5/8 ____	7.	1/25 ____	11.	4/7 ____
4.	1/4 ____	8.	47/64 ____	12.	3/8 ____

13. In figure 8-1, what decimal fraction of distance B is distance A? Give answer to 4 decimal places. ____

A = 1"
B = 3"

Fig. 8-1

C = 3"
12"
D

Fig. 8-2

14. In figure 8-2, what decimal fraction of distance D is distance C? ____

15. In figure 8-3, what decimal fraction of distance F is distance E? ____

16. In figure 8-3, what decimal fraction of distance H is distance G? Give answer to 4 decimal places. ____

F = 6 ft.
E = 9 in.
H = 3 ft. 3 in.
G

Fig. 8-3

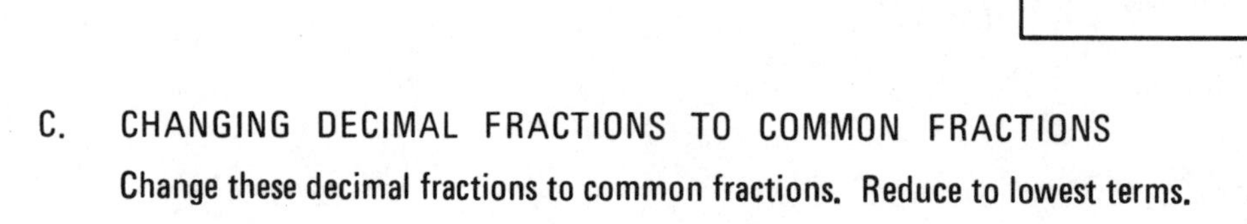

C. CHANGING DECIMAL FRACTIONS TO COMMON FRACTIONS

Change these decimal fractions to common fractions. Reduce to lowest terms.

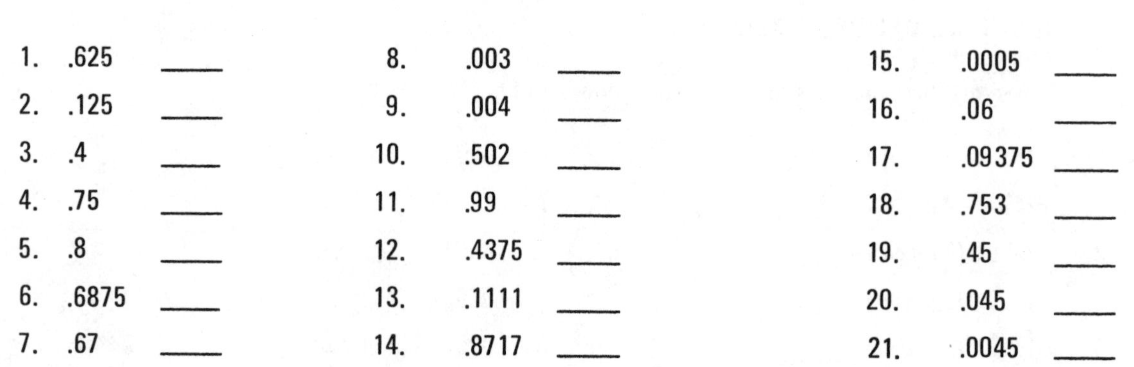

1.	.625 ____	8.	.003 ____	15.	.0005 ____
2.	.125 ____	9.	.004 ____	16.	.06 ____
3.	.4 ____	10.	.502 ____	17.	.09375 ____
4.	.75 ____	11.	.99 ____	18.	.753 ____
5.	.8 ____	12.	.4375 ____	19.	.45 ____
6.	.6875 ____	13.	.1111 ____	20.	.045 ____
7.	.67 ____	14.	.8717 ____	21.	.0045 ____

22. In figure 8-4, what common fractional part of distance B is distance A?

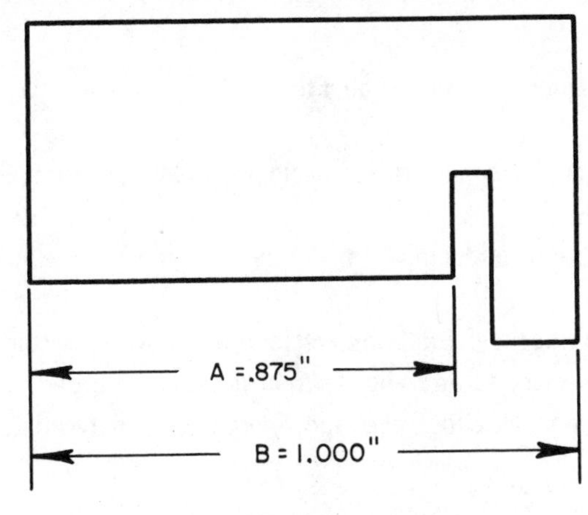

Fig. 8-4

23. In figure 8-5, what common fractional part of diameter C is diameter D?

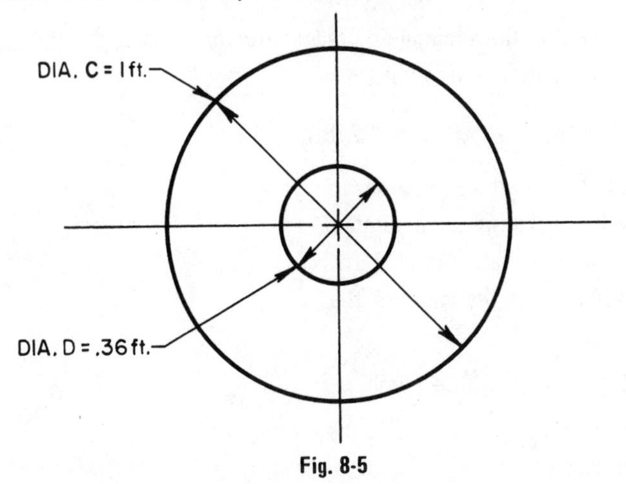

Fig. 8-5

24. In figure 8-6, what common fractional part of distance A is

a. Distance B? ____

b. Distance C? ____

c. Distance D? ____

d. Distance E? ____

e. Distance F? ____

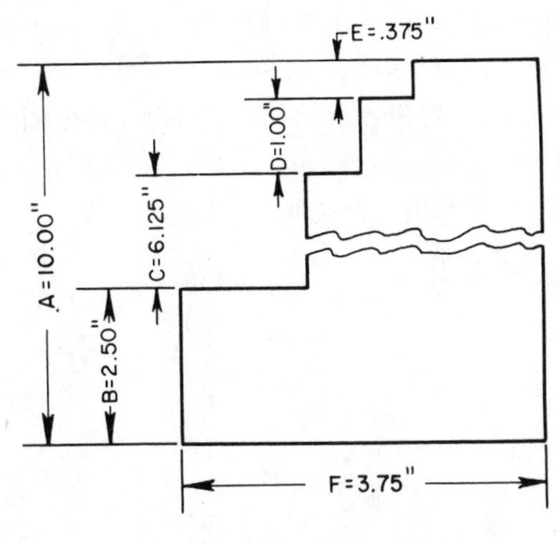

Fig. 8-6

Unit 9 Addition and Subtraction of Decimal Fractions

OBJECTIVES

After studying this unit, the student should be able to

- Add decimals.
- Add combinations of decimals, mixed decimals, and whole numbers.
- Subtract decimals.
- Subtract combinations of decimals, mixed decimals, and whole numbers.

Adding and subtracting decimal fractions are required at various stages in the production of most products and parts. It is necessary to add and subtract decimals in order to estimate machining costs and production times, to compute stock allowances and tolerances, to determine locations and lengths of cuts, and to inspect finished parts.

ADDING DECIMALS

Rule: To add decimals

- Arrange the numbers so that the decimal points are directly under each other.
- Proceed with addition as with whole numbers.

```
   7.35
 114.075
    .3422
    .003
+218.7
```

Example 1: Add 7.35, 114.075, .3422, .003, and 218.7

Note: To reduce the possibility of error, add zeros to decimals so that all the values have the same number of places to the right of the decimal point. Zeros added in this manner do not affect the value of the number.

▲ Proceed with addition as with whole numbers.

Note: The decimal point of the sum is placed in the same position as the other decimal points.

```
   7.3500
 114.0750
    .3422
    .0030
+218.7000
 340.4702
```

The decimal point location of a whole number is directly to the right of the last digit. For example, 1 = 1., 217 = 217., 200,000 = 200,000.

Example 2: Find the length of the swivel bracket shown.

▲ Arrange numbers so the decimal points are directly under each other.

▲ Add:

```
 .350
1.000
 .512
1.200
 .156
 .370
3.588
```

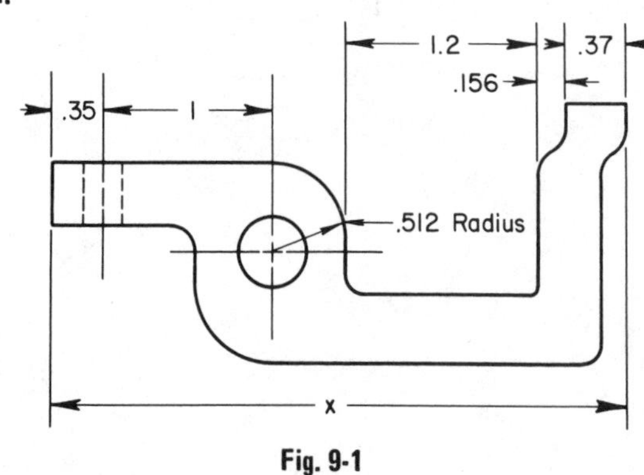

Fig. 9-1

SUBTRACTING DECIMALS

Rule: To subtract decimals

- Arrange the numbers so that the decimal points are directly under each other.
- Proceed with subtraction as with whole numbers.

Example 1: Subtract 13.261 from 25.6.

▲ Arrange the numbers so that the decimal points are directly under each other.

$$\begin{array}{r} 25.6 \\ -\ 13.261 \\ \hline \end{array}$$

Add two zeros to 25.6 so that it has the same number of decimal places as 13.261. Place the decimal point of the answer in the same position as the other decimal points.

$$\begin{array}{r} 25.600 \\ -\ 13.261 \\ \hline 12.339 \end{array}$$

▲ Subtract.

▲ Check: 12.339 + 13.261 = 25.600

Example 2: Determine dimensions A, B, C, and D of the support bracket shown in figure 9-2.

Solve for A:

A = .505 − .18

$$\begin{array}{rr} & .505 \\ - & .180 \\ \hline A = & .325 \end{array}$$

Solve for B:

B = 1.4 − .301

$$\begin{array}{rr} & 1.400 \\ - & .301 \\ \hline B = & 1.099 \end{array}$$

Solve for C:

C = 1.74 − .365

$$\begin{array}{rr} & 1.740 \\ - & .365 \\ \hline C = & 1.375 \end{array}$$

Solve for D:

D = .746 − .46

$$\begin{array}{rr} & .746 \\ - & .460 \\ \hline D = & .286 \end{array}$$

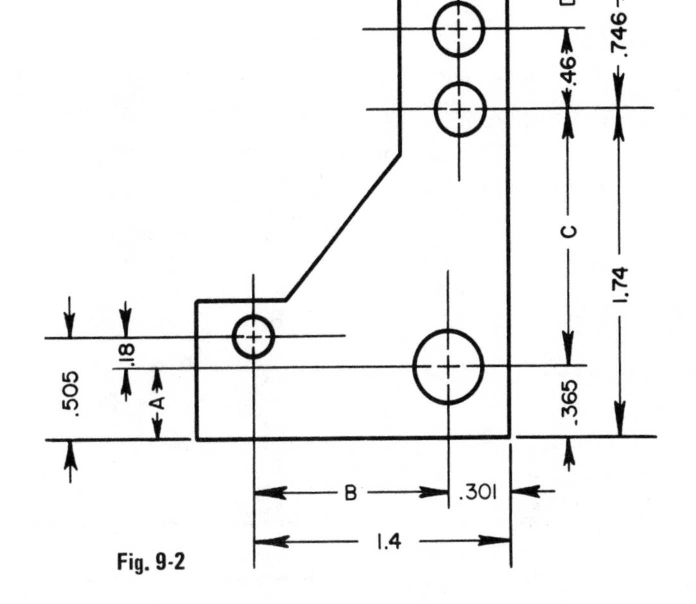

Fig. 9-2

APPLICATION

A. ADDING DECIMALS

1. Add these numbers

a.	.132 + 12.9 + 5	____	f.	4 + .4 + .04 + .004	____
b.	.003 + .13795	____	g.	87 + .0239 + 7.23	____
c.	.375 + .8 + .12	____	h.	.0001 + .1 + .01	____
d.	4.187 + .932 + .01	____	i.	4.705 + .0937 + .98	____
e.	873.14 + 19.3 + .137	____	j.	.057 + 5.7 + 570	____

2. Determine dimensions A, B, C, D, E, and F of this profile gage, figure 9-3.

   A = ____
   B = ____
   C = ____
   D = ____
   E = ____
   F = ____

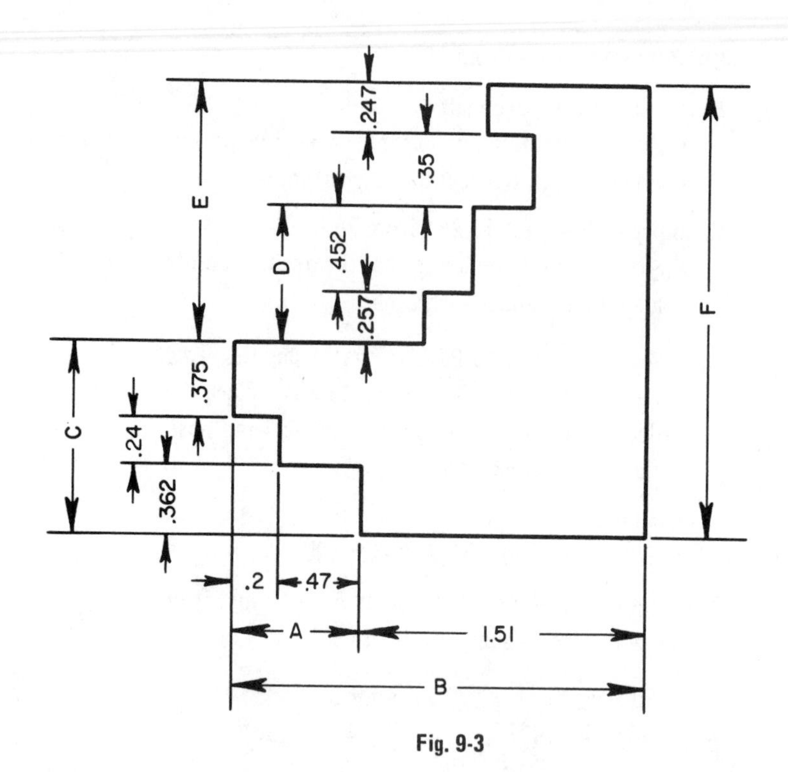

Fig. 9-3

3. A sine plate is to be set to a desired angle by using size blocks of the following thicknesses: 2.000 inches, .500 inch, .250 inch, .125 inch, .100 inch, .1007 inch, and .1001 inch. Determine the total height that the sine plate is raised. ____

4. Three cuts are required to turn a steel shaft. The depths of the cuts in inches are .250, .125, and .0055. How much stock has been removed per side? Round-off answer to 3 decimal places. ____

5. Determine dimensions A, B, C, D, and E of the gear arm shown.

   A = ____ B = ____ C = ____ D = ____ E = ____

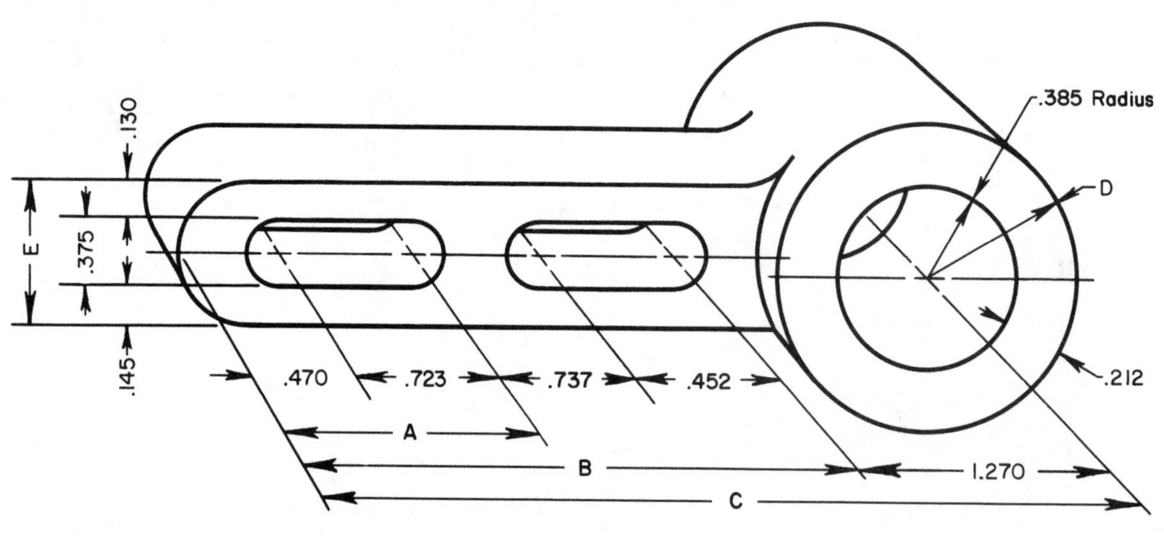

Fig. 9-4

B. SUBTRACTING DECIMALS

1. Subtract these numbers. Where necessary, round-off answers to 3 decimal places.

a. .617 - .4136 ____	d. .3 - .299 ____	g. .313 - .2323 ____
b. .319 - .0127 ____	e. .4327 - .232 ____	h. 3.872 - .0002 ____
c. 2.308 - .7859 ____	f. 23.062 - .973 ____	i. 5.923 - 3.923 ____

2. The front and right side views of a sliding shoe are shown. Determine dimensions A, B, C, D, E, and F.

A = ____
B = ____
C = ____
D = ____
E = ____
F = ____

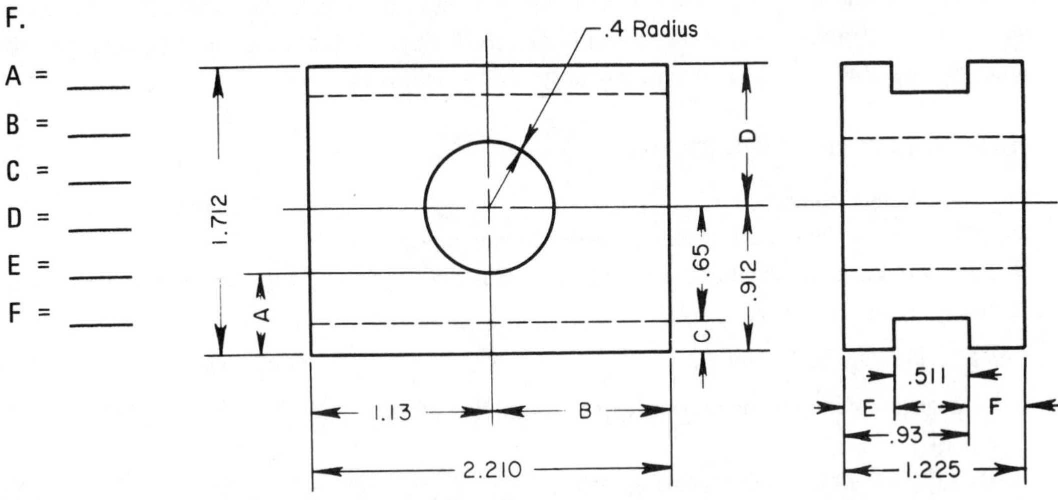

Fig. 9-5

3. Refer to the plate shown in figure 9-6 and determine the following distances:

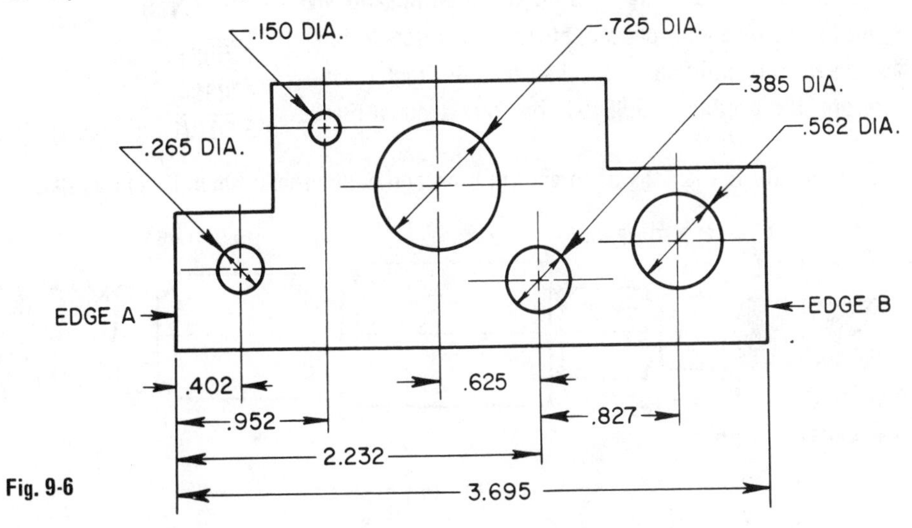

Fig. 9-6

a. The center distance between the .265 dia. hole and the .150 dia. hole. ____
b. The center distance between the .385 dia. hole and the .150 dia. hole. ____
c. The distance between edge A and the center of the .725 dia. hole. ____
d. The distance between edge B and the center of the .385 dia. hole. ____
e. The distance between edge B and the center of the .562 dia. hole. ____

Unit 10 Multiplication of Decimal Fractions

OBJECTIVES

After studying this unit, the student should be able to

- Multiply decimals.
- Multiply combinations of decimals, mixed decimals, and whole numbers.

A machinist must readily be able to multiply decimals for computing machine feeds and speeds, for determining tapers, and for determining lengths and stock sizes. Multiplication of decimals is also required in order to solve problems which involve geometry and trigonometry.

MULTIPLYING DECIMAL FRACTIONS

Rule: To multiply decimals

- Multiply using the same procedure as with whole numbers.
- Point off the same number of decimal places in the product as there are in the multiplicand and the multiplier combined.

Example 1: Multiply 50.123 by .87.

- ▲ Arrange the numbers the same as when multiplying whole numbers.

```
(multiplicand) 50.123
(multiplier)   x  .87
```

- ▲ Multiply the same as with whole numbers.

```
            50.123
            x  .87
            350861
           400984
(product)  4360701
```

- ▲ Beginning at the right of the product, point off as many decimal places as there are in both the multiplicand and the multiplier. The multiplicand, 50.123, has three decimal places, and the multiplier, .87, has two decimal places. Therefore, the product, 43.60701, has five decimal places.

```
  50.123  (3 places)
     .87  (2 places)
  350861
 400984
43.60701  (5 places)
```

Example 2: Compute the lengths of thread on each end of the shaft shown in figure 10-1.

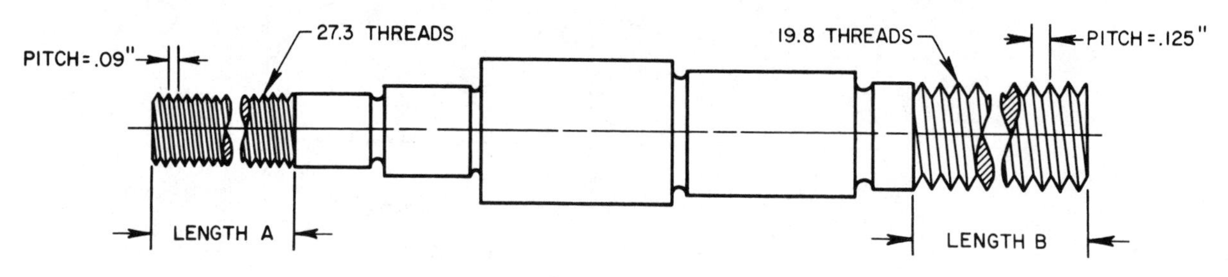

Fig. 10-1

Compute Length A:

```
 27.3 (1 place)
  .09 (2 places)
2.457 (3 places)
```

Compute Length B:

```
  19.8   (1 place)
 x.125   (3 places)
   990
  396
 198
2.4750   (4 places)
```

Note: When multiplying certain decimals, if the product has a smaller number of digits than the number of decimal places required, add as many zeros to the left of the product as are necessary to give the required number of decimal places.

Example: Multiply .0237 by .04. Round-off answer to five decimal places.

(multiplicand) .0237
(multiplier) x .04
948

▲ Multiply

The multiplicand, .0237, has four decimal places, and the multiplier, .04, has two decimal places. Therefore, the product must have six decimal places.

.0237 (4 places)
x .04 (2 places)
.000948 (6 places)

▲ Add three zeros to the left of the product.

▲ Round-off .000948 to five places. .00095

APPLICATION

1. Multiply these numbers. Where necessary, round-off answers to 4 decimal places.

   a. 3.876 x .012 ____ b. 2.2 x 1.5 ____ c. 40 x .15 ____ d. 8.93 x .32 ____

2. A section of a spur gear is shown. Given the circular pitches for various gear sizes, determine the working depths, clearances, and tooth thicknesses. Round-off answers to 4 decimal places.

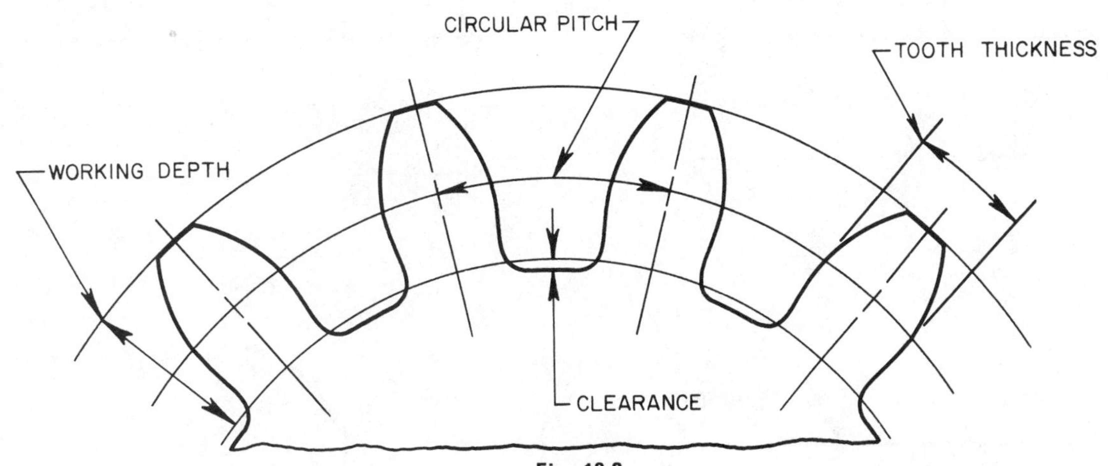

Fig. 10-2

Working depth = .6366 x Circular Pitch
Clearance = .05 x Circular Pitch
Tooth thickness = .5 x Circular Pitch

	Circular Pitch	Working Depth	Clearance	Tooth Thickness
a.	.3925			
b.	.1582			
c.	.8759			
d.	1.2378			
e.	1.5931			

3. Determine diameters A, B, C, D, and E of this shaft

Dia. A = ____

Dia. B = ____

Dia. C = ____

Dia. D = ____

Dia. E = ____

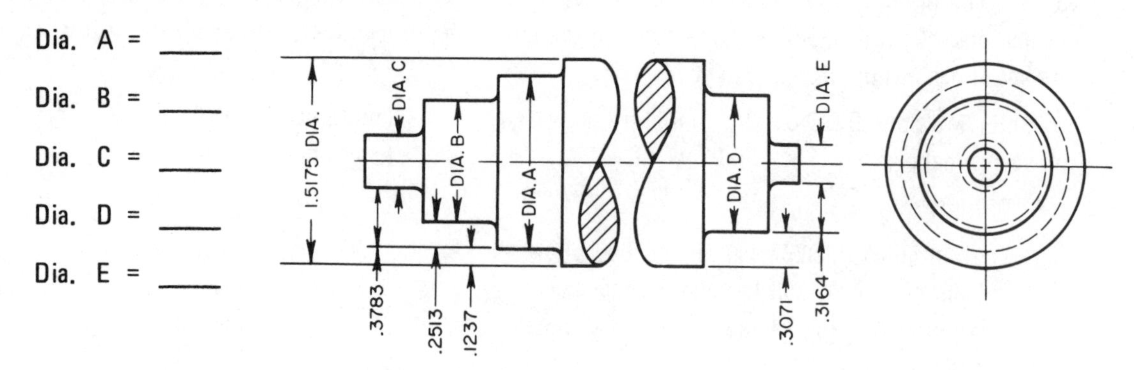

Fig. 10-3

4. Determine dimension x for each of the figures below.

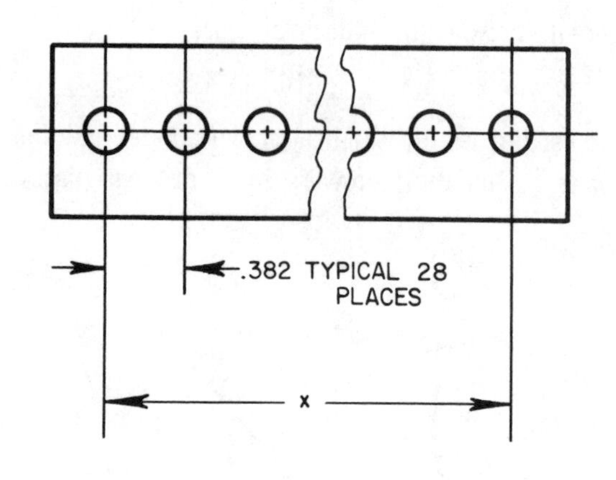

a. x = ________

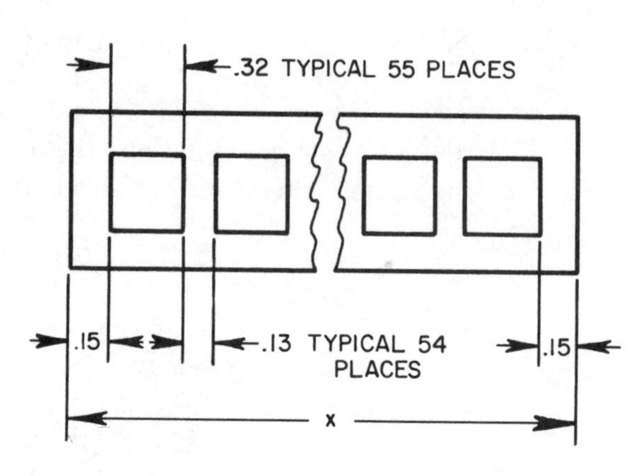

b. x = ________

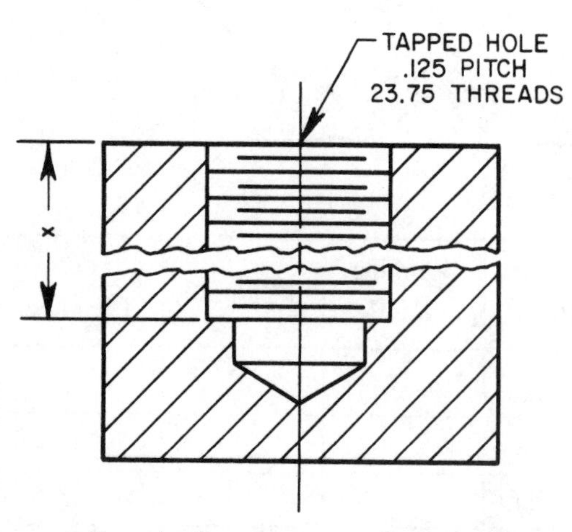

c. Round-off answers to 3 decimal places.
x = ________

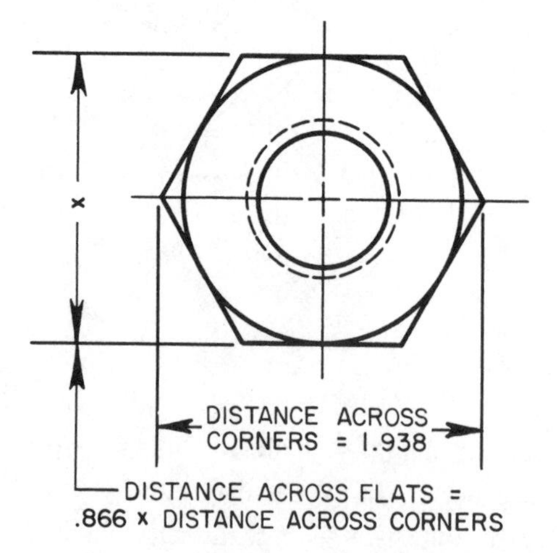

d. Round-off answers to 3 decimal places.
x = ________

Fig. 10-4

Unit 11 Division of Decimal Fractions

OBJECTIVES

After studying this unit, the student should be able to

- Divide decimals.
- Divide decimals with whole numbers.
- Divide decimals with mixed decimals.

Division with decimals is used for computing the manufacturing cost and time per piece after total production costs and times have been determined. Division with decimals is also required in order to compute thread pitches, gear tooth thicknesses and depths, cutting speeds, and depths of cut.

DIVIDING DECIMAL FRACTIONS

Moving a decimal point to the right is the same as multiplying the decimal by a power of 10.

$.237 \times 10 = 2.37$ $\qquad$ $.237 \times 100 = 23.7$

$.237 \times 1{,}000 = 237.$ $\qquad$ $.237 \times 10{,}000 = 2370.$

When dividing decimals, the value of the answer (quotient) is not changed if the decimal points of both the divisor and the dividend are moved the same number of places to the right. It is the same as multiplying both divisor and dividend by the same number.

$$.9375 \div .612 = (.9375 \times 1000) \div (.612 \times 1000) = 937.5 \div 612.$$

$$14.203 \div 6.87 = (14.203 \times 100) \div (6.87 \times 100) = 1420.3 \div 687.$$

Rule: To divide decimals

- Move the decimal point of the divisor as many places to the right as are necessary to make the divisor a whole number.
- Move the decimal point of the dividend the same number of places as were moved in the divisor.
- Place the decimal point in the quotient directly above the decimal point in the dividend.
- Divide as with whole numbers.

Example 1: Divide .643 by .28 (Answer is to be correct to three decimal places.)

- To make the divisor a whole number move the decimal point 2 places, 28.

  .28) .643
  (Divisor) (Dividend)

- The decimal point in the dividend is also moved 2 places, 64.3

  28.) 64.3

- Add two zeros to the dividend in order to obtain an answer to 3 decimal places.

  28) 64.300

Note: If a greater degree of accuracy is desired, extra zeros may be added and the resulting decimal fraction can be rounded-off at the desired number of decimal places.

- ▲ Place the decimal point of the quotient directly above the decimal point of the dividend.
- ▲ Divide as with whole numbers.

Note: The remainder 12/28 is less than 1/2; therefore, the answer is 2.296 (or, an extra zero may be added as noted earlier).

```
(Quotient)  2.296
        28) 64.300
            56
             83
             56
             270
             252
              180
              168
               12
```

Example 2: 3.19 ÷ .072 (Answer is to be correct to 2 decimal places.)

- ▲ Move the decimal point 3 places in the divisor, and 3 places in the dividend.

```
072. ) 3190.
```

- ▲ Add 2 zeros to the dividend.

```
72 ) 3190.00
```

- ▲ Place the decimal point of the quotient directly above the decimal point of the dividend.
- ▲ Divide.

Note: The remainder 40/72 is greater than 1/2, therefore, the answer is 44.31.

$$44.30\frac{40}{72} = 44.31$$

```
      44.30 40/72
72 ) 3190.00
     288
      310
      288
       220
       216
        40
```

When dividing a decimal fraction or a mixed decimal by a whole number, it is not necessary to move the decimal point of either the divisor or the dividend. Add zeros to the right of the dividend, if necessary, to obtain the desired number of decimal places in the answer.

Examples: Divide .63 by 12 to 4 decimal places

```
     .0525
12) .6300
```

Divide 33.97 by 5 to 3 decimal places

```
    6.794
5) 33.970
```

APPLICATION

1. Divide these numbers. Give answers to the indicated number of decimal places.

a.	.72 ÷ .432	(3 places)	____		e.	1.017 ÷ .07	(3 places)	____
b.	.92 ÷ .36	(2 places)	____		f.	16.3/3.8	(2 places)	____
c.	.001 ÷ .1	(4 places)	____		g.	37/.273	(2 places)	____
d.	10 ÷ .001	(3 places)	____		h.	.003/.78	(4 places)	____

2. Rack sizes are given according to diametral pitch. Given 4 different diametral pitches, find the linear pitch and the whole depth of each rack to 4 decimal places.

$$\text{Linear Pitch} = \frac{3.1416}{\text{Diametral Pitch}} \qquad \text{Whole Depth} = \frac{2.157}{\text{Diametral Pitch}}$$

	Diametral Pitch	Linear Pitch	Whole Depth
a.	6.75		
b.	2.5		
c.	7.25		
d.	16.125		

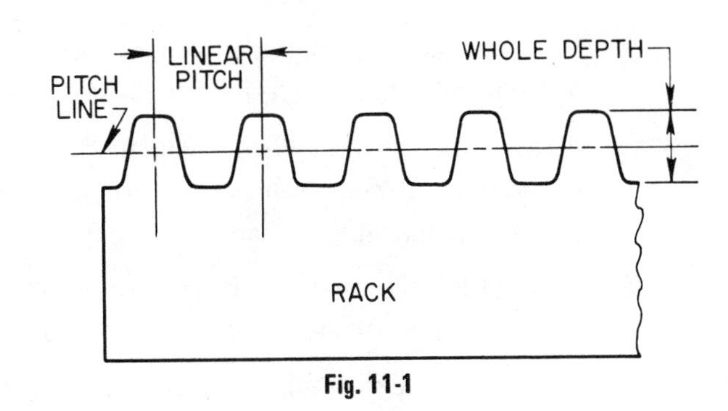

Fig. 11-1

3. Four sets of equally spaced holes are shown in this machined plate. Determine dimensions A, B, C, and D to 3 decimal places.

   a. Dim. A = ____

   b. Dim. B = ____

   c. Dim. C = ____

   d. Dim. D = ____

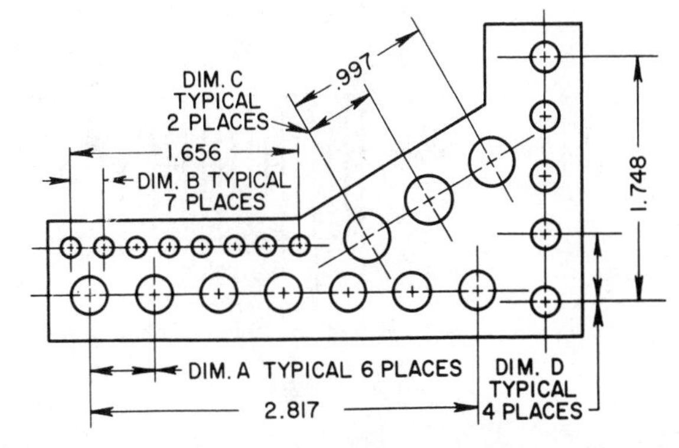

Fig. 11-2

4. A cross-sectional view of a bevel gear is shown. Given the diametral pitch and the number of gear teeth, determine the pitch diameter, the addendum, and the dedendum. Give answers to 4 decimal places.

$$\text{Pitch Diameter} = \frac{\text{Number of Teeth}}{\text{Diametral Pitch}}$$

$$\text{Dedendum} = \frac{1.157}{\text{Diametral Pitch}}$$

$$\text{Addendum} = \frac{1}{\text{Diametral Pitch}}$$

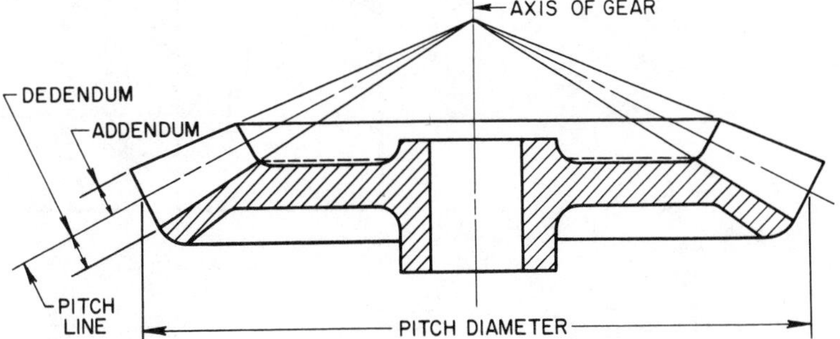

Fig. 11-3

	Diametral Pitch	Number of Teeth	Pitch Diameter	Addendum	Dedendum
a.	4	45			
b.	6	75			
c.	8	44			
d.	3	54			

5. How many complete bushings each .637 inch long can be cut from a bar of bronze which is 18.750 inches long? Allow .125 inch waste for each piece. __________

6. A shaft is being cut in a lathe. The tool feeds (advances) .015 inch each time the shaft turns once (1 revolution). How many revolutions will the shaft turn when the tool advances 3.120 inches? Give answer to 2 decimal places. __________

7. How much stock per stroke is removed by the wheel of a surface grinder if a depth of .1875 inch is reached after 75 strokes? Give answer to 4 decimal places. __________

8. An automatic screw machine is capable of producing one piece in .025 minutes. How many pieces can be produced in 1.375 hours? __________

9. This bolt has 7.7 threads. Determine the pitch to 3 decimal places. __________

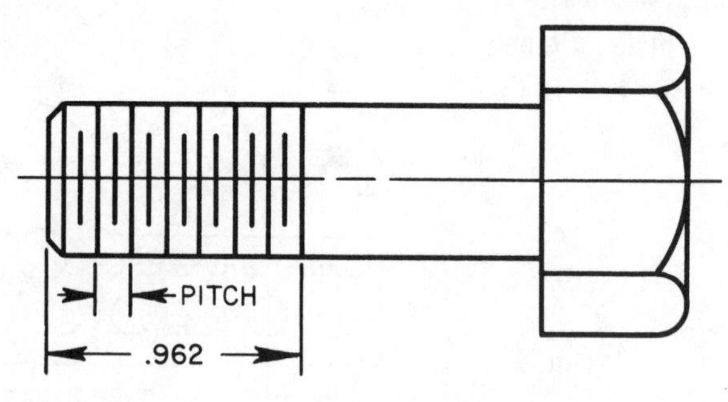

Fig. 11-4

10. This block has a threaded hole with a .0625 pitch. Determine the number of threads for the given depth to 2 decimal places.

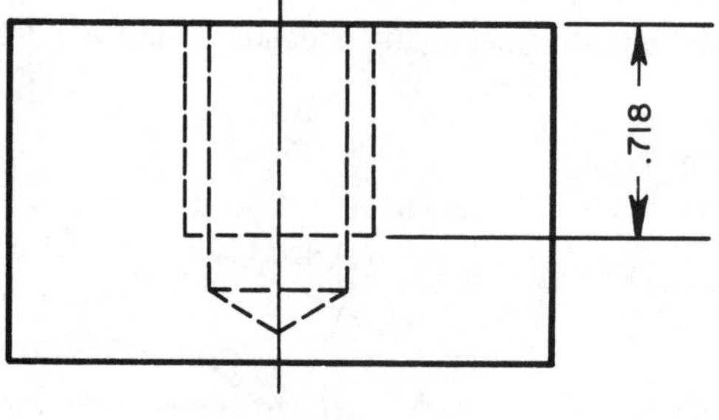

Fig. 11-5

11. The length of a side of a square equals the distance from point A to point B divided by 1.4142. Determine the length of a side of this square plate to 3 decimal places. __________

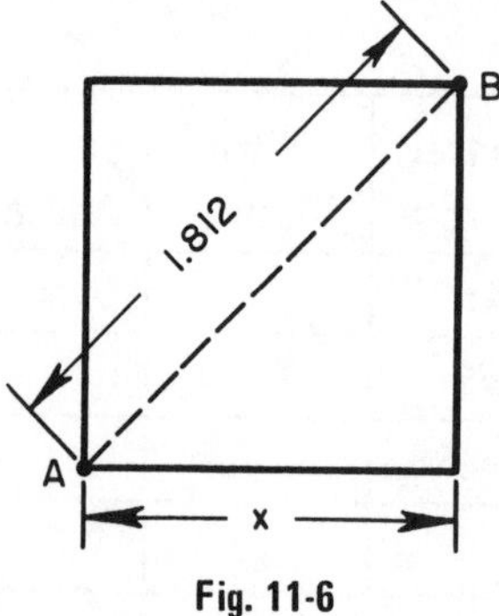

Fig. 11-6

Unit 12 Powers

OBJECTIVES

After studying this unit, the student should be able to

- Raise numbers to designated powers.
- Solve problems which involve combinations of powers with other basic operations.
- Solve problems requiring the use of powers in computing areas and volumes.

Powers of numbers are used to compute areas of square plates and circular sections and to compute volumes of cubes, cylinders, and cones. Use of powers is particularly helpful in determining distances in problems which require applications of geometry and trigonometry.

DESCRIPTION OF POWERS

Two or more numbers multiplied together to produce a given number are *factors* of the given number.

2 and 4 are factors of 8; $2 \times 4 = 8$

5 and 6 are factors of 30; $5 \times 6 = 30$

3 and 25 are factors of 75; $3 \times 25 = 75$

A *power* is the product of two or more equal factors.

3 x 3 is the second power of 3

5 x 5 x 5 is the third power of 5

.027 x .027 x .027 x .027 is the fourth power of .027

An *exponent* shows the number of times a number is to be taken as a factor. It is written above and to the right of the number.

3^2 means 3 x 3. The exponent 2 shows that 3 is taken as a factor twice. It is read as "3 to the second power" or "3 squared." $3^2 = 9$

2^5, two to the fifth power, means 2 x 2 x 2 x 2 x 2. $2^5 = 32$

$(3/4)^3$, three-quarters cubed, means 3/4 x 3/4 x 3/4. $(3/4)^3 = 27/64$

$.72^2$, .72 squared, means .72 x .72. $.72^2 = .5184$

3.5^4, 3.5 to the fourth power, means 3.5 x 3.5 x 3.5 x 3.5. $3.5^4 = 150.0625$

Example 1: Determine the areas of each of the two squares shown in figures 12-1 and 12-2. The area of a square equals the length of a side squared. The answer is given in square units.

Area = side^2

A = s^2

A = $(7/8'')^2$

A = $7/8'' \times 7/8''$

A = 49/64 square inches

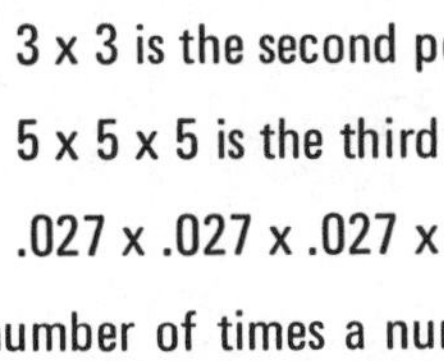

SIDE = $\frac{7''}{8}$

SIDE = $\frac{7''}{8}$

Fig. 12-1

Area = side^2

A = s^2

A = $(.75)^2$

A = $(.75') \times (.75')$

A = .5625 square feet

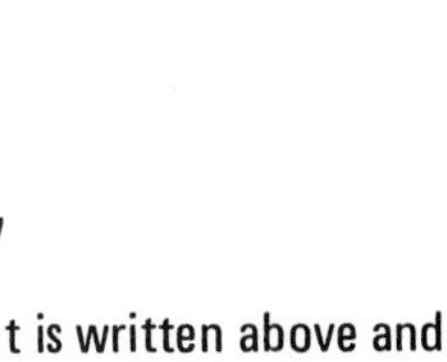

SIDE = .75'

SIDE = .75'

Fig. 12-2

Note: $A = s^2$ is called a *formula*. A formula is a short method of expressing an arithmetic relationship by the use of symbols. Known values may then be substituted for the symbols.

Example 2: Find the volume of the cube shown in figure 12-3. The volume of a cube equals the length of a side cubed. ($V = s^3$) The answer is given in cubic units.

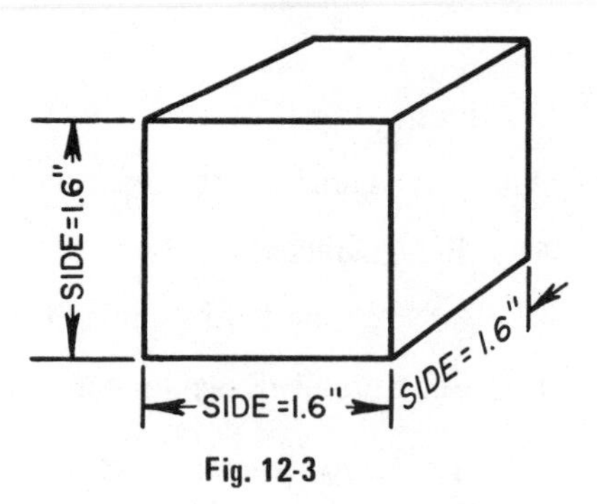

Fig. 12-3

$$\text{Volume} = \text{side}^3$$
$$V = s^3$$
$$V = (1.6'')^3$$
$$V = 1.6'' \times 1.6'' \times 1.6''$$
$$V = 4.096 \text{ cubic inches}$$

USE OF PARENTHESES

The parenthesis indicates that both the numerator and the denominator of a fraction are raised to the given power.

$$\left(\frac{2}{3}\right)^2 = \frac{2^2}{3^2} = \frac{2 \times 2}{3 \times 3} = \frac{4}{9}$$

Note: In this example, only the numerator is squared.,

$$\frac{2^2}{3} = \frac{2 \times 2}{3} = \frac{4}{3} = 1\frac{1}{3}$$

In this example, only the denominator is squared.

$$\frac{2}{3^2} = \frac{2}{3 \times 3} = \frac{2}{9}$$

Rule: To solve problems which involve operations within a parenthesis

- Perform the operations within the parenthesis first.
- Raise to the indicated power.

Examples:

$$(1.2 \times .6)^2 = .72^2 = .72 \times .72 = .5184$$
$$(.5 + 2.4)^2 = 2.9^2 = 2.9 \times 2.9 = 8.41$$
$$(.75 - .32)^2 = .43^2 = .43 \times .43 = .1849$$
$$\left(\frac{14.4}{3.2}\right)^2 = 4.5^2 = 4.5 \times 4.5 = 20.25$$

When solving power problems which also require addition, subtraction, multiplication, or division perform the power operation first.

Examples:

$$5 \times 3^2 - 12 = 5 \times 9 - 12 = 45 - 12 = 33$$
$$33.5 - 5.5^2 + 8.7 = 33.5 - 30.25 + 8.7 = 11.95$$
$$\frac{2.2^3 - 5.608}{1.4} = \frac{10.648 - 5.608}{1.4} = \frac{5.040}{1.4} = 3.6$$

Example: Compute the volume of the cylinder shown in figure 12-4 to 4 decimal places.

Volume = 3.14 x radius squared x height.

$V = \pi \times r^2 \times H$

$V = 3.14 \times .85^2 \times 1.25$

$V = 3.14 \times .7225 \times 1.25$

$V = 2.8358$ cubic inches

Note: The symbol π (pi) equals 3.14, 3 1/7, or 3.1416. It is a constant value used in arithmetic relationships involving circles.

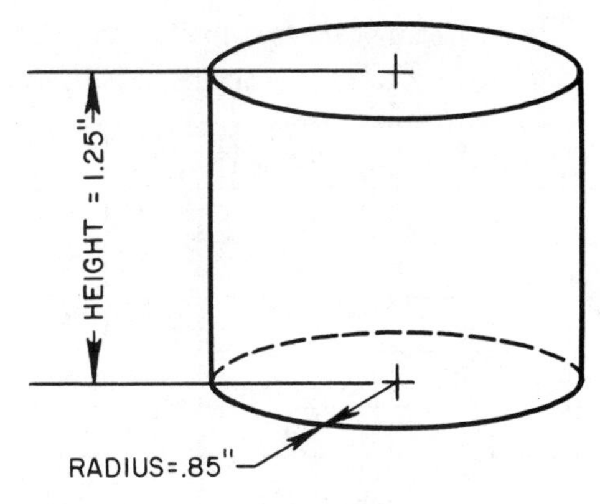

Fig. 12-4

APPLICATION

A. Raise the following numbers to the indicated power.

1. 3.4^3 ____
2. 1^8 ____
3. 100^4 ____
4. $(\frac{3}{4})^3$ ____
5. $\frac{3^3}{4}$ ____
6. $\frac{3}{4^3}$ ____
7. $(.3 \times 7)^2$ ____
8. $(18.6 + 9.5)^2$ ____
9. $(\frac{28.8}{7.2})^3$ ____

B. In the table below, the lengths of the sides of squares are given. Determine the areas of the squares. (Area = side^2, or $A = s^2$)

	S (in.)	A (sq. in.)
1.	1.25	
2.	23.07	
3.	.19	
4.	10.7	
5.	.02	

	S (in.)	A (sq. in.)
6.	1/4	
7.	7/8	
8.	3 1/2	
9.	13/16	
10.	13 3/4	

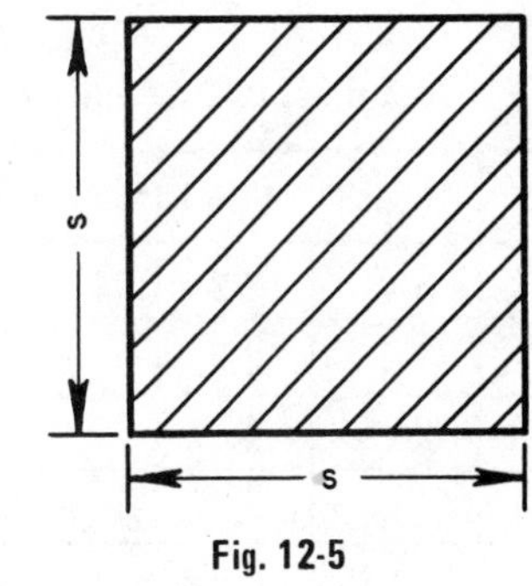

Fig. 12-5

C. In the table below, the lengths of the sides of cubes are given. Determine the volumes of the cubes. (Volume = side^3, or $V = s^3$) Round-off answers to 4 decimal places where necessary.

	S (in.)	V (cu. in.)
1.	.37	
2.	20.6	
3.	3.93	
4.	12	
5.	.075	

	S (in.)	V (cu. in.)
6.	1/3	
7.	7/8	
8.	1 1/2	
9.	10 1/4	
10.	3/4	

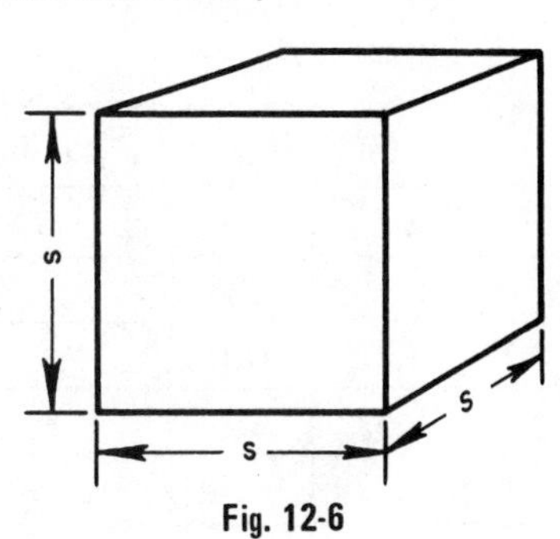

Fig. 12-6

D. In the table below, the radii of circles are given. Determine the areas of the circles. (Area = $3.14 \times R^2$, or $A = \pi \times R^2$) Give all answers in decimal form and round-off to 3 decimal places where necessary.

	R (in.)	A (sq. in.)
1.	6.2	
2.	18.	
3.	.07	
4.	9.28	
5.	.35	

	R (in.)	A (sq. in.)
6.	1/2	
7.	3/16	
8.	1 3/8	
9.	2 1/4	
10.	7/8	

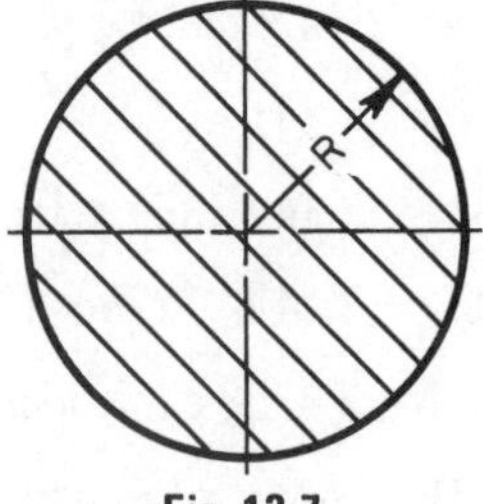

Fig. 12-7

E. In the table below, the diameters of spheres are given. Determine the volumes of the spheres. (Volume = $\frac{\pi \times D^3}{6}$; π = 3.14; D = Diameter.) Give all answers in decimal form and round-off to 2 decimal places where necessary.

	D (in.)	V (cu. in.)
1.	.45	
2.	.5	
3.	.75	
4.	.8	
5.	1.6	

	D (in.)	V (cu. in.)
6.	1/4	
7.	3/8	
8.	3 1/2	
9.	7/8	
10.	3/4	

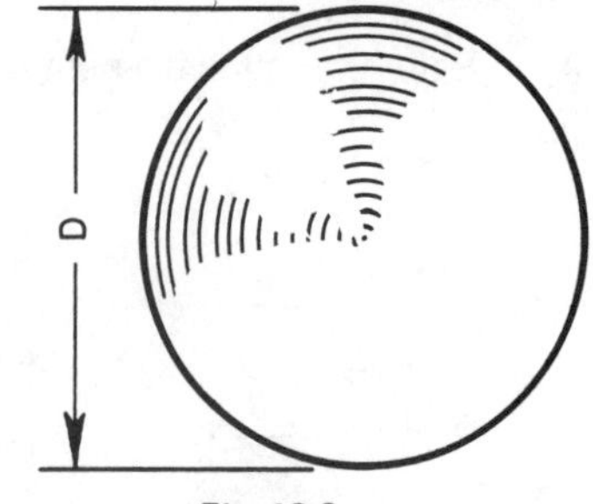

Fig. 12-8

F. In the table below, the radii and heights of cylinders are given. Determine the volumes of the cylinders. ($V = \pi \times R^2 \times H$; π = 3.14; H = height; R = radius.) Give all answers in decimal form and round-off to 2 decimal places where necessary.

	R (in.)	H (in.)	V (cu. in.)
1.	5	3.2	
2.	1.5	2.3	
3.	2.25	4.	
4.	.7	6.7	
5.	.81	.72	

	R (in.)	H (in.)	V (cu. in.)
6.	1/2	3	
7.	3/4	1 1/2	
8.	1 1/4	3	
9.	1/8	7 3/4	
10.	1	7/8	

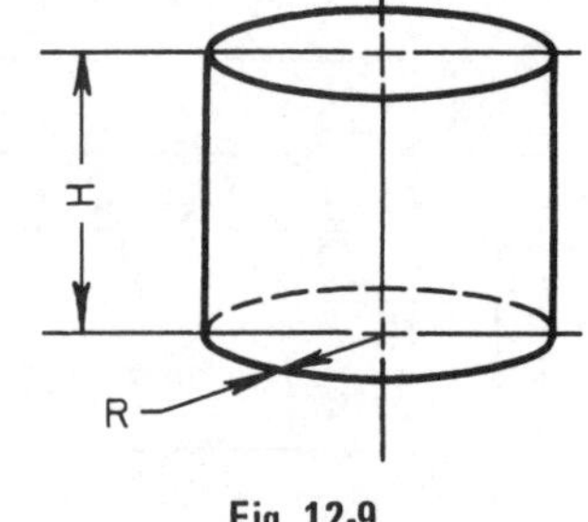

Fig. 12-9

G. In the table below, the diameters and heights of cones are given. Find the volumes of the cones. ($V = .2618 \times D^2 \times H$) Give all answers in decimal form and round-off to 2 decimal places where necessary.

	D (in.)	H (in.)	V (cu. in.)
1.	3.2	4	
2.	3	5	
3.	1.72	3.1	
4.	.92	6.2	
5.	.17	.88	

	D (in.)	H (in.)	V (cu. in.)
6.	1/4	4	
7.	3	7 1/8	
8.	1 3/4	5 1/2	
9.	2 3/8	1 5/8	
10.	1/2	9 1/2	

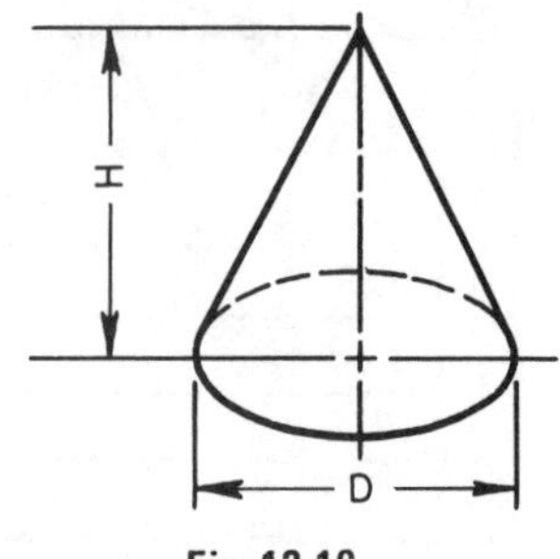

Fig. 12-10

H. Solve the following problems. Round-off answers to 3 decimal places where necessary.

1. Find the area of this square plate. A = ____
   ($A = S^2$)

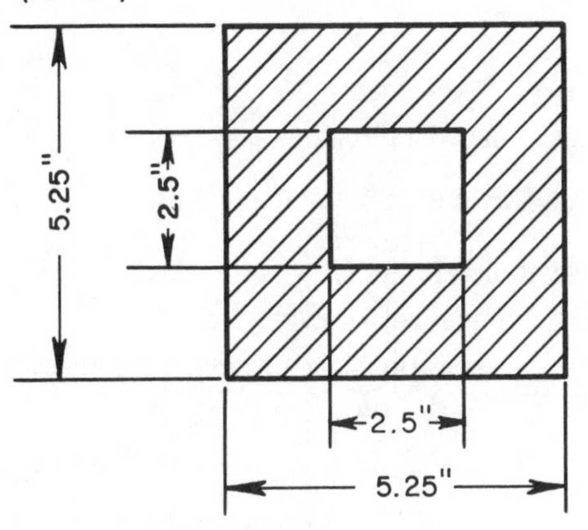

Fig. 12-11

2. Find the area of this washer. A = ____
   ($A = \pi R^2$)

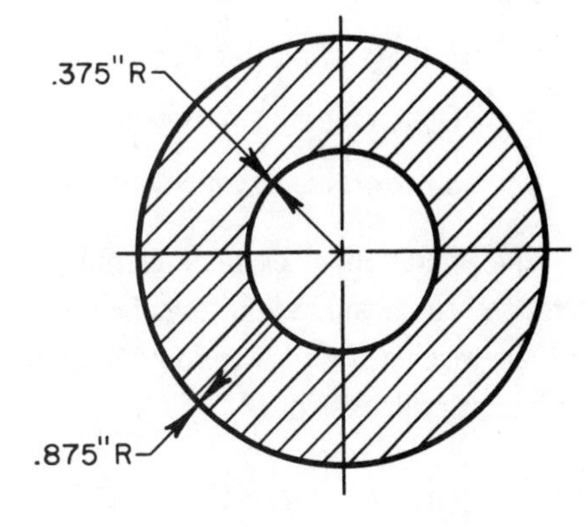

Fig. 12-12

3. Find the area of this spacer. A = ____

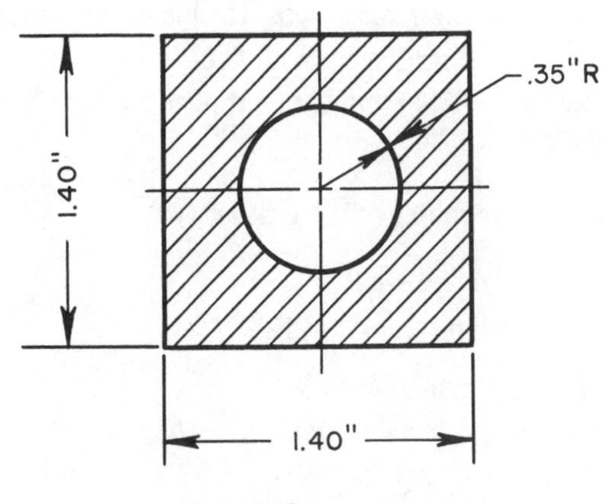

Fig. 12-13

4. Find the area of this plate. A = ____

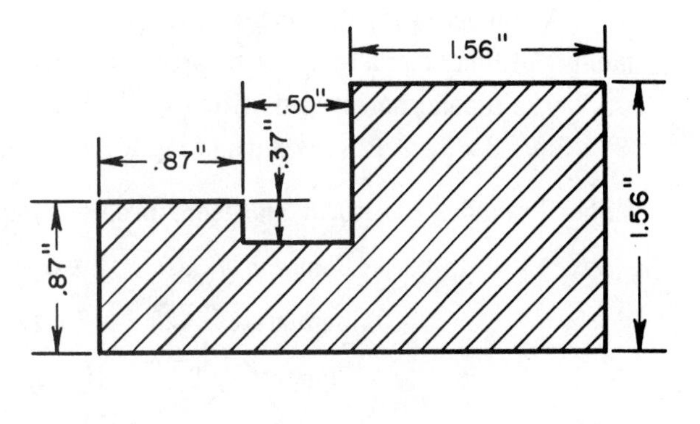

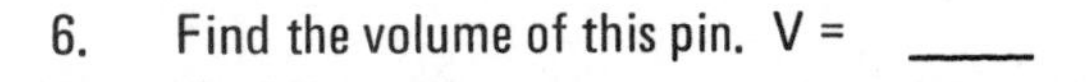

Fig. 12-14

5. Find the volume of this bushing. V = ____

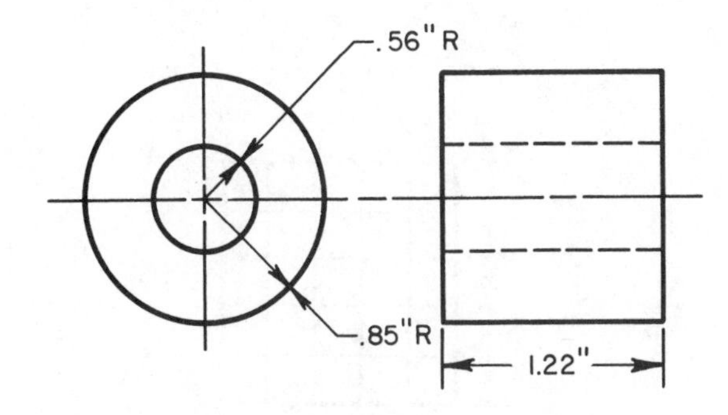

Fig. 12-15

6. Find the volume of this pin. V = ____

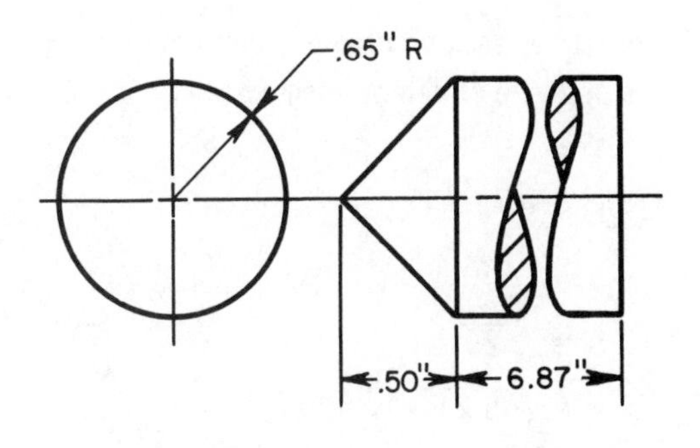

Fig. 12-16

Unit 13 Roots

OBJECTIVES

After studying this unit, the student should be able to

- Extract whole number roots.
- Determine square roots to any designated number of decimal places.
- Solve problems which involve combinations of roots with other basic operations.

The operation of extracting roots of numbers is used to determine lengths of sides and heights of squares and cubes and radii of circular sections when areas and volumes are known.

The machinist also uses roots in computing distances between various parts of machined pieces from given blueprint dimensions.

DESCRIPTION OF ROOTS

The *root* of a number is a quantity which is taken two or more times as an equal factor of the number. Determining a root is the opposite operation of determining a power.

The symbol $\sqrt{}$, called a *radical sign,* indicates a root of a number.

A number called the *index* is written to the left and above the radical sign. The index indicates the number of times that a root is to be taken as an equal factor to produce the given number.

The square root of 9 is written $\sqrt[2]{9}$. It asks the question, "What number multiplied by itself equals 9?" Since 3 x 3 = 9, 3 is the square root of 9. Therefore, $\sqrt[2]{9} = 3$.

Note: The index 2 is usually omitted in problems requiring square roots; $\sqrt[2]{9}$ is written $\sqrt{9}$.

Compute $\sqrt{36}$.	6 x 6 = 36; therefore, $\sqrt{36} = 6$
Compute $\sqrt{144}$.	12 x 12 = 144; therefore, $\sqrt{144} = 12$
Compute $\sqrt[3]{8}$	(Read, "compute the cube root of 8.") Since 2 x 2 x 2 = 8, the $\sqrt[3]{8} = 2$
Compute $\sqrt[3]{125}$.	5 x 5 x 5 = 125; therefore, $\sqrt[3]{125} = 5$
Compute $\sqrt[4]{81}$.	(Read, "compute the fourth root of 81.") Since 3 x 3 x 3 x 3 = 81, the $\sqrt[4]{81} = 3$

Example 1: Compute the length of the side of the square shown in figure 13-1. This square has an area of 25 square inches. Since $A = S^2$, the length of a side of the square equals the square root of the area.

Note: It has been pointed out that determining a root is the opposite of determining a power.

$$A = S^2$$
$$25 = S^2$$
$$S = \sqrt{A}$$
$$S = \sqrt{25}$$
$$S = \sqrt{5 \times 5}$$
$$S = 5 \text{ inches}$$

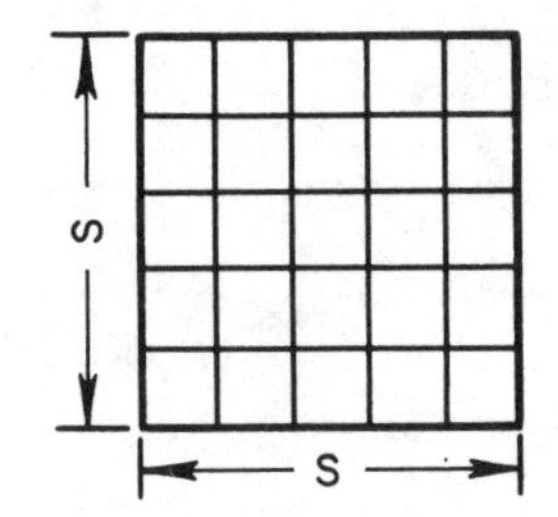

Fig. 13-1

Example 2: Compute the length of the side of the cube shown in figure 13-2. The volume of this cube equals 64 cubic inches. The length of a side of a cube equals the cube root of the volume. ($S = \sqrt[3]{V}$)

$$S = \sqrt[3]{V}$$

$$S = \sqrt[3]{64}$$

$$S = \sqrt[3]{4 \times 4 \times 4} = 4 \text{ inches}$$

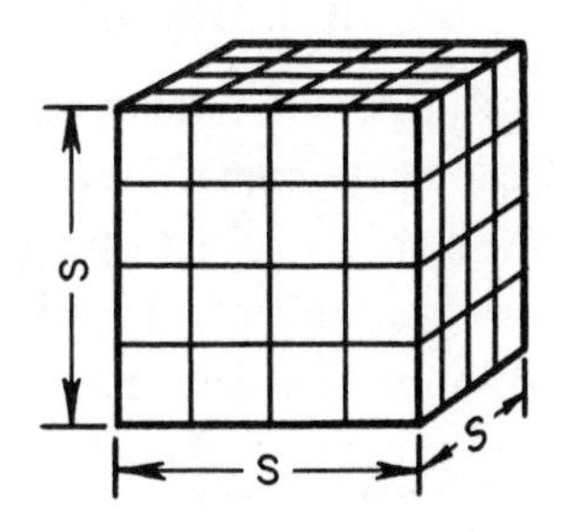

Fig. 13-2

If a root of a fraction is to be computed, both the numerator and the denominator must be enclosed with the radical sign.

Examples:

$$\sqrt{\frac{16}{25}}\,;\ \frac{4}{5} \times \frac{4}{5} = \frac{16}{25}\,;\ \text{Therefore, } \sqrt{\frac{16}{25}} = \frac{4}{5}$$

$$\sqrt[3]{\frac{27}{64}}\,;\ \frac{3}{4} \times \frac{3}{4} \times \frac{3}{4} = \frac{27}{64}\,;\ \text{Therefore, } \sqrt[3]{\frac{27}{64}} = \frac{3}{4}$$

Note: In this example, only the root of the numerator is taken.

$$\frac{\sqrt{16}}{25} = \frac{\sqrt{4 \times 4}}{25} = \frac{4}{25}$$

Note: In this example, only the root of the denominator is taken.

$$\frac{16}{\sqrt{25}} = \frac{16}{\sqrt{5 \times 5}} = \frac{16}{5} = 3\frac{1}{5}$$

Rule: To solve problems which involve operations within the radical sign

- Perform the operations within the radical sign first
- Then compute the root.

Examples:

$$\sqrt{3 \times 12} = \sqrt{36} = \sqrt{6 \times 6}; \qquad \text{therefore, } \sqrt{3 \times 12} = 6$$

$$\sqrt{5 + 59} = \sqrt{64} = \sqrt{8 \times 8}; \qquad \text{therefore, } \sqrt{5 + 59} = 8$$

$$\sqrt{128 - 7} = \sqrt{121} = \sqrt{11 \times 11}; \qquad \text{therefore, } \sqrt{128 - 7} = 11$$

$$\sqrt{\frac{32}{2}} = \sqrt{16} = \sqrt{4 \times 4}; \qquad \text{therefore, } \sqrt{\frac{32}{2}} = 4$$

Example: Compute the length of the chord C of the circular segment shown in figure 13-3.

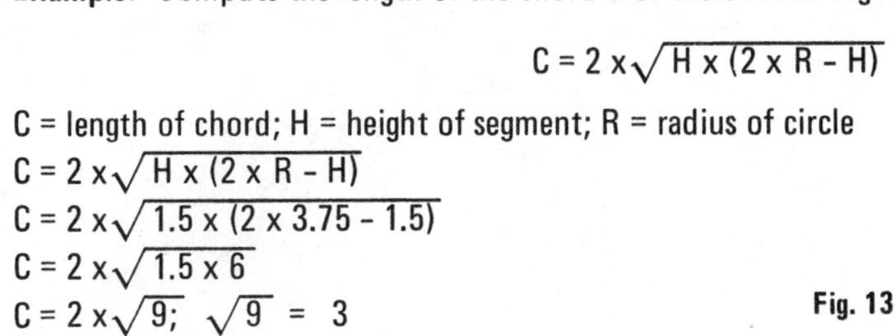

$$C = 2 \times \sqrt{H \times (2 \times R - H)}$$

C = length of chord; H = height of segment; R = radius of circle

$$C = 2 \times \sqrt{H \times (2 \times R - H)}$$

$$C = 2 \times \sqrt{1.5 \times (2 \times 3.75 - 1.5)}$$

$$C = 2 \times \sqrt{1.5 \times 6}$$

$$C = 2 \times \sqrt{9};\ \sqrt{9} = 3$$

$$C = 2 \times 3 = 6 \text{ inches}$$

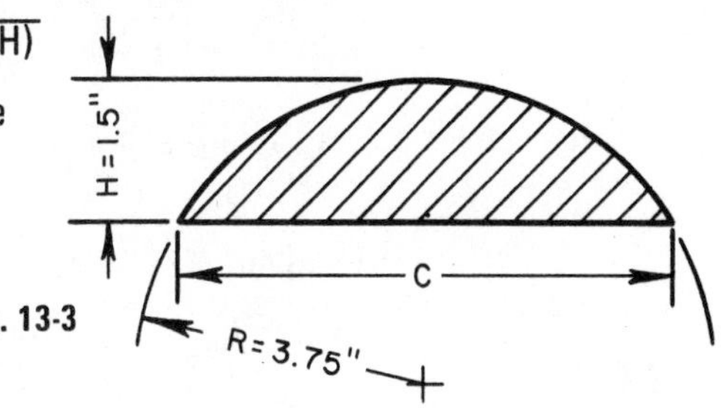

Fig. 13-3

GENERAL METHOD OF COMPUTING SQUARE ROOTS

The square root examples shown have all consisted of perfect squares. *Perfect squares* are numbers which have whole number square roots. These roots are relatively easy to determine by observation.

Most numbers do not have whole number square roots; therefore, a definite procedure must be used in computing square roots of most numbers.

The following examples illustrate the procedure used in determining square roots.

Example 1: Find the square root of 5410.218 to 2 decimal places.

▲ Beginning at the decimal point, group the digits in pairs.

Annex a zero to the 8 in order to form a pair of digits

▲ Place the decimal point directly above the decimal point of the number.

$$\sqrt{54\ 10\ .\ 21\ 80}$$

▲ Find the largest perfect square that can be subtracted from the first digit or pair of digits. The largest perfect square is 49. Write the square root of this perfect square above the first digit or group of digits. The square root of 49 is 7.

```
   7   .
 √54 10 . 21 80
  49
   5
```

▲ Bring down the next pair of digits (10) and place by the remainder (5).

```
   7   .
 √54 10 . 21 80
  49
   5 10
```

▲ Double the partial root and use this number as a trial divisor.

$2 \times 7 = 14$

```
     7   .
   √54 10 . 21 80
    49
14) 5 10
```

▲ Divide this trial divisor into the remainder, disregarding the last digit of the remainder.

$51 \div 14 = 3$

▲ Annex the quotient as the next figure in the square root. Also annex the same digit to the trial divisor.

Annex 3 to the root and to the trial divisor.

```
      7  3 .
    √54 10 . 21 80
     49
143) 5 10
```

▲ Multiply the complete divisor by the digit which was annexed to the root, and subtract.

Multiply $3 \times 143 = 429$

Subtract $510 - 429 = 81$

```
      7  3 .
    √54 10 . 21 80
     49
143) 5 10
     4 29
       81
```

▲ Repeat the process until the desired number of decimal places is obtained.

Bring down 21

Double 73; $2 \times 73 = 146$

Divide 146 into 812; $812 \div 146 = 5$.

Annex 5 to the root and to the trial divisor.

Multiply: 5 x 1465 = 7325.

Subtract: 8121 - 7325 = 796.

Bring down 80.

Double 735; 2 x 735 = 1470.

Divide 1470 into 7968; 7968 ÷ 1470 = 5.

Annex 5 to the root and to the trial divisor.

Multiply: 5 x 14705 = 73525.

Subtract: 79680 - 73525 = 6155.

```
           7  3 .  5  5
        √ 54 10 . 21 80
          49
  143 )  5 10
         4 29
 1465 )   81 21
          73 25
14705 )   7 96 80
          7 35 25
            61 55
```

Note: The remainder $\frac{6155}{14705}$ is less than 1/2; therefore, the answer is 73.55

Example 2: Determine the square root of 923.7 to 3 decimal places.

```
          3  0 .  3  9  2
        √ 9 23 . 70 00 00
          9
   60 )   0 23
            00
  603 )     23 70
            18 09
 6069 )      5 61 00
             5 46 21
60782 )        14 79 00
               12 15 64
                2 63 36
```

Note: The remainder $\frac{26336}{60782}$ is less than 1/2, therefore the answer is 30.392.

Example 3: Determine the square root of .0039 to 4 decimal places.

```
         .0  6  2  4
      √ .00 39 00 00
            36
 122 )       3 00
             2 44
1244 )         56 00
               49 76
                6 24
```

Note: The remainder $\frac{624}{1244}$ is greater than therefore the answer is .0625.

APPLICATION

A. The following problems have either whole number roots or numerators and denominators which have whole number roots. Determine these roots.

1. $\sqrt[3]{216}$ ____
2. $\sqrt{\frac{9}{16}}$ ____
3. $\frac{\sqrt{4}}{9}$ ____
4. $\frac{25}{\sqrt{36}}$ ____
5. $\sqrt{\frac{3}{4} \times \frac{3}{4}}$ ____
6. $\sqrt{.5 \times 8}$ ____
7. $\sqrt{56.7 + 87.3}$ ____
8. $\sqrt{16.4 - 7.4}$ ____
9. $\sqrt[3]{\frac{428.8}{6.7}}$ ____

B. The following problems have whole number square roots. Solve for the missing values in the tables.

1. The areas of squares are given in this table. Determine the lengths of the sides.

$$S = \sqrt{A}$$

	A (sq.in.)	S (in.)
a.	225	
b.	121	
c.	100	
d.	81	
e.	1	

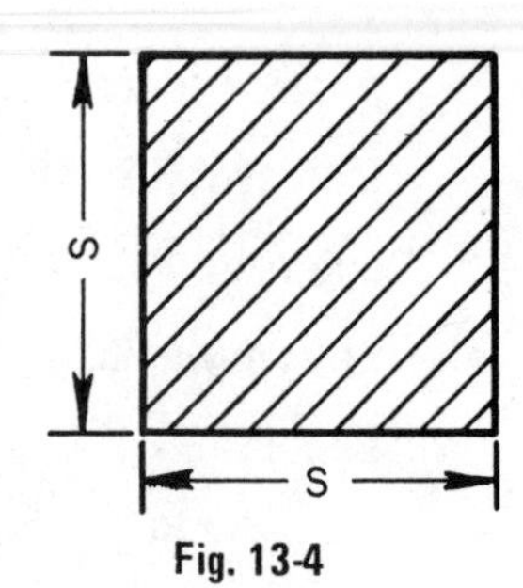

Fig. 13-4

2. The volumes of cubes are given in this table. Determine the lengths of the sides.

$$S = \sqrt[3]{V}$$

	V (cu.in.)	S (in.)
a.	216	
b.	64	
c.	343	
d.	1000	
e.	1	

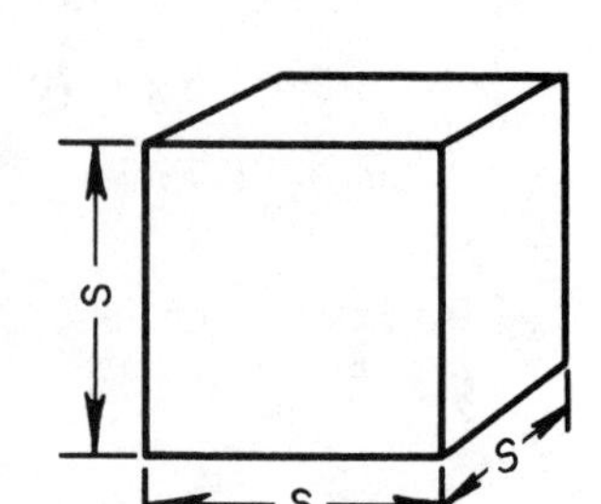

Fig. 13-5

3. The areas of circles are given in this table. Determine the lengths of the radii.

$$R = \sqrt{\frac{A}{\pi}}, \quad \pi = 3.14$$

	A (sq.in.)	R (in.)
a.	50.24	
b.	12.56	
c.	314	
d.	28.26	
e.	153.86	

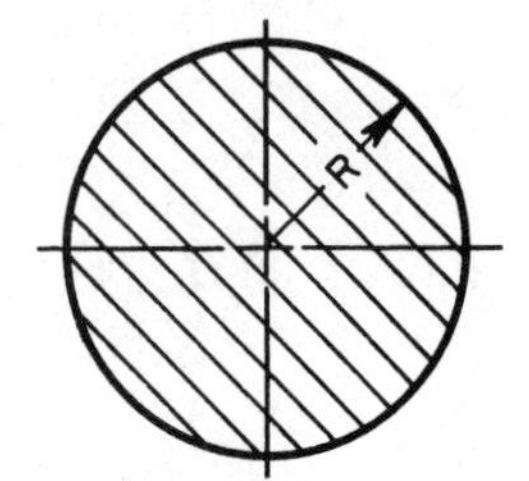

Fig. 13-6

4. The volumes of spheres are given in this table. Determine the lengths of the diameters.

$$D = \sqrt[3]{\frac{V}{.5236}}$$

	V (cu.in.)	D (in.)
a.	14.1372	
b.	113.0976	
c.	4.1888	
d.	.5236	
e.	523.6	

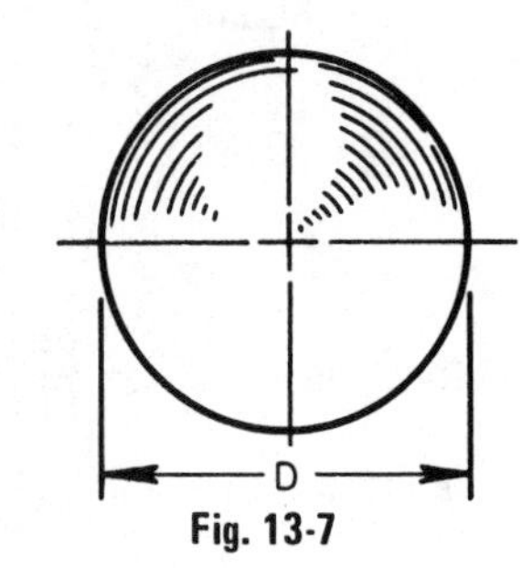

Fig. 13-7

C. The following problems have square roots that are not whole numbers. Compute these roots to the indicated number of decimal places.

1. $\sqrt{12.54}$ (3) ____
2. $\sqrt{391}$ (2) ____
3. $\sqrt{\frac{4}{5}}$ (3) ____
4. $\sqrt{3\frac{1}{2}}$ (3) ____
5. $\sqrt{.07 \times 32}$ (2) ____
6. $\sqrt{15.82 + 3.71}$ (2) ____
7. $\sqrt{178.5 - 163.7}$ (3) ____
8. $\sqrt{\frac{.441}{70}}$ (4) ____

D. The following problems have roots that are not whole numbers. Solve for the missing values in the tables.

1. The volumes of cylinders and their heights are given in this table. Find the lengths of the radii to 2 decimal places.

$$R = \sqrt{\frac{V}{\pi \times H}}; \quad \pi = 3.14$$

	V(cu.in.)	H(in.)	R(in.)
a.	249.896	5	
b.	132.634	12	
c.	14	29	
d.	10	28	

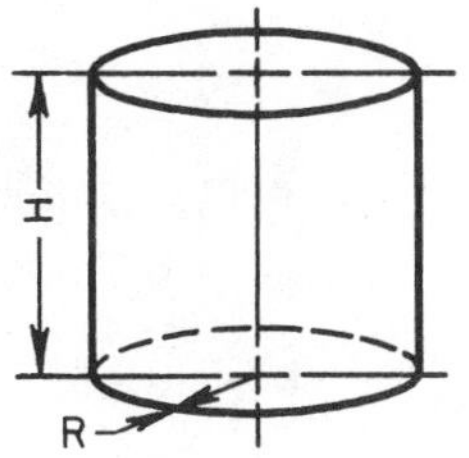

Fig. 13-8

2. The volumes of cones and their heights are given in this table. Compute the lengths of the diameters to 2 decimal places.

$$D = \sqrt{\frac{V}{.262 \times H}}$$

	V(cu.in.)	H(in.)	D(in.)
a.	116.328	6	
b.	19.388	2	
c.	1257.6	10	
d.	15	55	

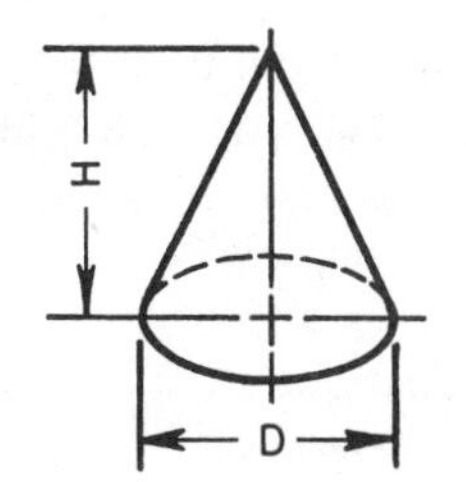

Fig. 13-9

E. Solve the following problems.

1. The pitch of broach teeth depends upon the length of cut, the depth of cut, and the material being broached. The minimum pitch = $3 \times \sqrt{L \times d \times F}$; L = length of cut; d = depth of cut; F = a factor related to the type of material being broached. Find the minimum pitch to 3 decimal places for broaching cast iron where L = .825", d = .005", F = 5. __________

2. The dimensions of keys and keyways are determined in relation to the diameter of the shafts with which they are used.

$$D = \sqrt{\frac{L \times T}{.3}}$$

D = shaft diameter; L = key length; T = key thickness

What is the shaft diameter that would be used with a key where L = 2.7" and T = .25"?

____

Unit 14 Table of Decimal Equivalents and Combined Operations of Decimals

OBJECTIVES

After studying this unit, the student should be able to

- Determine fractional and decimal equivalents by using the decimal equivalent table.
- Determine nearest fractional equivalents of decimals by using the decimal equivalent table.
- Solve problems using any combination of operations.

Generally, fractional blueprint dimensions are given in multiples of 64ths of an inch. A machinist is often required to convert these fractional dimensions to decimal equivalents for machine settings. When laying out parts such as castings that have ample stock allowances, it is sometimes convenient to use a fractional steel scale and to convert decimal dimensions to the nearest equivalent fractions. The amount of computation and the chances of error can be reduced by using the decimal equivalent table.

DECIMAL EQUIVALENT TABLE

Fraction	Decimal	Fraction	Decimal	Fraction	Decimal	Fraction	Decimal
1/64	.015625	17/64	.265625	33/64	.515625	49/64	.765625
1/32	.03125	9/32	.28125	17/32	.53125	25/32	.78125
3/64	.046875	19/64	.296875	35/64	.546875	51/64	.796875
1/16	.0625	5/16	.3125	9/16	.5625	13/16	.8125
5/64	.078125	21/64	.328125	37/64	.578125	53/64	.828125
3/32	.09375	11/32	.34375	19/32	.59375	27/32	.84375
7/64	.109375	23/64	.359375	39/64	.609375	55/64	.859375
1/8	.125	3/8	.375	5/8	.625	7/8	.875
9/64	.140625	25/64	.390625	41/64	.640625	57/64	.890625
5/32	.15625	13/32	.40625	21/32	.65625	29/32	.90625
11/64	.171875	27/64	.421875	43/64	.671875	59/64	.921875
3/16	.1875	7/16	.4375	11/16	.6875	15/16	.9375
13/64	.203125	29/64	.453125	45/64	.703125	61/64	.953125
7/32	.21875	15/32	.46875	23/32	.71875	31/32	.96875
15/64	.234375	31/64	.48438	47/64	.734375	63/64	.984375
1/4	.250	1/2	.500	3/4	.750	1	1.000

The following examples illustrate the use of the decimal equivalent table.

Example 1: Find the decimal equivalent of 23/32". The decimal equivalent .71875" is shown directly to the right of 23/32".

Example 2: Find the fractional equivalent of .3125". The fractional equivalent 5/16" is shown directly to the left of .3125".

Example 3: Find the nearest fractional equivalents of the decimal dimensions given on the casting shown in figure 14-1.

Compute dimension A. The decimal .757 lies between .750 and .765625. The difference between .757 and .750 is .007. The difference between .757 and .765625 is .008625. Since .007 is less than .008625, the .750 value is closer to .757. The fractional equivalent of .750" is 3/4".

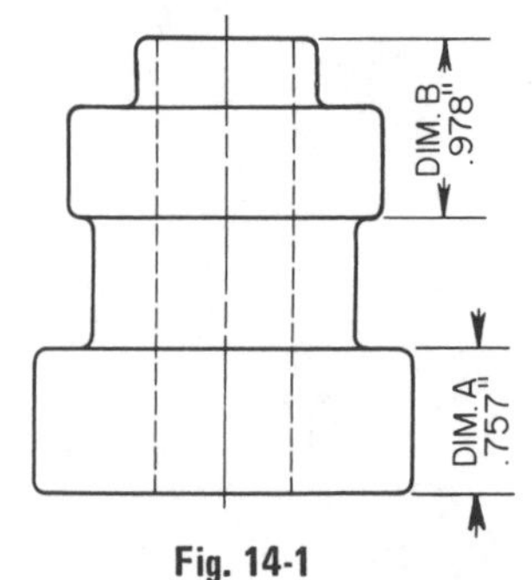

Fig. 14-1

Compute dimension B. The decimal .978 lies between .96875 and .984375. The difference between .978 and .96875 is .00925. The difference between .978 and .984375 is .006375. Since .006375 is less than .00925, the .984375 value is closer to .978. The fractional equivalent of .984375" is 63/64".

COMBINED OPERATIONS OF DECIMALS

In the process of completing a job, a machinist must determine stock sizes, cutter sizes, feeds and speeds, and roughing allowances as well as cutting dimensions. Usually most and sometimes all of the fundamental operations of mathematics must be used for computations in the manufacture of a part.

Determination of powers and roots must also be considered in the order of operations which was discussed in unit 6. The following rule incorporates all six fundamental operations.

Rule for order of six fundamental operations:

- All operations within parentheses. If there are parentheses within parentheses, perform the operations within the innermost parenthesis first.
- Powers and roots. Powers and roots are determined as they occur.
- Multiplication and division. Perform multiplication and division operations in the order in which they occur.
- Addition and subtraction. Addition and subtraction can be taken in any order.

Example 1: Find the value of $7.875 + 3.2 \times 4.3 - 2.73$.

- ▲ Multiply: $3.2 \times 4.3 = 13.76$
- ▲ Add: $7.875 + 13.76 = 21.635$
- ▲ Subtract: $21.635 - 2.73 = 18.905$

Example 2: Find the value of $(27.34 - 4.82) \div (2.41 \times 1.78 + 7.89)$.

- ▲ Perform operations within parentheses.
  Subtract: $27.34 - 4.82 = 22.52$ (first parenthesis)
  Multiply: $2.41 \times 1.78 = 4.2898$ (second parenthesis)
  Add: $4.2898 + 7.89 = 12.1798$ (second parenthesis)
- ▲ Divide: $22.52 \div 12.1798 = 1.85$ (to 2 decimal places)

Example 3: Find the value of $\dfrac{13.79 + (27.6 \times .3)^2}{\sqrt{23.04} + .875 - 3.76}$

- ▲ Consider the numerator and the denominator as if each were within parentheses. All of the operations are performed in the numerator and in the denominator before the division is performed.

$$\frac{13.79 + (27.6 \times .3)^2}{\sqrt{23.04} + .875 - 3.76} = (13.79 + (27.6 \times .3)^2) \div (\sqrt{23.04} + .875 - 3.76)$$

- ▲ In the numerator:
  - ▲ Multiply: $27.6 \times .3 = 8.28$
  - ▲ Square: $8.28^2 = 68.5584$.
  - ▲ Add: $13.79 + 68.5584 = 82.3484$
- ▲ In the denominator:
  - ▲ Extract the square root: $\sqrt{23.04} = 4.8$
  - ▲ Add: $4.8 + .875 = 5.675$
  - ▲ Subtract: $5.675 - 3.76 = 1.915$
  - ▲ Divide: $82.3484 \div 1.915 = 43.002$ (to 3 decimal places)

Example 4: Blanks in the shape of regular pentagons (5-sided figures) are punched from strip stock as shown in figure 14-2. Determine the width of strip stock required, using the given dimensions and the formula for dimension R.

$$R = \sqrt{r^2 + s^2 \div 4}$$

$$\text{Width} = R + .980 + 2 \times .125$$

$$\text{Width} = \sqrt{.980^2 + 1.424^2 \div 4} + .980 + 2 \times .125$$

▲ Compute the operations under the radical sign.

- ▲ Square: $.980^2 = .9604$
- ▲ Square: $1.424^2 = 2.027776$
- ▲ Divide: $2.027776 \div 4 = .506944$
- ▲ Add: $\sqrt{.9604 + .506944} = \sqrt{1.467344}$
- ▲ Extract the square root: $\sqrt{1.467344} = 1.211$

▲ Multiply: $2 \times .125 = .250$

▲ Add: $1.211 + .980 + .250 = 2.441$ inches

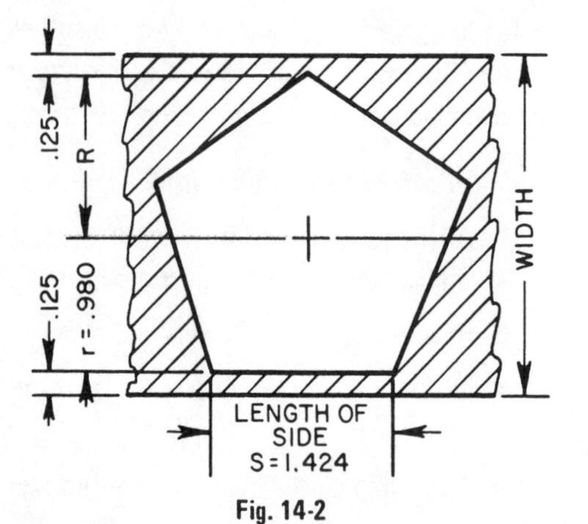

Fig. 14-2

Note: In solving expressions which consist of numerous multiplication and power operations, it is often necessary to carry out the work to two or three more decimal places than the number of decimal places required in the answer.

APPLICATION

A. Find the fractional or decimal equivalents of these numbers using the decimal equivalent table.

1. 25/32 ____	4. 15/16 ____	7. .3125 ____
2. 9/32 ____	5. 5/64 ____	8. .28125 ____
3. 11/32 ____	6. .671875 ____	9. .203125 ____

B. Find the nearest fractional equivalents of these decimals using the decimal equivalent table.

1. .541 ____	3. .459 ____	5. .209 ____
2. .762 ____	4. .498 ____	6. .805 ____

C. Solve these examples of combined operations. Round-off answers to 2 decimal places where necessary.

1. $.5231 + 11.664 \div 4.32 \times .521$ ____	6. $27.16 \div \sqrt{1.76 + 12.32}$ ____
2. $81.07 \div 12.1 + 2 \times 3.7$ ____	7. $(\sqrt{3.98 + .87 \times 3.9})^2$ ____
3. $\frac{56.050}{3.8} \times .875 - 3.92$ ____	8. $(3.29 \times 1.7)^2 \div (3.82 - .37)$ ____
4. $(24.78 - 19.32) \times 4.6$ ____	9. $.25 \times (\frac{\sqrt{64} \times 3.87}{8.32 - 5.13}) + 18.3^2$ ____
5. $(13.5 \div 4 - 1.76)^2 \times 4.5$ ____	10. $18.32 - \sqrt{\frac{7.88 + 13.5}{3.5^2 - .52}} \times .7$ ____

D. Solve the following problems which require combined operations.

1. The figure shows the three-wire method of checking screw threads. With proper diameter wires and a micrometer, very accurate pitch diameter measurements can be made. Using the formula given, determine the micrometer dimension over wires of the American Standard threads in the table below. Give answer to 4 decimal places. $M = D - (1.5155 \times P) + (3 \times W)$

	Major Diameter D	Pitch P	Wire Diameter W	Dimension Over Wires M
a.	.8750	.1250	.0900	
b.	.2500	.0500	.0350	
c.	.6250	.1000	.0700	
d.	1.3750	.16667	.1500	
e.	2.5000	.2500	.1500	

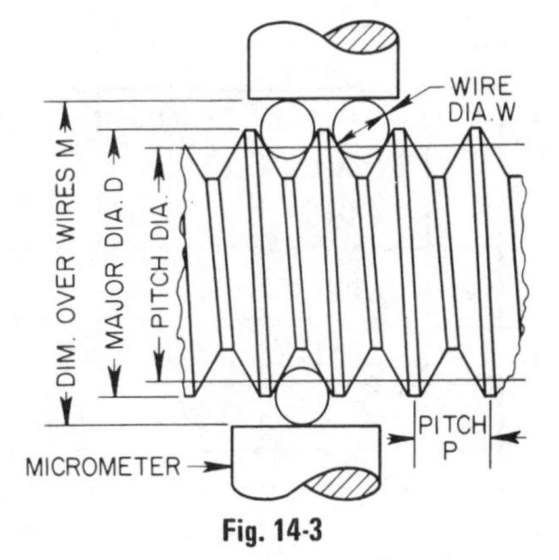

Fig. 14-3

2. A bronze bushing with a diameter of .8750 inch is to be pressed into a mounting plate. The assembly print calls for a bored hole in the plate to be .0015 inch less in diameter than the bushing diameter. The hole diameter in the plate checks .8702 inch. How much must the diameter of the plate hole be increased in order to meet the print specification? ________

3. A stamped sheet steel plate is shown, figure 14-4. Compute dimensions A - F to 3 decimal places. ________

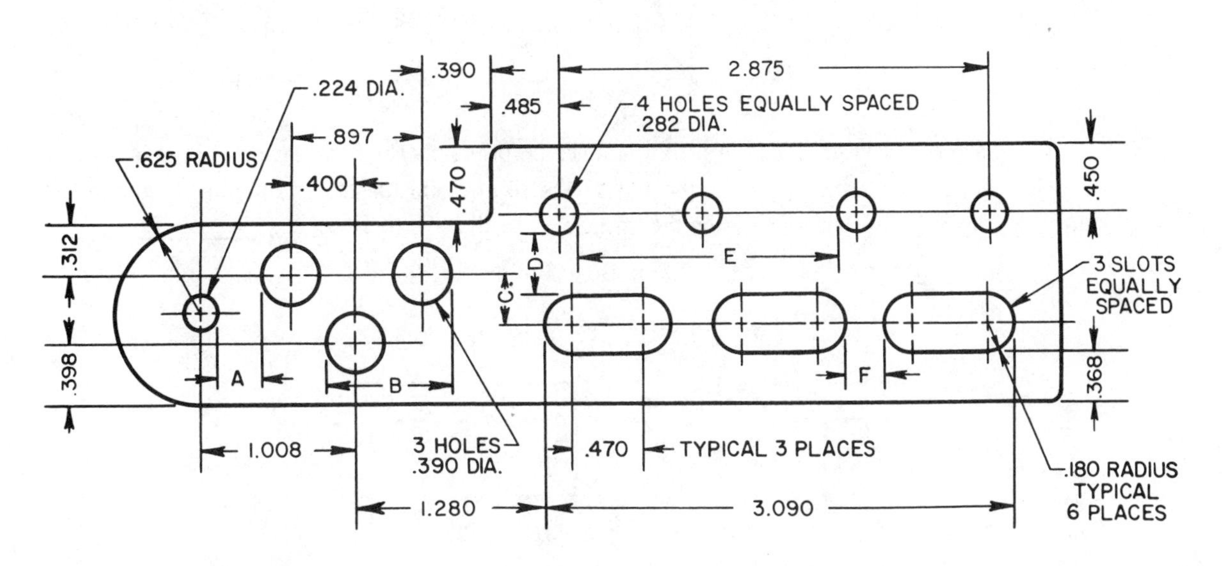

Fig. 14-4

A = ____ B = ____ C = ____ D = ____ E = ____ F = ____

4. A flat is to be milled in three pieces of round stock each of a different diameter. The length of the flat is determined by the diameter of the stock and the depth of cut. The table gives the required length of flat and the stock diameter for each piece. Determine the depth of cut for each piece to 3 decimal places using the formula

$$C = \frac{D}{2} - .5 \times \sqrt{4 \times (\frac{D}{2})^2 - F^2}$$

	Diameter D	Length of Flat F	Depth of Cut C
a.	1.400"	1.200"	
b.	2.200"	1.600"	
c.	3.600"	1.800"	

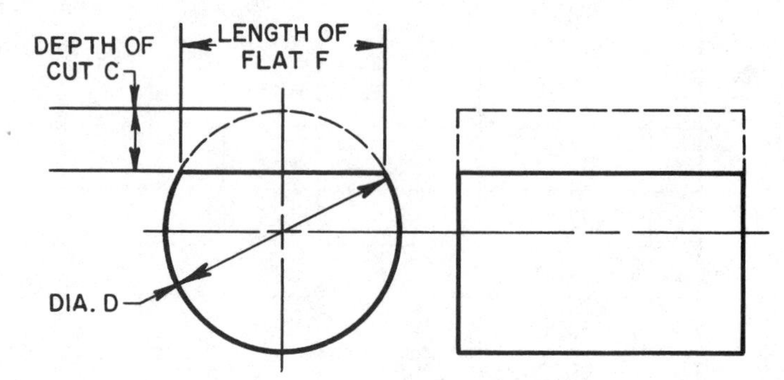

Fig. 14-5

5. A groove is machined in a circular plate with a 1.625-inch diameter. Two milling cuts, one .250 inch deep and the second .125 inch deep, are made. A grinding operation then removes .016 inch. What is the distance from the center of the plate to the bottom of the groove?

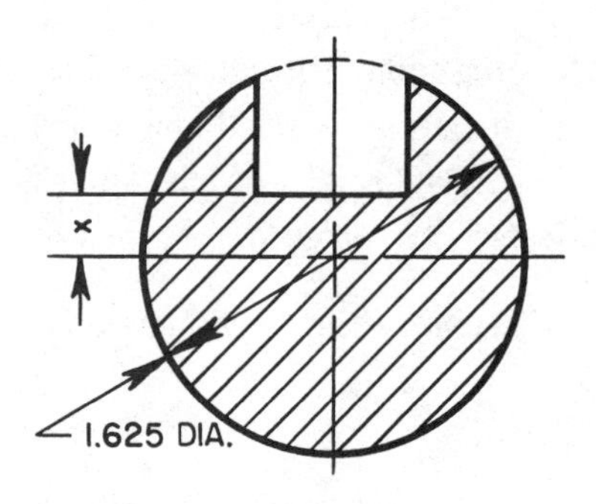

Fig. 14-6

6. A 60° groove has been machined in a fixture. The groove is checked by placing a pin in the groove and indicating the distance between the top of the fixture and the top of the pin. Compute this distance H to 3 decimal places by using the formula

$$H = 1.5 \times D - .866 \times W$$

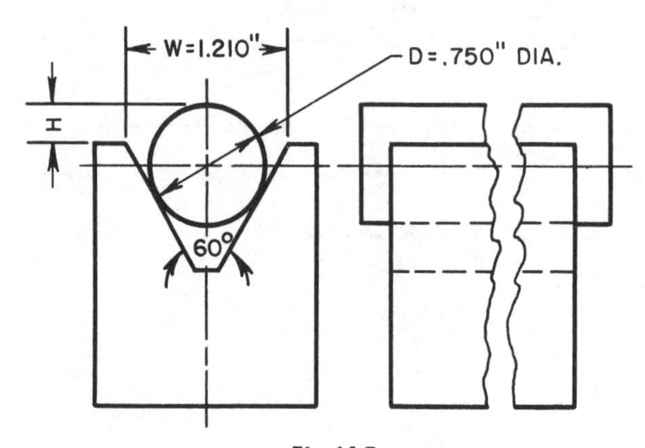

Fig. 14-7

SECTION II

Measurement

Unit 15 Degree of Precision, Tolerance and Clearance

OBJECTIVES

After studying this unit, the student should be able to

- Compute the range of numbers with reference to their degree of precision.
- Compute maximum and minimum limits of dimensions using unilateral and bilateral tolerances.
- Compute maximum and minimum clearances and interferences using unilateral and bilateral tolerances.

The basic units of measure used in the United States are established and maintained by the Bureau of Standards. These units of measure are based on international standards. In 1960 metal bars which were used as international standards were replaced by light waves which afford a greater degree of accuracy and precision.

In the United States, the American Standard Association determines industrial standards. This association, with the cooperation of other similar organizations throughout the world, establishes and maintains international standards of measure, specification, and practice.

Most countries use the metric system of measurement; at present the United States uses the decimal-inch system. The decimal-inch system is characterized by ease of computation and convenient conversion with metric units. It is expected that the U.S. will gradually adopt the metric system over the next decade.

DEGREE OF PRECISION OF MEASURING INSTRUMENTS

The cost of producing a part increases with the degree of precision called for; therefore, no greater precision than is actually required should be specified on a drawing.

The degree of precision specified for a particular machining operation dictates the type of machine, the machine setup, and the measuring instrument used for that operation.

Limitations of Measuring Instruments

Various measuring instruments have differing limitations on the degree of precision possible.

- steel rules or scales, 1/64" for fractional rules, .010" for decimal rules
- micrometers and calipers, .001", with vernier attachment .0001"
- precision gage blocks, (manufactured up to .000002" precision, but the degree of precision of measurement is only as precise as the measuring instrument that is used with the blocks)
- dial indicators (comparison measurement), .001" to .0001"
- electronic comparators and air gages, .0001" to .000001"

The degree of precision achieved not only depends on the limitations of the measuring instrument but can also be affected by errors in measurement. These errors can be caused by defects in the instruments, environmental changes such as differences in temperature, or the inaccuracy of the person using them.

DEGREE OF PRECISION OF NUMBERS

The degree of precision of a number depends upon the unit of measurement. The degree of precision of a number increases as the number of decimal places increases.

Example 1: The degree of precision of 2″ is to the closest inch as shown in figure 15-1. The range of values includes all numbers equal to or greater than 1.5″ or less than 2.5″.

Example 2: The degree of precision of 2.0″ is to closest 10th of an inch as shown in figure 15-2. The range of values includes all numbers equal to or greater than 1.95″ and less than 2.05″.

Example 3: The degree of precision of 2.00″ is to the closest 100th of an inch as shown in figure 15-3. The range of values includes all numbers equal to or greater than 1.995″ and less than 2.005″.

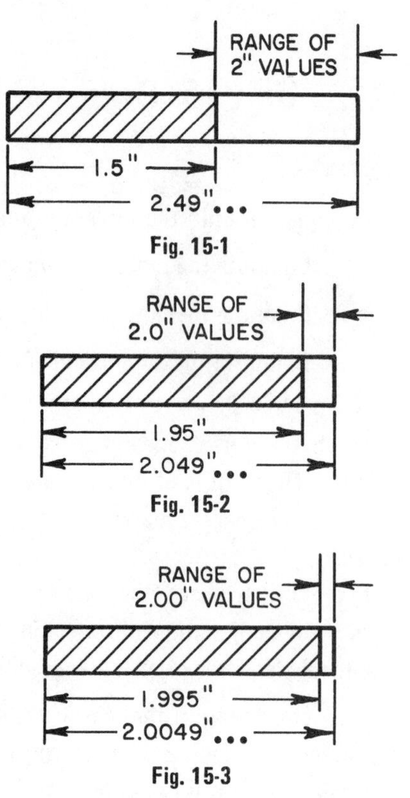

Fig. 15-1

Fig. 15-2

Fig. 15-3

The degree of precision of 2.000″ is to the closest 1000th of an inch. The range of values includes all numbers equal to or greater than 1.9995″ and less than 2.0005″.

The degree of precision of .1753″ is to the closest 10,000th of an inch. The range of values includes all numbers equal to or greater than .17525″ and less than .17535″.

TOLERANCE

Tolerance is the amount of variation permitted on a dimension of a part. Tolerance is equal to the difference between the maximum and minimum dimensions (limits) of a part.

For example, if the maximum limit of a hole diameter is .878″ and the minimum limit is .872″, the tolerance is .878″ - .872 = .006″.

A *basic dimension* is the standard size from which the maximum and minimum limits are made. *Unilateral tolerance* means that the total tolerance is taken in one direction only from the basic dimension. Figure 15-4 shows three applications of unilateral tolerances where clearance is required between shafts and holes.

Bilateral tolerance means that the tolerance is divided, partly plus and partly minus from the basic dimension. Figure 15-5 shows three applications of bilateral tolerances where clearance is required between shafts and holes.

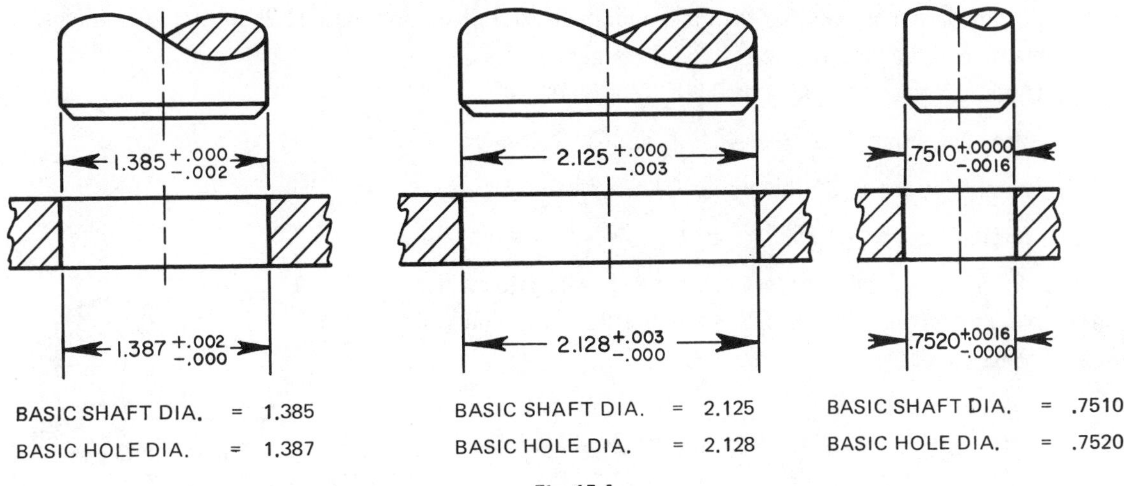

Fig. 15-4

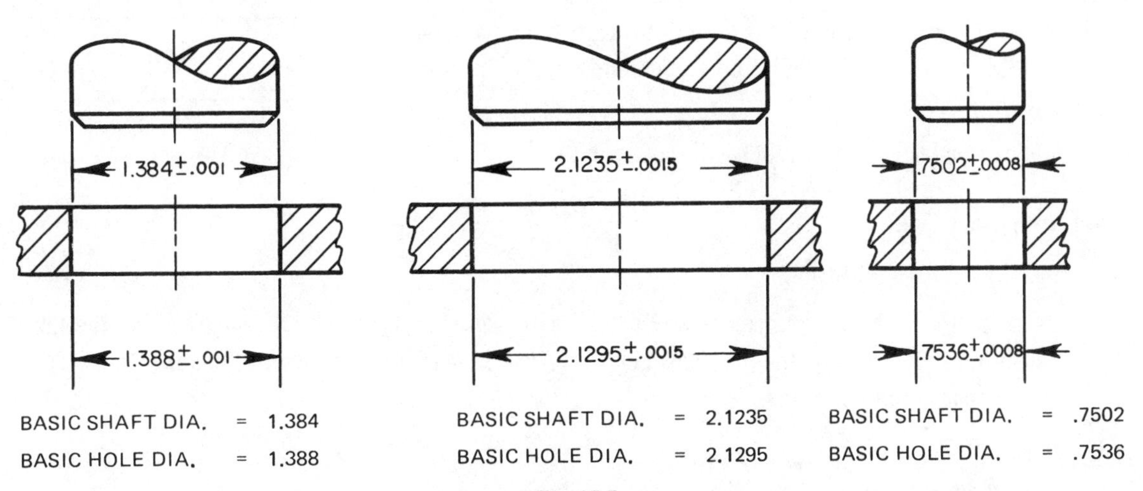

Fig. 15-5

A *mean dimension* is a value which is midway between the maximum and minimum limits. Where bilateral dimensioning is used and the plus and minus values are the same, the mean dimension is equal to the basic dimension.

Examples of Tolerance, Clearance, and Interference

Example 1: Figure 15-6 shows mating parts which require a clearance fit.

- ▲ Compute the maximum and minimum clearances between the parts.
- ▲ Convert the given unilateral tolerance to bilateral tolerance using the mean dimension as the basic dimension.
- ▲ Maximum clearance; subtract the minimum dimension of the top piece (1.2580 - .0028 = 1.2552) from the maximum dimension of the bottom piece. (1.2590 + .0028 = 1.2618). Maximum clearance = 1.2618 - 1.2552 = .0066.

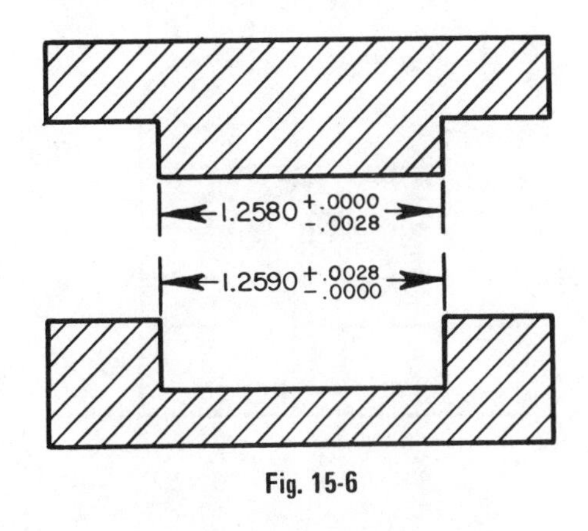

Fig. 15-6

- ▲ Minimum clearance; subtract the maximum dimension of the top piece (1.2580) from the minimum dimension of the bottom piece (1.2590). Minimum clearance = 1.2590 - 1.2580 = .0010.
- ▲ Bilateral tolerance on top piece; .0028 ÷ 2 = .0014

  Mean dimension = 1.2580 - .0014 = 1.2566. Tolerance = 1.2566 ± .0014
- ▲ Bilateral tolerance on bottom piece; .0028 ÷ 2 = .0014

  Mean dimension = 1.2590 + .0014 = 1.2604. Tolerance = 1.2604 ± .0014

Example 2: Figure 15-7 shows a pin which is to be pressed into a hole. This is a *forced or interference fit.* Compute the maximum and minimum interference.

- ▲ Maximum interference; subtract the minimum hole diameter (1.3850 - .0005 = 1.3845) from the maximum pin diameter (1.3865 + .0005 = 1.3870).

  Maximum interference = 1.3870 - 1.3845 = .0025
- ▲ Minimum interference; subtract the maximum hole diameter (1.3850 + .0005 = 1.3855) from the minimum pin diameter (1.3865 - .0005 = 1.3860).

  Minimum interference = 1.3860 - 1.3855 = .0005

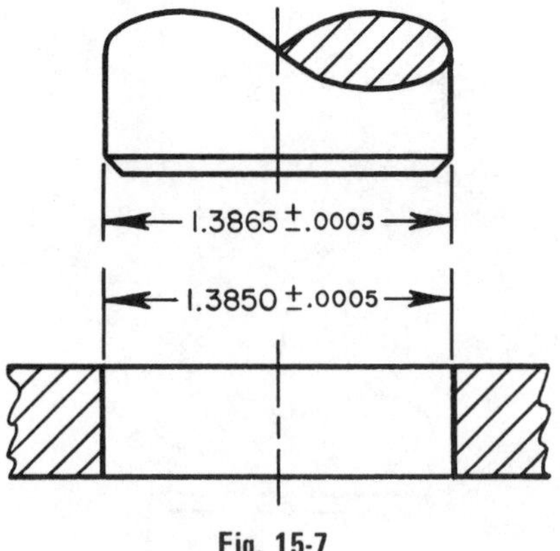

Fig. 15-7

APPLICATION

A. DEGREE OF PRECISION

Refer to the following statement and compute the missing dimensions B and C, using the given dimension A for each of the problems below. "The range of values of A are all numbers equal to or greater than B and less than C."

1. A = 5.077″ B = ____ C = ____
2. A = 3.6″ B = ____ C = ____
3. A = 10″ B = ____ C = ____
4. A = .09″ B = ____ C = ____
5. A = .9″ B = ____ C = ____
6. A = 1.00″ B = ____ C = ____

B. UNILATERAL TOLERANCES

Figures 15-8 and 15-9 show clearance fits between mating parts. Given either dimension A or B compute the missing dimensions in the tables. The solution to problem "a." is shown.

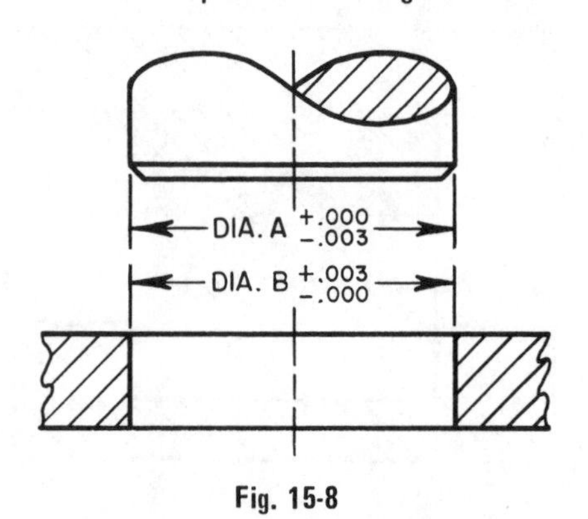

Fig. 15-8

	Dia. A	Dia. B	Max. Dia.	Min. Dia.	Basic Dia.	Mean Dia.
a.	1.6370	—	1.6370	1.6340	1.6370	1.6355
	—	1.6400	1.6430	1.6400	1.6400	1.6415
b.	2.7320	—				
	—	2.7380				
c.	.8635	—				
	—	.8655				
d.	.1872	—				
	—	.1880				

2.

	Dim. A	Dim. B	Max. Dim.	Min. Dim.	Basic Dim.	Mean Dim.
a.	4.7856	——				
	——	4.7860				
b.	2.3492	——				
	——	2.3501				
c.	1.8934	——				
	——	1.8938				
d.	.9992	——				
	——	.9998				

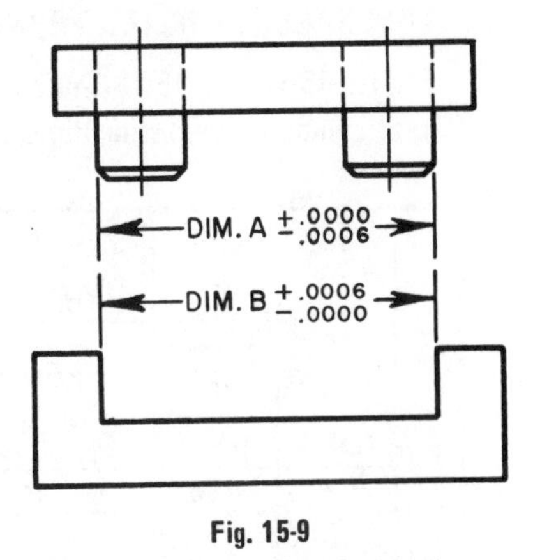

Fig. 15-9

C. BILATERAL TOLERANCES

Figures 15-10 and 15-11 show clearance fits between mating parts. Given either dimension A or B compute the missing dimensions in the tables. The solution to problem "a." is shown.

1.

	Dia. A	Dia. B	Max. Dia.	Min. Dia.	Basic Dia.	Mean Dia.
a.	2.4350	——	2.4370	2.4330	2.4350	2.4350
	——	2.4400	2.4420	2.4380	2.4400	2.4400
b.	5.0275	——				
	——	5.0330				
c.	.5760	——				
	——	.5806				
d.	1.0325	——				
	——	1.0375				

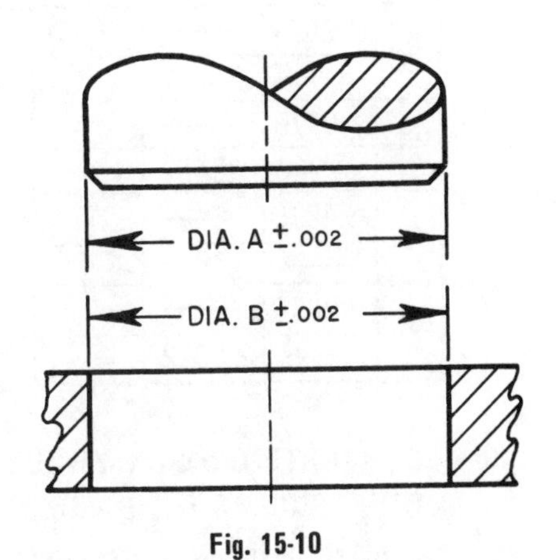

Fig. 15-10

2.

	Dim. A	Dim. B	Max. Dim.	Min. Dim.	Basic Dim.	Mean Dim.
a.	1.0554	——				
	——	1.0580				
b.	.7392	——				
	——	.7412				
c.	3.2377	——				
	——	3.2405				
d.	1.9995	——				
	——	2.0020				

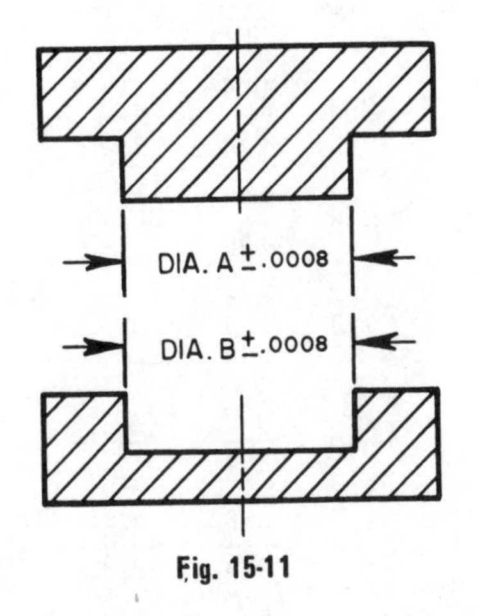

Fig. 15-11

D. TOLERANCES APPLIED TO INTERFERENCE FITS (Press fits)

Figures 15-12 and 15-13 show interference fits between a pin and a hole: Given either dimension A or B, compute the missing dimensions in the tables.

1.

	Dia. A	Dia. B	Max. Dia.	Min. Dia.	Basic Dia.	Mean Dia.
a.	.7853	——				
	——	.7838				
b.	.1870	——				
	——	.1862				
c.	2.9075	——				
	——	2.9064				
d.	1.5136	——				
	——	1.5120				

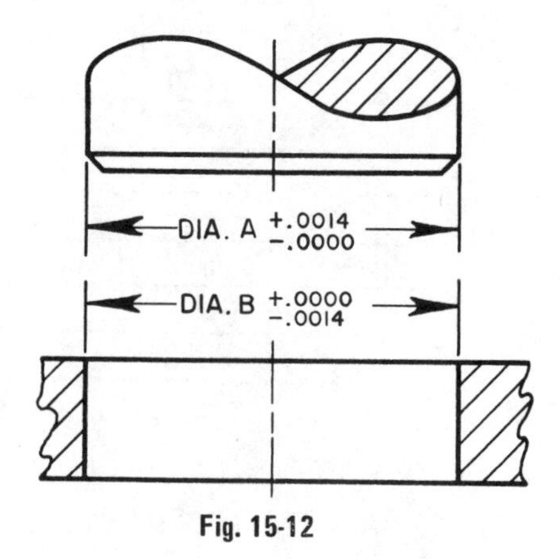

Fig. 15-12

2.

	Dia. A	Dia. B	Max. Dia.	Min. Dia.	Basic Dia.	Mean Dia.
a.	1.3874	——				
	——	1.3852				
b.	.9470	——				
	——	.9452				
c.	3.4896	——				
	——	3.4870				
d.	2.0094	——				
	——	2.0072				

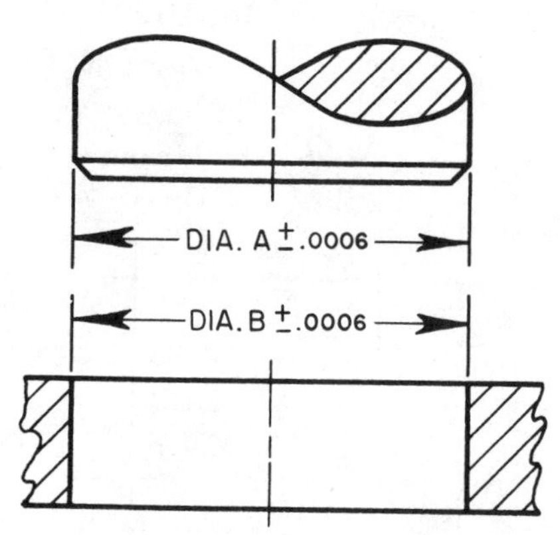

Fig. 15-13

E. CONVERTING UNILATERAL TOLERANCES TO BILATERAL TOLERANCES USING THE MEAN DIMENSION AS THE BASIC DIMENSION

Convert the following from unilateral to bilateral tolerances.

1. $8.1702 \pm \frac{.0000}{.0016}$ ______
2. $.064 \pm \frac{.000}{.016}$ ______
3. $.938 \pm \frac{.010}{.000}$ ______
4. $1.734 \pm \frac{.002}{.000}$ ______
5. $3.000 \pm \frac{.000}{.004}$ ______
6. $2.385 \pm \frac{.006}{.000}$ ______
7. $.073 \pm \frac{.000}{.008}$ ______
8. $.0080 \pm \frac{.0000}{.0012}$ ______
9. $4.1873 \pm \frac{.0014}{.0000}$ ______
10. $6.0707 \pm \frac{.0006}{.0000}$ ______
11. $.9998 \pm \frac{.0018}{.0000}$ ______
12. $1.0012 \pm \frac{.0000}{.0074}$ ______
13. $6.3573 \pm \frac{.0000}{.0010}$ ______
14. $.3007 \pm \frac{.0032}{.0000}$ ______
15. $2.5390 \pm \frac{.0054}{.0000}$ ______
16. $.1873 \pm \frac{.000}{.003}$ ______
17. $.0010 \pm \frac{.0000}{.0008}$ ______
18. $8.4649 \pm \frac{.0022}{.0000}$ ______

F. CLEARANCE AND INTERFERENCE

Compute the answers to the following problems.

1.

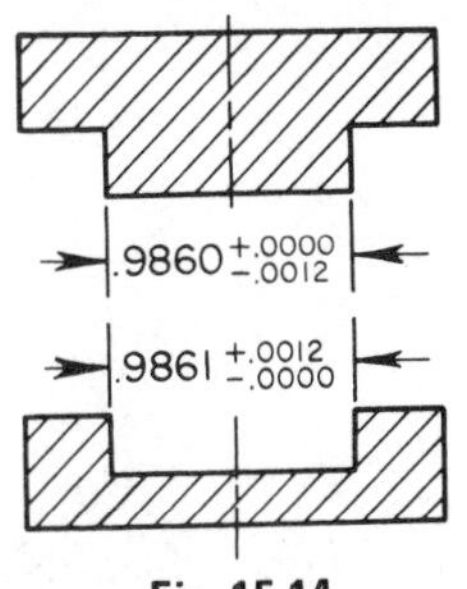

Fig. 15-14

a. Maximum clearance = ____

b. Minimum clearance = ____

2.

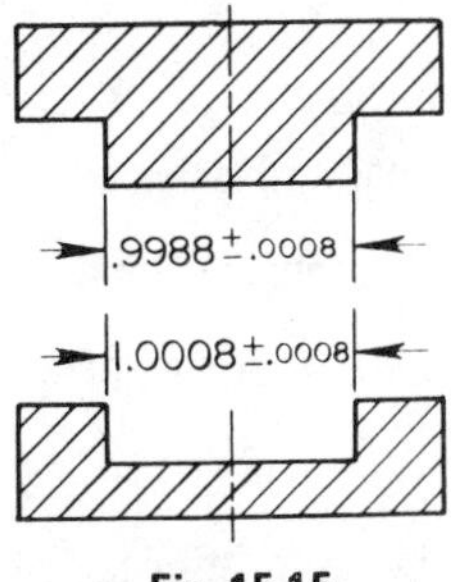

Fig. 15-15

a. Maximum clearance = ____

b. Minimum clearance = ____

3.

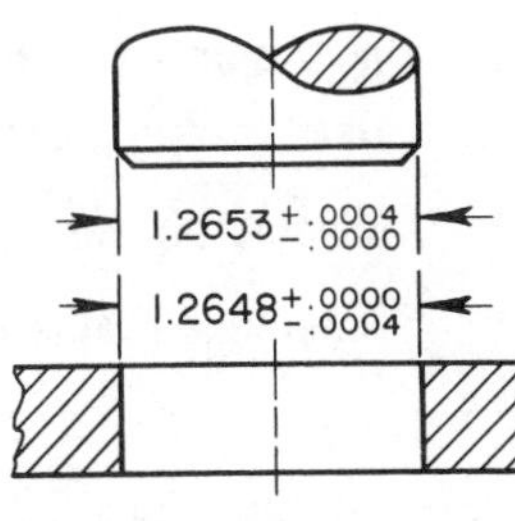

Fig. 15-16

a. Maximum interference = ____

b. Minimum interference = ____

4.

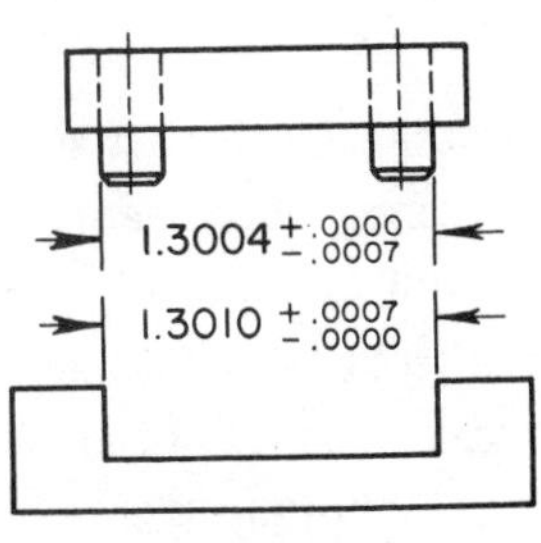

Fig. 15-17

a. Maximum clearance = ____

b. Minimum clearance = ____

5.

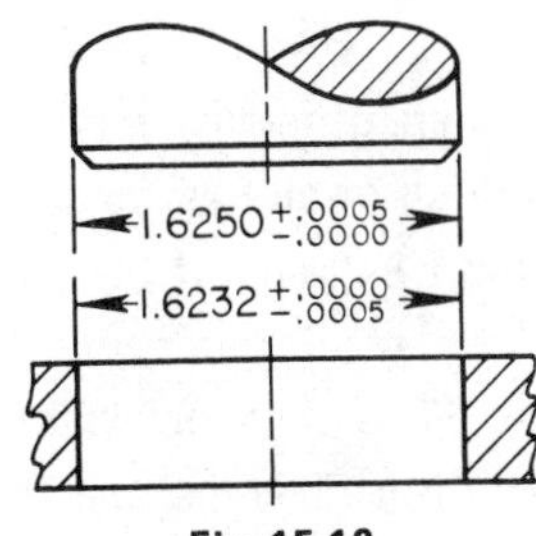

Fig. 15-18

a. Maximum interference = ____

b. Minimum interference = ____

6.

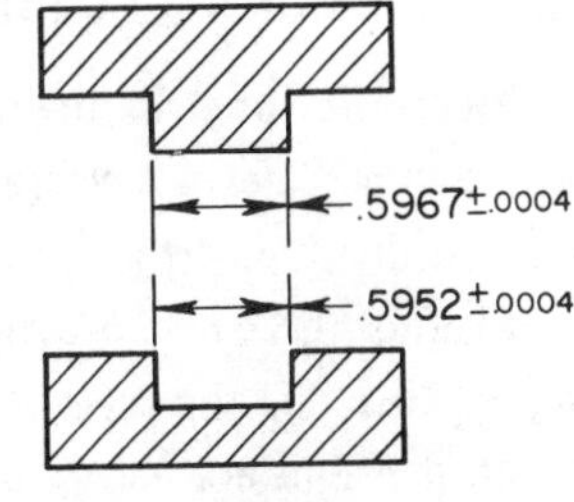

Fig. 15-19

a. Maximum interference = ____

b. Minimum interference = ____

7. Refer to figure 15-20.

a. the mean pin diameters = ____

b. the mean hole diameters = ____

c. the maximum dim. A = ____

d. the minimum dim. A = ____

e. the maximum dim. B = ____

f. the minimum dim. B = ____

g. the maximum total clearance between dim. C and dim. D = ____

h. the minimum total clearance between dim. C and dim. D = ____

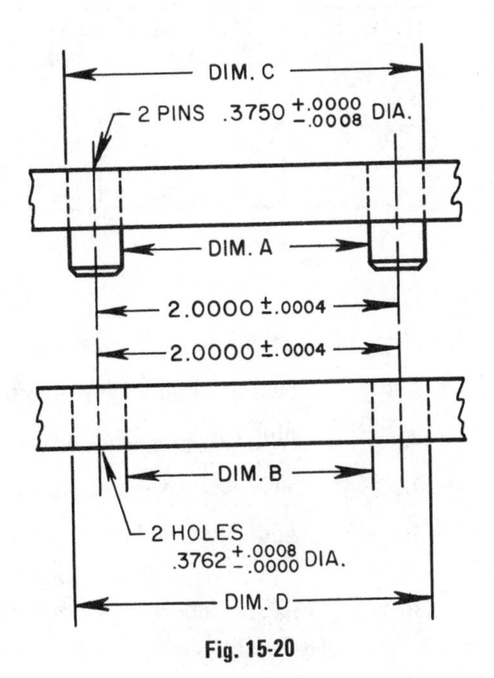

Fig. 15-20

Unit 16 Steel Rules and Gage Blocks

OBJECTIVES

After studying this unit, the student should be able to

- Read measurements from fractional and decimal-inch steel rules.
- Measure dimensions using fractional and decimal-inch steel rules.
- Determine proper gage blocks to be used for specified dimensions.

STEEL RULES

Steel rules are widely used for machine shop applications which do not require a high degree of precision. The steel rule is often the most practical measuring instrument to use for checking dimensions where stock allowances for finishing are provided. Steel rules are also used for locating roughing cuts on machined pieces and for determining the approximate locations of parts for machine setups.

DESCRIPTION OF STEEL RULES

Steel rules used in the machine shop are generally fractional-inch and decimal-inch rules six inches long, although rules anywhere from a fraction of an inch to several inches in length are sometimes used. The finest division of the fractional rules is 1/64" and of the decimal rule, 1/100".

Figures 16-1 and 16-2 show examples of reading measurements from the fractional and decimal rules. In figure 16-1, the top scale is graduated in 64ths of an inch and the bottom scale in 32nds of an inch. The common divisions are halves, quarters, eighths, sixteenths, and thirty-seconds. Sixty-fourths are added on the top scale.

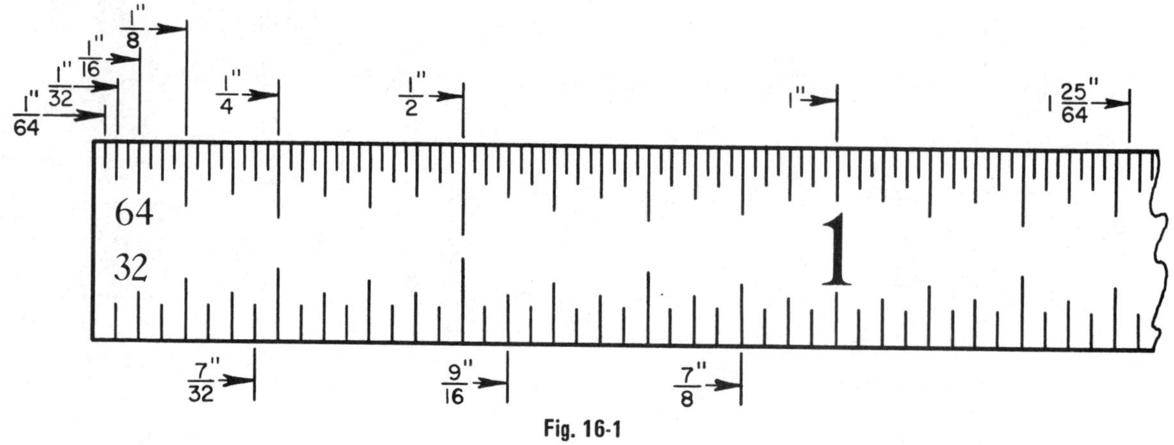

Fig. 16-1

Examples: Locate the following values on the scales in figure 16-1.

▲ 7/8" – Subtract 1 one-eighth division from 1". (1" = 8/8")

▲ 9/16" - Add 1 one-sixteenth division to 1/2". (1/2" = 8/16")

▲ 7/32" - Subtract 1 one-thirty-second from 1/4". (1/4" = 8/32")

▲ 1 25/64 - Add 1/64" to 1 3/8". (1 3/8" = 1 24/64")

In figure 16-2 the top scale is graduated in 100ths and the bottom scale in 50ths. The common divisions are halves, tenths, and fiftieths. Hundredths are added on the top scale.

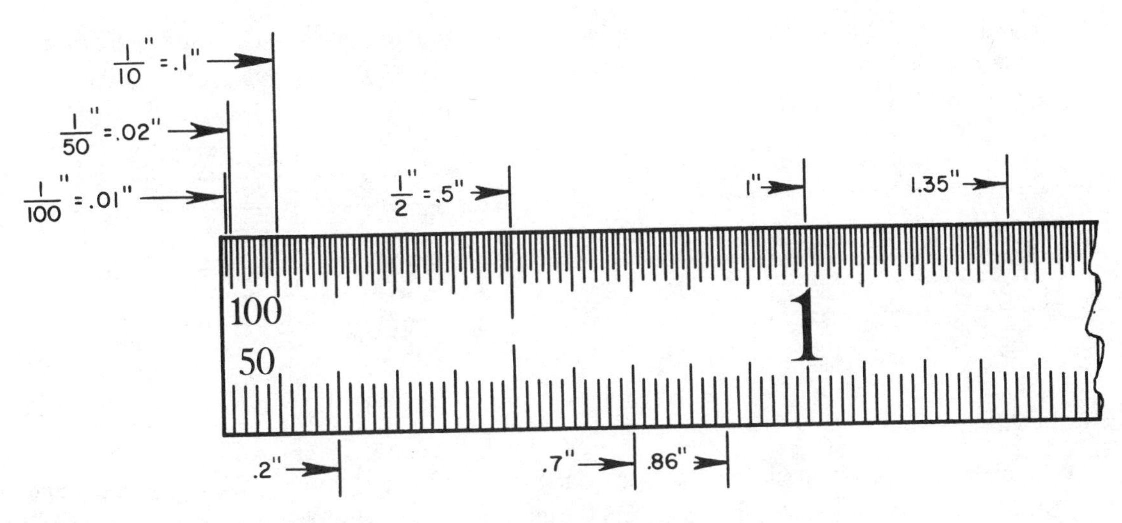

Fig. 16-2

Examples: Locate the following values on the scales in figure 16-2.

▲ .2" – Count 2 one-tenth divisions. 2 x .1" = .2"

▲ .7" – Add 2 one-tenth divisions to the .5 division, .5" + 2 x .1" = .7"

▲ .86" – Add 3 one-fiftieth divisons to 8 one-tenth divisions, 8 x .1" + 3 x .02" = .86

▲ 1.35" – Count 1 one-inch division, plus 3 one-tenth divisions plus 5 one-hundredth divisions, 1" + 3 x .1" + 5 x .01" = 1.35"

CORRECT PROCEDURE IN THE USE OF A STEEL RULE

- The end of the rule can be used as a reference point provided it is used with a knee (a straight ground block), as shown in figure 16-3. If a knee is not used, the end of the rule should not be used as a reference point, as shown in figure 16-4. Instead, the 1" graduation should be used as the reference point and the 1" must be subtracted from the measurement obtained, as shown in figure 16-5.

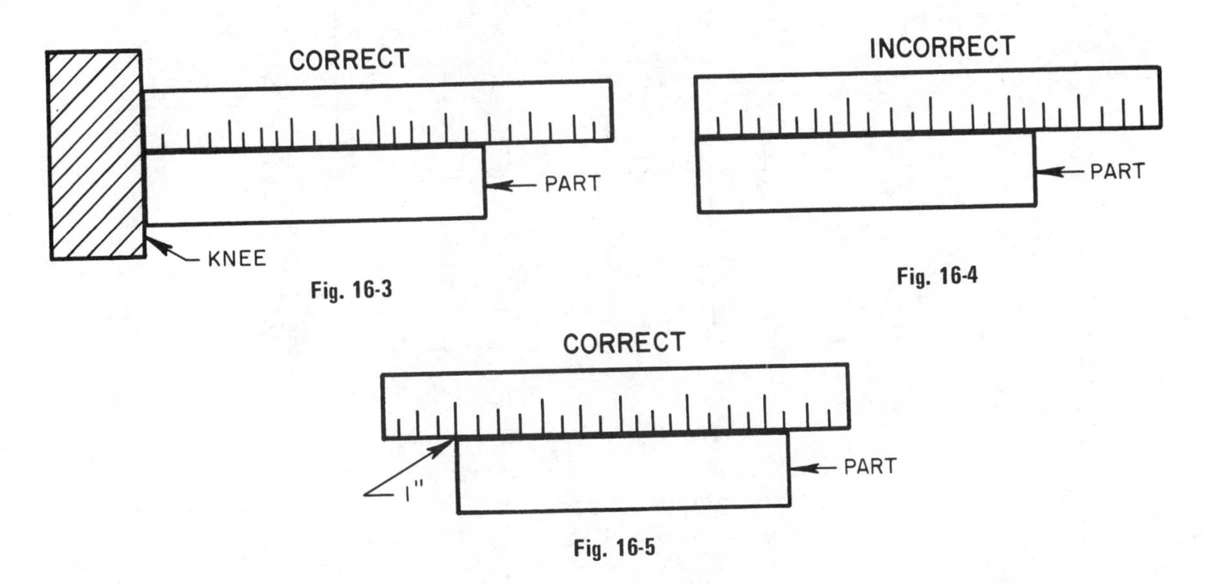

Fig. 16-3

Fig. 16-4

Fig. 16-5

- The scale edge of the rule should be put on the part to be measured, as shown in figure 16-6. Do not lay the rule flat on the part, as shown in figure 16-7. Following the correct procedure eliminates *parallax error* (error caused by the scale and the part being in different planes).

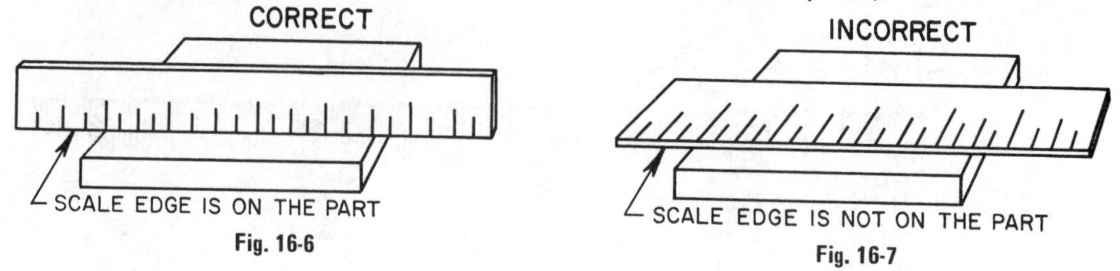

Fig. 16-6

Fig. 16-7

GAGE BLOCKS

Gage blocks are used in machine shops as standards for checking and setting (calibration) of micrometers, calipers, dial indicators, and other measuring instruments. Other applications of gage blocks are for layout, machine setups, and surface plate inspection.

DESCRIPTION OF GAGE BLOCKS

Gage blocks are either rectangular or square blocks, which when properly used, provide a degree of precision to the millionths. By *wringing* the blocks (slipping the blocks one over the other using light pressure), a combination of the proper blocks can be achieved which provides a desired length. Wringing the blocks produces a very thin air gap that acts similarly to a liquid film in holding the blocks together.

There is a variety of different gage block sets available. Figure 16-8 lists the thicknesses of blocks of a gage block set which is frequently used in machine shops.

9 BLOCKS .0001" SERIES

.1001	.1002	.1003	.1004	.1005	.1006	.1007	.1008	.1009

49 BLOCKS .001" SERIES

.101	.102	.103	.104	.105	.106	.107	.108	.109
.110	.111	.112	.113	.114	.115	.116	.117	.118
.119	.120	.121	.122	.123	.124	.125	.126	.127
.128	.129	.130	.131	.132	.133	.134	.135	.136
.137	.138	.139	.140	.141	.142	.143	.144	.145
.146	.147	.148	.149					

19 BLOCKS .050" SERIES

.050	.100	.150	.200	.250	.300	.350	.400	.450
.500	.550	.600	.650	.700	.750	.800	.850	.900
.950								

4 BLOCKS 1.000" SERIES

1.000	2.000	3.000	4.000

Fig. 16-8

Usually there is more than one combination of blocks which will give a desired length. The most efficient procedure for determining block combinations is to eliminate the digits of the desired measurement going from right to left. Application of this procedure saves time and minimizes the number of blocks and the chances of error.

Example 1: Determine a combination of gage blocks for 3.6453".

- ▲ Choose the block which eliminates the last digit to the right of the decimal point, the 3. Choose the .1003 block. (3.6453 - .1003 = 3.545)
- ▲ Eliminate the next digit, the 5. Choose the .145 block. (3.545 - .145 = 3.400)
- ▲ Eliminate the next digit, the 4. Choose the .400 block. (3.400 - .400 = 3.000)
- ▲ The 3.000 block completes the required dimension as shown in figure 16-9. (.1003" + .145" + .400" + 3.000" = 3.6453")

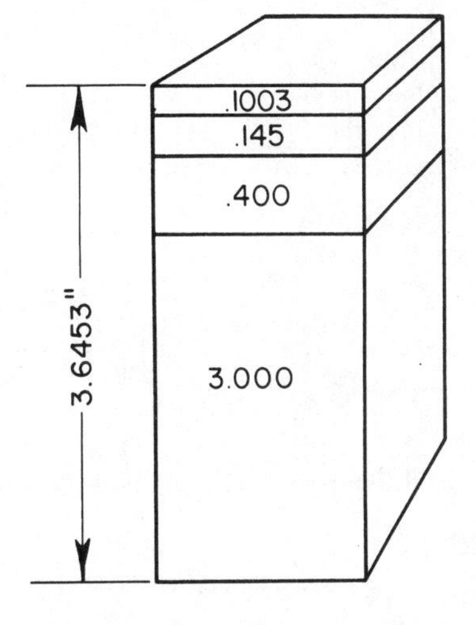

Fig. 16-9

Example 2: Determine a combination of gage blocks for 6.8792".

- ▲ Eliminate the 2. Choose the .1002 block. (6.8792 - .1002 = 6.779)
- ▲ Eliminate the 9. Choose the .129 block. (6.779 - .129 = 6.650)
- ▲ Eliminate the 5. Choose the .650 block. (6.650 - .650 = 6.000)
- ▲ The 2.000 and 4.000 blocks complete the required dimension as shown in figure 16-10. (.1002" + .129" + .650" + 2.000" + 4.000" = 6.8792")

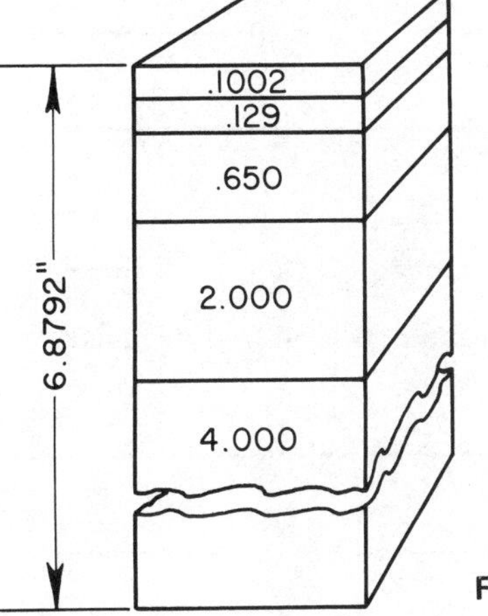

Fig. 16-10

APPLICATION

A. STEEL RULES

1. Read measurements a-p on the fractional-inch rule shown below.

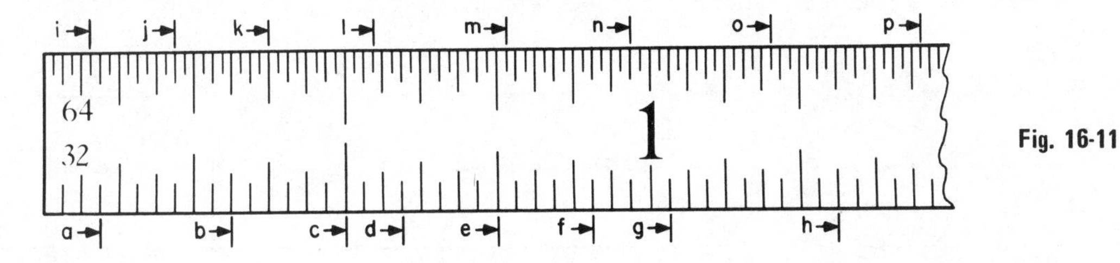

Fig. 16-11

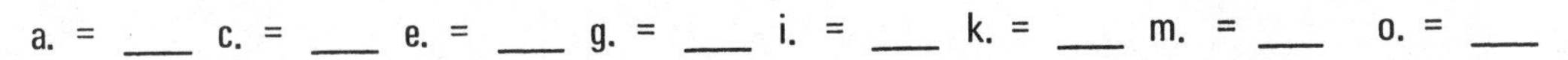

a. = ___ c. = ___ e. = ___ g. = ___ i. = ___ k. = ___ m. = ___ o. = ___

b. = ___ d. = ___ f. = ___ h. = ___ j. = ___ l. = ___ n. = ___ p. = ___

2. Read measurements a-p on the decimal-inch rule shown below.

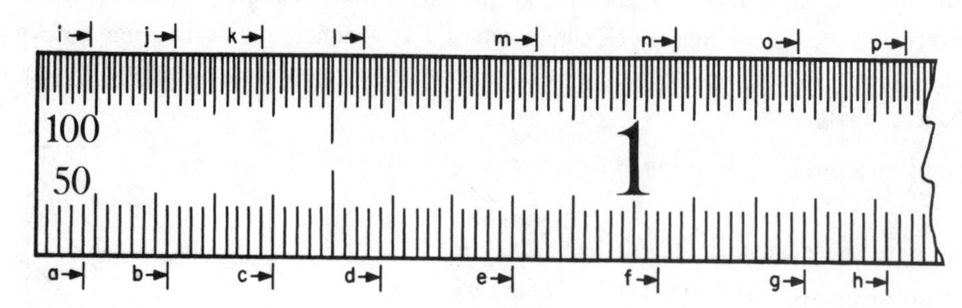

Fig. 16-12

a. = ____ e. = ____ i. = ____ m. = ____
b. = ____ f. = ____ j. = ____ n. = ____
c. = ____ g. = ____ k. = ____ o. = ____
d. = ____ h. = ____ l. = ____ p. = ____

3. Measure the lengths of these lines to the nearest sixteenth of an inch.

a b c
d e f g
h i j

a. = ____ c. = ____ e. = ____ g. = ____ i. = ____
b. = ____ d. = ____ f. = ____ h. = ____ j. = ____

4. Measure the lengths of these lines to the nearest thirty-second of an inch.

a b c d
e f g
h i j

a. = ____ c. = ____ e. = ____ g. = ____ i. = ____
b. = ____ d. = ____ f. = ____ h. = ____ j. = ____

5. Measure the lengths of these lines to the nearest sixty-fourth of an inch.

a b c
d e f g
h i j

a. = ____ c. = ____ e. = ____ g. = ____ i. = ____
b. = ____ d. = ____ f. = ____ h. = ____ j. = ____

6. Measure the diameters of the holes shown in this plate to the nearest .020".

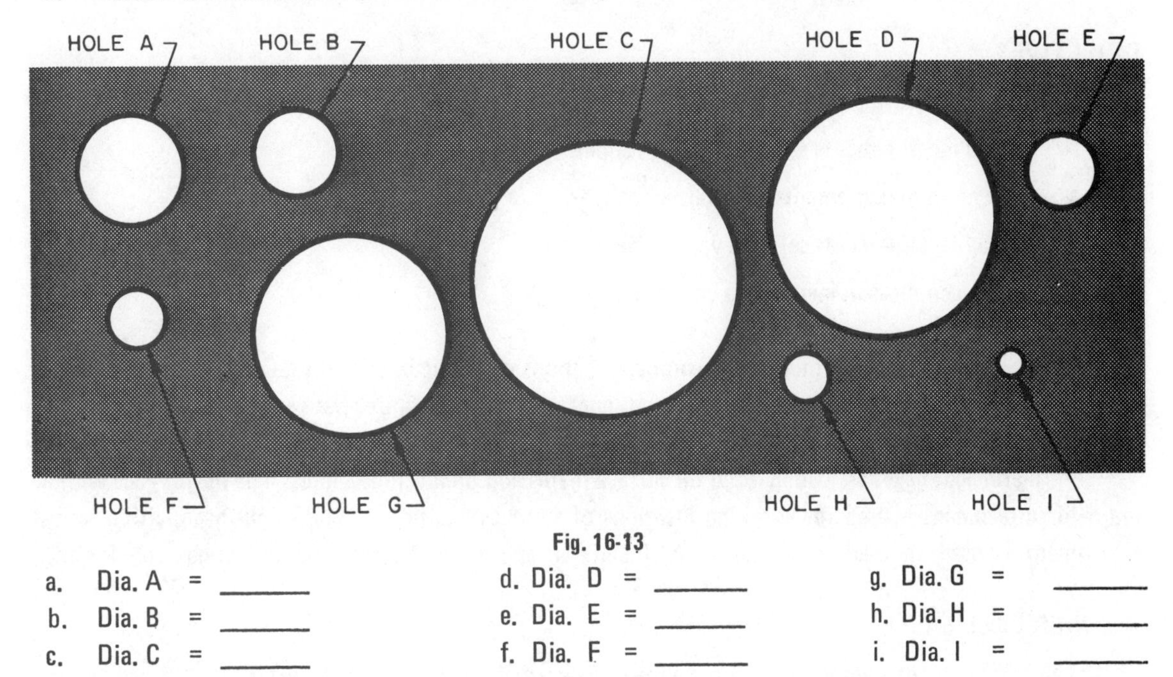

Fig. 16-13

a. Dia. A = ______	d. Dia. D = ______	g. Dia. G = ______
b. Dia. B = ______	e. Dia. E = ______	h. Dia. H = ______
c. Dia. C = ______	f. Dia. F = ______	i. Dia. I = ______

7. Using a fractional-inch and a decimal-inch steel rule and a sharp-pointed pencil, mark off the following dimensions.

a. 2 1/8"	e. 4 3/32"	i. 5 5/8"	m. 1.7"	q. .52"	u. 5.76"
b. 3/8"	f. 3 51/64"	j. 4 19/32"	n. 3.3"	r. 2.88"	v. .44"
c. 9/16"	g. 1 17/32"	k. 9/64"	o. .8"	s. 3.02"	w. 2.92"
d. 13/32"	h. 15/16"	l. 21/64"	p. 4.4"	t. .98"	x. 3.08"

B. GAGE BLOCKS

Using the gage block sizes given in figure 16-8, determine a combination of gage blocks for each of the dimensions given below. Note: Usually more than one combination of blocks will give the desired dimension.

1. 4.8638" ______	12. 1.0001" ______
2. 1.8702" ______	13. .2731" ______
3. 3.1222" ______	14. 2.7311" ______
4. .6333" ______	15. 5.090" ______
5. .3759" ______	16. 6.0907" ______
6. 5.8002" ______	17. 2.9789" ______
7. 7.973" ______	18. .8754" ______
8. .9999" ______	19. 7.7777" ______
9. 10.375" ______	20. 10.0101" ______
10. 9.050" ______	21. 9.4346" ______
11. 4.8757" ______	22. 3.9208" ______

Unit 17 Vernier Instruments: Caliper and Height Gage

OBJECTIVES

After studying this unit, the student should be able to

- Read measurements set on a vernier caliper.
- Set given measurements on a vernier caliper.
- Read measurements set on a vernier height gage.
- Set given measurements on a vernier height gage.

Vernier calipers are used in machine shop applications when the degree of precision to thousandths of an inch is adequate. They are used for checking lengths of parts, distances between holes, and both inside and outside diameters of cylinders.

Vernier height gages are widely used on surface plates and on machine tables. The height gage with an indicator attachment is used for checking locations of surfaces and holes. The height gage with a scriber attachment is used to mark reference lines, locations, and stock allowances on castings and forgings.

VERNIER CALIPER

The basic parts of a vernier caliper are a main scale which is similar to a steel rule with a fixed jaw and a sliding jaw with a vernier scale. The vernier scale slides parallel to the main scale and provides a degree of precision to .001″. Calipers are available in a wide range of lengths with different types of jaws and scale graduations. Figure 17-1 shows a vernier caliper which is commonly used in machine shops. The main scale is divided into inches and the inches are divided into 10 divisions each equal to .1″. The .1″ divisions are divided into 4 parts each equal to .025″. The vernier scale consists of 25 divisions.

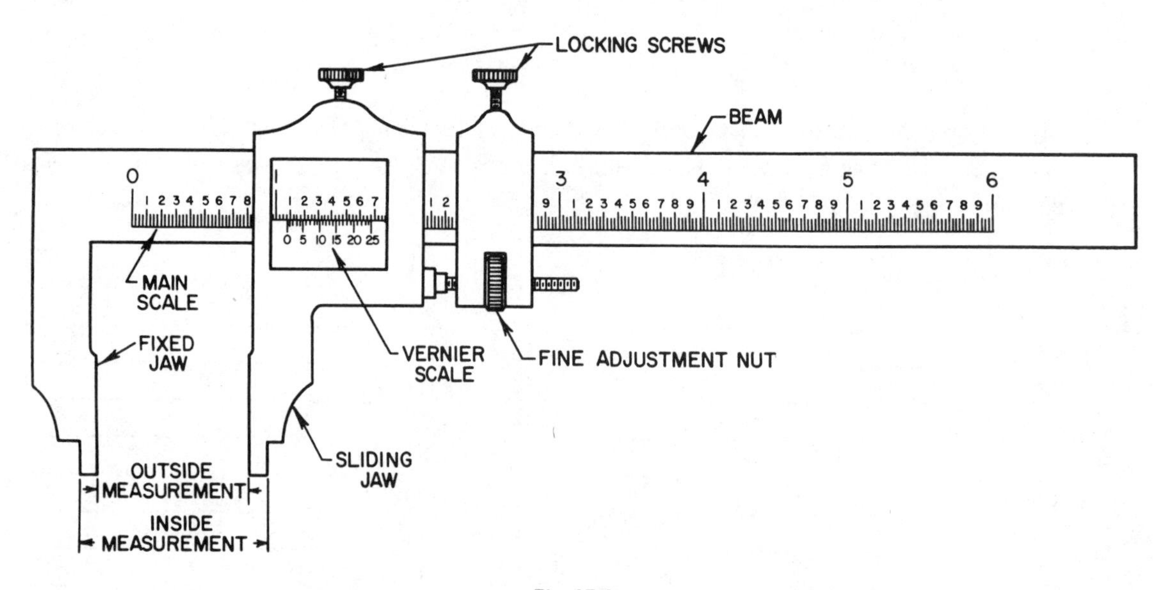

Fig. 17-1

The vernier scale has 25 divisions in a length equal to a length on the main scale that has 24 divisions. The difference between a main scale division and a vernier division is 1/25 of .025″ or .001″, figure 17-2.

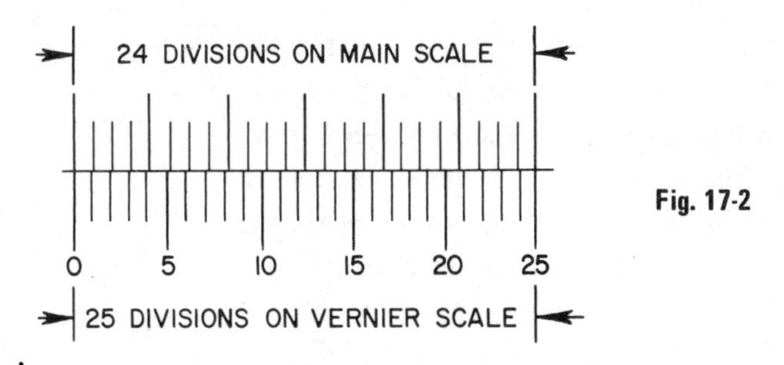

Fig. 17-2

PROCEDURE FOR READING AND SETTING A MEASUREMENT ON A VERNIER CALIPER

A measurement is read by adding the thousandths reading on the vernier scale to the reading from the main scale.

On the main scale the user reads the number of inch divisions, .1" divisions, and .025" divisions that are to the left of the zero graduation on the vernier scale. On the vernier scale, he finds the graduation that most closely coincides with a graduation on the main scale. This vernier graduation indicates the number of thousandths that are added to the main scale reading.

Setting a given measurement is the reverse procedure of reading a measurement on the vernier caliper.

Example 1: Read the measurement set on the vernier caliper shown in figure 17-3.

- ▲ In reference to the zero division on the vernier scale read two 1" divisions, seven .1" divisions, and three .025" divisions on the main scale: 2" + .7" + .075" = 2.775".
- ▲ Observe which vernier scale graduation most closely coincides with a main scale graduation. The eighth vernier scale graduation coincides; therefore, .008" is added to 2.775". Measurement = 2.775" + .008" = 2.783".

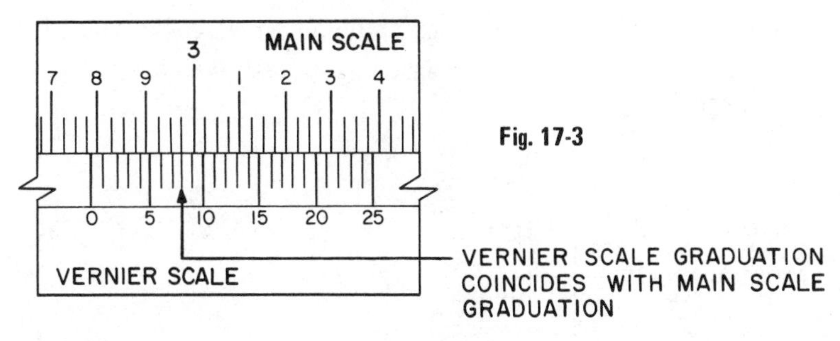

Fig. 17-3

Example 2: Set 1.237" on a vernier caliper as shown in figure 17-4.

- ▲ Move the vernier zero graduation to 1" + .2" + .025" on the main scale.
- ▲ An additional .012" (1.237" - 1.225") is set by adjusting the sliding jaw until the 12 graduation on the vernier scale coincides with a graduation on the main scale.

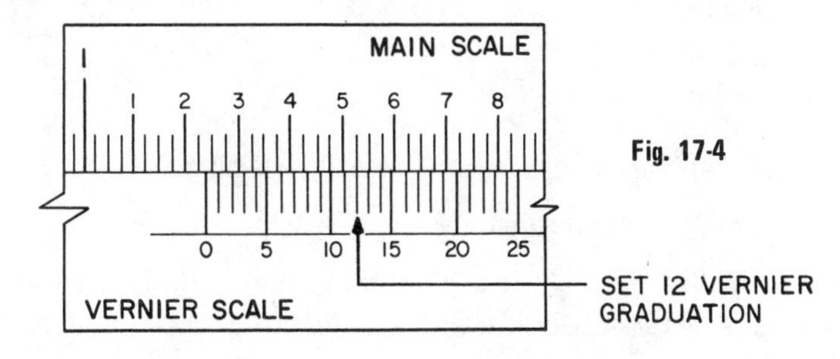

Fig. 17-4

The accuracy of measurement obtainable with a vernier caliper depends on the user's ability to align the caliper with the part which is being measured and the user's "feel" when measuring. The line of measurement must be parallel to the beam of the caliper and lie in the same plane as the caliper. Care must be used to prevent too loose or too tight a caliper setting.

VERNIER HEIGHT GAGE

The vernier height gage and vernier caliper are similar in operation. The height gage also has a sliding jaw; the fixed jaw is the surface plate with which the height gage is usually used. The gage can be used with a scriber, a depth gage attachment, or an indicator. The indicator is the most widely used and, generally, the most accurate attachment. Figure 17-5 shows the parts of a vernier height gage.

Measurements on the vernier gage are read and set using the same procedure as with the vernier caliper.

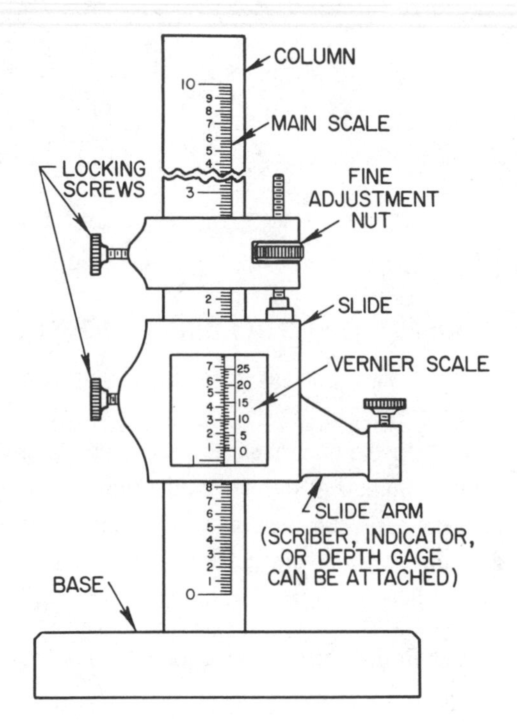

Fig. 17-5

Example 1: Read the measurement set on the vernier height gage shown in figure 17-6.

- ▲ In reference to the zero division on the vernier scale read 5", four .1" divisions, and two .025" divisions on the main scale: 5" + .4" + .050" = 5.450".
- ▲ Observe which vernier scale graduation most closely coincides with the main scale graduation. The twenty-first vernier scale graduation coincides; therefore, .021" is added to 5.450". Measurement = 5.450" + .021" = 5.471".

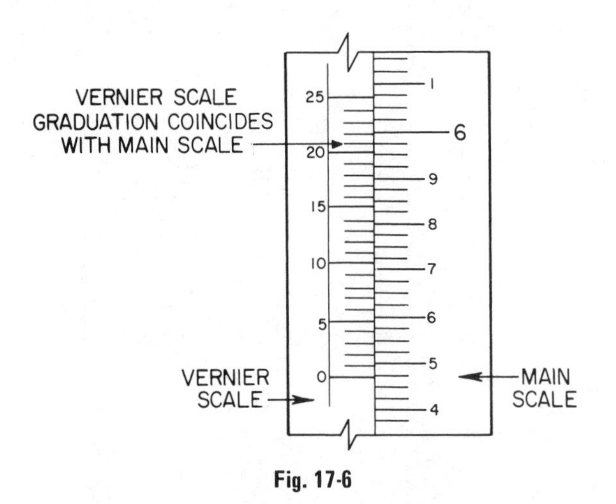

Fig. 17-6

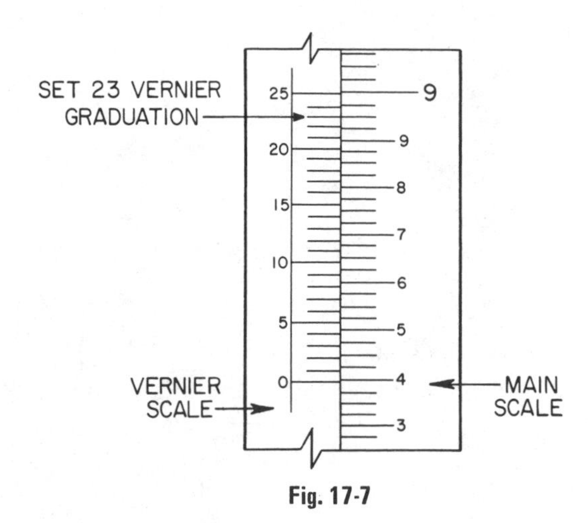

Fig. 17-7

Example 2: Set 8.398" on a vernier height gage as shown in figure 17-7.

- ▲ Move the vernier zero graduation to 8" + .3" + .075" = 8.375".
- ▲ An additional .023" (8.398" - 8.375") is set by turning the fine adjustment screw until the 23 graduation on the vernier scale coincides with a graduation on the main scale.

APPLICATION

A. VERNIER CALIPER

1. Read the vernier caliper measurements a – h for the following settings.

a.

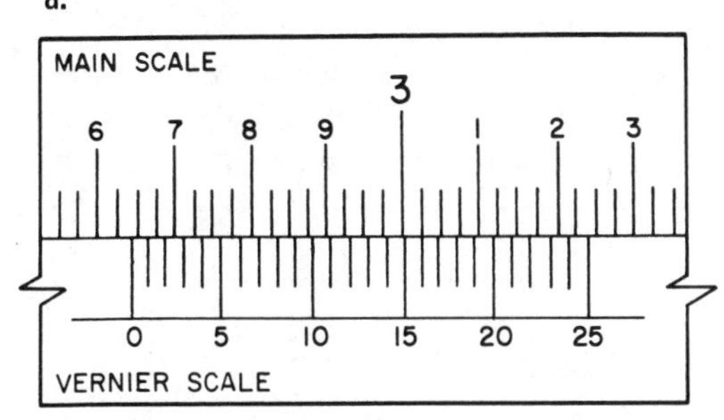

b.

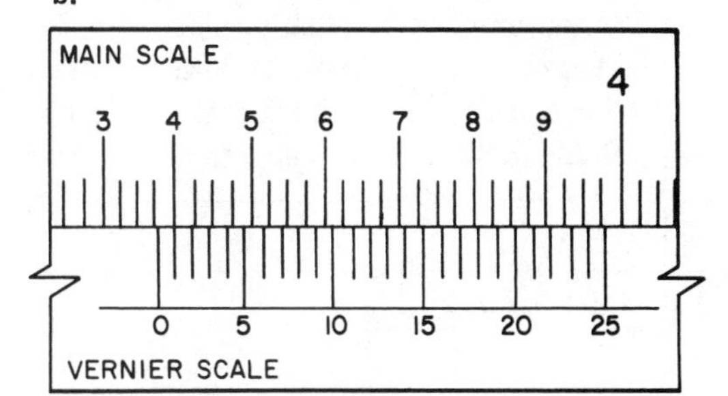

c.

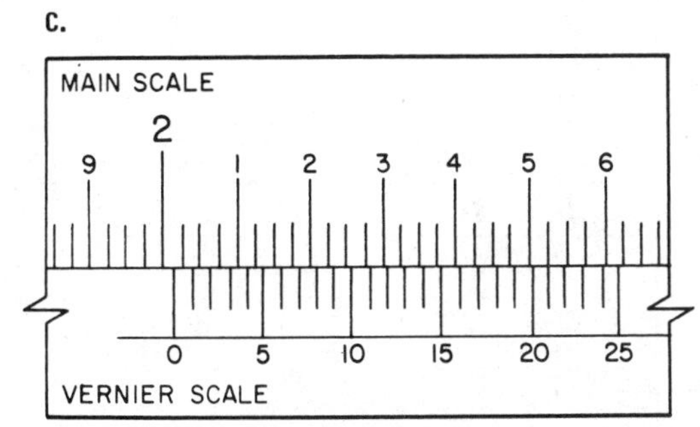

d.

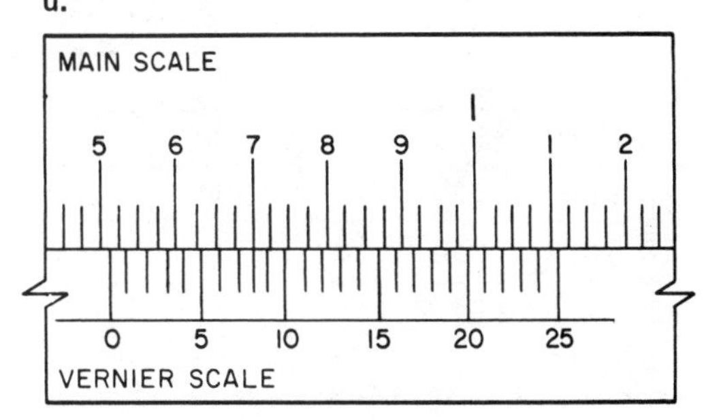

e.

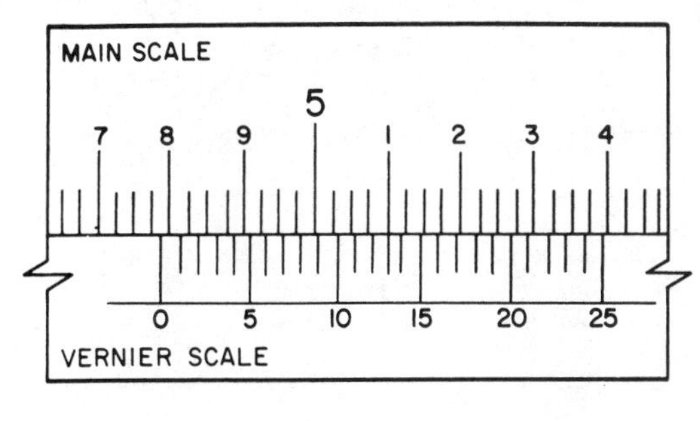

f.

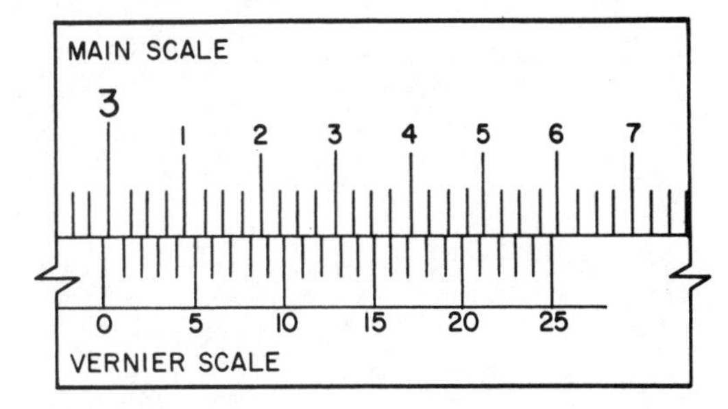

g.

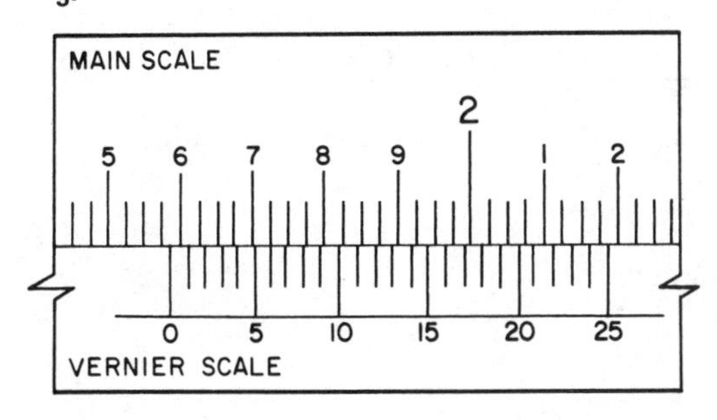

h.

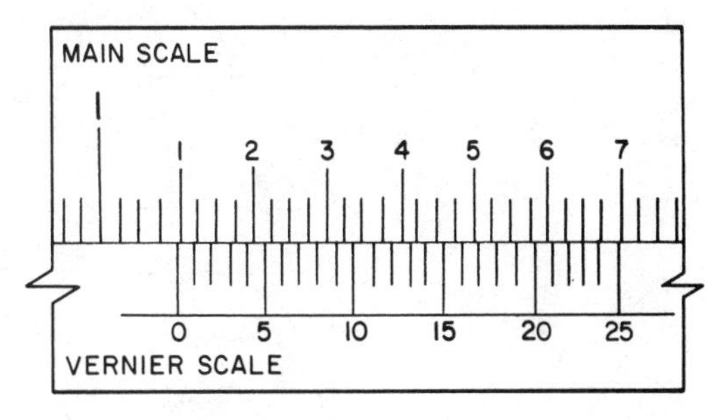

2. The tables below give the position of the zero graduation on the vernier scale in reference to the main scale and the vernier scale graduation that coincides with a main scale graduation. Determine the vernier caliper settings. The answer to the first problem is given.

	Zero Vernier Graduation Lies Between These Main Scale Graduations	Vernier Graduation That Coincides With A Main Scale Graduation	Vernier Caliper Setting
a.	1.875 - 1.900	19	1.894
b.	3.025 - 3.050	21	
c.	.050 - .075	3	
d.	5.775 - 5.800	11	
e.	1.225 - 1.250	7	
f.	.075 - .100	16	
g.	4.000 - 4.025	4	
h.	2.650 - 2.675	9	
i.	1.000 - 1.025	13	
j.	5.975 - 6.000	24	
k.	2.825 - 2.850	8	
l.	4.950 - 4.975	1	

	Zero Vernier Graduation Lies Between These Main Scale Graduations	Vernier Graduation That Coincides With A Main Scale Graduation	Vernier Caliper Setting
m.	.000 - .025	5	
n.	.825 - .850	15	
o.	3.550 - 3.575	23	
p.	5.075 - 5.100	20	
q.	3.325 - 3.350	17	
r.	2.075 - 2.100	6	
s.	4.400 - 4.425	10	
t.	1.025 - 1.050	14	
u.	.675 - .700	18	
v.	.050 - .075	2	
w.	3.000 - 3.025	19	
x.	2.925 - 2.950	22	

3. Refer to the following sentence and to the given vernier caliper settings to find values A, B, and C. "The zero vernier scale graduation lies between A and B on the main scale and the vernier graduation C coincides with the main scale graduation." The answer to the first problem is given.

		A	B	C
a.	3.242"	A = 3.225"	B = 3.250"	C = 17
b.	2.877"	A = ______	B = ______	C = ______
c.	5.939"	A = ______	B = ______	C = ______
d.	.611"	A = ______	B = ______	C = ______
e.	4.369"	A = ______	B = ______	C = ______
f.	.094"	A = ______	B = ______	C = ______
g.	7.857"	A = ______	B = ______	C = ______
h.	1.646"	A = ______	B = ______	C = ______
i.	6.024"	A = ______	B = ______	C = ______
j.	.022"	A = ______	B = ______	C = ______
k.	3.333"	A = ______	B = ______	C = ______
l.	5.999"	A = ______	B = ______	C = ______
m.	.388"	A = ______	B = ______	C = ______
n.	.965"	A = ______	B = ______	C = ______

4. Distances between centers of holes are often checked by measuring the inside distances between the holes and adding the radius of each hole to the vernier caliper measurement obtained. The following problems give the hole diameters and the distances between centers. Determine the settings on the caliper main and vernier scales.

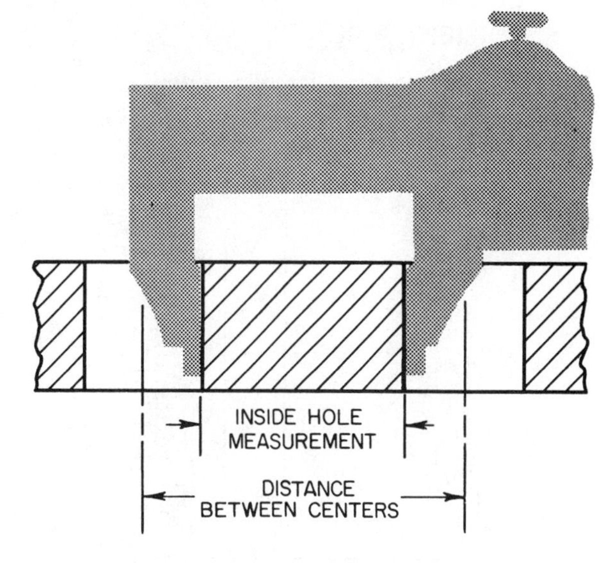

a.

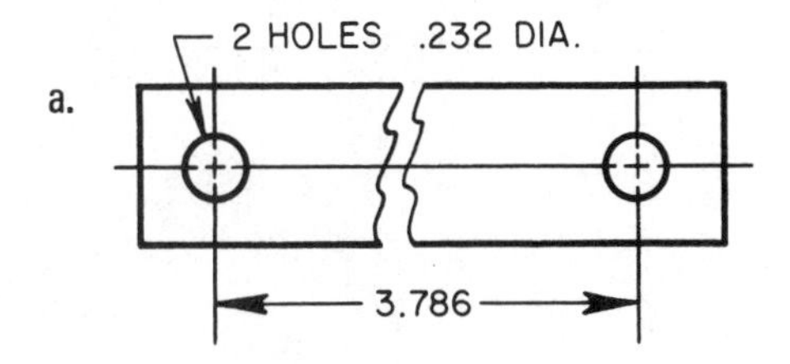

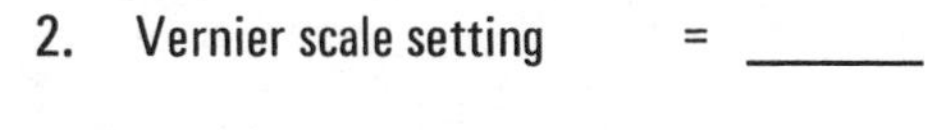

1. Main scale setting = ______
2. Vernier scale setting = ______

b.

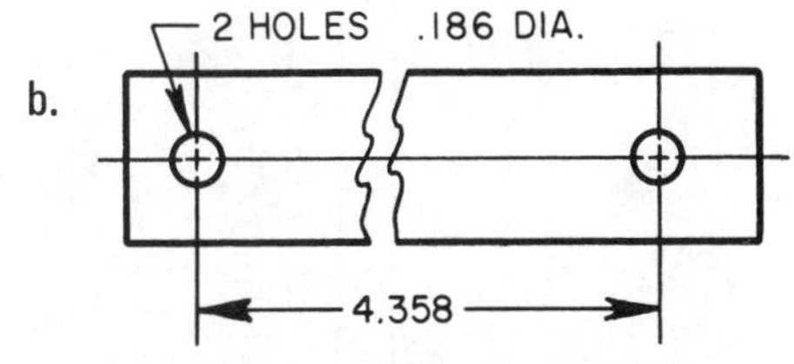

1. Main scale setting = ______
2. Vernier scale setting = ______

c.

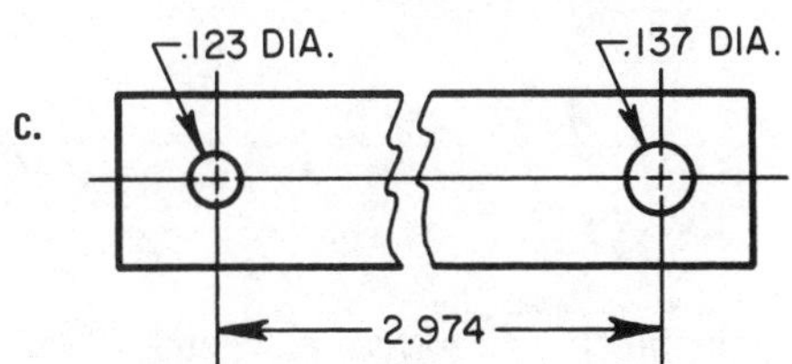

1. Main scale setting = ______
2. Vernier scale setting = ______

d.

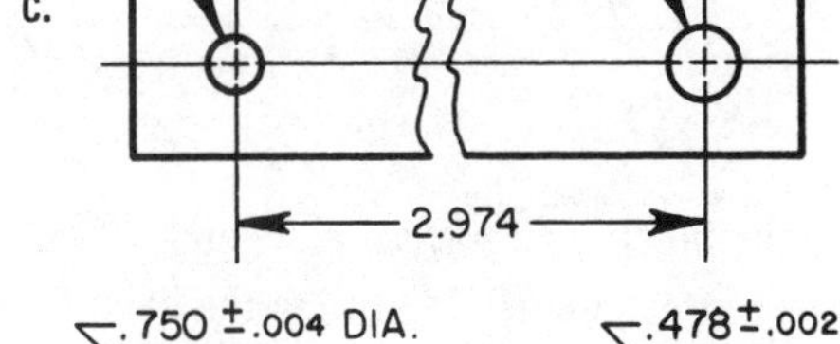

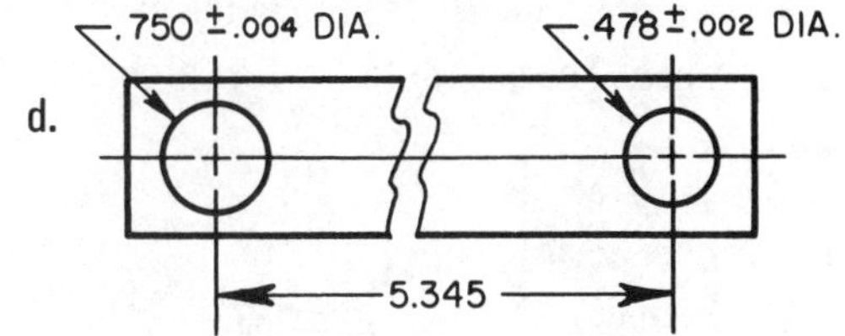

1. Main scale setting = ______
2. Max. vernier scale setting = ______
3. Min. vernier scale setting = ______

e.

.375±.003 DIA. .327±.005 DIA.

3.262±.003

1. Main scale setting = ______
2. Max. vernier scale setting = ______
3. Min. vernier scale setting = ______

B. HEIGHT GAGE

1. Read height gage measurements a – h for the following settings:

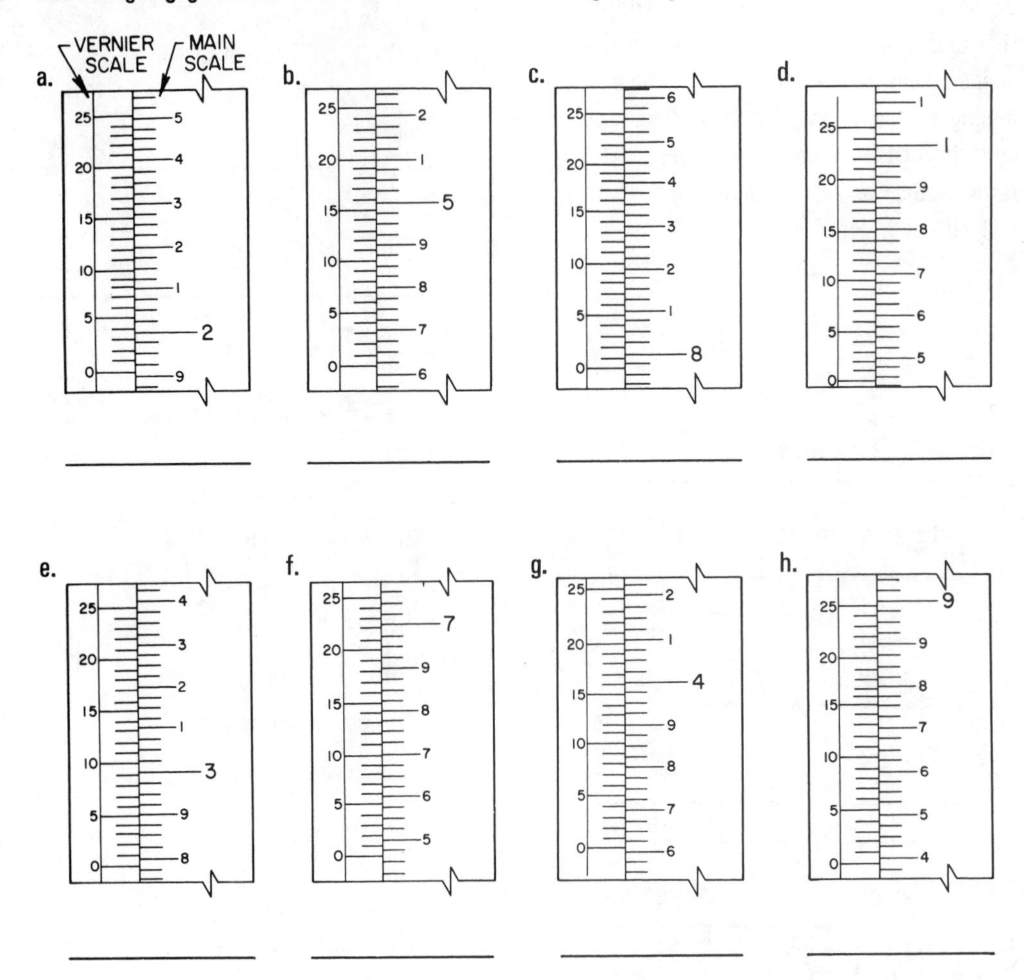

2. The hole locations of this block are checked by placing the block on a surface plate and indicating the bottom of each hole using a height gage with an indicator attachment. Determine the height gage settings from the bottom of the part to the bottom of the holes. The answer to the first problem is given.

Hole Number	Hole Diameter	Print Locations to Centers of Holes	Height Gage Main Scale Setting	Height Gage Vernier Scale Setting
1	.376″	A = .640″	.450″ - .475″	2
2	.250″	B = 1.008″		
3	.188″	C = .514″		
4	.496″	D = .312″		
5	.132″	E = .810″		

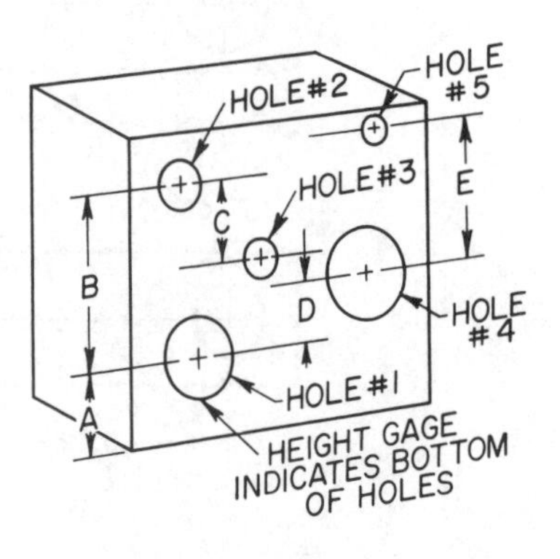

Unit 18 Micrometers

OBJECTIVES

After studying this unit, the student should be able to

- Read settings from the barrel and thimble scales of a .001-inch micrometer.
- Set given dimensions on the scales of .001-inch and .0001-inch micrometers.
- Read settings from the barrel, thimble, and vernier scales of .0001-inch micrometers.

Micrometers are basic measuring instruments used by machinists. Micrometers are available in a wide range of sizes and types. Ordinary micrometers are used to measure dimensions between parallel surfaces of parts and outside diameters of cylinders. Other types, depth micrometers, screw thread micrometers, disc and blade micrometers, bench micrometers, and inside micrometers, also have wide application in the machine shop.

THE MICROMETER

Figure 18-1 shows a .001-inch micrometer with its principal parts labeled.

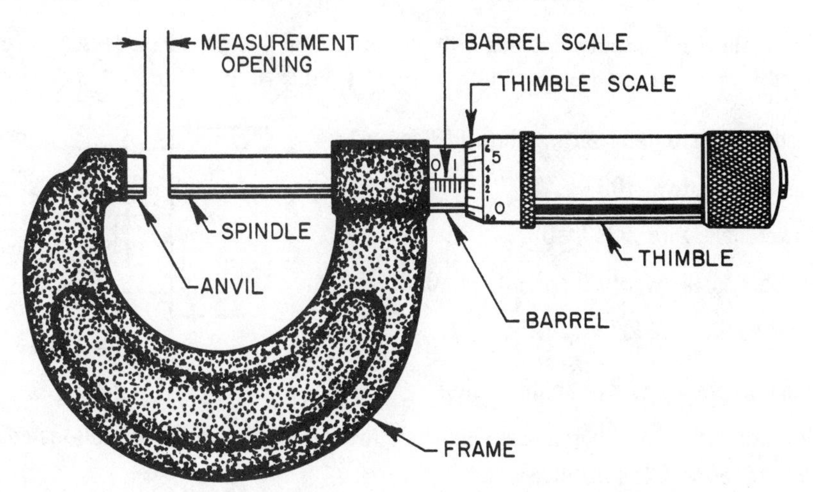

Fig. 18-1

The part to be measured is placed between the anvil and the spindle. The barrel of a micrometer consists of a scale which is one inch long. The one-inch length is divided into ten divisions each equal to .100 inch. The .100-inch divisions are further divided into four divisions each equal to .025-inch.

The thimble has a scale which is divided into twenty-five parts. One revolution of the thimble moves .025 inch on the barrel scale. Therefore, a movement of one graduation on the thimble equals 1/25 of .025 inch or .001 inch along the barrel.

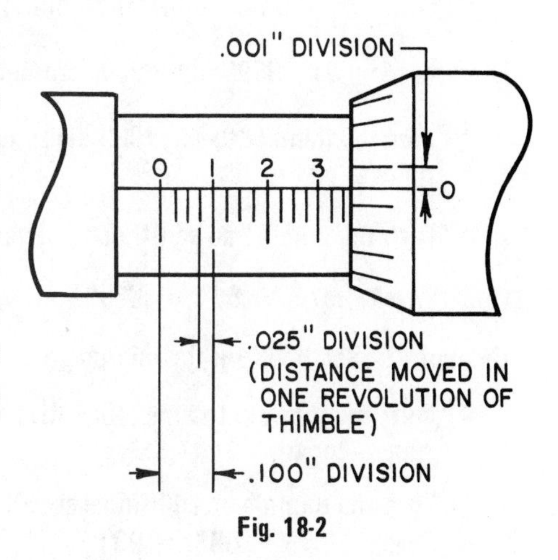

Fig. 18-2

READING AND SETTING THE MICROMETER

A micrometer is read by observing the position of the bevel edge of the thimble in reference to the scale on the barrel. The user observes the greatest .100-inch division and the number of .025-inch divisions on the barrel scale. To this barrel reading, he adds the number of the .001-inch divisions on the thimble that coincide with the horizontal line on the barrel scale.

Rule: To read the micrometer

- Observe the greatest .100-inch division on the barrel scale.
- Observe the number of .025-inch divisions on the barrel scale.
- Add the thimble reading that coincides with the horizontal line on the barrel scale.

Example 1: Read the micrometer setting shown in figure 18-3.

▲ Observe the greatest .100-inch division on the barrel scale. (three .100-inch divisions = .300 inch)

▲ Observe the number of .025-inch divisions between the .300-inch mark and the thimble. (two .025-inch divisions = .050 inch)

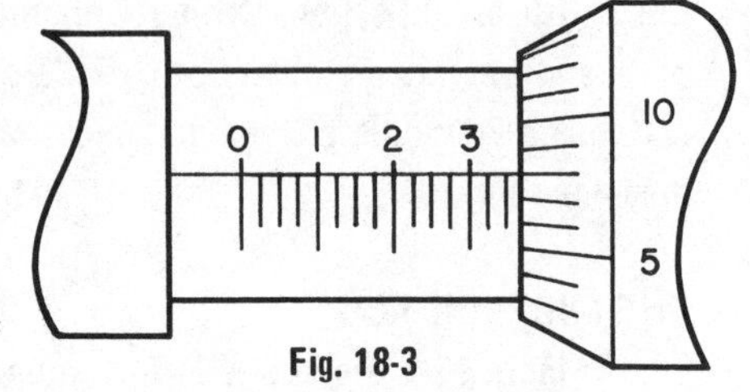

Fig. 18-3

▲ Add the thimble reading that coincides with the horizontal line on the barrel scale. (Eight .001" divisions = .008") Micrometer reading = .300" + .050" + .008" = .358".

Example 2: Read the micrometer setting shown in figure 18-4.

▲ On the barrel scale, two .100" = .200."

▲ On the barrel scale, zero .025" = 0".

▲ On the thimble scale, twenty-three .001" = .023".

Micrometer reading = .200" + .023" = .223".

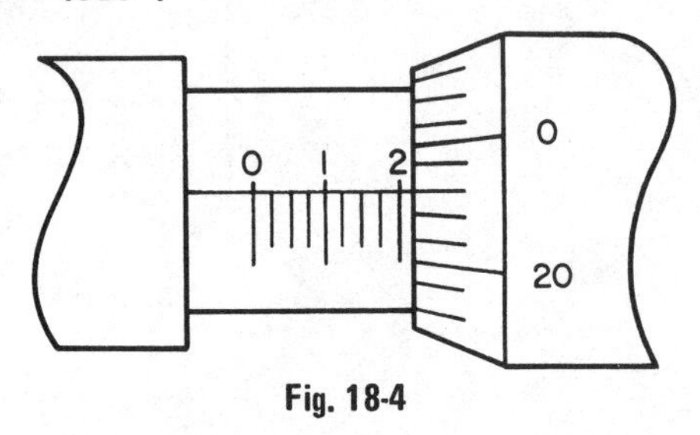

Fig. 18-4

Rule: To set the micrometer to a given dimension

- Turn the thimble until the barrel scale indicates the required number of .100-inch divisions plus the necessary number of .025-inch divisions.
- Turn the thimble until the thimble scale indicates the required additional .001-inch divisions.

Example 1: Set .689" on the micrometer.

▲ Turn the thimble to six .100" divisions plus three .025" divisions on the barrel scale. (6 x .100" + 3 x .025" = .675")

▲ Turn the thimble an additional fourteen .001" divisions.

Check: .675" + .014" = .689"

Example 2: Set .931" on the micrometer.

▲ Turn the thimble to nine .100" divisions plus one .025" division on the barrel scale. (9 x .100" + 1 x .025" = .925")

▲ Turn the thimble an additional six .001" divisions.
Check: .925" + .006" = .931"

THE VERNIER (.0001") MICROMETER

The addition of a vernier scale on the barrel of a .001-inch micrometer increases the degree of precision of the instrument to .0001 inch. The barrel scale and thimble scale of a vernier micometer are identical to that of a .001-inch micrometer.

Figure 18-5 shows the relative positions of the barrel scale, thimble scale, and vernier scale of a .0001-inch vernier micrometer.

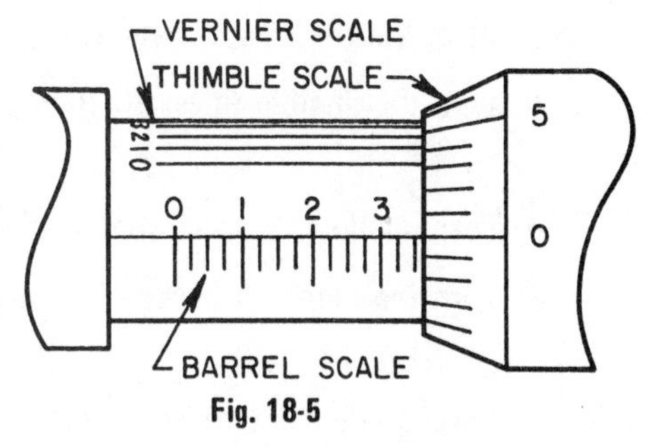

Fig. 18-5

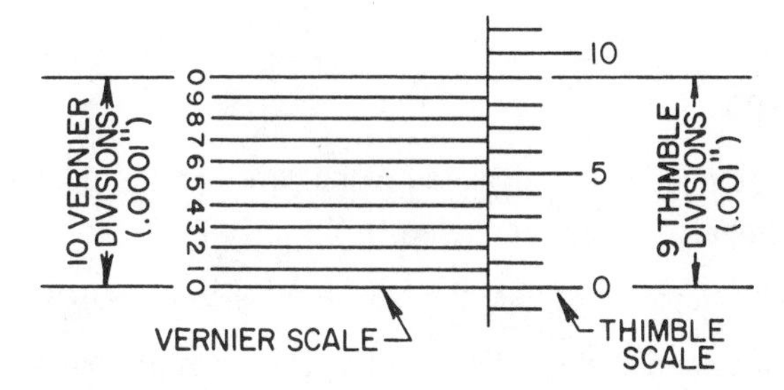

Fig. 18-6

The vernier scale consists of ten divisions. Ten vernier divisions on the circumference of the barrel are equal in length to nine divisions of the thimble scale. The difference between one vernier division and one thimble division is .0001 inch. Figure 18-6 shows a flattened view of a vernier and a thimble scale.

READING AND SETTING THE VERNIER MICROMETER

Reading a vernier micrometer is the same as reading an ordinary micrometer except for the addition of reading the vernier scale. A particular vernier graduation coincides with a thimble scale graduation. This vernier graduation gives the number of .0001-inch divisions that are added to the barrel and thimble scale readings.

Examples of Vernier Micrometer Readings:

Figure 18-7 shows a flattened view of a vernier micrometer setting. Read this setting.

- ▲ Read the barrel scale reading.
  Three .100" divisions plus
  three .025" divisions = .375"
- ▲ Read the thimble scale. The reading is between the .009" and .010" divisions, therefore, the thimble reading is .009".

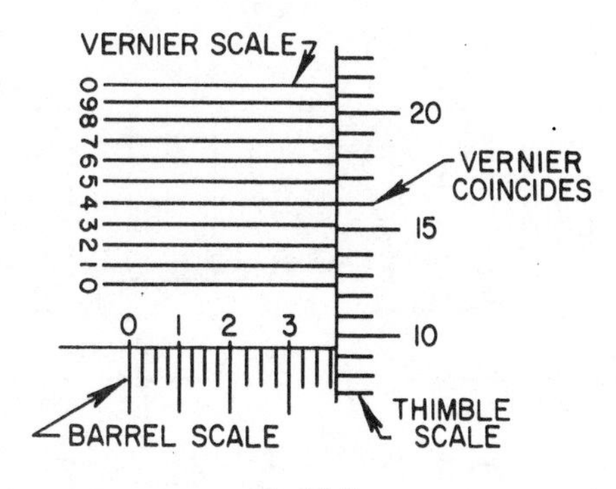

Fig. 18-7

- ▲ Read the vernier scale. The .004" division of the vernier scale coincides with a thimble division.
  Vernier micrometer reading = .375" + .009" + .0004" = .3844"

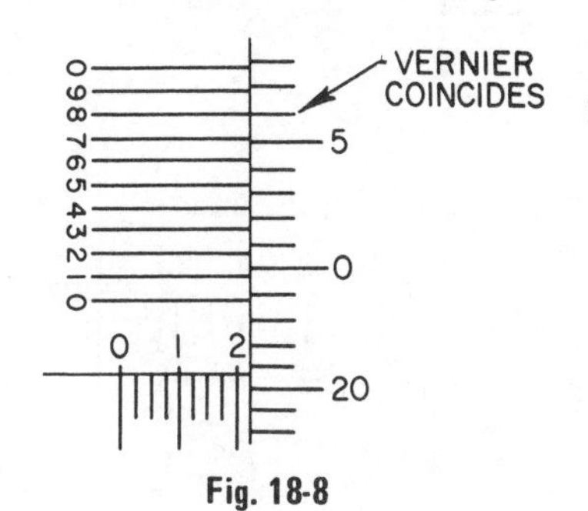

Fig. 18-8

Figure 18-8 shows a flattened view of a vernier micrometer setting. Read this setting.

- ▲ On the barrel scale read .200"
- ▲ On the thimble scale read .020"
- ▲ On the vernier scale read .0008"
  Vernier micrometer reading = .200" + .020" + .0008 = .2208"

Example of Vernier Micrometer Setting: Set .7693″ on the vernier micrometer.

▲ Turn the thimble to seven .100″ divisions plus two .025″ divisions on the barrel scale: 7 x .100″ + 2 x .025″ = .750″

▲ Turn the thimble an additional nineteen .001″ divisions.

▲ Turn the thimble carefully until a graduation on the thimble scale coincides with the .0003″ division on the vernier scale.

Check: .750″ + .019″ + .0003″ = .7693″

PROPER PROCEDURE IN THE USE OF MICROMETERS

- Micrometers should not be adjusted too loose or too tight while being used to measure a part; proper "feel" is extremely important and generally is developed through experience.
- The micrometer must be held perpendicular to the surfaces of the part which is being measured.
- The micrometer should be "rocked" across the diameter and along the axis of a cylinder to prevent a false reading when the diameter of the cylinder is being measured.
- More than one reading should be taken in measuring a part, and the readings averaged.
- Micrometers must be kept clean and periodically calibrated with gage blocks.

APPLICATION

A. Read the settings on the .001-inch micrometer scales shown below.

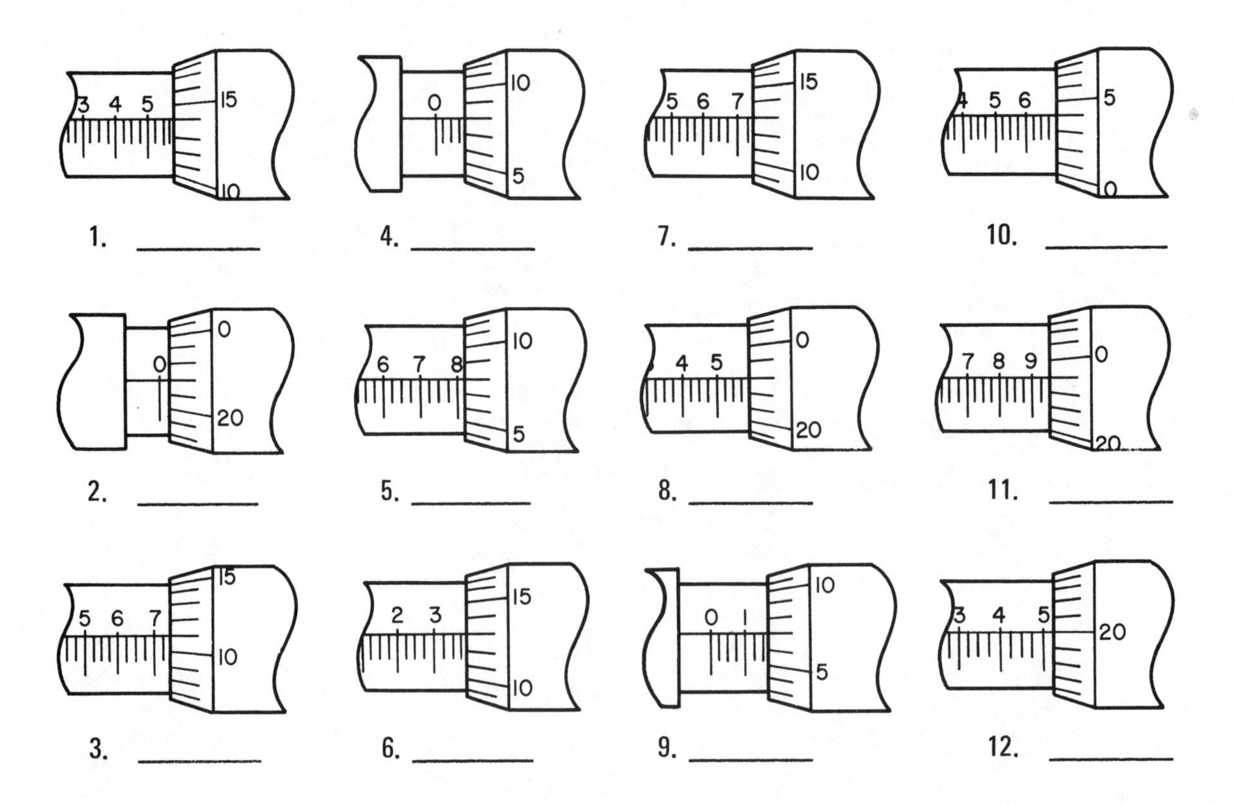

B. Given the following barrel scale and thimble scale settings of a .001-inch micrometer, determine the readings in the tables. The answer to the first problem is given.

	Barrel Scale Setting is Between:	Thimble Scale Setting	Micrometer Reading
1.	.425 - .450	.016	.441
2.	.075 - .100	.007	
3.	.150 - .175	.003	
4.	.875 - .900	.012	
5.	.300 - .325	.024	

	Barrel Scale Setting is Between:	Thimble Scale Setting	Micrometer Reading
6.	.000 - .025	.021	
7.	.025 - .050	.013	
8.	.750 - .775	.017	
9.	.975 - 1.000	.006	
10.	.625 - .650	.016	

C. Given the following .001-inch micrometer readings, determine the barrel scale and thimble scale settings. The answer to the first problem is given.

	Micrometer Reading	Barrel Scale Setting	Thimble Scale Setting
1.	.387	.375 - .400	.012
2.	.839		
3.	.973		
4.	.002		
5.	.059		

	Micrometer Reading	Barrel Scale Setting	Thimble Scale Setting
6.	.998		
7.	.036		
8.	.281		
9.	.517		
10.	.666		

D. Read the settings on the .0001-inch micrometer scales shown below. The vernier, thimble, and barrel scales are shown in flattened views.

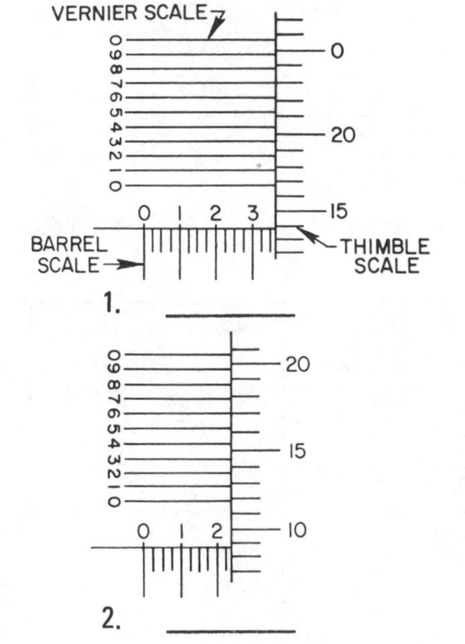

1. ________

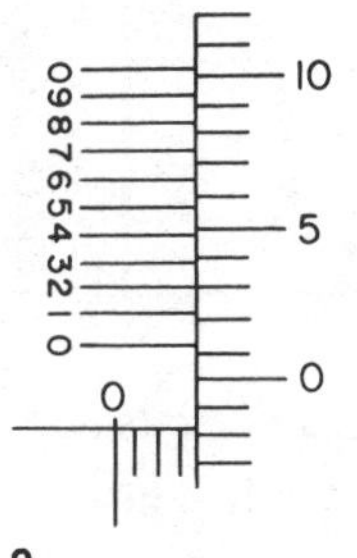

3. ________

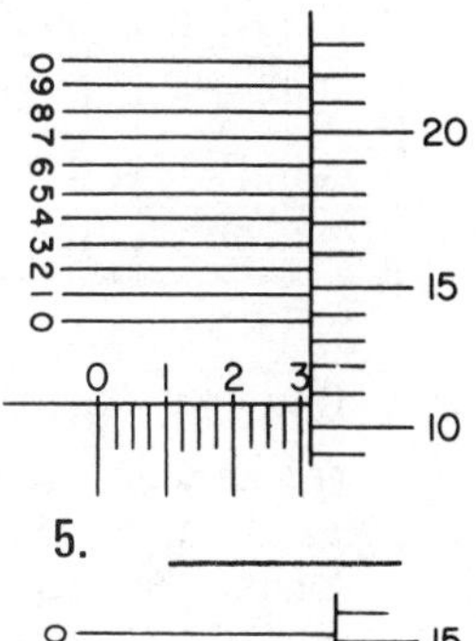

5. ________

2. ________

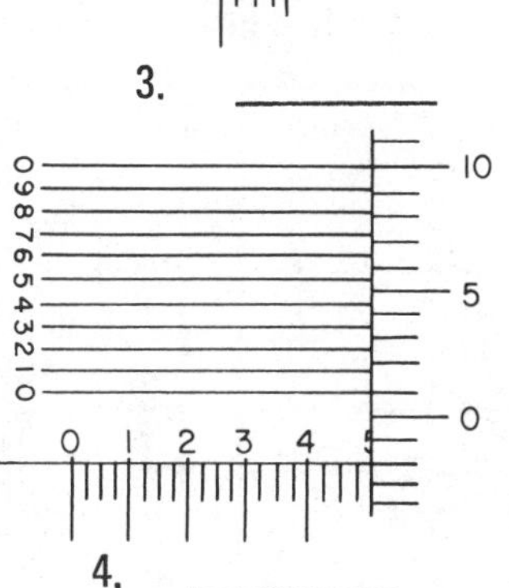

4. ________

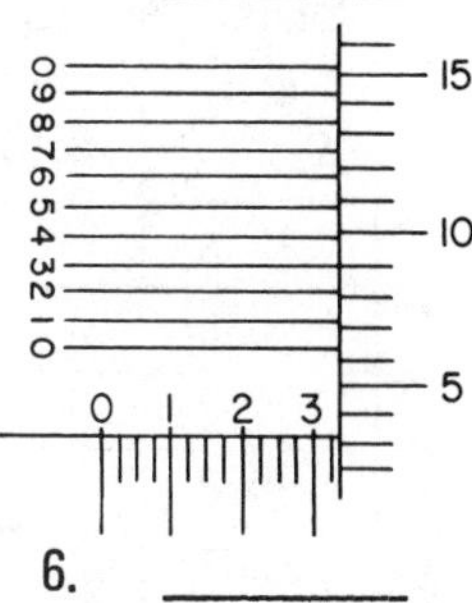

6. ________

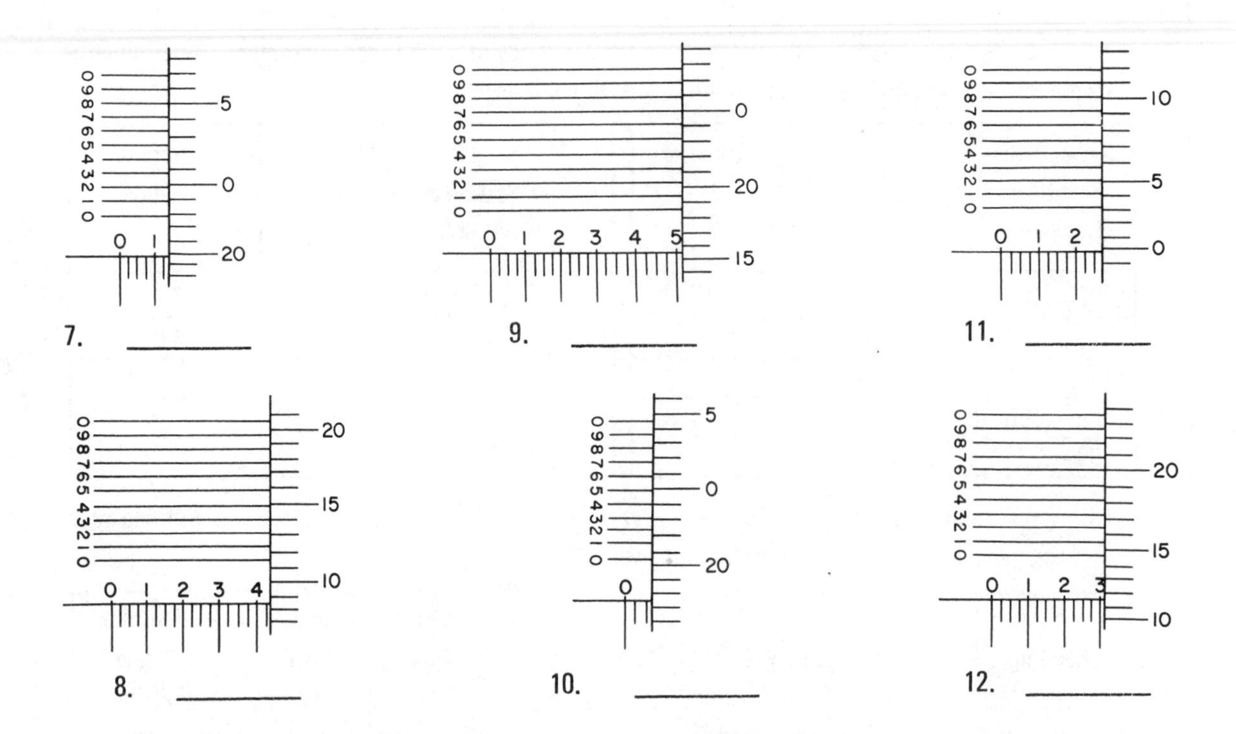

E. Given the following barrel scale, thimble scale, and vernier scale settings of a .0001-inch micrometer, determine the micrometer readings in the tables below. The answer to the first problem is given.

	Barrel Scale Setting is Between:	Thimble Scale Setting is Between:	Vernier Scale Setting	Microm-eter Reading
1.	.375-.400	.017-.018	.0008	.3928
2.	.125-.150	.008-.009	.0003	
3.	.950-.975	.021-.022	.0009	
4.	.075-.100	.011-.012	.0005	
5.	.200-.225	.000-.001	.0004	

	Barrel Scale Setting is Between:	Thimble Scale Setting is Between:	Vernier Scale Setting	Microm-eter Reading
6.	.625-.650	.021-.022	.0003	
7.	.000-.025	.000-.001	.0009	
8.	.275-.300	.020-.021	.0007	
9.	.850-.875	.009-.010	.0004	
10.	.125-.150	.014-.015	.0008	

F. Given the following .0001-inch micrometer readings, determine the barrel scale, thimble scale and, vernier scale settings. The answer to the first problem is given.

	Microm-eter Reading	Barrel Scale Setting	Thimble Scale Setting	Vernier Scale Setting
1.	.7846	.775-.800	.009-.010	.0006
2.	.1035			
3.	.0079			
4.	.9898			
5.	.3001			

	Microm-eter Reading	Barrel Scale Setting	Thimble Scale Setting	Vernier Scale Setting
6.	.0008			
7.	.8008			
8.	.3135			
9.	.9894			
10.	.0379			

Unit 19 Metric System

OBJECTIVES

After studying this unit, the student should be able to

- Compute problems using metric length units.
- Change English units of length to metric units.
- Change metric units of length to English units.
- Compute problems by converting values from one system to the other.

The metric system is the system of measurement which is used by most countries. Both the English and Metric systems have certain advantages and disadvantages. The metric system is superior to the English system for computational purposes. The metric system has been used for many years in the United States in the manufacture of guns, ammunition, and optics. With increased imports and exports of production machinery, tools, and products, the metric system is being more widely used in the United States, and it appears that the trend toward this increased use will continue. Therefore, it is important that the machine technician be able to compute with metric units.

METRIC UNITS

The metric system is used for all units of measure such as length, area, volume, and weight. The part of the system which is generally used by the machinist and draftsman is length measurement.

The standard unit of length is the *meter.* One meter equals 39.37 inches. Dimensions on engineering drawings are usually given in millimeters, regardless of the size of the dimensions. One *millimeter* equals 1/1000 meter.

Common metric units of length with their abbreviations are listed in table 19-1.

Metric Units of Linear Measure					
1 decimeter (dm.)	=	.1 meter (m.)	10 decimeters (dm.)	=	1 meter (m.)
1 centimeter (cm.)	=	.01 meter (m.)	100 centimeters (cm.)	=	1 meter (m.)
1 millimeter (mm.)	=	.001 meter (m.)	1000 millimeters (mm.)	=	1 meter (m.)
1 kilometer (km.)	=	1000 meters (m.)	.001 kilometers (km.)	=	1 meter (m.)

Table 19-1

Examples of changing units. (Refer to table 19-1.)

▲ Change 4 centimeters (cm.) to meters (m.). Since 1 cm. = .01 m., 4 cm. = 4 x .01 m. = .04 m.

▲ Change .0127 meters (m.) to millimeters (mm.). Since 1 m. = 1000 mm., .0127 m. = .0127 x 1000 mm. = 12.7 mm.

▲ Add .5 meters (m.), 3.7 decimeters (dm.), 12 centimeters (cm.), and 218 millimeters (mm.). Give the answer in millimeters.

a. Change each value to the equivalent value in meters:

.5 m. = .5 m., 3.7 dm. = 3.7 x .1 m. = .37 m., 12 cm. = 12 x .01 m. = .12 m.; 218 mm. = 218 x .001 m. = .218 m.

b. Combine .5 m.
.37 m.
.12 m.
.218 m.
1.208 m.

c. Change meters to millimeters:
Since 1m. = 1000 mm.,
1.208 m. = 1.208 x 1000 mm. =
1208 mm.

CHANGING UNITS FROM ONE SYSTEM TO THE OTHER

It is often necessary to convert values between the English system and the metric system. The conversion is easily accomplished by the substitution of English and metric equivalents shown in table 19-2.

Metric to English Units			English to Metric Units		
1 meter (m.)	=	39.37 inches	1 inch	=	.0254 meter (m.)
	=	3.2808 feet	1 foot	=	.3048 meter (m.)
1 centimeter (cm.)	=	.3937 inch	1 inch	=	2.54 centimeters (cm.)
1 millimeter (mm.)	=	.03937 inch	1 inch	=	25.4 millimeters (mm.)

Table 19-2

Figure 19-1 shows the relationship of English and metric units by comparing decimal-inch and metric scales.

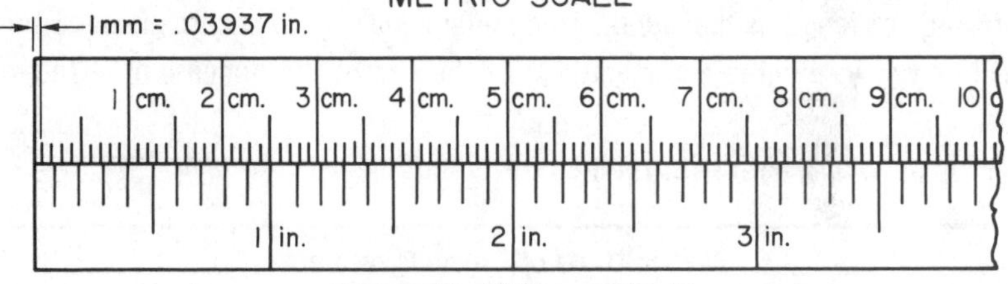

Fig. 19-1

Examples of converting values between systems. (Refer to table 19-2.)

▲ Change .25 meters to inches.

Since 1 m. = 39.37 in., .25 m. = 25 x 39.37 in. = 9.8425 in.

▲ Change 3.800 inches to centimeters.

Since 1 in. = 2.54 cm., 3.800 in. = 3.800 x 2.54 cm. = 9.652 cm.

▲ Change the dimensions given in figure 19-2 to millimeters.

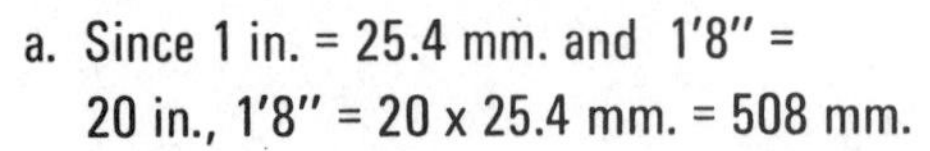

a. Since 1 in. = 25.4 mm. and 1'8" = 20 in., 1'8" = 20 x 25.4 mm. = 508 mm.

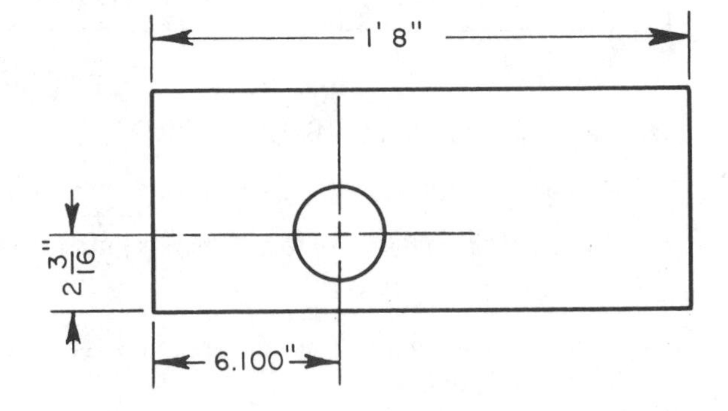

Fig. 19-2

b. Since 1 in. = 25.4 mm., 6.100" = 6.100 x 25.4 mm. = 154.94 mm.

c. Since 1 in. = 25.4 mm. and 2 3/16" = 2.1875", 2 3/16 in. = 2.1875 x 25.4 mm. = 55.5625 mm.

APPLICATION

A. METRIC UNITS

1. Change the following lengths to meters.

   a. 30 decimeters ____ c. 200 centimeters ____ e. 8,000 millimeters ____

   b. .75 decimeters ____ d. 6,250 centimeters ____ f. 21,000 millimeters ____

2. Change the following lengths to millimeters.

   a. 1 meter ____ c. 6 meters ____ e. 12 centimeters ____

   b. 2.75 decimeters ____ d. 47.3 centimeters ____ f. .004 kilometers ____

3. Addition: Give answers in millimeters.

   a. 13.7 m. + 96.6 m. ____ c. 10.9 cm. + .37 dm. ____

   b. .5 mm. + .028 mm. + 17.6 mm. ____ d. 2 cm. + .7 dm. + .003 m. ____

4. Subtraction: Give answers in millimeters.

   a. 8 m. - 23 cm. ____ c. 120 dm. - 87 cm. ____

   b. 720 mm. - 2 dm. ____ d. .25 m. - 3.5 cm. ____

5. Determine, in millimeters, dimensions F, G, and H on the plate shown.

   a. a = 78.3 mm., b = 100 mm., c = 210.6 mm.

   d = 65 mm. e = 104.4 mm.

   F = ____ G = ____ H = ____

   b. a = .325 dm., b = 4.3 cm., c = 10.35 cm.

   d = 83.42 mm., e = 13.4 cm.

   F = ____ G = ____ H = ____

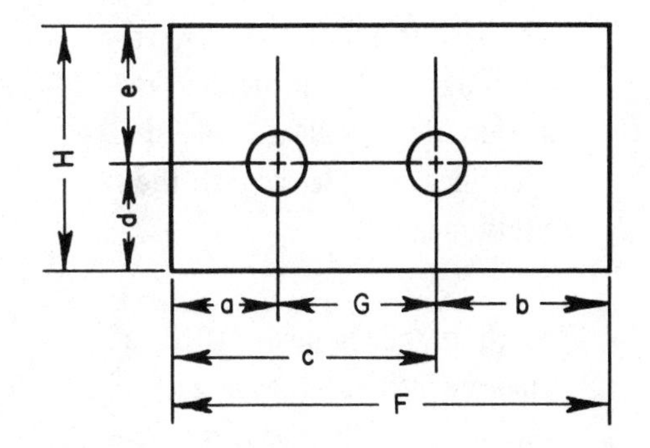

6. Determine, in millimeters, dimensions C and D on the part shown.

   a. a = .85 m. b = 28 cm.

   C = ____ D = ____

   b. a = 66.5 dm. b = 32.45 cm.

   C = ____ D = ____

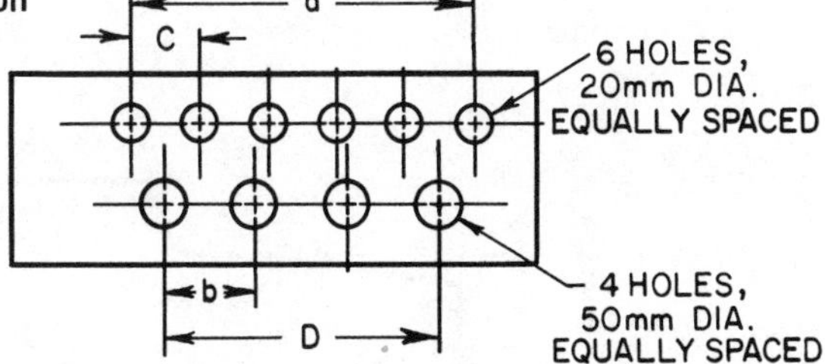

7. A piece of sheet metal is 1.12 m. wide. Strips each 3.4 cm. wide are cut, allowing 3 mm. for each strip. Determine the number of complete strips cut and the width of the waste strip in millimeters. ________

8. An aluminum slab .076 m. thick is machined with three equal cuts, each cut 10 mm. deep. Determine the finished thickness of the slab in centimeters. ________

B. CHANGING UNITS

1. Change the following metric lengths to inches. Where necessary round-off answers to 3 decimal places.

a. 10 mm.	____	e. 100 mm.	____	i. 10 cm.	____
b. .1 m.	____	f. 20 cm.	____	j. 110 mm.	____
c. .06 m.	____	g. 55.2 mm.	____	k. 27.7 cm.	____
d. .003 m.	____	h. 7.8 cm.	____	l. 9.4 mm.	____

2. Change the following lengths to the indicated metric units. Where necessary, round-off answers to three decimal places.

a. 1 in. to mm.	____	f. 2.5 in. to mm.	____	k. 1 in. to cm.	____
b. 12 in. to cm.	____	g. 1 in. to m.	____	l. 100 in. to m.	____
c. .25 in. to mm.	____	h. 8.2 in. to cm.	____	m. 33 in. to m.	____
d. 17.87 in. to mm.	____	i. 212 in. to m.	____	n. .014 in. to mm.	____
e. 5/16 in. to mm.	____	j. 9 3/8 in. to cm.	____	o. 2 13/16 in. to mm.	____

3. The part shown on the right is dimensioned in millimeters. Change each dimension to inches and redimension the drawing. Round-off dimensions to the closest thousandths of an inch.

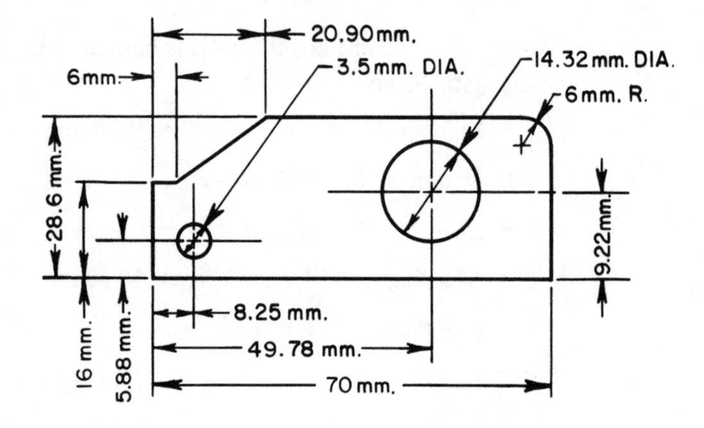

4. The shaft shown below is dimensioned in inches. Change each dimension to millimeters and redimension the drawing. Round-off dimensions to the closest hundredth of a millimeter.

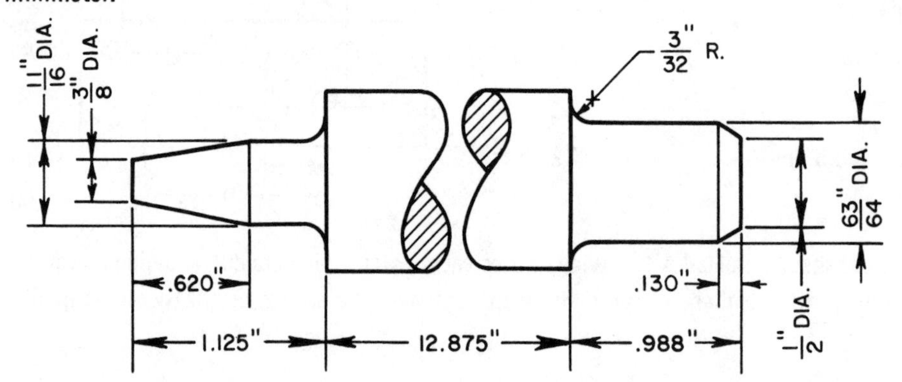

5. Determine the total length of stock in inches required to make 35 bushings, each 38.1 mm. long allowing 3/16" waste for each bushing.

SECTION III

Fundamentals of Algebra

Unit 20 Introduction to Symbolism

OBJECTIVES

After studying this unit, the student should be able to

- Translate word statements into algebraic expressions.
- Convert diagram dimensions into algebraic expressions.
- Evaluate algebraic expressions by substituting numbers for symbols.

Algebra is a branch of mathematics in which letters are used to represent numbers. By the use of letters, general rules called *formulas* can be stated mathematically. Algebra is an extension of arithmetic; therefore, the rules and procedures which apply to arithmetic also apply to algebra. Many problems which are difficult or impossible to solve by arithmetic can be solved by algebra.

The basic principles of algebra discussed in this text are intended to provide a practical background for machine shop applications. A knowledge of algebraic fundamentals is essential in the use of trade handbooks and for the solutions of many geometric and trigonometric problems.

SYMBOLISM

Symbols are the language of algebra. Both arithmetic numbers and literal numbers are used in algebra.

- *Arithmetic numbers* are numbers which have definite numerical values, such as 4, 5.17, and 7/8.
- *Literal numbers* are letters which represent arithmetic numbers, such as a, x, V, and P. Depending on how it is used, a literal number can represent one particular arithmetic number, a wide range of numerical values, or all numerical values.
- An *algebraic expression* is a word statement put into mathematical form by using literal numbers, arithmetic numbers, and signs of operation.

Customarily the multiplication sign (x) is not used in algebra because it can be misinterpreted as the letter x. When a literal number is multiplied by a numerical value, or when two or more literal numbers are multiplied, no sign of operation is required;

5 x a is written 5a; 17 x c, 17c; a x b, ab; V x P = VP; 6 x a x b x c = 6 abc

Parentheses () or raised dots · are often used in place of the multiplication sign (x) when numerical values are multiplied; 3 x 4 is written 3(4) or 3 · 4; 6 x 17/2, 6(17/2) or 6 · 17/2; 18 x 3.4 x 5^2 is written 18(3.4) (5^2) or 18 · 3.4 · 5^2.

Examples of Algebraic Expressions

▲ Figure 20-1 shows a dimension which is increased by 1/2 inch. If x is the original dimension, the increased dimension is x + 1/2 inch.

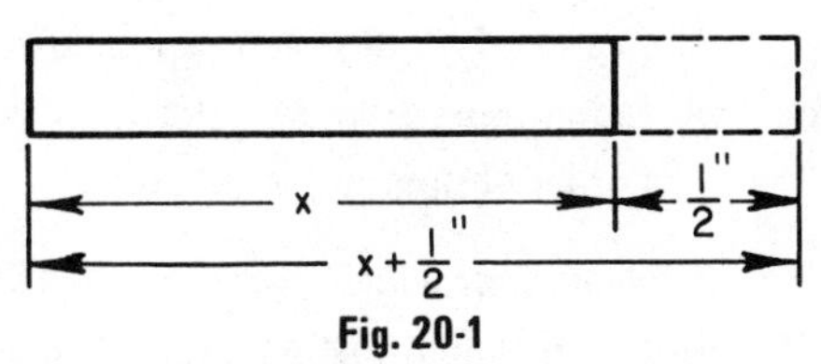

Fig. 20-1

▲ The production rate of a new machine is 4 times as great as an old machine. If the old machine produced y parts per hour, the new machine produces 4y parts per hour.

▲ A length of drill rod shown in figure 20-2 is cut into 3 equal pieces. If L is the length of the drill rod, the length of each piece is L/3.

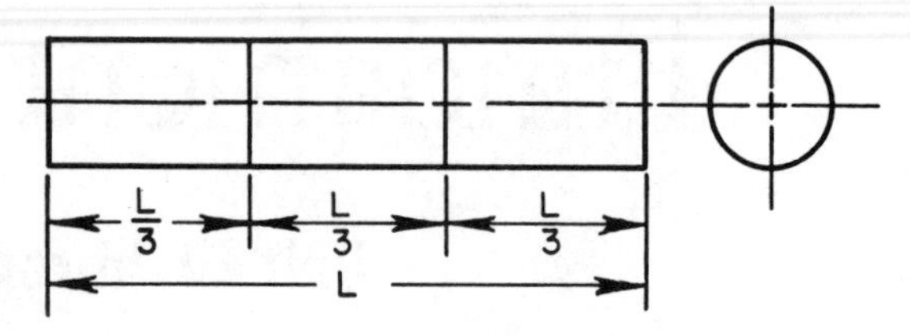

Fig. 20-2

▲ A step block is shown in figure 20-3. Dimension B equals 3/4 of dimension A and dimension C is twice dimension A. If dimension A is d inches, dimension B is 3/4d inches and dimension C is 2d inches. The total height is d inches + 3/4d inch + 2d inches or 3 3/4d inches.

Note: If no arithmetic number appears before a literal number, it is assumed that the value is the same as if a one(1) appeared before the letter, d = 1d.

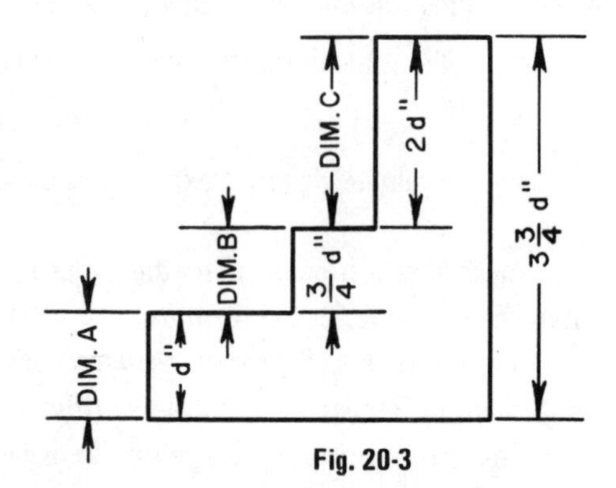

Fig. 20-3

▲ Figure 20-4 shows a plate with 8 drilled holes. The distance from the left edge of the plate to hole 1 and the distance from the right edge of the plate to hole 8 are each equal to a inches. The distances between holes 1 and 2, holes 2 and 3, and holes 3 and 4 are each equal to b inches. The distances between holes 4 and 5, holes 5 and 6, holes 6 and 7, and holes 7 and 8 are equal to c inches.

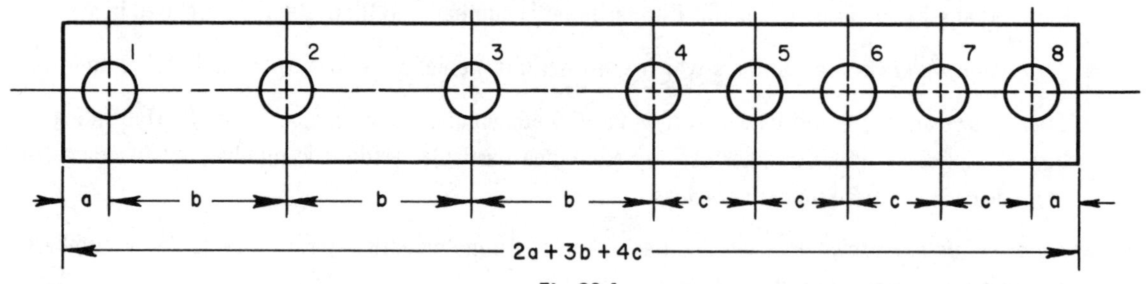

Fig. 20-4

The total length of the plate is a + b + b + b + c + c + c + c + a, or 2a + 3b + 4c. The distance from hole 2 to hole 7 is b + b + c + c + c, or 2b + 3c. The distance from hole 1 to the right edge of the plate minus the distance from hole 4 to hole 7 is (b + b + b + c + c + c + c + a) - (c + c + c) or, (3b + 4c + a) - 3c or, 3b + c + a.

Note: Only like literal numbers may be arithmetically added.

EVALUATION OF ALGEBRAIC EXPRESSIONS

The value of an algebraic expression is found by substituting given numerical values for literal values and solving the expression by following the order of operations as in arithmetic.

The order of operations follows:

- *All operations within parentheses.* If there are parentheses within parentheses, perform the operations within the innermost parenthesis first.

- *Powers and roots.*
- *Multiplication and division.* Perform multiplication and division operations in the order in which they occur.
- *Addition and subtraction.* Addition and subtraction can be taken in any order.

Examples of Evaluation of Algebraic Expressions

▲ A rectangle is shown in figure 20-5. The perimeter of a rectangle = 2L + 2W; therefore, P = 2(5) + 2(3) = 10 + 6 = 16 inches.

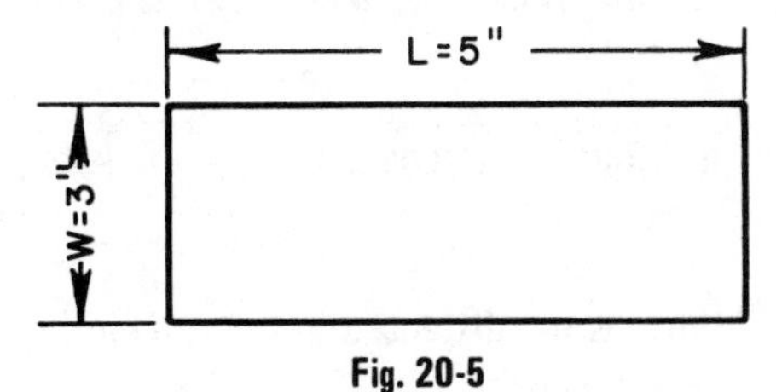

Fig. 20-5

▲ Figure 20-6 shows a ring. The area of the ring = $\pi R^2 - \pi r^2$; ($\pi = 3.14$). $A = 3.14 \cdot 5^2 - 3.14 \cdot 2^2 = 3.14 \cdot 25 - 3.14 \cdot 4 = 78.50 - 12.56 = 65.94$ square inches.

▲ An ellipse is shown in figure 20-7. The approximate perimeter of an ellipse is shown by

$\pi \sqrt{2(a^2 + b^2)}$; ($\pi = 3.14$).

$P = 3.14 \sqrt{2(8^2 + 6^2)} = 3.14 \sqrt{2(64 + 36)} =$

$3.14\sqrt{2(100)} = 3.14\sqrt{200} = 3.14\,(14.14) =$ 44.40 inches

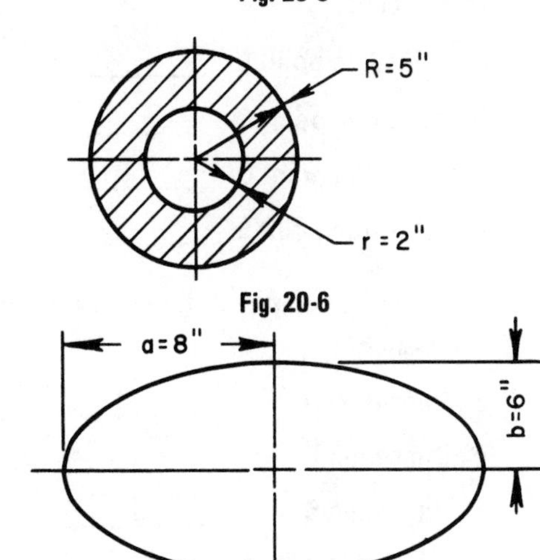

Fig. 20-6

Fig. 20-7

▲ If b = 6, d = 4, and y = 2, then $\frac{3(2b + 3dy)}{4(7d - bd)} = \frac{3(2 \cdot 6 + 3 \cdot 4 \cdot 2)}{4(7 \cdot 4 - 6 \cdot 4)} = \frac{3(12 + 24)}{4(28 - 24)} = \frac{3 \cdot 36}{4 \cdot 4} = \frac{108}{16} = 6.75$

▲ If m = 2, p = 3, x = 8, then $3m(4p + 5(x - m) + p)^2 = 3 \cdot 2(4 \cdot 3 + 5(8 - 2) + 3)^2 = 6(12 + 5 \cdot 6 + 3)^2 = 6(45)^2 = 6(2025) = 12,150$

▲ If a = 5, b = 10, and c = 8, then $\frac{6a}{b} + \frac{abc}{20}(a^3 - 12b) = \frac{6 \cdot 5}{10} + \frac{5 \cdot 10 \cdot 8}{20}(5^3 - 12 \cdot 10) =$

$\frac{30}{10} + \frac{400}{20}(125 - 120) = 3 + 20(5) = 3 + 100 = 103$

APPLICATION

A. ALGEBRAIC EXPRESSIONS

Express the following problems in algebraic form.

1. The product of 6 and x increased by y. __________
2. The sum of a and 12. __________
3. Subtract b from 25. __________
4. Subtract 25 from b. __________

5. Divide r by s. ________
6. Twice L minus one-half P. ________
7. The product of x and y divided by the square of m. ________

8. a. What is the total length of this part? ________

   b. What is the length from point A to point B? ________

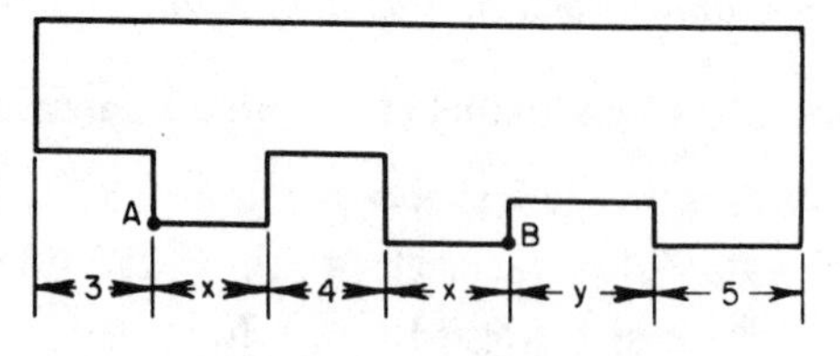

9. What is the distance between the indicated points?

   a. Point A to point B ________

   b. Point F to point C ________

   c. Point B to point C ________

   d. Point D to point E ________

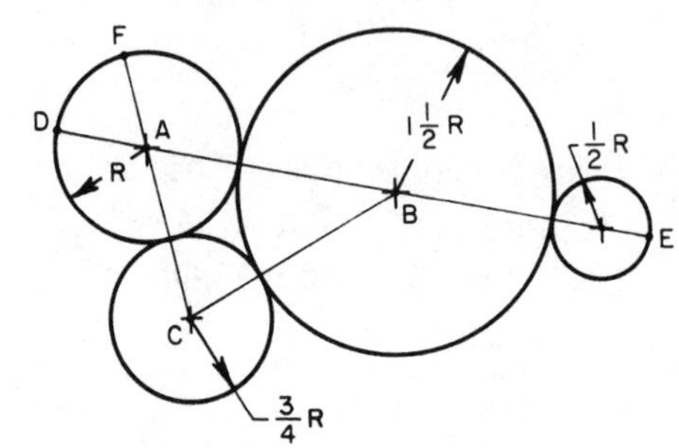

10. What is the length of the following dimensions?

    a. Dimension A ________

    b. Dimension B ________

    c. Dimension C ________

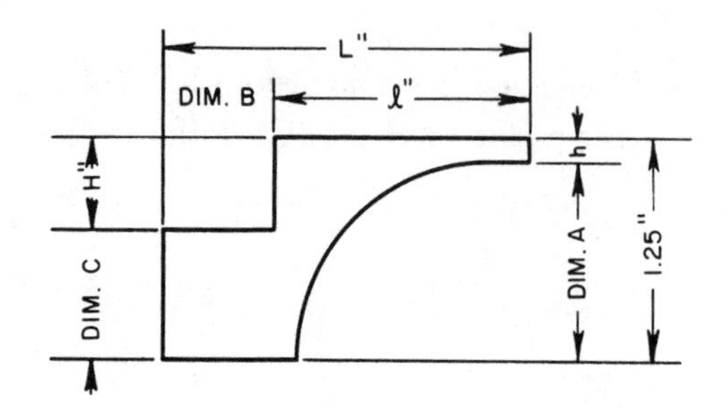

11. Stock is removed in two operations from a block n inches thick. A milling operation removes p inches and a grinding operation removes t inches. What is the final thickness of the block? ________

12. Given: s as the length of a side of a hexagon, r as the radius of the inside circle, and R as the radius of the outside circle.

    a. What is the length of r if r equals the product of .866 and the length of a side of the hexagon? ________

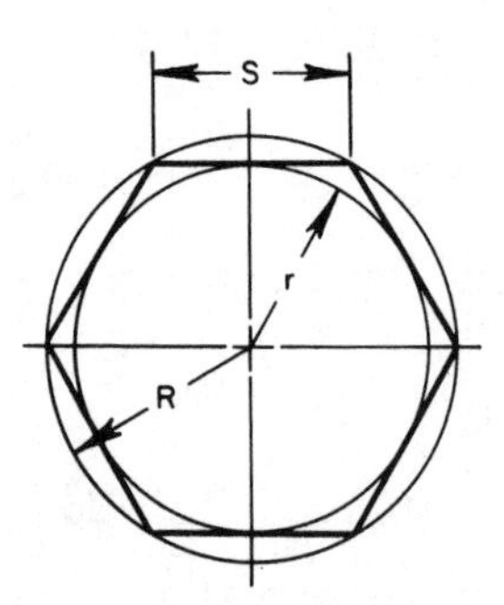

    b. What is the length of R if R equals the product of 1.155 and the radius of the inside circle? ________

    c. What is the area of the hexagon if the area equals the product of 2.598 and the square of the radius of the outside circle? ________

B. EVALUATION OF ALGEBRAIC EXPRESSIONS

Substitute the given numbers for letters and find the values of the following expressions.

1. If $a = 4$ and $c = 2$
   a. $5a + 3c^2$ __________
   b. $5c + a$ __________
   c. $\frac{10c}{a}$ __________
   d. $\frac{a + c}{a - c}$ __________
   e. $\frac{a + 5c}{ac + a}$ __________

2. If $b = 8$, $d = 4$, and $e = 2$
   a. $\frac{b}{d} + e - 3$ __________
   b. $bd\,(3 + 4d - b)$ __________
   c. $5b - (bd + 3)$ __________
   d. $3e\,(b - e) - d(\frac{b}{2})$ __________
   e. $\frac{12d}{e} - (3b - (d + e) + 4)$ __________

3. If $x = 6$ and $y = 3$
   a. $2\,xy + 7$ __________
   b. $3x - 2y + xy$ __________
   c. $\frac{5xy - 2y}{8\,x - xy}$ __________
   d. $\frac{4x - 4y}{3}$ __________
   e. $6\,x - 3y + xy$ __________

4. If $m = 5$, $p = 4$, $r = 3$
   a. $m + mp^2 - r^3$ __________
   b. $(p + 2)^2\,(m - r)^2$ __________
   c. $\frac{(pr)^2}{2} - pr + m^3$ __________
   d. $\frac{p^3 + 3p - 12}{m^2 + 15}$ __________
   e. $\frac{r^3}{3p - 9} + m^2\,(mp - 6r)^2$ __________

5. a. Area = $1/2\,d^2$, Area = __________
   b. $S = .7071d$, S = __________

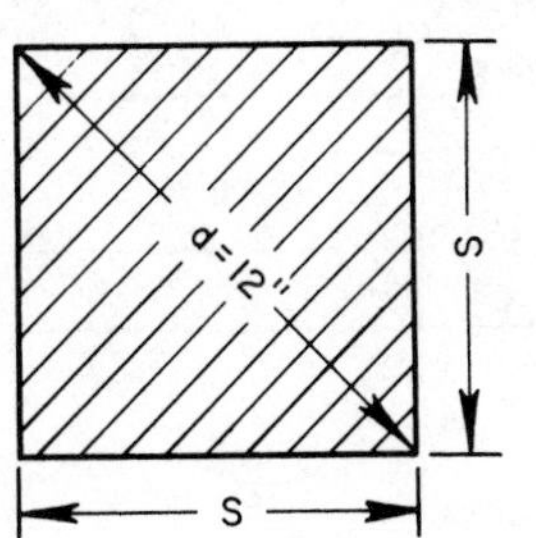

6. a. $\ell = \frac{\pi R\alpha}{180}$, ℓ = __________
   b. Area = $1/2\,R\ell$, Area = __________

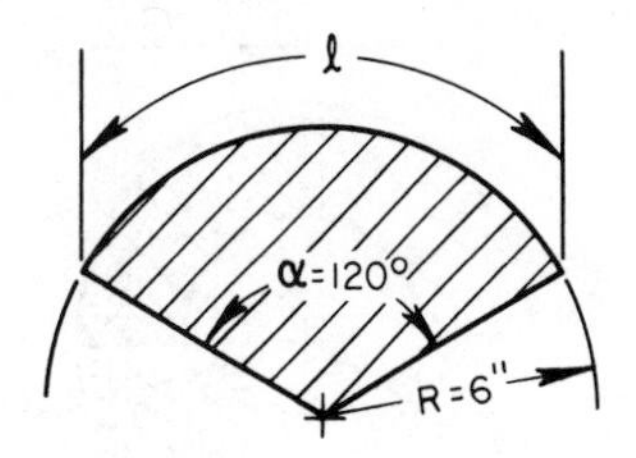

7. a. $S = 1/2\,(a + b + c)$, S = __________
   b. Area = $\sqrt{s\,(s - a)\,(s - b)\,(s - c)}$,
   Area = __________

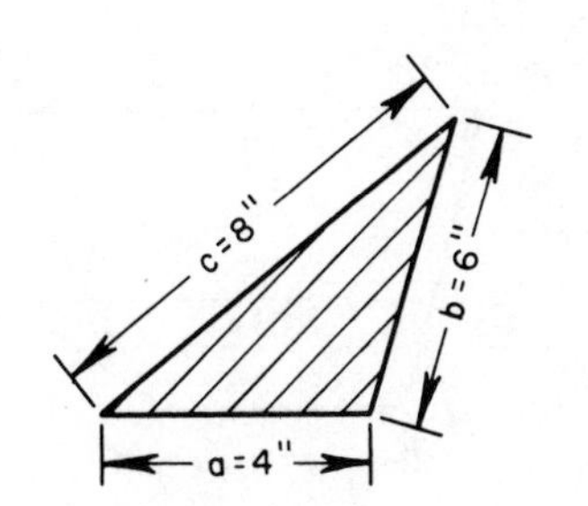

8. a. $r = \frac{c^2 + 4h^2}{8h}$, r = __________
   b. $\ell = .0175\,r\alpha$, ℓ = __________

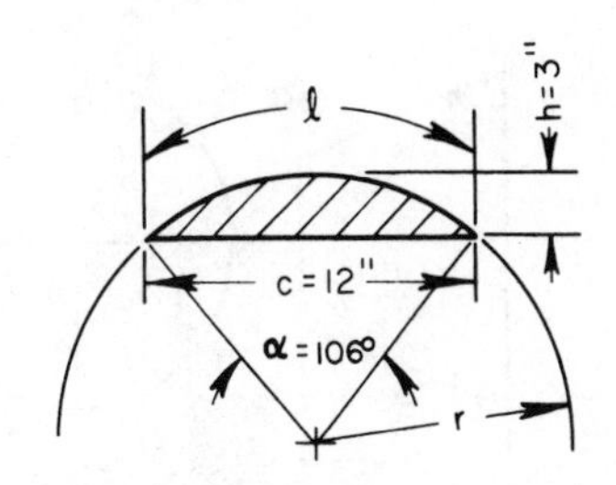

9. Area $= \dfrac{(H + h)\,b + ch + aH}{2}$

Area = ______________

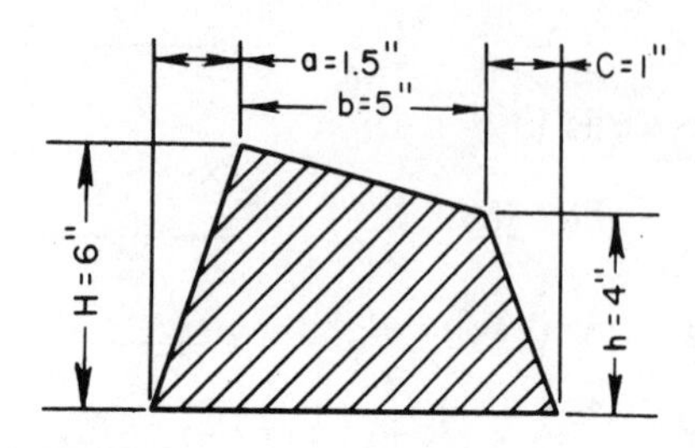

10. Length of belt on pulleys =

$$2\,C + \frac{11D + 11d}{7} + \frac{(D - d)^2}{4\,C}$$

Length of belt = ______________

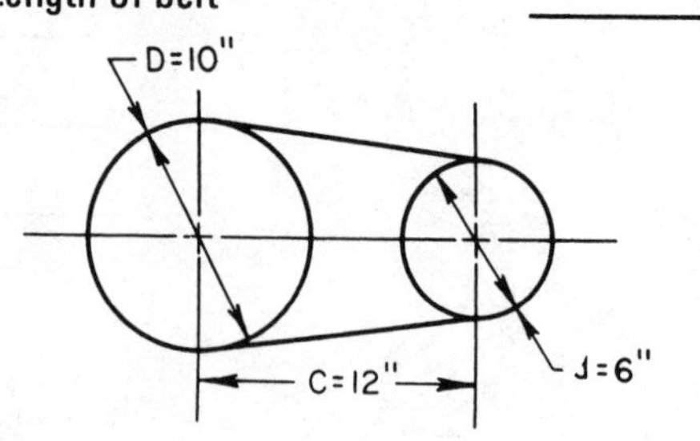

11. Area = $dt + 2a\,(s + n)$

Area = ______________

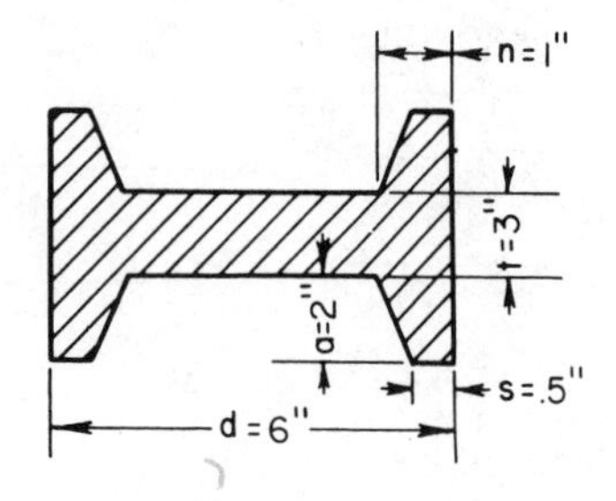

12. Area = $\pi\,(ab - cd)$

Area = ______________

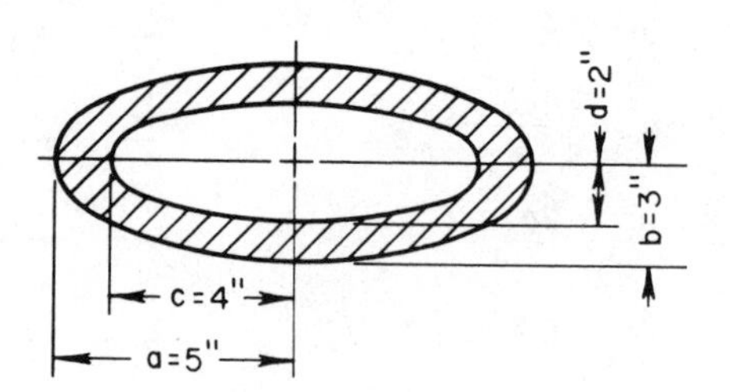

13. Area $= \dfrac{\pi\,(R^2 - r^2)}{2}$

Area = ______________

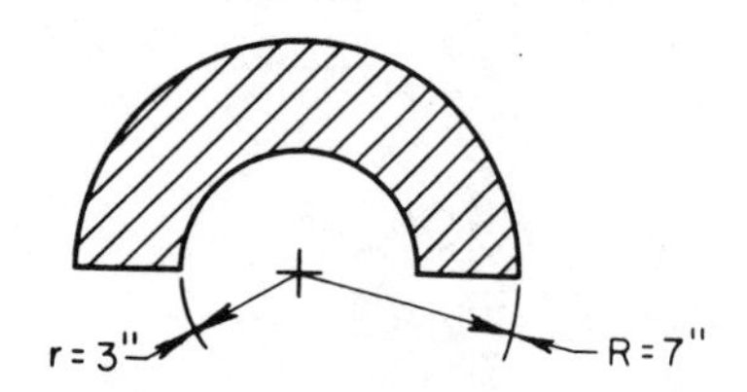

14. Area = $t[b + 2\,(a - t)]$

Area = ______________

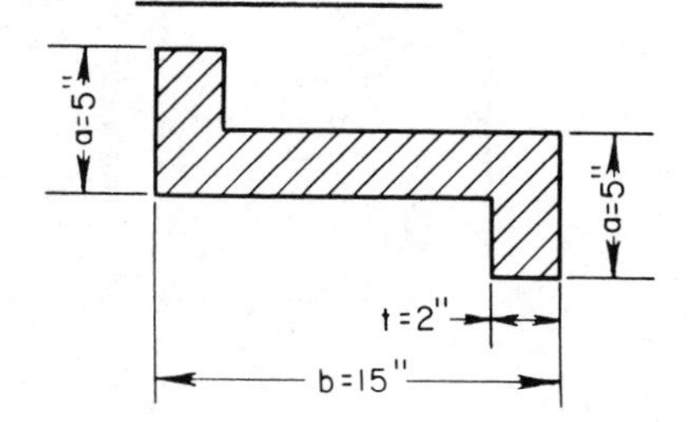

15. a. $S = \sqrt{(R-r)^2 + h^2}$, S = ______________

b. Volume = $1.05\,h\,(R^2 + Rr + r^2)$

Volume = ______________

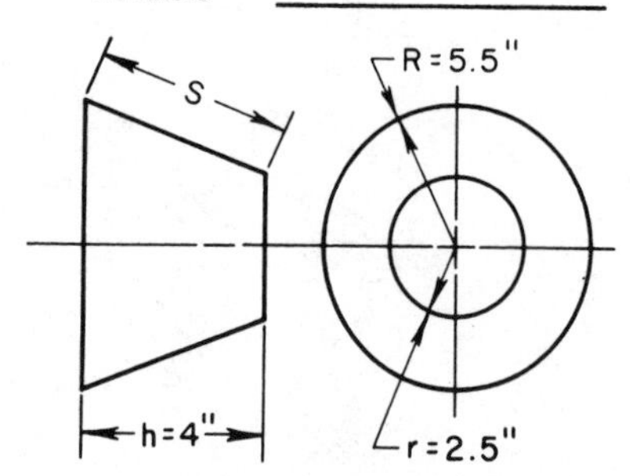

16. Volume $= \dfrac{(2a + c)\,bh}{6}$

Volume = ______________

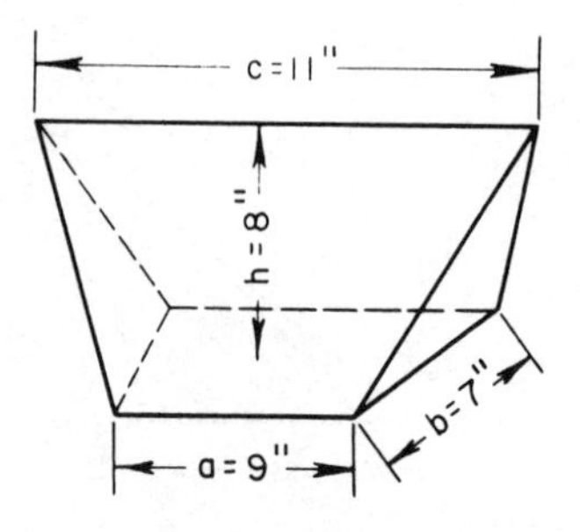

Unit 21 Signed Numbers

OBJECTIVES

After studying this unit, the student should be able to

- Compare signed numbers according to size and direction using the number scale.
- Determine absolute values of signed numbers.
- Perform basic operations of addition, subtraction, multiplication, division, powers, and roots, using signed numbers.
- Solve expressions which involve combined operations of signed numbers.

Signed numbers are required for solving problems in mechanics and trigonometry. Positive and negative numbers express direction, such as machine table movement from a reference point. Signed numbers are particularly useful in programming machining operations for numerical control.

SIGNED NUMBERS

Numbers that are preceded by a plus or a minus sign are called *signed numbers.* The plus sign indicates that the number is greater than zero, and the minus sign indicates that the number is less than zero.

A number which has no sign or one which is preceded by a plus sign is a *positive number.* The numbers used in arithmetic are all positive numbers. For example, +7 is the same as 7; they both mean 7 units greater than 0.

A number which is preceded by a minus sign is a *negative number.* A negative number must be preceded by a minus sign. For example, -5 means 5 units less than 0.

THE NUMBER SCALE

The number scale in figure 21-1 shows the relationship of positive and negative numbers. It shows distance and direction between numbers. Considering a number as a starting point and counting to a second number to the right represents positive (+) direction. Counting to the left represents negative (-) direction.

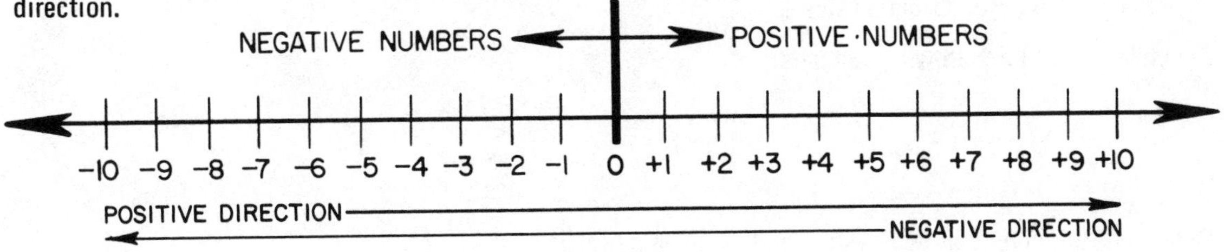

Fig. 21-1

Examples:

▲ Starting at -2 and counting to the right to +6 represents 8 units in a positive (+) direction; +6 is 8 units greater than -2.

▲ Starting at +6 and counting to the left to -2 represents 8 units in a negative (-) direction; -2 is 8 units less than +6.

▲ Starting at -3 and counting to the left to -10 represents 7 units in a (-) direction; -10 is 7 units less than -3.

▲ Starting at -9 and counting to the right to 0 represents 9 units in a (+) direction; 0 is 9 units greater than -9.

OPERATIONS USING SIGNED NUMBERS

The following rules and examples show how to perform operations of addition, subtraction, multiplication, division, powers, and roots with signed numbers.

- The *absolute value* of a number is the number without a sign. For example, the absolute value of +4 is 4, the absolute value of -4 is also 4. Therefore, the absolute value of +4 and -4 is the same value, 4.

  The absolute value of -20 is 15 greater than the absolute value of +5; 20 is 15 greater than 5.

Addition

Rule: To add two or more positive numbers

- Add the numbers as in arithmetic.

Examples: Add the following numbers

$$\begin{array}{r} +3 \\ \underline{+5} \\ +8 \end{array} \qquad \begin{array}{r} 15 \\ \underline{7} \\ 22 \end{array} \qquad 2 + 9 + 13 = 24 \qquad +12 + (+15) = +27$$

Rule: To add two or more negative numbers

- Add their absolute values and prefix a minus sign.

Examples: Add the following numbers

$$\begin{array}{r} -5 \\ \underline{-2} \\ -7 \end{array} \qquad \begin{array}{r} -13 \\ -4 \\ \underline{-15} \\ -32 \end{array} \qquad \begin{array}{l} -6 + (-5) = -11 \\ -8 + (-10) + (-4) + (-3) = -25 \end{array}$$

Rule: To add a positive and a negative number

- Subtract the smaller absolute value from the larger absolute value and prefix the sign of the number having the larger absolute value.

Examples: Add the following numbers

$$1.\ \begin{array}{r} +5 \\ \underline{-3} \\ +2 \end{array} \qquad 2.\ \begin{array}{r} -5 \\ \underline{+3} \\ -2 \end{array} \qquad 3.\ \begin{array}{r} -17 \\ \underline{+17} \\ 0 \end{array} \qquad \begin{array}{l} 4.\ +12 + (-8) = +4 \\ \\ 5.\ -12 + (+8) = -4 \end{array}$$

Rule: To add more than two positive and negative numbers

- Add all the positive numbers.
- Add all the negative numbers.
- Add the sums.

Examples: Add the following numbers

▲ $-2 + 4 + (-10) + 5 = 9 + (-12) = -3$

▲ $8 + 7 + (-6) + 4 + (-3) + (-5) + 10 = 29 - 14 = 15$

▲ $4 + (-6) + 12 + 3 + (-7) + 1 + (-5) + (-2) = 20 - 20 = 0$

Subtraction

Rule: To subtract signed numbers

- Change the sign of the number subtracted (subtrahend) to the opposite sign.
- Follow the rules for addition.

Note: When the sign of the subtrahend is changed the problem becomes one in addition. Therefore, subtracting a negative number is the same as adding a positive number, and subtracting a positive number is the same as adding a negative number.

Examples:

▲ Subtract 5 from 8; $8 - (+5) = 8 + (-5) = 3$

▲ Subtract 8 from 5; $5 - (+8) = 5 + (-8) = -3$

▲ Subtract -5 from 8; $8 - (-5) = 8 + (+5) = 13$

▲ Subtract -5 from -8; $-8 - (-5) = -8 + (+5) = -3$

▲ $-3 - (+7) = -3 + (-7) = -10$

▲ $0 - (-14) = 0 + (+14) = 14$

▲ $0 - (+14) = 0 + (-14) = -14$

▲ $-14 - (-14) = -14 + (+14) = 0$

Multiplication

Rule: To multiply two or more signed numbers

- Find the product of the numbers.
- If the number of negative signs is even, the product is positive.
- If the number of negative signs is odd, the answer is negative.
- If the problem consists of all positive numbers, the product is positive.

Note: It is not necessary to count the number of positive values in a problem consisting of both positive and negative numbers. The number of negative values determines the sign of the product.

Examples:

▲ $4 \times (-3)$ consists of 1 negative; 1 is an odd number; therefore, $4 \times (-3) = -12$

▲ $-4 \times (-3)$ consists of 2 negatives; 2 is an even number; therefore, $-4 \times (-3) = +12$

▲ $(-2)(-4)(-3)(-1)(-2)(-1) = +48$; (6 negatives, even number)

▲ $(-2)(-4)(-3)(-1)(-2) = -48$; (5 negatives, odd number)

▲ $(2)(4)(3)(1)(2) = +48$; (all positives)

▲ $(2)(-4)(-3)(1)(-2) = -48$; (3 negatives, odd number)

▲ $(-2)(4)(-3)(-1)(-2) = +48$; (4 negatives, even number)

Note: The product of any number and 0 = 0; for example, $0 \times 9 = 0$, $0 \times -9 = 0$.

Division

Rule: To divide two numbers that have the same sign (both positive or both negative)

- Divide as in arithmetic.
- Prefix a plus sign.

Examples:

$$\frac{-8}{-2} = +4; \ \frac{8}{2} = +4; \ 15 \div 3 = +5; \ -3\overline{)-15} = +5$$

Rule: To divide a positive and a negative number

- Divide as in arithmetic.
- Prefix a minus sign.

Examples:

$$-\frac{30}{5} = -6; \ \frac{30}{-5} = -6; \ -21 \div 3 = -7; \ -3\overline{)21} = -7$$

Note: Zero divided by any number = 0; for example, $0 \div +3 = 0$, $0 \div -3 = 0$. A number divided by 0 is not allowed. For the purposes of this text, division by 0 has no meaning; for example, $14 \div 0$ and $-14 \div 0$ have no meaning.

Powers

Rule: Since the power operation is a multiplication operation, the rules of multiplication of signed numbers apply.

- A positive number raised to any power is positive.

Examples:

$$3^2 = +9; \ 3^3 = +27; \ 2^4 = +16; \ 2^5 = +32$$

▲ A negative number raised to an even power is positive, and raised to an odd power is negative.

Examples:

$$-3^2 = (-3)(-3) = +9; -3^3 = (-3)(-3)(-3) = -27; -2^4 = (-2)(-2)(-2)(-2) = +16;$$

$$-2^5 = (-2)(-2)(-2)(-2)(-2) = -32$$

▲ The value of a number raised to a negative power is equal to the number inverted and raised to a positive power.

Examples:

▲ $3 = \frac{3}{1}; \ 3^{-2} = \frac{3^{-2}}{1}; \ \frac{3^{-2}}{1} = \frac{1}{3^2}; \ \frac{1}{3^2} = \frac{1}{9}$. Therefore, $3^{-2} = \frac{1}{9}$

▲ $2 = \frac{2}{1}; \ 2^{-3} = \frac{2^{-3}}{1}; \ \frac{2^{-3}}{1} = \frac{1}{2^3}; \ \frac{1}{2^3} = \frac{1}{8}$. Therefore, $2^{-3} = \frac{1}{8}$

▲ $-4 = \frac{-4}{1}; \ -4^{-3} = \frac{-4^{-3}}{1}; \ \frac{-4^{-3}}{1} = \frac{1}{-4^3} = \frac{1}{-64}$. Therefore, $-4^{-3} = \frac{1}{-64}$

Rule: When a number is raised to a fractional power

- The numerator indicates the exponent of the number
- The denominator indicates the root of the number.

Examples:

▲ $25^{1/2} = \sqrt[2]{25^1}$; $\sqrt{25} = 5$. Therefore, $25^{1/2} = 5$.

▲ $8^{1/3} = \sqrt[3]{8^1}$; $\sqrt[3]{8} = 2$. Therefore, $8^{1/3} = 2$.

▲ $8^{2/3} = \sqrt[3]{8^2}$; $\sqrt[3]{8^2} = \sqrt[3]{64}$; $\sqrt[3]{64} = 4$. Therefore, $8^{2/3} = 4$.

▲ $36^{-1/2} = \frac{1}{36^{1/2}}$; $\frac{1}{36^{1/2}} = \frac{1}{\sqrt{36}}$; $\frac{1}{\sqrt{36}} = \frac{1}{6}$. Therefore, $36^{-1/2} = \frac{1}{6}$.

Roots

Rule: A positive number which has an even index has a positive and a negative root.

Example:

$\sqrt{25} = \sqrt{(+5)(+5)} = +5$; $\sqrt{25} = \sqrt{(-5)(-5)} = -5$. Therefore, $\sqrt{25} = +5$ and -5.

Note: In computing problems in this text which have both a positive and a negative root, disregard the negative root and use only the positive root.

Rule: If the index of the number is odd, the sign of the root is the same as the sign of the number.

Examples:

$\sqrt[3]{27} = 3$; $\sqrt[3]{-27} = -3$; $\sqrt[5]{32} = 2$; $\sqrt[5]{-32} = -2$

Note: The square root of a negative number has no solution in the real number system.

Example:

$\sqrt{-4}$ does not equal -2; $\sqrt{(-2)(-2)} = \sqrt{+4}$

$\sqrt{-4}$ does not equal $+2$; $\sqrt{(+2)(+2)} = \sqrt{+4}$

APPLICATION

A. THE NUMBER SCALE

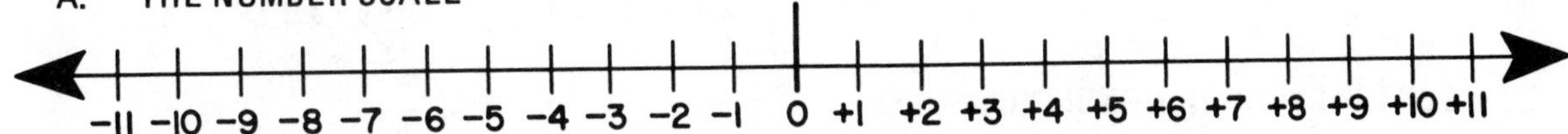

Refer to the number scale and give the direction (+ or −) and the number of units counted going from

1. − 11 to − 2 ________
2. − 8 to − 1 ________
3. − 6 to 0 ________
4. −2 to − 8 ________
5. + 2 to − 8 ________
6. + 3 to + 9 ________
7. + 10 to − 10 ________
8. + 10 to 0 ________
9. + 2 to + 7 ________
10. + 9 to + 1 ________
11. + 11 to 0 ________
12. 0 to − 4 ________
13. − 7.5 to + 10 ________
14. + 10 to − 7.5 ________
15. − 10.8 to − 2.3 ________
16. − 2.3 to − .8 ________
17. + 7½ to 2¼ ________
18. + 6⅞ to 0 ________

B. COMPARING SIGNED NUMBERS

1. Select the greater of the two signed values and indicate the number of units by which it is greater.

a. + 5, - 14 ______ d. + 8, + 13 ______ g. + 14.3, + 21 ______

b. + 7, - 3 ______ e. + 20, - 22 ______ h. - 1.8, + 1.8 ______

c. - 6, - 2 ______ f. - 18, - 4 ______ i. + 17.6, - 21.9 ______

2. List the following signed numbers in order of value starting with the smallest number.

a. + 17, - 1, + 2, 0, - 18, + 4, - 22 ______

b. - 5, + 5, 0, + 13, + 27, - 21, - 2, - 19 ______

c. + 10, - 10, - 7, + 7, 0, + 25, - 25, + 14 ______

d. 0, 15, - 3.6, - 2.5, - 14.9, + 17, + .3 ______

e. - 16, + 14 1/8, - 13 7/8, + 6, - 4 3/8 ______

C. ABSOLUTE VALUE

Convert the following pairs of signed numbers to absolute values and subtract the smaller absolute value from the larger absolute value.

1. + 23, - 14 ____ ____
2. - 17, + 11 ____ ____
3. - 6, + 6 ____ ____
4. + 25, + 13 ____ ____
5. - 16, + 16 ____ ____
6. - 32.1, - 29.7 ____ ____

D. ADDITION OF SIGNED NUMBERS

Add the following signed numbers as indicated.

1. + 4 + (- 17) ______
2. - 17 + (+ 4) ______
3. - 17 + (- 4) ______
4. + 17 + (+ 4) ______
5. - 23 + (- 5) ______
6. + 13.5 + .6 ______
7. - 16 + (- 16.2) ______
8. $\frac{3}{8} + (-\frac{9}{16})$ ______
9. $-\frac{1}{3} + (-\frac{1}{4})$ ______
10. - 17 + 3 + 1 + (- 1) + (- 31) ______

E. SUBTRACTION OF SIGNED NUMBERS

Subtract the following signed numbers as indicated.

1. - 8 - (- 4) ______
2. + 8 - (+ 4) ______
3. + 4 - (- 8) ______
4. + 4 - (+ 8) ______
5. - 23 - 23 ______
6. 20 - (- 3) ______
7. - 14 - (- 33) ______
8. .75 - .87 ______
9. - 15.8 - (- .9) ______
10. 16 3/16 - (- 1/8) ______
11. - 3/4 - 3/16 ______
12. (5 + 8) - (2 + 7) ______
13. (2 + 7) - (5 + 8) ______
14. (7 - 9) - (- 3 + 12) ______

F. MULTIPLICATION OF SIGNED NUMBERS

Multiply the following signed numbers as indicated.

1. $(-7)(5)$ ____
2. $(-8)(-3)$ ____
3. $(6)(7)$ ____
4. $(0)(-14)$ ____
5. $(-1/4)(-1/2)$ ____
6. $(-1/8)(3/4)$ ____
7. $(15)(-3/8)$ ____
8. $(-2\ 1/2)(-1\ 3/8)$ ____
9. $(-2)(3)(-5)$ ____
10. $(-2)(-1)(-3)$ ____
11. $(-2)(2)(3)(-3)(1)$ ____
12. $(-4)(-1)(-1)(-1)$ ____

G. DIVISION OF SIGNED NUMBERS

Divide the following signed numbers as indicated.

1. $\frac{-15}{3}$ ____
2. $\frac{15}{-3}$ ____
3. $\frac{-15}{-3}$ ____
4. $12 \div 4$ ____
5. $-23 \div 23$ ____
6. $-5 \div -5$ ____
7. $-1 \div 5$ ____
8. $5 \div -1$ ____
9. $\frac{1}{2} \div -\frac{1}{2}$ ____
10. $-\frac{3}{4} \div -\frac{1}{8}$ ____
11. $6.2\overline{)3.1}$ ____
12. $-3.1\overline{)6.2}$ ____

H. POWERS OF SIGNED NUMBERS

Raise the following signed numbers to the indicated powers.

1. -3^3 ____
2. -3^2 ____
3. -2^3 ____
4. -5^2 ____
5. -5^3 ____
6. $(\frac{1}{5})^3$ ____
7. 3^{-2} ____
8. -3^{-2} ____
9. -2^{-3} ____
10. $4^{1/2}$ ____
11. $16^{1/2}$ ____
12. $-27^{1/3}$ ____

I. ROOTS OF SIGNED NUMBERS

Determine the indicated root of the following signed numbers.

1. $\sqrt[3]{-64}$ ____
2. $\sqrt[3]{64}$ ____
3. $\sqrt{100}$ ____
4. $\sqrt[3]{-1000}$ ____
5. $\sqrt[3]{-64}$ ____
6. $\sqrt[3]{125}$ ____
7. $\sqrt[3]{-125}$ ____
8. $\sqrt{16}$ ____
9. $\sqrt[7]{-128}$ ____
10. $\sqrt{1/4}$ ____
11. $\sqrt[3]{8/27}$ ____
12. $\sqrt[3]{-8/27}$ ____
13. $\sqrt[3]{\frac{-8}{-27}}$ ____
14. $\frac{\sqrt[5]{32}}{-4}$ ____
15. $\frac{-1}{\sqrt[3]{-27}}$ ____

J. COMBINED OPERATIONS OF SIGNED NUMBERS

Solve the following problems using the proper order of operations.

1. $17 - (3)(-2) + (-5)^2$ ______
2. $4 - 5(8 - 10)$ ______
3. $-2(4 + 2) + 3(5 - 7)$ ______
4. $5 - 3(8 - 6) - (1 - 4)$ ______
5. $-1(8 - 3)(3 - 8)(1 - 2)$ ______
6. $8 - 4(19 - 5 + 3 - 6) + 10$ ______
7. $-3^3 + 3^3 - (-6)(3) - (\frac{-6}{2})$ ______
8. $4^2 + \sqrt[3]{-8} + (-4)(0)(-3)$ ______
9. $\sqrt[3]{-21(2) - (-15)} + (16)^{1/2}$ ______
10. $[(15 - 40)(5)]^{-1/3} + 10^{-2}$ ______

Unit 22 Algebraic Operations of Addition, Subtraction, and Multiplication

OBJECTIVES

After studying this unit, the student should be able to

- Perform the basic algebraic operations of addition, subtraction, and multiplication.

A knowledge of basic algebraic operations is essential in order to solve equations. For certain applications, formulas given in machine trade handbooks cannot be used directly as given, but must be rearranged. Formulas are rearranged by using the principles of algebraic operations which are presented in this unit.

DEFINITIONS

It is important to understand the following definitions in order to apply procedures which are required for solving problems involving basic operations.

- A *term* of an algebraic expression is that part of the expression which is separated from the rest by a plus or a minus sign. For example, $4x + ab/2x - 12 + 3ab^2x - 8a\sqrt{b}$ is an expression that consists of five terms: $4x$, $ab/2x$, 12, $3ab^2x$, and $8a\sqrt{b}$.
- A *factor* is one of two or more literal and numerical values of a term that are multiplied. For example, 4 and x are each factors of $4x$; 3, a, b^2 and x are each factors of $3ab^2x$; 8, a, and $\sqrt{b}$ are each factors of $8a\sqrt{b}$.

Note: It is of primary importance to distinguish between factors and terms.

- A *numerical coefficient* is the number factor of a term. The letter factors of a term are the *literal factors.* For example, in the term $5x$, 5 is the numerical coefficient; x is the literal factor.

  In the term $\frac{1}{3}ab^2c^3$, $\frac{1}{3}$ is the numerical coefficient; a, b^2, and c^3 are the literal factors.
- *Like terms* are terms that have identical literal factors; the numerical coefficients do not have to be the same. For example, $6x$ and $13x$ are like terms; $15ab^2c^3$, $3.2ab^2c^3$, and $1/8\ ab^2c^3$ are like terms.
- *Unlike terms* are terms which have different literal factors. For example, $12x$ and $12y$ are unlike terms; $15xy$, $3x^2y$, and $4x^2y^2$ are unlike terms.

Note: Although the literal factors are x and y, they are raised to different powers.

BASIC ALGEBRAIC OPERATIONS

Addition

Like terms can be added; the addition of unlike terms can only be indicated. As in arithmetic, like things can be added, but unlike things cannot be added. For example, 4 inches + 5 inches = 9 inches. Both values are inches; therefore, they can be added. But 4 inches + 5 pounds cannot be added because they are unlike things.

Rule: To add like terms

- Add the numerical coefficients.
- Leave the literal factors unchanged.

Note: If a term does not have a numerical coefficient, the coefficient 1 is understood: $x = 1x$, $abc = 1abc$, $n^2rs^3 = 1n^2rs^3$

Examples: Add the following like terms.

$3x$	x	$-5xy^2$	$6x^2y^3$	$2(a+b)$
$\underline{12x}$	$\underline{-14x}$	$\underline{+5xy^2}$	$\underline{-13x^2y^3}$	$-3(a+b)$
$15x$	$-13x$	0	$-7x^2y^3$	$\underline{7(a+b)}$
				$6(a+b)$

Rule: The addition of unlike terms can only be indicated.

Examples: Add the following unlike terms.

			$8a$
15	$7x$	$3x$	$-6b$
$\underline{x}$	$\underline{8y}$	$\underline{-7x^2}$	$\underline{2c}$
$15 + x$	$7x + 8y$	$3x + (-7x^2)$	$8a + (-6b) + 2c$

Rule: To add two or more expressions that consist of more than one term

- Group the like terms.
- Add.

Examples: Add the following expressions.

- Add: $12x - 2xy + 6x^2y^3$ and $-4x - 7xy + 5x^2y^3$
- Group like terms

$$\begin{array}{r} 12x - 2xy + 6x^2y^3 \\ \underline{-4x - 7xy + 5x^2y^3} \\ 8x - 9xy + 11x^2y^3 \end{array}$$

- Add: $6a - 7b$ and $18b - 3ab + a$ and $-14a + ab^2 - 5ab$
- Group like terms

$$\begin{array}{rrrr} 6a & -7b & & \\ a & +18b & -3ab & \\ \underline{-14a} & & \underline{-5ab} & \underline{+ab^2} \\ -7a & +11b & -8ab & +ab^2 \end{array}$$

Subtraction

Like terms can be subtracted; the subtraction of unlike terms can only be indicated. The same principles apply in arithmetic. For example, 8 feet - 3 feet = 5 feet, but 8 feet - 3 ounces cannot be subtracted.

Rule: To subtract terms

- Subtract the numerical coefficients.
- Leave the literal factors unchanged.

Examples: Subtract the following like terms as indicated.

$18ab - 7ab = 11ab$

$-5x^2y - 8x^2y = -13x^2y$

$bx^2y^3 - 13bx^2y^3 = -12bx^2y^3$

$-24\,dmr - (-24dmr) = 0$

Rule: The subtraction of unlike terms can only be indicated.

Examples: Subtract the following unlike terms as indicated.

$3x^2$	$-13abc$	$-2xy$
$\underline{-(+2x)}$	$\underline{-(+8abc^2)}$	$\underline{-(-7y)}$
$3x^2 - 2x$	$-13abc - 8abc^2$	$-2xy + 7y$

Rule: To subtract expressions that consist of more than one term.

- Group like terms.
- Subtract.

Note: Each term of the subtrahend is subtracted following the procedure for subtraction of signed numbers.

Examples: Subtract 7 a + 3b - 3d from 8a - 7b + 5d

$$\begin{array}{r} 8a - 7b + 5d \\ \underline{-(7a + 3b - 3d)} \end{array} = \begin{array}{r} 8a - 7b + 5d \\ \underline{+(-7a - 3b + 3d)} \\ a - 10b + 8d \end{array}$$

▲ Subtract as indicated: $(3x^2 + 5x - 12xy) - (7x^2 - x - 3x^3 + 6y)$

$$\begin{array}{l} 3x^2 + 5x - 12xy \\ \underline{-(7x^2 - x \qquad - 3x^3 + 6y)} \end{array} = \begin{array}{l} 3x^2 + 5x - 12xy \\ \underline{+(-7x^2 + x \qquad + 3x^3 - 6y)} \\ -4x^2 + 6x - 12xy + 3x^3 - 6y \end{array}$$

Multiplication

Rule: To multiply two or more terms

- Multiply the numerical coefficients following the procedure for multiplication of signed numbers.
- Add the exponents of the same letter factors.
- Show the product as a combination of all numerical and literal factors.

Examples: Multiply $(-3x^2)\ (6x^4)$

▲ Multiply numerical coefficients $(-3)\ (6) = -18$

▲ Add exponents of like letter factors $(x^2)\ (x^4) = x^{2+4} = x^6$

▲ Show product as combination of $-18x^6$ (all numerical and literal factors)

Note: $x^2 = x \cdot x$ and $x^4 = x \cdot x \cdot x \cdot x$. Therefore, $(x^2)\ (x^4) = (x \cdot x)\ (x \cdot x \cdot x \cdot x) = x^6$ (x is used as a factor 6 times.)

$(3a^2b^3)\ (7ab^3) = (3)\ (7)\ (a^{2+1})\ (b^{3+3}) = 21a^3b^6$

$(-4a)\ (-7b^2c^2)\ (-2ac^3d^3) = (-4)\ (-7)\ (-2)\ (a^{1+1})\ (b^2)\ (c^{2+3})d^3 = -56a^2b^2c^5d^3$

Rule: To multiply expressions that consist of more than one term within an expression

- Multiply each term of one expression by each term or the other expression.
- Combine like terms.

Note: This multiplication procedure is consistent with arithmetic.

Examples in arithmetic: $3(4 + 2) = 3(6) = 18$ or

▲ Multiply each term of one expression by one of the other expressions

$3(4 + 2) =$
$3(4) + 3(2) = 12 + 6 = 18$

▲ Combine like terms.

$(5 + 3)(2 + 4) = 8(6) = 48$ or

$(5 + 3)(2 + 4) = \underset{(1)}{5(2)} + \underset{(2)}{5(4)} + \underset{(3)}{3(2)} + \underset{(4)}{3(4)} = 10 + 20 + 6 + 12 = 48$

Examples in algebra:

$3a(6 + 2a^2) = (3a)(6) + 3a(2a^2) = 18a + 6a^3$

$-5x^2y(3xy - 4x^3y^2 + 5y) = -5x^2y(3xy) - 5x^2y(-4x^3y^2) - 5x^2y(5y) = -15x^3y^2 + 20x^5y^3 - 25x^2y^2$

▲ Multiply each term of one expression by each term of the other expression.

$(3c + 5d^2)(4d^2 - 2c) = \underset{(1)}{3c(4d^2)} + \underset{(2)}{3c(-2c)} + \underset{(3)}{5d^2(4d^2)} + \underset{(4)}{5d^2(-2c)} = 12cd^2 - 6c^2 + 20d^4 - 10cd^2$

▲ Combine like terms. $12cd^2 - 10cd^2 - 6c^2 + 20d^4 = 2cd^2 - 6c^2 + 20d^4$

APPLICATION

A. ADDITION OF SINGLE TERMS

Add the following terms.

1. $-8x^2y$, $5x^2y$
2. $-4x$, $6x$
3. $-7c^2d$, $-c^2d$
4. $3(a + b)$, $-8(a + b)$
5. d^2, 0
6. $-\frac{3}{4}mn$, $\frac{1}{2}mn$
7. $14x$, x^2
8. $-3xy^2$, $8xy^2$, $-7xy^2$
9. $-5.6dt^3$, $-2.3dt^3$, $-3.4dt^3$
10. $3.2a^2bc$, $6.7ab^2c$, $-4.4a^2bc$

B. ADDITION OF EXPRESSIONS WITH TWO OR MORE TERMS

Add the following expressions.

1. $-5x + 7xy - 8y$
   $-9x - 12xy + 13y$

2. $3a - 11d - 8m$
   $-a + 11d - 3m$

3. $-6ab - 7a^2b^2 - 3a^3b$
   $-5ab + 14a^2b^2 - 12a^3b$
   $-9ab - 7a^2b^2 + a^3b$
   $ab - 2a^3b$

4. $(3xy^2 + x^2y - x^2y^2), (2x^2y + x^2y^2)$ ________

5. $(10a - 5b), (-12a - 7b), (17a + b)$ ________

6. $(x^3 + 5), (3x - 7x^2 + 7), (x - 3x^3)$ ________

7. $(b^4 + 4b^3c - 2b^2c), (4b^3c - 7bc)$ ________

8. $(x^2 - 4xy), (4xy - y^2), (-x^2 + y^2)$ ________

C. SUBTRACTION OF SINGLE TERMS

Subtract the following terms as indicated.

1. $7xy^2 - (-13xy^2)$ ______
2. $3xy - xy$ ______
3. $-3xy - xy$ ______
4. $-3xy - (-xy)$ ______
5. $9ab - (-9ab)$ ______
6. $-5a^2 - (5a^2)$ ______
7. $.7a^2b^2 - 1.5a^2b^2$ ______
8. $0 - (-8mn^3)$ ______
9. $-8mn^3 - 0$ ______
10. $7/8\ x^2 - (-3/8\ x^2)$ ______
11. $13a - 7a^2$ ______
12. $-13a - (-7a^2)$ ______
13. $.6xy - .9xy^2$ ______
14. $-ax^2 - ax^2$ ______
15. $1/2dt - (-3/8dt)$ ______
16. $1/2d^2t^2 - (-1/2d^2t^2)$ ______
17. $18 - 3x$ ______
18. $3x - 18$ ______
19. $-3.2d - 6.4d$ ______
20. $-1.4xy - (-1.4xy)$ ______

D. SUBTRACTION OF EXPRESSIONS WITH TWO OR MORE TERMS

Subtract the following expressions as indicated.

1. $(2a^2 - 3a) - (7a^2 - 8a)$ ______
2. $(4x^2 + 8xy) - (3x^2 + 5xy)$ ______
3. $(9b^2 + 1) - (9b^2 - 1)$ ______
4. $(9b^2 - 1) - (9b^2 - 1)$ ______
5. $(xy^2 - x^2y^2 + x^3y^2) - 0$ ______
6. $(2a^3 - .5a^2) - (-a^3 + a^2 - a)$ ______
7. $(5x + 3xy - 7y) - (3y^2 - x^2y)$ ______
8. $(-d^2 - dt + dt^2) - (-4 + dt)$ ______

E. MULTIPLICATION OF SINGLE TERMS

Multiply the following terms as indicated.

1. $(-5b^2c)\ (3b^3)$ ____
2. $(x)\ (x^2)$ ____
3. $(-3a^2)\ (-4a^3)$ ____
4. $(8ab^2c)\ (7a^3bc^2)$ ____
5. $(-x^3y^3)\ (3x^2y^4)$ ____
6. $(-3xy)\ (0)$ ____
7. $(7ab^4)\ (3a^4b)$ ____
8. $(-3d^5r^4)\ (-d^3)$ ____
9. $(-3d^5r^4)\ (-d^3)\ (-1)$ ____
10. $(.3x^2y^4)\ (.4x^5)$ ____
11. $(1/4\ a^3)\ (3/8\ a^2)$ ____
12. $(-5x)\ (0)\ (-5x)$ ____
13. $(m^2t)\ (st)$ ____
14. $(-1.6bc)\ (2.1)$ ____
15. $(abc^3)\ (c^3d)$ ____
16. $(2x^6y^6)\ (-2x^2)$ ____
17. $(-2/3\ mt)\ (t^4)$ ____
18. $(7ab^3)\ (-7a^3b)$ ____
19. $(-.3a^3b^2)\ (-5b^4)$ ____
20. $(-x^2y)\ (-xy)\ (-x)$ ____
21. $(d^4m^2)(-1)(-m^3)$ ____

F. MULTIPLICATION OF EXPRESSIONS WITH TWO OR MORE TERMS

Multiply the following expressions as indicated.

1. $-7x^2y^3\ (2xy^2 - 3x^4)$ ______
2. $3a^2\ (-a^2 + a^3b)$ ______
3. $-2a^3b^2\ (4ab^3 - b^2 - 4)$ ______
4. $xy^2\ (x^2 + y^3 + xy)$ ______
5. $-4(dt + t^2 - 1)$ ______
6. $(m^2t^3s^4)\ (-m^4s^2 + m - s^5)$ ______
7. $(3x + 7)\ (x^2 + 8)$ ______
8. $(7x^2 - y^3)\ (-2x^3 + y^2)$ ______
9. $(5ax^3 + bx)\ (2a^2x^3 + b^2x)$ ______
10. $(-4a^2b^3 + 5xy^2)\ (4a^2b^3 - 5xy)$ ______

Unit 23 Algebraic Operations of Division, Powers, and Roots

OBJECTIVES

After studying this unit, the student should be able to

- Perform the basic algebraic operations of division, powers, and roots.
- Remove parentheses which are preceded by a plus or minus sign.
- Simplify algebraic expressions which involve combined operations.

BASIC ALGEBRAIC OPERATIONS

Division

Rule: To divide two terms

- Divide the numerical coefficients, following the procedure for signed numbers.
- Subtract the exponents of letter factors of the divisor from the exponents of the same letter factors of the dividend.

Note: This division procedure is consistent with arithmetic.

Example: $\frac{2^5}{2^2} = \frac{\cancel{2} \cdot \cancel{2} \cdot 2 \cdot 2 \cdot 2}{\cancel{2} \cdot \cancel{2}} = 2 \cdot 2 \cdot 2 = 2^3 = 8$ or $\frac{2^5}{2^2} = 2^{5-2} = 2^3 = 8$

Example 1 in algebra: Divide $-16x^3$ by $8x$

- ▲ Divide the numerical coefficients $\frac{-16}{8} = -2$
- ▲ Subtract the exponents of the letter factor in the divisor from the exponent of the like factor in the dividend.

$$\frac{x^3}{x} = x^{3-1} = x^2$$

- ▲ Combine: $-2x^2$, or $\frac{-16x^3}{8x} = \left(\frac{-16}{8}\right)(x^{3-1}) = -2x^2$

Example 2: $\frac{-30a^3b^5c^2}{-5a^2b^3} = \left(\frac{-30}{-5}\right)(a^{3-2})(b^{5-3})(c^2) = 6ab^2c^2$

Example 3: $\frac{3x}{x} = \left(\frac{3}{1}\right) x^{1-1} = 3x^0 = 3$

Note: In arithmetic any number divided by itself equals 1. For example, $4/4 = 1$, $4/4 = 4^{1-1} = 4^0$. Therefore, $4^0 = 1$. Any number raised to the zero power equals 1: $5^3/5^3 = 5^{3-3} = 5^0 = 1$; $\frac{a^3b^2c}{a^3b^2c} = (a^{3-3})(b^{2-2})(c^{1-1}) = a^0b^0c^0 = (1)(1)(1) = 1$

Rule: To divide when the dividend consists of more than one term

- Divide each term of the dividend by the divisor.
- Combine. This division procedure is consistent with arithmetic.

Example: $\frac{6+8}{2} = \frac{14}{2} = 7$; $\frac{6+8}{2} = \frac{6}{2} + \frac{8}{2} = 3 + 4 = 7$

Example in algebra: $\frac{-20xy^2 + 15x^2y^3 + 35x^3y}{-5xy} = \frac{-20xy^2}{-5xy} + \frac{15x^2y^3}{-5xy} + \frac{35x^3y}{-5xy} = 4y - 3xy^2 - 7x^2$

Powers

Rule: To raise a term to a power

- Raise the numerical coefficient to the indicated power following the procedure for powers of signed numbers.
- Multiply each of the literal factor exponents by the exponent of the power to which it is to be raised.

Note: This power procedure is consistent with arithmetic.

Example: $(2^2)^3 = 4^3 = 64$; $(2^2)^3 = (2 \cdot 2)(2 \cdot 2)(2 \cdot 2) = 2^6 = 64$ (2 is used as a factor 6 times.)

Examples in algebra: Find $(x^3)^2$

▲ Raise the numerical coefficient to the indicated power $(1)^2 = 1$

▲ Multiply each literal factor exponent by the exponent to which it is to be raised. $(x^3)^2$ $3 \times 2 = 6$

▲ Combine x^6

Note: $(x^3)^2$ is not the same as x^3x^2. $(x^3)^2 = (x^3)(x^3) = (x \cdot x \cdot x)(x \cdot x \cdot x) = x^6$. (x is a factor 6 times.) $x^3x^2 = (x^3)(x^2) = (x \cdot x \cdot x)(x \cdot x) = x^5$ (x is a factor 5 times.)

Solve: $(-3a^2b^4c)^3$

▲ Raise the numerical coefficient to the indicated power. $(-3)^3 = -27$

▲ Multiply each literal factor exponent by the exponent to which it is to be raised.

$$(a^2)^3 = a^6;\ (b^4)^3 = b^{12};\ (c^1)^3 = c^3$$

▲ Combine: $-27a^6b^{12}c^3$

Example: $(\frac{-1}{2}x^3(yd^2)^3r^4)^2 = (\frac{-1}{2})^2\ (x^{3 \cdot 2})(y^{1 \cdot 3}d^{2 \cdot 3})^2r^{4 \cdot 2} = \frac{1}{4}\ x^6(y^{3 \cdot 2})\ (d^{6 \cdot 2})r^8 = \frac{1}{4}\ x^6y^6d^{12}r^8$

Rule: If an expression consists of more than one term raised to a power, solve as a multiplication problem.

Example: $(2x + 4y^3)^2 = (2x + 4y^3)(2x + 4y^3)$;

(1) (2)

$(2x + 4y^3)\ (2x + 4y^3) = 2x(2x) + 2x(4y^3) + 4y^3(2x) + 4y^3(4y^3) = 4x^2 + 8xy^3 + 8xy^3 + 16y^6 =$
$4x^2 + 16xy^3 + 16y^6$.

(3) (4)

Roots

Rule: To extract the root of a term

- Determine the root of the numerical coefficient following the procedure for roots of signed numbers.
- The roots of the literal factors are determined by dividing the exponent of each literal factor by the index of the root.

  This procedure for extracting roots is consistent with arithmetic.

Example: $\sqrt[2]{2^6} = \sqrt[2]{64} = 8$; $\sqrt[2]{2^6} = 2^{6/2} = 2^3 = 8$

Examples in algebra: $\sqrt{25a^6b^4c^8} = \sqrt{25}(a^{6/2})(b^{4/2})\ (c^{8/2}) = 5a^3b^2c^4$

$$\sqrt[3]{-27dx^9y^2} = \sqrt[3]{-27}(d^{1/3})(x^{9/3})(y^{2/3}) = -3d^{1/3}\ x^3y^{2/3}$$

$$\sqrt[4]{16/81\ d^8t^{12}y^2} = \sqrt[4]{16/81}\ (d^{8/4})(t^{12/4})(y^{2/4}) = 2/3\ d^2t^3y^{1/2}$$

Note: Roots of expressions that consist of two or more terms cannot be extracted by this procedure. This fact is consistent with arithmetic.

Example: $\sqrt{3^2 + 4^2} = \sqrt{9 + 16} = \sqrt{25} = 5$, the correct answer; but $\sqrt{3^2 + 4^2}$ does not equal $\sqrt{3^2} + \sqrt{4^2}$; $\sqrt{3^2} + \sqrt{4^2} = 3 + 4 = 7$; 7 does not equal 5.

Removal of Parentheses

Rule: To remove parentheses, if they are preceded by a plus (+) sign

- The parentheses are removed without changing the signs of any terms.

Example: $5a + (4b + 7 - 3d) = 5a + 4b + 7 - 3d$

Rule: To remove parentheses if they are preceded by a negative (-) sign

- Remove the parentheses.
- Change the signs of each of the terms.

Example: $3c^2 - (8x + 3y - 7) = 3c^2 - 8x - 3y + 7$

Combined Operations

Rule: To solve expressions requiring combined operations

- Apply the proper order of operations as discussed in units 14 and 20.

Examples:

$$10x - 3x(2 + x - 4x^2) = 10x - 6x - 3x^2 + 12x^3 = 4x - 3x^2 + 12x^3$$

$$15a^6b^3 + (2a^2b)^3 - \frac{a^7(b^3)^2}{ab^3} = 15a^6b^3 + 8a^6b^3 - \frac{a^7b^6}{ab^3} = 15a^6b^3 + 8a^6b^3 - a^6b^3 = 22a^6b^3$$

$$-4a(15 - 3(2a + ab) + a) - 2a^2b = -4a(15 - 6a - 3ab + a) - 2a^2b =$$
$$-60a + 24a^2 + 12a^2b - 4a^2 - 2a^2b = -60a + 20a^2 + 10a^2b$$

APPLICATION

A. DIVISION OF SINGLE TERMS

Divide the following terms as indicated.

1. $\frac{-18a^4b^5}{6ab^2}$ ____
2. $\frac{21x^3y^2}{3xy}$ ____
3. $\frac{14x^3}{14x^3}$ ____
4. $\frac{xy^2}{-1}$ ____
5. $\frac{0}{13ab^2}$ ____
6. $\frac{-30a^5d^2}{-6a^2d^2}$ ____
7. $(-12x^2y^8) \div (-3x^2y^5)$ ____
8. $(-6d^3t^2) \div (6d^3t^2)$ ____
9. $0 \div x^3y^7$ ____
10. $(9a^3bc^2y) \div (-a^3)$ ____
11. $(4.5xy) \div (-.5y)$ ____

B. DIVISION OF EXPRESSIONS WITH TWO OR MORE TERMS

Divide the following expressions as indicated.

1. $(9x^6y^3 - 6x^2y^5) \div (3xy^2)$ ____
2. $(15a^2 + 25a^5) \div (-a)$ ____
3. $(2x - 4y) \div 2$ ____
4. $(-18a^2b^7 - 12a^5b^5) \div (-6a^2b^5)$ ____
5. $(7cd^2 - 35c^2d - 7) \div (-7)$ ____
6. $(.8x^5y^6 + .2x^4y^7) \div (2x^2y^4)$ ____
7. $(-.9a^2x - .3ax^2 + .6) \div (-.3)$ ____
8. $(5y^2 - 25xy^2 - 10y^5) \div 5y^2$ ____

C. POWERS OF SINGLE TERMS

Raise the following terms to indicated powers.

1. $(-4a^3b^2c^4)^3$ ____
2. $(4a^4b^3)^2$ ____
3. $(-5x^4y^5)^2$ ____
4. $(-2a^2b^2c^3)^4$ ____
5. $(-d^4m^5x)^3$ ____
6. $(.5x^3y)^3$ ____
7. $(-6(a^2b^3)^2c)^2$ ____
8. $(-3b^2(m^3)^2x^3)^3$ ____
9. $(-2a^2b)^{-3}$ ____

D. POWERS OF EXPRESSIONS OF TWO OR MORE TERMS

Raise the following terms to the indicated powers and combine like terms where possible.

1. $(3x^2 - 5y^3)^2$ ____
2. $(a^3 + b^4)^2$ ____
3. $(5t^2 - 6x)^2$ ____
4. $(a^2b^3 + ab^3)^2$ ____
5. $(.6d^3t^2 - .2t)^2$ ____
6. $(-.4x^2y - y^4)^2$ ____
7. $(\frac{2}{3}c^2d + \frac{3}{4}cd^2)^2$ ____
8. $((x^2)^3 - (y^3)^2)^2$ ____
9. $((-a^4b)^2 + (x^2y)^3)^2$ ____

E. ROOTS

Determine the roots of the following terms.

1. $\sqrt{36a^6b^2c}$ ____
2. $\sqrt{4x^2y^4}$ ____
3. $\sqrt{25x^8y^6}$ ____
4. $\sqrt{81c^{10}d^2x^6}$ ____
5. $\sqrt[5]{-32x^{10}}$ ____
6. $\sqrt{100ab^4}$ ____
7. $\sqrt[3]{-64dt^9}$ ____
8. $\sqrt[3]{-8cd}$ ____
9. $\sqrt[3]{c^2dt^9}$ ____
10. $\sqrt[3]{-x^2d^2t}$ ____
11. $\sqrt{1/16xy^6}$ ____
12. $\sqrt{64/81xy^6}$ ____

F. REMOVAL OF PARENTHESES

Remove parentheses and combine like terms where possible.

1. $7x-(2x - 3x^2 - x^3)$ ____
2. $8x - (3d + c)$ ____
3. $-2a^2 + (3 - 6a)$ ____
4. $-15a^2b-(-2a^2b-a + b^2)$ ____
5. $-(x^2 + y^2 - xy)$ ____
6. $+(x^2 + y^2 - xy)$ ____

G. COMBINED OPERATIONS

Simplify the following expressions and combine like terms where possible.

1. $15 - 2(3xy)^2 + x^2y^2 - 3$ ____
2. $5(a^2 - b) + a^2 - b$ ____
3. $(2 - c^2)(2 + c^2) + c$ ____
4. $\frac{ab}{a} - (\frac{-a^2b}{a^2} - \frac{a^3b}{a^3})$ ____
5. $\frac{4 - 8x + 16x^2}{2} + \frac{3x^4}{x^2}$ ____
6. $\frac{16xy^8}{2xy^2} - (y^2)^3 + 15$ ____
7. $\frac{\sqrt{25x^2}(3xy^3)}{-5} - (-4)$ ____
8. $\sqrt{\frac{64d^6}{9}} \div d^2$ ____
9. $\frac{12x^6 + 16x^4y}{(2x)^2} - (16x^4y^2)^{1/2}$ ____
10. $-4a(-8 + (ab^2)^3 - 12)$ ____

Unit 24 Introduction to Equations

OBJECTIVES

After studying this unit, the student should be able to

- Convert word statements into mathematical statements.
- Convert graphic information into equations.
- Solve simple equations using logical reasoning.

It is essential that the skilled machine technician understand equations and their applications. The solution of equations is required to compute problems using trade handbook formulas. Often machine shop problems are solved using a combination of equations, with elements of geometry and trigonometry.

EXPRESSION OF EQUALITY

An equation is a mathematical expression of equality between two or more quantities and always contains the equal sign (=). The value of all quantities on the left side of the equal sign equals the value of all quantities on the right side of the equal sign.

The following are examples of simple equations:

$$7 + 2 = 5 + 4 \qquad \frac{12}{3} + 2 \times 5 = 18 - 4$$

$$3 \times 5\frac{1''}{2} = 16\frac{1''}{2} \qquad 360^\circ = 5 \times 80^\circ - 40^\circ$$

$$a + b = c - d \qquad \frac{xy}{2} = x + y$$

Because it expresses the equality of the quantities on the left and on the right of the equal sign, an equation is a balanced mathematical statement.

In figure 24-1, the total weight on the left side of the scale equals the total weight on the right side; therefore, the scale balances.

3 pounds + 5 pounds + 2 pounds = 4 pounds + 6 pounds; 10 pounds = 10 pounds

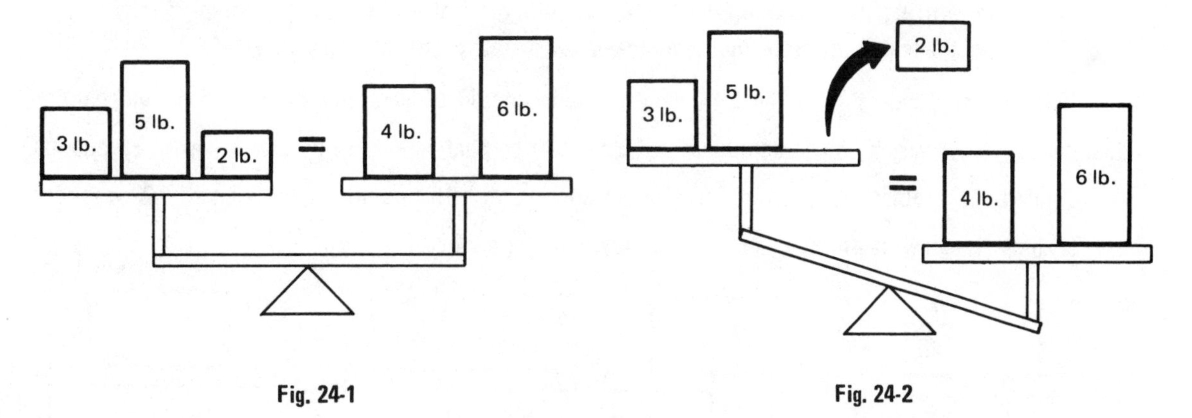

Fig. 24-1 Fig. 24-2

Figure 24-2 shows that when the 2-pound weight is removed from the scale, the scale is no longer in balance.

3 pounds + 5 pounds does not equal 4 pounds + 6 pounds, 8 pounds does not equal 10 pounds.

THE UNKNOWN QUANTITY

In general, an equation is used to determine the numerical value of an unknown quantity. Although any letter or symbol can be used to represent the unknown quantity, the letter x is commonly used.

The first letter of the unknown quantity is often used. Some common letter designations are

l to represent length	p to represent pressure
A to represent area	f to represent feed of cutter
t to represent time	w to represent weight

FROM WORD QUESTIONS TO EQUATIONS

An equation asks a question. It asks for the value of the unknown which makes the left side of the equation equal to the right side. The question asked may not be in equation form; instead it may be expressed in words.

It is important to develop the ability to convert word questions into mathematical questions, or equations. A problem must be fully understood before it can be written as an equation.

Whether the word problem is simple or complex, a definite logical procedure should be followed to analyze the problem. A few or all of the following steps may be required, depending on the complexity of the particular problem.

- Carefully read the entire problem, several times if necessary.
- Break the problem down into simpler parts.
- It is sometimes helpful to draw a simple picture as an aid in visualizing the various parts of the problem.
- Identify and list the unknowns. Give each unknown a letter name, such as x.
- Decide where the equal sign should be, and group the parts of the problem on the proper side of the equal sign.
- Check. Are the statements on the left equal to the statements on the right of the equal sign?
- After writing the equation, check it against the original problem, step-by-step. Does the equation state mathematically what the problem states in words?

Note: The method used to solve the following examples should be carefully followed and understood.

Example 1: What weight must be added to 15 pounds so that it will be in balance with 22 pounds?

▲ Convert the problem from word form to equation form: 15 pounds + what weight = 22 pounds?

▲ To help visualize the problem, a picture is shown in figure 24-3.

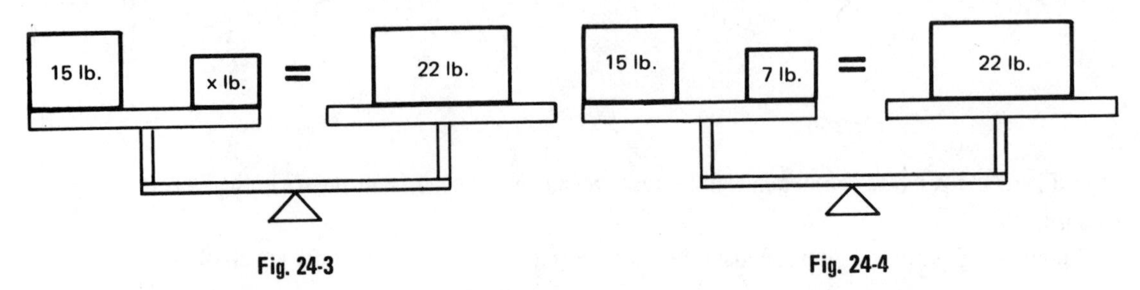

Fig. 24-3

Fig. 24-4

▲ Let x represent the unknown weight. Omit the weight units. Therefore, $15 + x = 22$.

▲ Ask the question: What number added to 15 = 22? Since 7 added to 15 = 22, $x = 7$.

▲ Figure 24-4 shows that 7 pounds must be added to 15 pounds to equal 22 pounds and make the scale balance.

▲ Check the answer by substituting 7 for x in the original equation. $15 + 7 = 22$

▲ The equation is balanced. $22 = 22$

Example 2: If a finished piece 9½ inches long is required, how many inches must be cut from a 12-inch length of bar stock? Make no allowance for the thickness of the cut.

▲ Convert the problem from word form to equation form

12 inches - the number of inches cut off = 9½ inches.

▲ A picture of the problem is shown in figure 24-5.

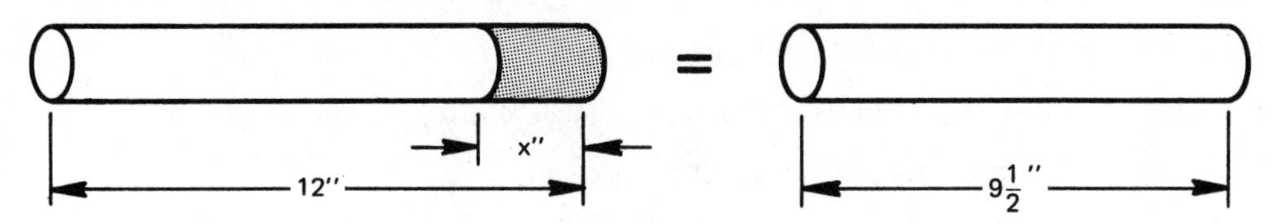

Fig. 24-5

▲ Let x represent the number of inches cut off. Omit the inch units. $12 - x = 9\frac{1}{2}$.

▲ Ask the question: What number subtracted from 12 = 9½? Since $12 - 2\frac{1}{2} = 9\frac{1}{2}$, $x = 2\frac{1}{2}$.

▲ Figure 24-6 shows that 2½ inches must be cut off 12 inches to balance the equation or equal 9½ inches.

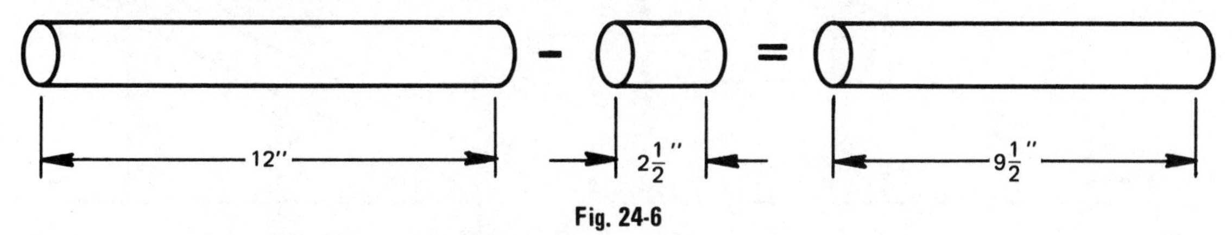

Fig. 24-6

▲ Check the answer by substituting 2½ inches for x in the original equation. 12" - 2½" = 9½"

12" - 2½" = 9½"   The equation is balanced. 9½" = 9½"

Example 3: The sum of two angles equals 90°. One angle is twice as large as the other. What is the size of the smaller angle?

▲ Convert the problem from word to equation form: An angle + an angle twice as large = 90°.

▲ A picture of the problem is shown in figure 24-7.

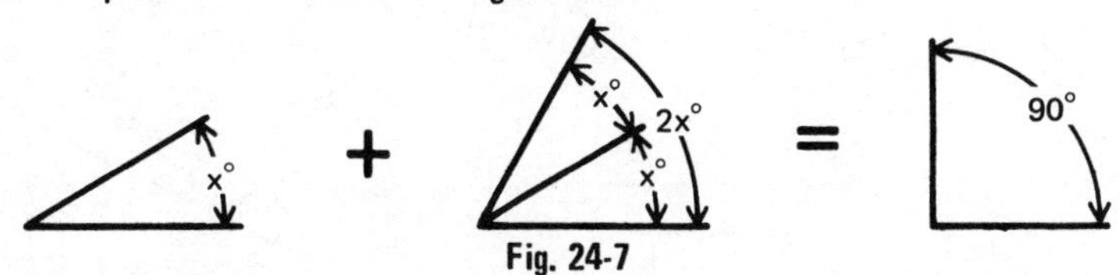

Fig. 24-7

▲ Let x represent the smaller angle. The larger angle is twice as large or equal to 2x. Therefore, $x + 2x = 90$, or $3x = 90$.

▲ Ask the question: What number multiplied by 3 = 90? Since 3 multipled by 30 = 90, $x = 30$.

▲ The smaller angle $x = 30^\circ$ and the larger angle $= 2x$ or 60° as shown in figure 24-8.

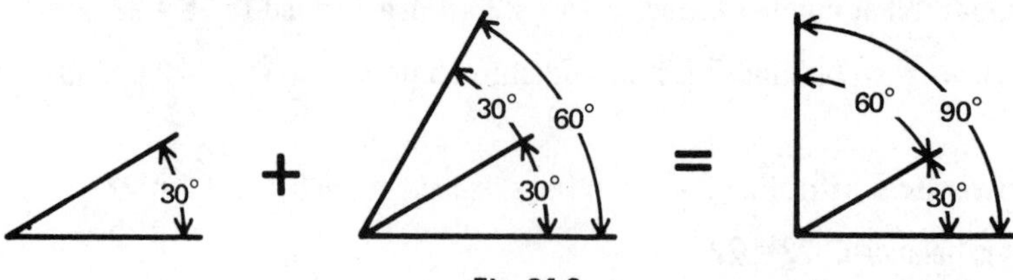

Fig. 24-8

▲ Check the answer by substituting 30° for x in the original equation.

$$30^\circ + 2\,(30^\circ) = 90^\circ;\ 30^\circ + 60^\circ = 90^\circ.\ \text{The equation is balanced. } 90^\circ = 90^\circ$$

Example 4: Three gage blocks are used to tilt a sine plate. The total height of the three blocks is 2.75 inches. The bottom block is 4 times as thick as the middle block. The middle block is twice as thick as the top block. How thick is each block?

▲ Convert the problem from word form to equation form:

a. Let x represent the thickness of the thinnest block, the top block.

b. The middle block is twice as thick as the top block, or $2x$.

c. The bottom block is four times as thick as the middle block, or $(4)\,(2x) = 8x$.

d. The sum of the three blocks = 2.75″. Therefore, $x + 2x + 8x = 2.75''$, or $11x = 2.75''$.

▲ A picture of the problem is shown in figure 24-9.

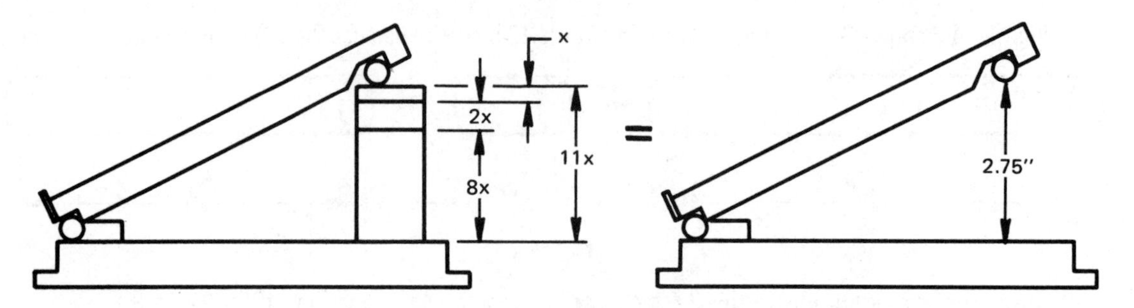

Fig. 24-9

▲ Ask the question: What number multiplied by 11 = 2.75? Since $11 \times .25 = 2.75$, $x = .25$.

▲ The thickness of each block is shown in figure 24-10:

a. The top block is x″ thick or .25″.

b. The middle block is 2x or 2(.25″) = .50″

c. The bottom block is 8x or 8(.25″) = 2.00″.

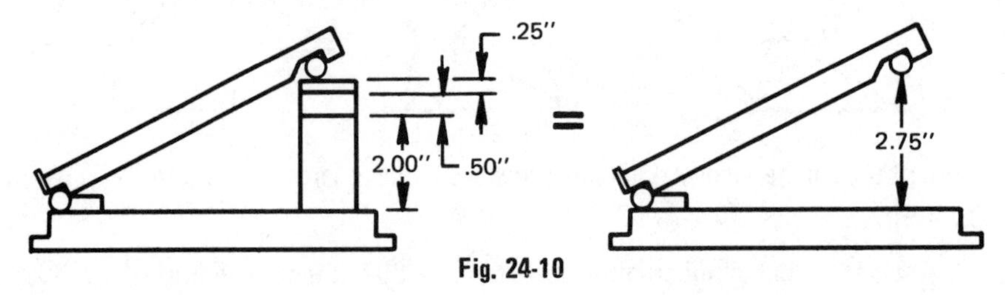

Fig. 24-10

▲ Check the answer by substituting .25" for x in the original equation: x" + 2x" + 4(2x") = 2.75"

.25" + 2(.25") + 4(2)(.25")	= 2.75"
.25 + .50" + 2.00"	= 2.75"
The equation is balanced. 2.75"	= 2.75"

In many cases the problems to be solved in actual machine shop applications will be more difficult than the preceding examples. It is essential, therefore, to be able to use the procedure shown to analyze the problem, determine the unknowns, and set up the equation.

CHECKING THE EQUATION

At the end of each of the preceding examples, the value found for the unknown was substituted in the original equation to prove that it was the correct value. If both sides of the equation are equal, the equation is balanced and the solution is correct.

Example: $11x + 4 = 70$, $11x = 66$, $x = 6$.

Substitute 6 for the value of x: $11(6) + 4 = 70$, $66 + 4 = 70$.

The equation is balanced. $70 = 70$.

All work in a machine shop should be checked and rechecked to prevent errors. This procedure saves time, labor, and money.

APPLICATION

A. Convert each of the following word problems to equations. Let the unknown number equal x and solve for the value of the unknown. Check the equation by comparing it to the word question. Does the equation state mathematically what the problem states in words? Check whether the equation is balanced by substituting the value of the unknown in the equation.

1. A number plus 18 equals 32. __________
2. A number less 7 equals 15. __________
3. Five times a number equals 55. __________
4. A number divided by 3 equals 9. __________
5. Thirty-two divided by a number equals 8. __________
6. A number plus twice the number equals 36. __________
7. Six times a number minus the number equals 45. __________
8. Seven times a number plus eight times the number equals 60. __________
9. Sixty divided by 3 times a number equals 4. __________
10. A piece of bar stock 16 inches long is cut into two unequal lengths. One piece is 3 times as long as the other. How long is each piece? __________
11. Three blocks are used to tilt a sine plate. The total height of the three blocks is 4.5". The first block is 3 times as thick as the second block. The second block is twice as thick as the third block. How thick is each block?

    a. __________

    b. __________

    c. __________

12. Five holes are drilled in a steel plate on a bolt circle. There are 300° between hole 1 and hole 5. The number of degrees between any two consecutive holes doubles in going from hole 1 to hole 5. Find the number of degrees between the indicated holes.

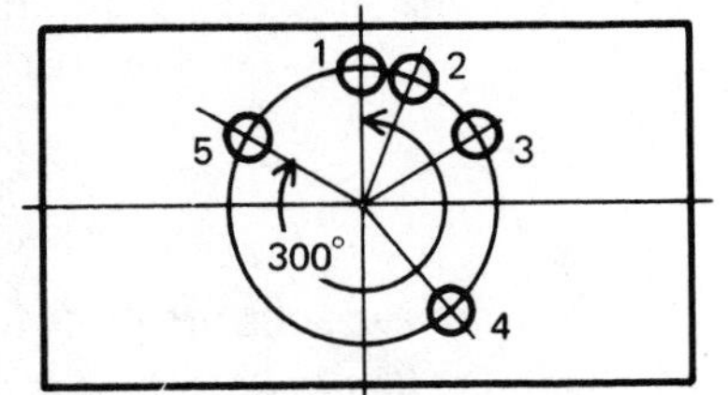

a. 1 and 2 ____ c. 3 and 4 ____

b. 2 and 3 ____ d. 4 and 5 ____

13. The total amount of stock milled off an aluminum casting in two cuts is .300 inch. The roughing cut is .250 inch greater than the finish cut. What is the depth of the finish cut? ____

B. In each of the following problems, refer to the corresponding figure. Write an equation, solve for x, and check.

1.  2.  3. 

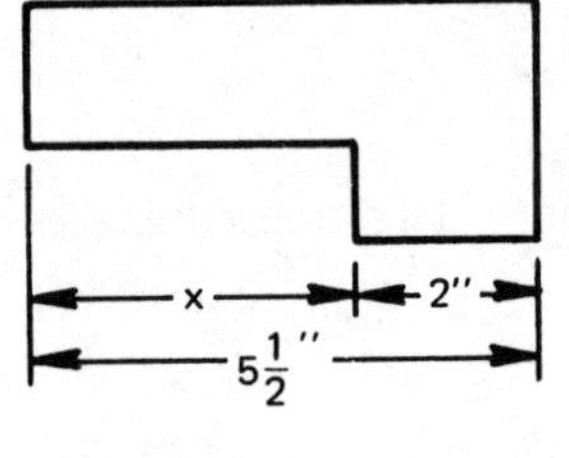

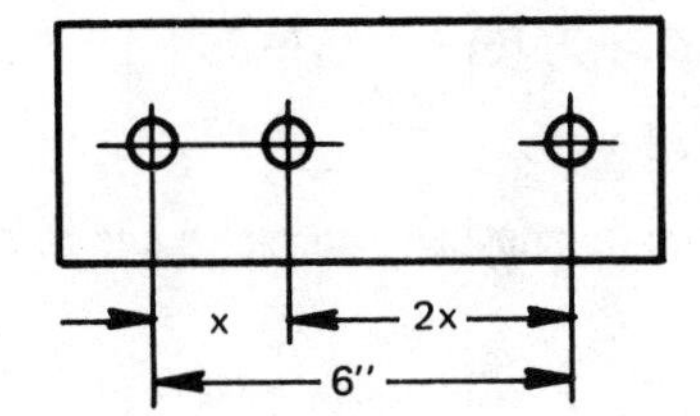

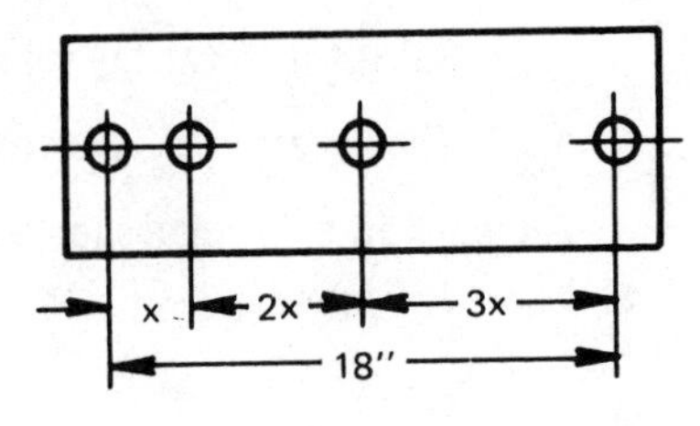

4. ____ 5. ____ 6. ____

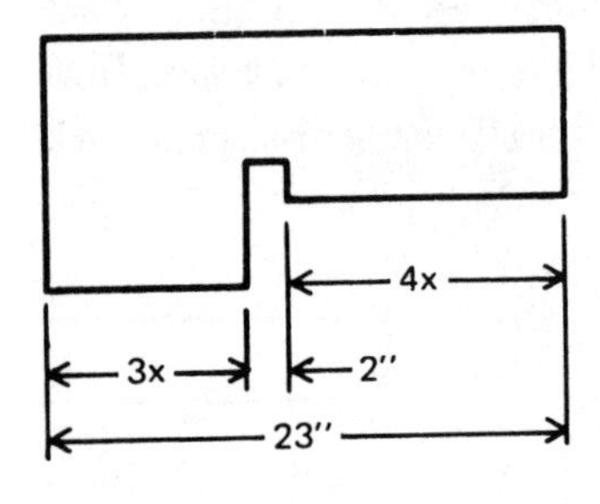

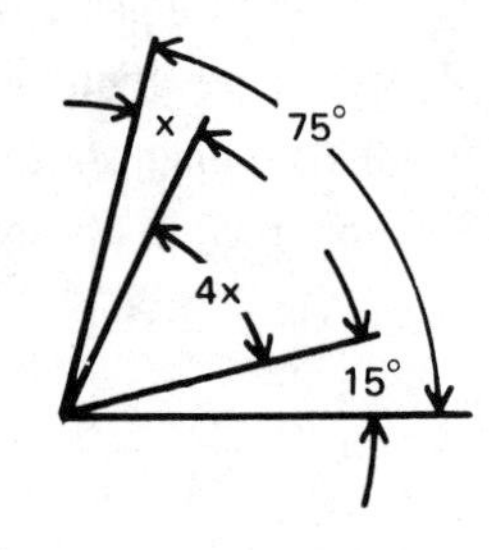

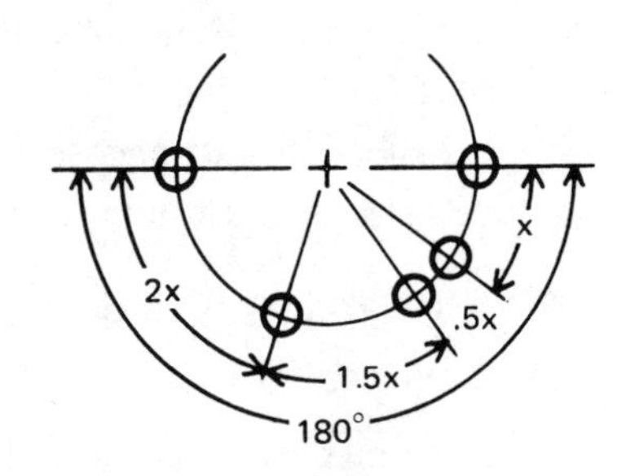

7. ____ 8. ____

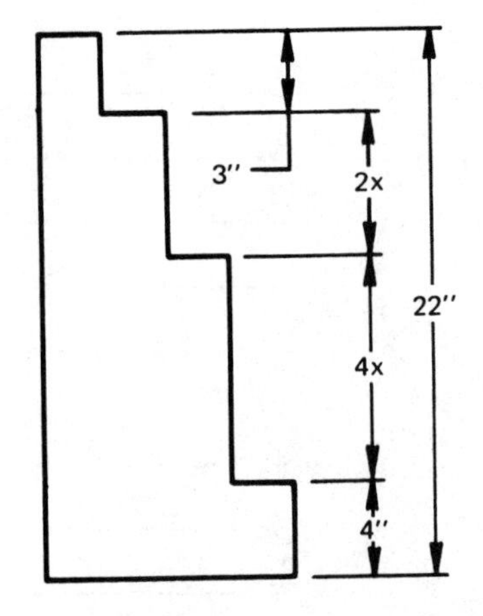

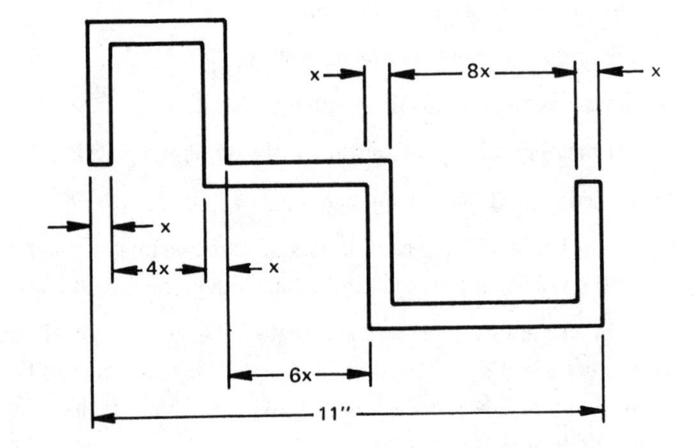

C. For each of the following problems, refer to the proper figure, solve for the unknowns, and check.

1. Find the distances between the following holes.

   a. Hole 1 to Hole 2 ________
   b. Hole 2 to Hole 3 ________
   c. Hole 3 to Hole 4 ________
   d. Hole 4 to Hole 5 ________
   e. Hole 5 to Hole 6 ________
   f. Hole 2 to Hole 4 ________
   g. Hole 3 to Hole 6 ________

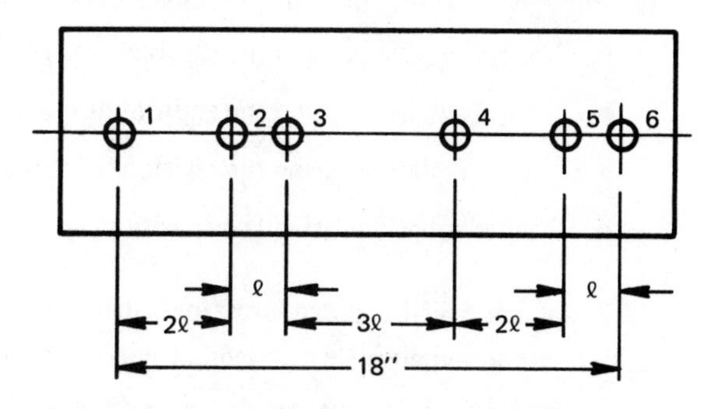

2. Find the distances between points.

   a. A and B ________
   b. C and D ________
   c. E and F ________

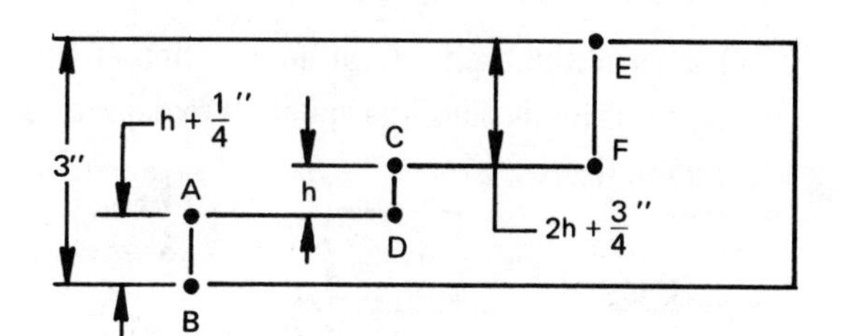

3. Find the value of each of the four angles.

   a. ∠1 = ________
   b. ∠2 = ________
   c. ∠3 = ________
   d. ∠4 = ________

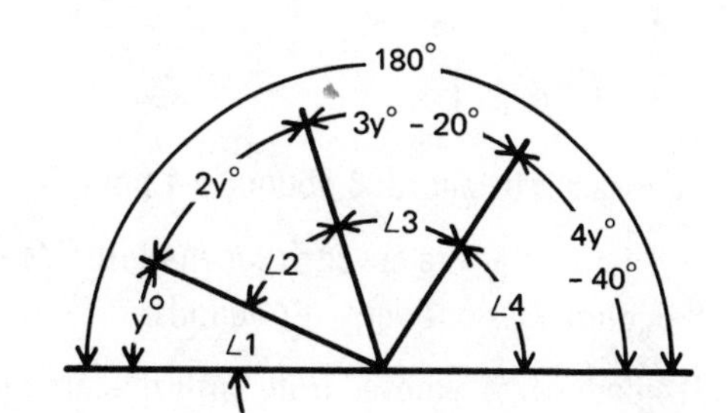

D. Solve for the unknown values in the following equations.

1. $x + 4x = 30, x =$ ____
2. $x + 3 = 12, x =$ ____
3. $2y + 5y + 3y = 70, y =$ ____
4. $32 = 21 + y, y =$ ____
5. $18 - a = 12, a =$ ____
6. $b - 13 = 80, b =$ ____
7. $3b + 5b - 2b = 48, b =$ ____
8. $3(5a) = \frac{6(30)}{2}, a =$ ____

9. $\frac{1}{2}x = 42, x =$ ____
10. $\frac{x}{4} = 12, x =$ ____
11. $\frac{27}{x} = 9, x =$ ____
12. $\frac{d}{6} + 4 = 9, d =$ ____
13. $.75x - .5x = \frac{18 + 30}{4}, x =$ ____
14. $6(2.5x) + 5x = 60, x =$ ____
15. $\frac{2y + 4y + 6y}{3} = 80, y =$ ____
16. $27 - (3)(6) = b + 7, b =$ ____

Unit 25 Solution of Equations by Subtraction, Addition, and Division

OBJECTIVES

After studying this unit, the student should be able to

- Solve equations using the subtraction principle of equality.
- Solve equations using the addition principle of equality.
- Solve equations using the division principle of equality.
- Solve equations using transposition.

There are specific procedures for solving equations using the fundamental principles of equality. Equations are solved more directly and efficiently by application of these principles.

SOLUTION OF EQUATIONS BY SUBTRACTION

The values on each side of an equation are equal and the equation is balanced. If the same value is subtracted from both sides, the equation remains balanced. The equation 8 pounds + 4 pounds = 12 pounds is pictured in figure 25-1.

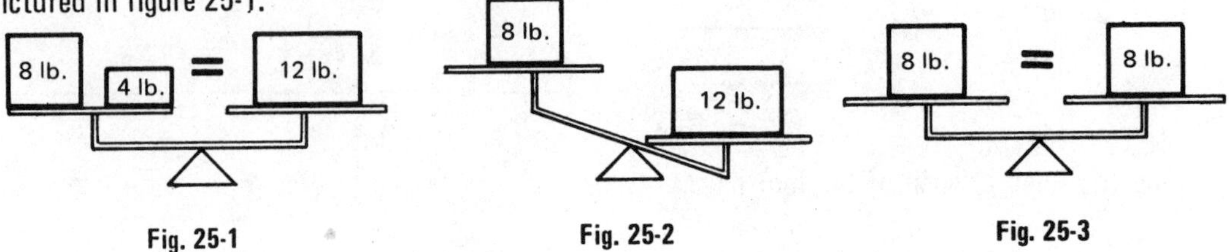

Fig. 25-1 Fig. 25-2 Fig. 25-3

- The scale is balanced; 8 pounds + 4 pounds = 12 pounds. 12 pounds = 12 pounds.
- If 4 pounds are removed from the left side only, the scale is not in balance, as shown in figure 25-2. 8 pounds *does not* equal 12 pounds.
- If 4 pounds are removed from both the left and right side, then the scale remains balanced, as shown in figure 25-3. 8 pounds + 4 pounds - 4 pounds = 12 pounds - 4 pounds. 8 pounds = 8 pounds.

The subtraction principle of equality states that if the same number is subtracted from both sides of an equation, the sides remain equal, or the equation remains balanced.

The subtraction principle is used to solve an equation in which a number is added to the unknown, such as $x + 15 = 20$.

Rule: To solve an equation in which a number is added to the unknown,

- Subtract the number which is added to the unknown from both sides of the equation.

Examples: Solve for the unknowns in the following problems using the subtraction principle.

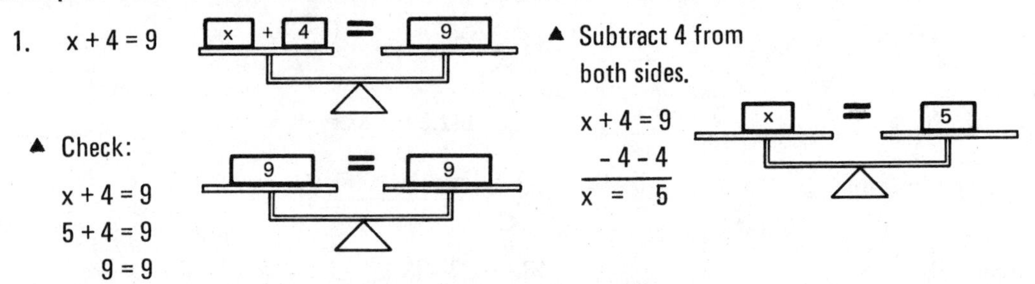

1. $x + 4 = 9$

▲ Subtract 4 from both sides.

$$\begin{aligned} x + 4 &= 9 \\ -4 & \;\; -4 \\ \hline x &= 5 \end{aligned}$$

▲ Check:

$x + 4 = 9$

$5 + 4 = 9$

$9 = 9$

2. ▲ Write an equation: $5.5 + y = 17$

▲ Subtract 5.5 from both sides:

$$\begin{array}{r} 5.5 + y = 17 \\ -\ 5.5 \quad -\ 5.5 \\ \hline y = 11.5 \end{array}$$

▲ Check: $5.5 + y = 17$

$5.5 + 11.5 = 17$

$17 = 17$

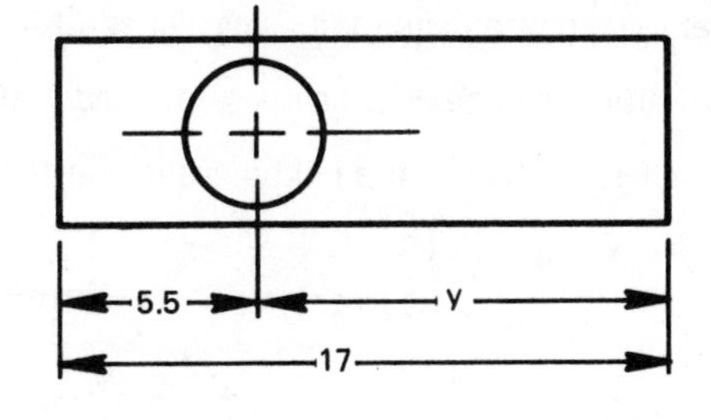

3. $-39 = P + 18$

▲ $$\begin{array}{r} -39 = P + 18 \\ -18 \quad -18 \\ \hline -57 = P \end{array}$$

▲ Check:

$-39 = P + 18$

$-39 = -57 + 18$

$-39 = -39$

4. $W + 4\frac{3}{4} = 12$

▲ $$\begin{array}{r} W + 4\frac{3}{4} = 12 \\ -4\frac{3}{4} \quad -\ 4\frac{3}{4} \\ \hline W \quad = 7\frac{1}{4} \end{array}$$

▲ Check:

$W + 4\frac{3}{4} = 12$

$7\frac{1}{4} + 4\frac{3}{4} = 12$

$12 = 12$

Transposition

Transposing a term means transferring, or moving, it from one side of an equation to the other. Transposition is an alternate method of solving equations.

Rule: To transpose a term from one side of an equation to the other,

- Change the operation of the term to the opposite operation.

Examples: Solve for the unknowns in the following problems by using transposition. Observe that solving these problems by transposition is actually a shortcut application of the subtraction principle.

1. $x + 7 = 10$

▲ Seven is added to x. Subtraction is the opposite operation of addition. Therefore, when 7 is moved to the right side of the equation it is subtracted.

$x = 10 - 7$

$x = 3$

▲ Check: $x + 7 = 10$

$3 + 7 = 10$

$10 = 10$

2. $y + 10.7 = 18$

▲ Move 10.7 from the left side of the equation to the right side and subtract.

$y = 18 - 10.7$

$y = 7.3$

▲ Check: $y + 10.7 = 18$

$7.3 + 10.7 = 18$

$18 = 18$

3. ▲ $T + 6\frac{1}{8} = -19$

$T = -19 - 6\frac{1}{8}$

$T = -25\frac{1}{8}$

▲ Check: $T + 6\frac{1}{8} = -19$

$-25\frac{1}{8} + 6\frac{1}{8} = -19$

$-19 = -19$

SOLUTION OF EQUATIONS BY ADDITION

The addition principle of equality states that if the same number is added to both sides of an equation, the sides remain equal, or the equation remains balanced.

The addition principle is used to solve an equation in which a number is subtracted from the unknown, such as $x - 17 = 30$.

Rule: To solve an equation using the additon principles of equality

- Add the number which is subtracted from the unknown to both sides of the equation.

Examples: Solve for the unknowns in the following problems using the addition principle:

1. $x - 6 = 15$

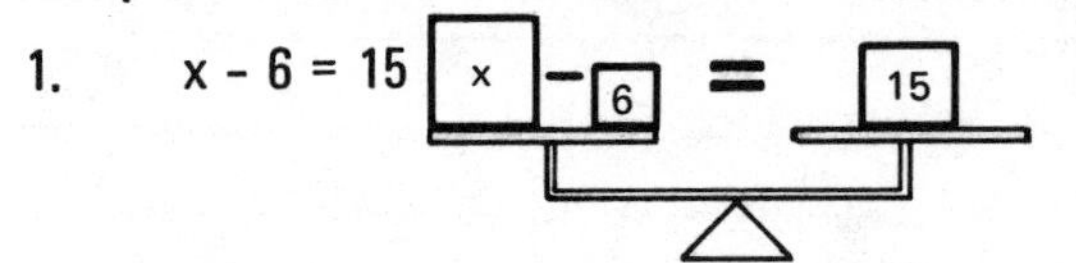

▲ Add 6 to both sides.

$$\begin{array}{r} x - 6 = 15 \\ +6 + \ 6 \\ \hline x = 21 \end{array}$$

▲ Check:

$$\begin{array}{r} x - \ 6 = 15 \\ 21 - \ 6 = 15 \\ 15 = 15 \end{array}$$

2. Seven inches are cut from a block. The remaining block is ten inches high. What was the height of the original block?

▲ Let y = the height of the original block.

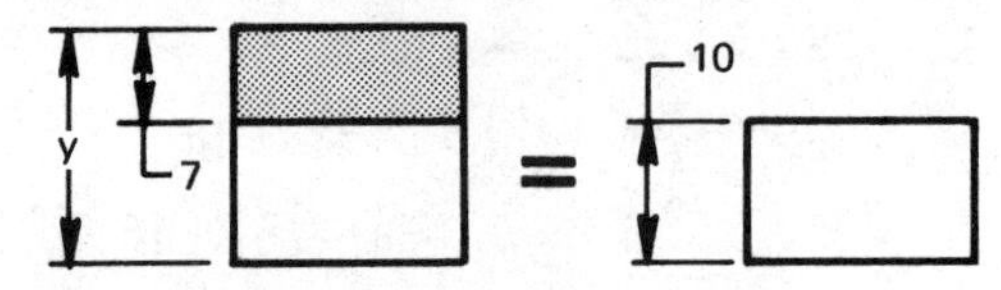

▲ Write an equation. $y - 7 = 10$

▲ Add 7 to both sides.

$$\begin{array}{r} y - 7 = 10 \\ +7 \quad +7 \\ \hline y = 17 \text{ inches} \end{array}$$

▲ Check:

$$\begin{array}{r} y - 7 = 10 \\ 17 - 7 = 10 \\ 10 = 10 \end{array}$$

3. $-35 = P - 20.4$

▲
$$\begin{array}{r} -35 \quad = P - 20.4 \\ +20.4 \qquad +20.4 \\ \hline -14.6 = P \end{array}$$

▲ Check:

$$\begin{array}{r} -35 = P - 20.4 \\ -35 = -14.6 - 20.4 \\ -35 = -35 \end{array}$$

Transposition

Examples: Solve for the unknown in the following problems using transposition. Observe that solving these problems by transposition is actually a shortcut application of the addition principle.

1. $x - 4 = 19$

▲ Four is subtracted from x. Addition is the opposite operation of subtraction. Therefore, when 4 is moved to the right side of the equation, it is added.

$$\begin{array}{r} x = 19 + 4 \\ x = 23 \end{array}$$

▲ Check:

$$\begin{array}{r} x - \ 4 = 19 \\ 23 - \ 4 = 19 \\ 19 = 19 \end{array}$$

2. $y - 16.9 = 30$

▲ Move 16.9 from the left side of the equation to the right side and add.

$$\begin{array}{r} y = 30 + 16.9 \\ y = 46.9 \end{array}$$

▲ Check:

$$\begin{array}{r} y - 16.9 = 30 \\ 46.9 - 16.9 = 30 \\ 30 = 30 \end{array}$$

3. $-14 = P - 27\frac{3}{4}$

▲
$$\begin{array}{r} -14 + 27\frac{3}{4} = P \\ 13\frac{3}{4} = P \end{array}$$

▲ Check:

$$\begin{array}{r} -14 = P - 27\frac{3}{4} \\ -14 = 13\frac{3}{4} - 27\frac{3}{4} \\ -14 = -14 \end{array}$$

SOLUTION OF EQUATIONS BY DIVISION

The division principle of equality states that if both sides of an equation are divided by the same number, the sides remain equal, or the equation remains balanced.

The division principle is used to solve an equation in which a number is multiplied by the unknown, such as $3x = 18$.

Rule: To solve an equation in which a number is multiplied by the unknown

- Divide the number which is multiplied by the unknown into both sides of the equation.

Example: Solve for the unknowns in the following problems using the division principle.

1. $6x = 24$

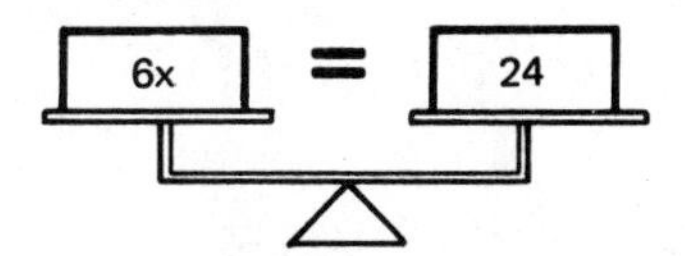

▲ Divide both sides by 6.

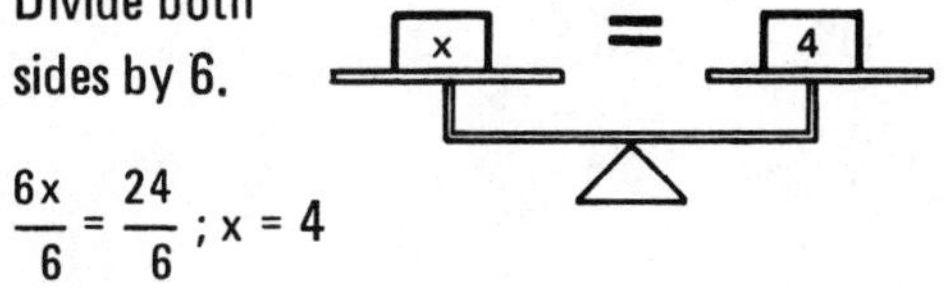

$\frac{6x}{6} = \frac{24}{6}$; $x = 4$

▲ Check:

$6x = 24$

$6(4) = 24$

$24 = 24$

2.

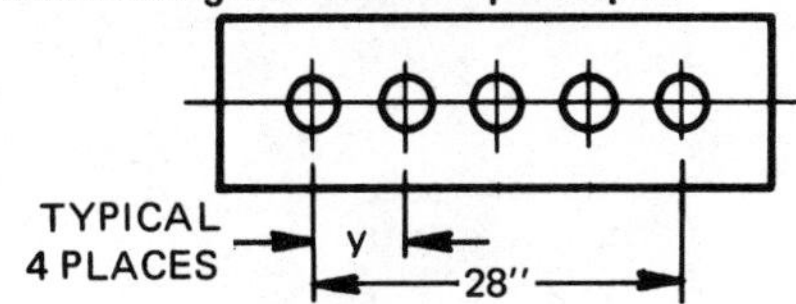

▲ Write an equation: $4y = 28$

▲ Divide both sides by 4; $\frac{4y}{4} = \frac{28}{4}$, $y = 7$

▲ Check: $4y = 28$

$4(7) = 28$

$28 = 28$

3. $-14.4 = 3.2F$

$\frac{-14.4}{3.2} = \frac{3.2F}{3.2}$

$-4.5 = F$

▲ Check:

$-14.4 = 3.2F$

$-14.4 = 3.2(-4.5)$

$-14.4 = -14.4$

4. $7\frac{1}{4}A = 21\frac{3}{4}$

$\frac{7\frac{1}{4}A}{7\frac{1}{4}} = \frac{21\frac{3}{4}}{7\frac{1}{4}}$

$A = 3$

▲ Check:

$7\frac{1}{4}A = 21\frac{3}{4}$

$7\frac{1}{4}(3) = 21\frac{3}{4}$

$21\frac{3}{4} = 21\frac{3}{4}$

Examples: Solve for the unknown in the following problems using transposition. Observe that solving these problems by transposition is actually a shortcut application of the division principle.

1. $10x = 70$

▲ The unknown x is multiplied by 10. Division is the opposite operation of multiplication. Therefore, when 10 is moved to the right side of the equation it is used as a divisor.

$x = \frac{70}{10}$

$x = 7$

▲ Check:

$10x = 70$

$10(7) = 70$

$70 = 70$

2. $3.6P = 18$

▲ Move 3.6 from the left side to the right side of the equation and use as a divisor.

$P = \frac{18}{3.6}$

$P = 5$

▲ Check:

$3.6P = 18$

$3.6(5) = 18$

$18 = 18$

3. $-\frac{1}{8}R = -3$

▲ $R = \frac{-3}{-\frac{1}{8}}$

$R = 24$

▲ Check:

$-\frac{1}{8}R = -3$

$-\frac{1}{8}(24) = -3$

$-3 = -3$

APPLICATION

A. SOLUTION BY SUBTRACTION

Solve each of the following equations by both the subtraction principle of equality and transposition. Check all answers.

1. $x + 12 = 27$, x = ____
2. $x + 32 = 56$, x = ____
3. $A + 17 = 43$, A = ____
4. $7 = y + 7$, y = ____
5. $-14 = V + 1$, V = ____
6. $W + \frac{3}{4} = \frac{1}{8}$, W = ____
7. $Y + 13\frac{7}{8} = 21\frac{1}{8}$, Y = ____
8. $37.75 = x + 29.55$, x = ____
9. $36\frac{11}{16} = y + 17\frac{5}{16}$, y = ____
10. $P + 18\frac{29}{32} = 50\frac{1}{32}$, P = ____

B. SOLVE BY SUBTRACTION

Write an equation for each of the following problems, solve for the unknown, and check.

1. a. Equation: ________
   b. x = ________

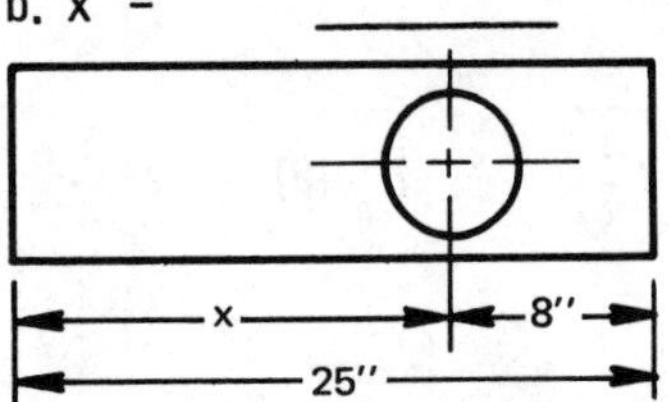

2. a. Equation: ________
   b. y = ________

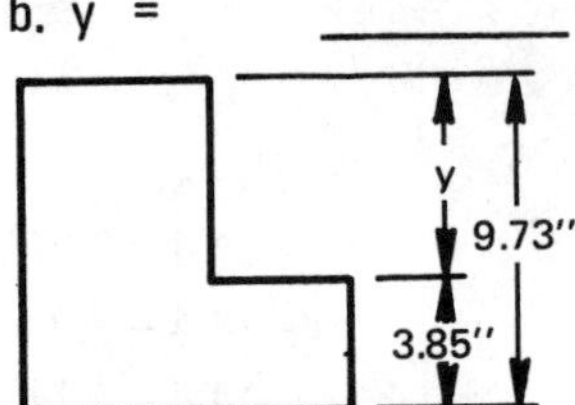

3. a. Equation: ________
   b. r = ________

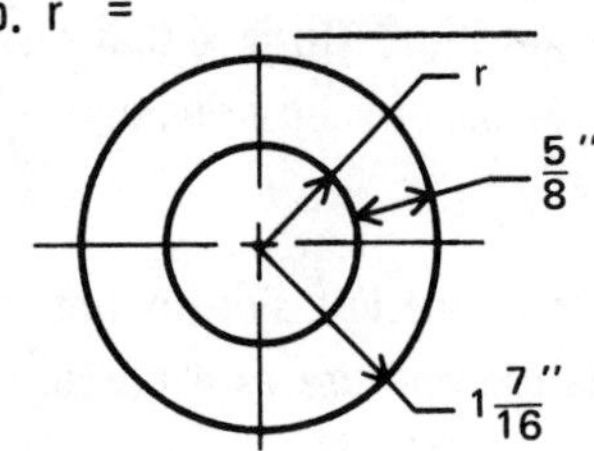

4. a. Equation: ________
   b. T = ________

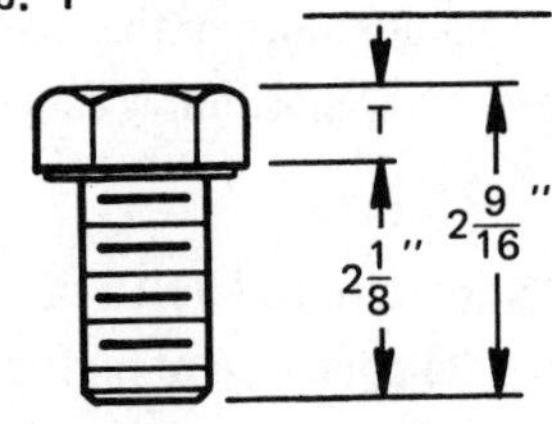

5. a. Equation: ________
   b. x = ________

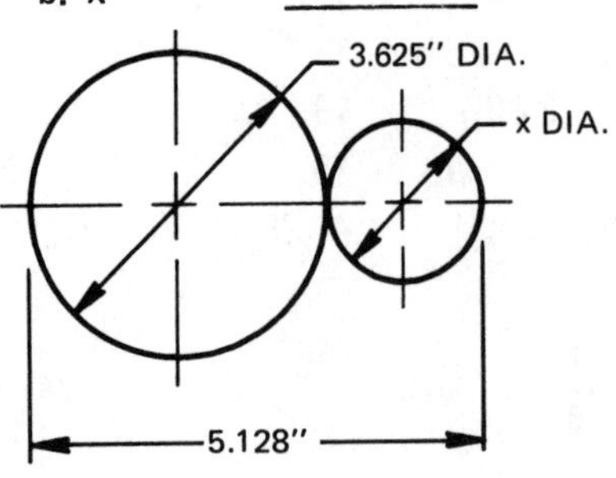

6. a. Equation: ________
   b. H = ________

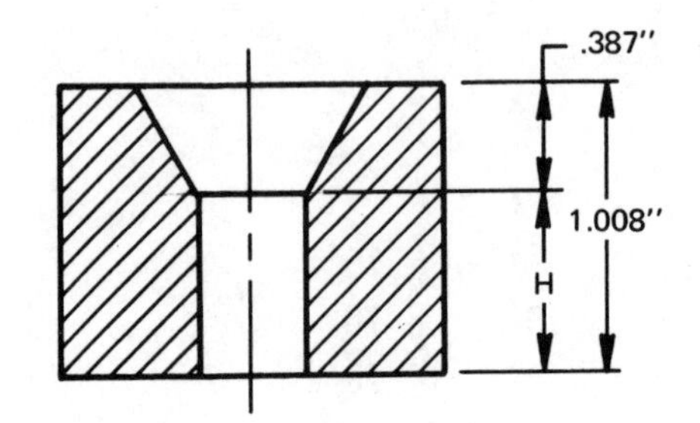

7. The height of 2 gage blocks is .8504 inch. One block is .750 inch thick. What is the thickness of the other block?

   a. Equation: __________ b. Thickness = __________

8. Three holes are drilled in a housing. The center distance between the first hole and the second hole is 7 3/4 inches and the center distance between the first hole and the third hole is 11 1/8 inches. What is the distance between the second hole and the third hole?

   a. Equation: __________ b. Distance = __________

9. A milling cut of 9/32 inch is required to provide a reference surface on a rough casting which is 7 5/8 inches high. What is the height of the casting after the milling cut?

   a. Equation: __________ b. Height = __________

10. A shaft rotates in a bearing which is .3873 inch in diameter. The total clearance between the shaft and bearing is .0008 inch. What is the diameter of the shaft?

    a. Equation: __________ b. Diameter = __________

C. SOLUTION BY ADDITION

Solve each of the following equations by both the addition principle of equality and transposition. Check all answers.

1. $y - 14 = 31$, y = ____
2. $x - 12 = 19$, x = ____
3. $F - 53 = 7$, F = ____
4. $-18 = M - 23$, M = ____
5. $6 = V - 12$, V = ____
6. $D - \frac{1}{8} = \frac{3}{8}$, D = ____
7. $\frac{7}{16} = r - \frac{7}{32}$, r = ____
8. $T - \frac{9}{64} = -\frac{13}{16}$, T = ____
9. $x - 3\frac{5}{8} = 17\frac{1}{16}$, x = ____
10. $-\frac{3}{4} = y - \frac{3}{32}$, y = ____

D. SOLVE BY ADDITION

Write an equation for each of the following problems, solve for the unknown, and check.

1. A bushing has a body diameter of 1 3/4 inches which is 9/16 inch less than the head diameter. What is the size of the head diameter?

   a. Equation: __________

   b. Size = __________

2. The flute length of a reamer is 1 1/8 inch, which is 3 3/8 inches less than the shank length. How long is the shank?

   a. Equation: __________

   b. Length = __________

3. A hole is countersunk to a depth of .275 inch. The depth of the countersink is 1.650 inches less than the depth of the .625-inch diameter hole.

   a. Equation: ________

   b. Depth = ________

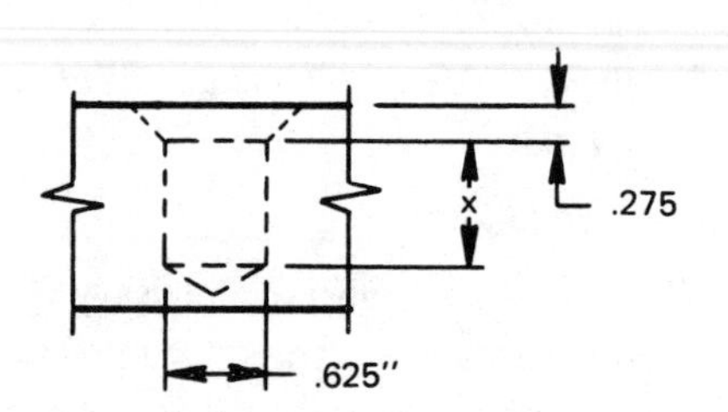

E. SOLUTION BY DIVISION

Solve each of the following equations by both the division principle of equality and transposition. Check.

1. $9x = 60$ $x =$ ____
2. $12P = 48$, $P =$ ____
3. $27 = 3m$, $m =$ ____
4. $18y = -54$, $y =$ ____
5. $8L = 2$, $L =$ ____
6. $-2.7y = 14.85$, $y =$ ____
7. $-20 = -60F$, $F =$ ____
8. $\frac{7}{8}R = 1\frac{3}{4}$, $R =$ ____
9. $\frac{3}{16}x = -\frac{3}{8}$, $x =$ ____
10. $\frac{3}{8}x = \frac{3}{16}$, $x =$ ____

F. SOLVE BY DIVISION

Write an equation for each of the following problems, solve for the unknown.

1. a. Equation: ________

   b. x = ________

2. a. Equation: ________

   b. x = ________

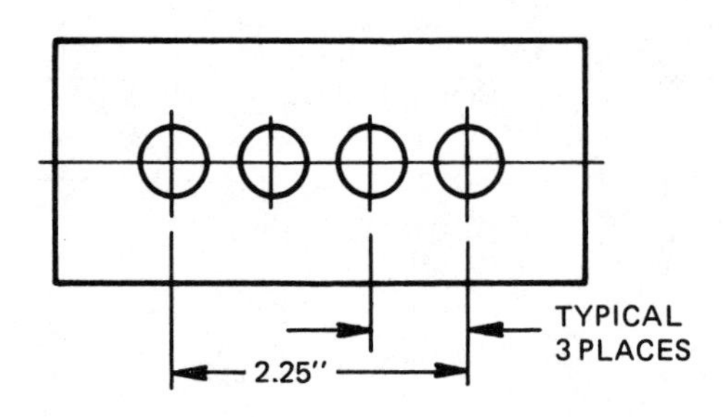

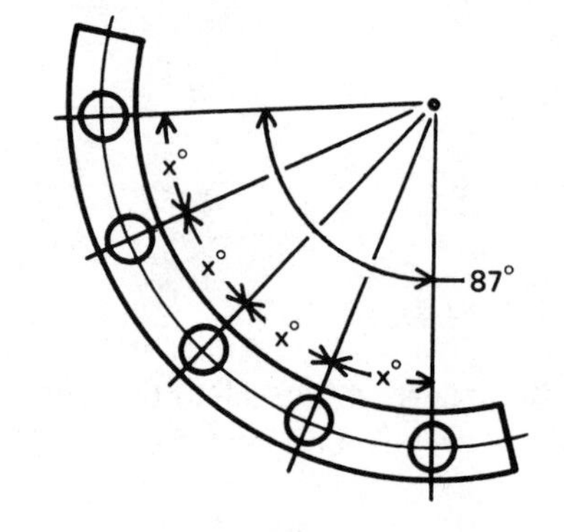

3. The feed of a drill is the depth of material that the drill penetrates in one revolution. The total depth of penetration equals the product of the number of revolutions and the feed. Compute the feed of a drill which cuts to a depth of 3.300 inches while turning 500 revolutions.

   a. Equation: ________

   b. Feed = ________

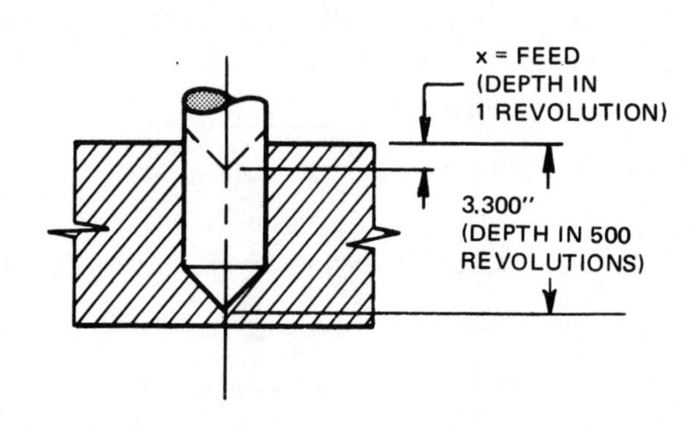

Unit 26 Solution of Equations by Multiplication, Roots, and Powers

OBJECTIVES

After studying this unit, the student should be able to

- Solve equations using the multiplication principle of equality.
- Solve equations using the root principle of equality.
- Solve equations using the power principle of equality.
- Solve equations using transposition

SOLUTION OF EQUATIONS BY MULTIPLICATION

The multiplication principle of equality states that if both sides of an equation are multiplied by the same number, the sides remain equal, or the equation remains balanced.

The multiplication principle is used to solve an equation in which a number is divided into the unknown, such as $\frac{x}{4} = 10$.

Rule: To solve an equation in which a number is divided into the unknown

- Multiply both sides of the equation by the number which is divided into the unknown.

Examples: Solve for the unknowns in the following problems using the multiplication principle.

1. $\frac{x}{3} = 7$

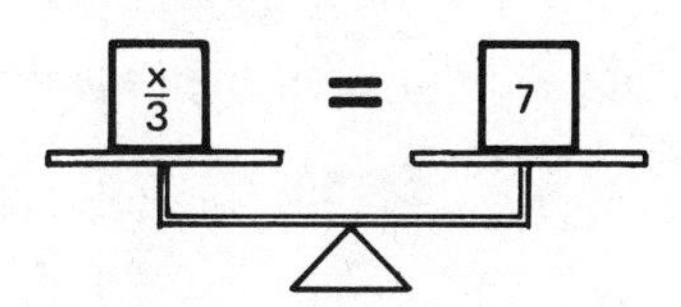

▲ Multiply both sides by 3. $\frac{x}{3} = 7$

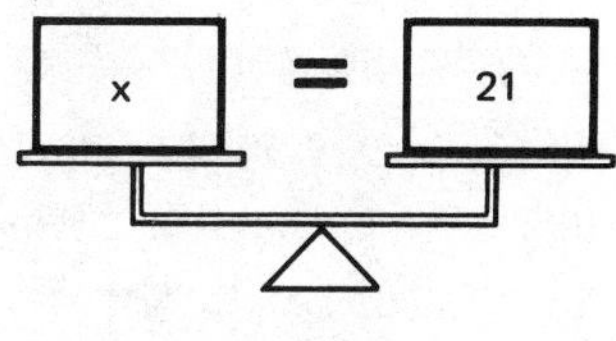

$3(\frac{x}{3}) = 3(7)$ $x = 21$

▲ **Check:**

$\frac{x}{3} = 7$

$\frac{21}{3} = 7$

$7 = 7$

2. A length of bar stock is cut into 5 equal pieces. Each piece is 4.5 inches long. What was the length of the bar before it was cut?

▲ Let y = the length of the bar before it was cut.

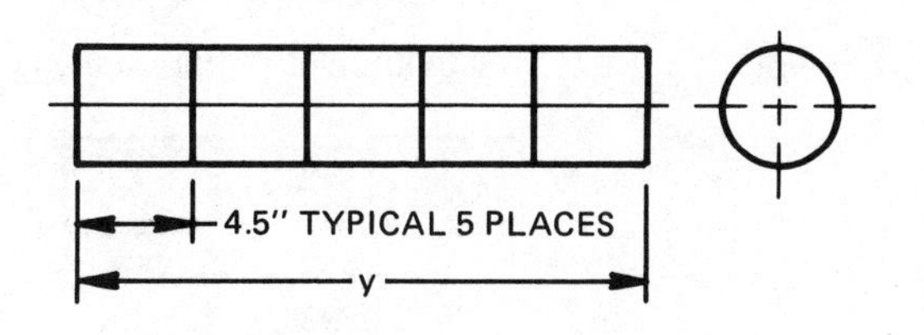

▲ Write an equation: $\frac{y}{5} = 4.5$

▲ Multiply both sides by 5. $5(\frac{y}{5}) = 5(4.5)$

$y = 22.5$ inches

▲ Check: $\frac{y}{5} = 4.5$

$\frac{22.5}{5} = 4.5$

$4.5 = 4.5$

3. $6\frac{1}{8} = \frac{F}{-5}$

$-5(6\frac{1}{8}) = -5(\frac{F}{-5})$

$-30\frac{5}{8} = F$

▲ Check: $6\frac{1}{8} = \frac{F}{-5}$

$6\frac{1}{8} = \frac{-30\frac{5}{8}}{-5}$

$6\frac{1}{8} = 6\frac{1}{8}$

Examples: Solve for the unknowns in the following problems using transposition. Observe that solving these problems by transposition is actually a shortcut application of the multiplication principle.

1. $\frac{x}{7} = 8$

▲ Seven is divided into x. Multiplication is the opposite operation of division, therefore, when 7 is moved to the right side of the equation, it is used as a multiplier.

$x = 7(8)$
$x = 56$

▲ Check: $\frac{x}{7} = 8$

$\frac{56}{7} = 8$

$8 = 8$

2. $8 = \frac{L}{4.2}$

▲ Move 4.2 from the right side of the equation to the left side and multiply.

$8(4.2) = L$
$33.6 = L$

▲ Check: $8 = \frac{L}{4.2}$

$8 = \frac{33.6}{4.2}$

$8 = 8$

3. $\frac{R}{6\frac{1}{4}} = -3$

▲ $R = -3(6\frac{1}{4})$

$R = -18\frac{3}{4}$

▲ Check: $\frac{R}{6\frac{1}{4}} = -3$

$\frac{-18\frac{3}{4}}{6\frac{1}{4}} = -3$

$-3 = -3$

SOLUTION OF EQUATIONS BY ROOTS

The root principle of equality states that if the same root of both sides of an equation is taken, the sides remain equal, or the equation remains balanced.

The root principle is used to solve an equation that contains an unknown which is raised to a power, such as $x^2 = 36$.

Rule: To solve an equation in which an unknown is raised to a power

▲ Extract the root of both sides which leaves the unknown with an exponent of one.

Examples: Solve for the unknowns in the following problems using the root principle.

1. $x^2 = 9$

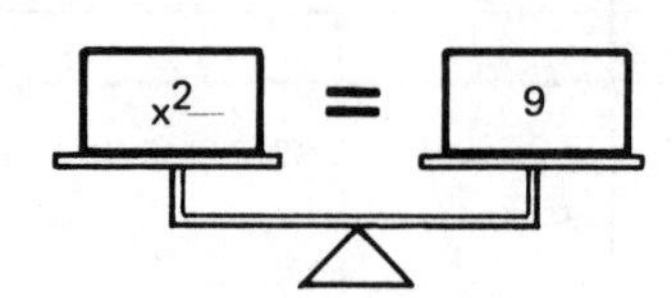

▲ Extract the square root of both sides.

$x^2 = 9$
$\sqrt{x^2} = \sqrt{9}$
$x = 3$

▲ Check:

$x^2 = 9$
$3^2 = 9$
$9 = 9$

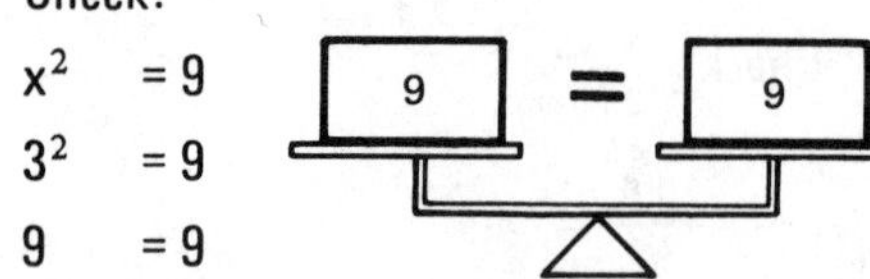

2. The area of a square piece of sheet steel equals 16 square feet. What is the length of each side?

▲ Let s = length of each side.

▲ Write an equation:

$s^2 = 16$

▲ Extract the square root of both sides.

$\sqrt{s^2} = \sqrt{16}$; $s = 4$

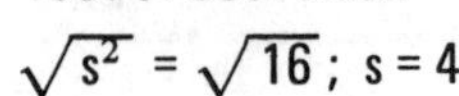

▲ Check: $s^2 = 16$
$4^2 = 16$
$16 = 16$

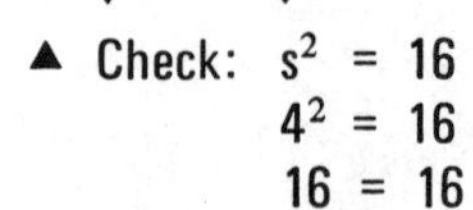

3. $T^3 = -64$
$\sqrt[3]{T^3} = \sqrt[3]{-64}$
$T = -4$

▲ Check: $T^3 = -64$
$(-4)^3 = -64$
$-64 = -64$

4. $V^2 = 9/64$
$\sqrt{V^2} = \sqrt{9/64}$
$V = \frac{3}{8}$

Check: $V^2 = \frac{9}{64}$
$(\frac{3}{8})^2 = \frac{9}{64}$
$\frac{9}{64} = \frac{9}{64}$

Examples: Solve for the unknowns in the following problems using transposition. Observe that solving these problems by transposition is actually a shortcut application of the root principle.

1. $x^2 = 25$

▲ The unknown x is squared. Extracting a square root is the opposite of squaring, therefore, when the square is moved to the right side of the equation, it is changed to a square root.

$x = \sqrt{25}$ ▲ Check: $x^2 = 25$

$x = 5$ $\quad 5^2 = 25$

$25 = 25$

2. $27 = T^3$

▲ Move the cubing operation from the right side of the equation to the left and change it to a cube root operation.

$\sqrt[3]{27} = T$ ▲ Check: $27 = T^3$

$3 = T$ $\quad 27 = 3^3$

$27 = 27$

SOLUTION OF EQUATIONS BY POWERS

The power principle of equality states that if both sides of an equation are raised to the same power, the sides remain equal, or the equation remains balanced.

The power principle is used to solve an equation that contains a root of the unknown, such as $\sqrt{x} = 8$

Rule: To solve an equation which contains a root of the unknown

- Raise both sides of the equation to the power which leaves the unknown with an exponent of one.

Examples: Solve for the unknowns in the following problems, using the power principle.

1. $\sqrt{x} = 8$

▲ Square both sides.

$\sqrt{x} = 8$

$(\sqrt{x})^2 = 8^2$

$x = 64$

▲ Check:

$\sqrt{x} = 8$

$\sqrt{64} = 8$

$8 = 8$

2. The length of a side of a cube equals 2 inches. The cube root of the volume equals the length of a side. Find the volume of the cube.

▲ Let V = the volume of the cube.

▲ Write an equation.

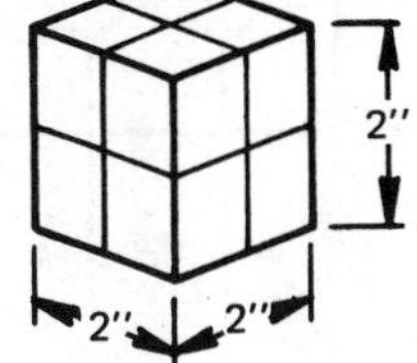

$\sqrt[3]{V} = 2$

▲ Cube both sides.

$(\sqrt[3]{V})^3 = 2^3$

$V = 8$ cubic inches

▲ Check: $\sqrt[3]{V} = 2$

$\sqrt[3]{8} = 2$

$2 = 2$

Examples: Solve for the unknowns in the following problems using transposition. Observe that solving these problems by transposition is actually a shortcut application of the power principle.

1. $\sqrt{x} = 12$

▲ The square root of the unknown is given. Squaring is the opposite operation of extracting the square root. Therefore, the square root is changed to squaring when moved to the right side of the equation.

$x = 12^2$ $\quad$ Check: $\sqrt{x} = 12$

$x = 144$ $\quad \sqrt{144} = 12$

$12 = 12$

2. $4 = \sqrt[3]{y}$

▲ Move the cube root operation from the right side of the equation to the left and change it to a cubing operation.

$4^3 = y$ ▲ Check:

$64 = y$ $\quad 4 = \sqrt[3]{y}$

$4 = \sqrt[3]{64}$

$4 = 4$

APPLICATION

A. SOLUTION BY MULTIPLICATION

Solve each of the following equations by both the multiplication principle of equality and transposition. Check all answers.

1. $\frac{x}{8} = 3.5,$ x = ______
2. $\frac{y}{2} = 7,$ y = ______
3. $\frac{P}{14} = 3,$ P = ______
4. $4 = \frac{L}{9},$ L = ______
5. $\frac{x}{3} = 12,$ x = ______
6. $H \div -5\frac{1}{8} = 10,$ H = ______
7. $N \div -\frac{3}{4} = -\frac{1}{4},$ N = ______
8. $X \div .05 = 7.8,$ x = ______
9. $\frac{P}{-.75} = .005,$ P = ______
10. $\frac{R}{100} = -.06,$ R = ______

B. SOLVE BY MULTIPLICATION

Write an equation for each of the following problems, solve for the unknown and check.

1. a. Equation: ______

   b. x = ______

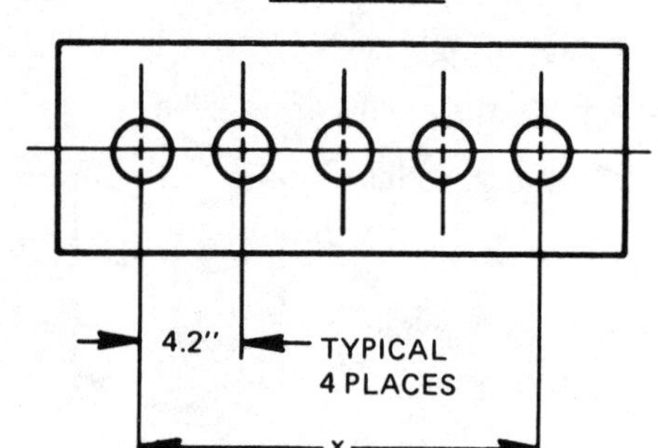

2. a. Equation: ______

   b. x = ______

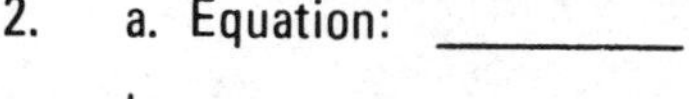

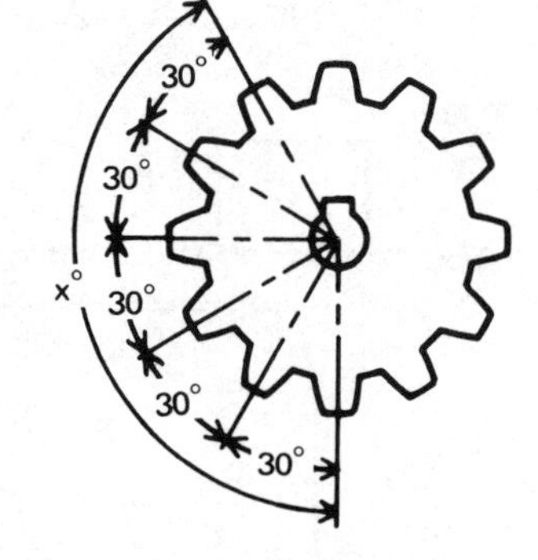

3. A 10-inch sine plate is tilted at an angle of 45°. The gage block height divided by 10 equals .70711 inch. Compute the height of the gage blocks.

   a. Equation: ______

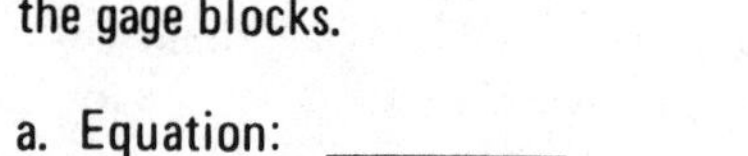

   b. Length = ______

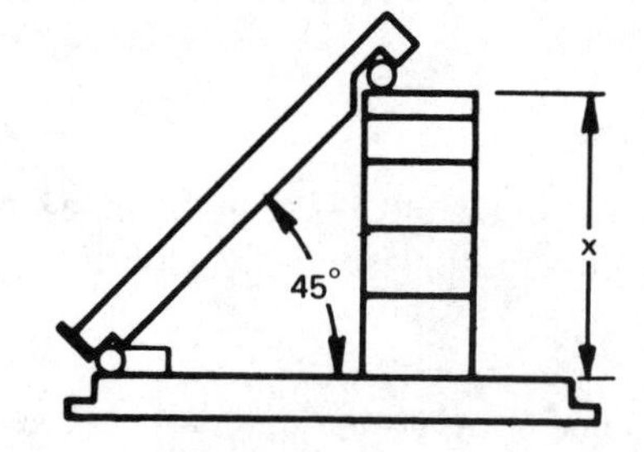

4. The width of a rectangular sheet of metal is equal to the area of the sheet divided by its length. Compute the area of a sheet which is 3¼ feet wide and 5½ feet long.

   a. Equation: ______

   b. Area = ______

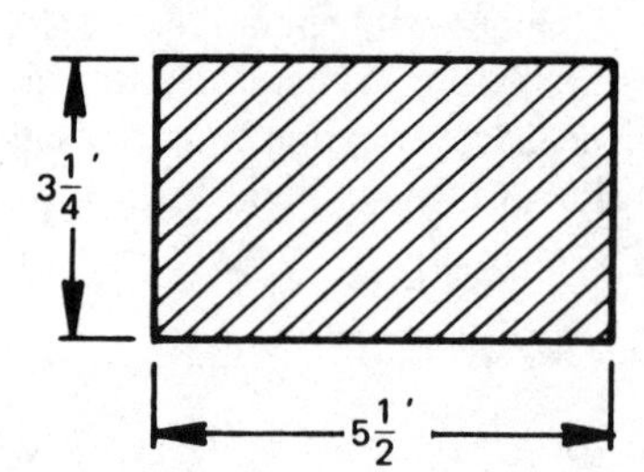

5. The depth of an American Standard thread divided by .6495 is equal to the pitch. Compute the depth of a thread with a .050-inch pitch.

   a. Equation: ________

   b. Depth = ________

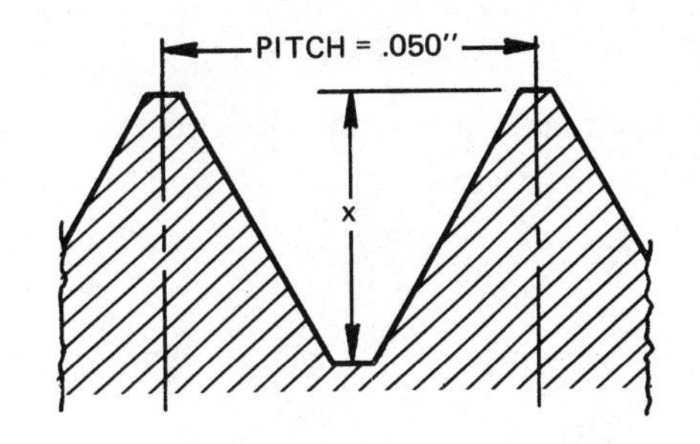

C. SOLUTION BY ROOTS

Solve each of the following equations by both the root principle of equality and transposition. Check all answers.

1. $x^3 = -64$ x = ______
2. $y^2 = 81$,   y = ______
3. $M^2 = 25$,   M = ______
4. $P^2 = 144$,   P = ______
5. $V^3 = 27$,   V = ______
6. $F^3 = .027$,   F = ______
7. $H^2 = 10{,}000$,   H = ______
8. $x^2 = \frac{4}{81}$,   x = ______
9. $y^3 = \frac{8}{27}$,   y = ______
10. $y^3 = \frac{-8}{27}$,   y = ______

D. SOLVE BY ROOTS

Write an equation for each of the following problems, solve for the unknown, and check.

1. The area of a square equals the length of a side squared. Given the areas of squares, compute the lengths of the sides.

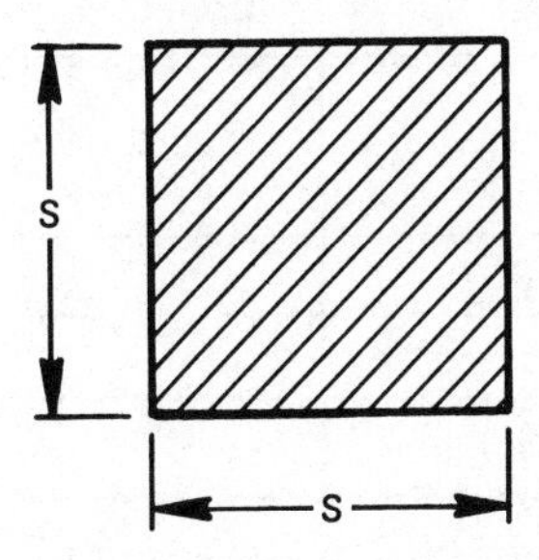

	Area	Equation	S
a.	36 square inches		
b.	25/81 square foot		
c.	1.44 square feet		
d.	49/121 square foot		
e.	.0049 square foot		

2. The volume of a cube equals the length of a side cubed. Given the volumes of cubes, compute the lengths of the sides.

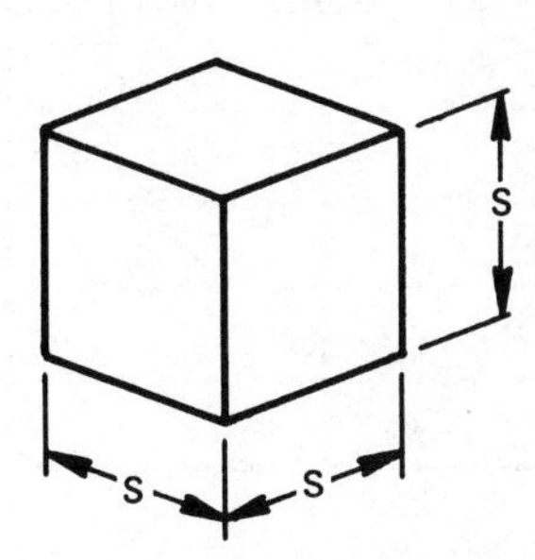

	Volume	Equation	S
a.	125 cubic inches		
b.	64/125 cubic foot		
c.	.064 cubic foot		
d.	1/216 cubic foot		
e.	.027 cubic foot		

E. SOLUTION BY POWERS

Solve each of the following equations by both the power principle of equality and transposition. Check all answers.

1. $\sqrt{x} = 5.6$, x = ______
2. $\sqrt{y} = 4$, y = ______
3. $\sqrt{P} = 9$, P = ______
4. $\sqrt[3]{V} = 5$, V = ______
5. $\sqrt[4]{S} = 3$, S = ______
6. $\sqrt[3]{T} = -4$, T = ______
7. $\sqrt{x} = 3.8$, x = ______
8. $\sqrt{y} = .01$, y = ______
9. $\sqrt{L} = \frac{5}{8}$, L = ______
10. $\sqrt[3]{V} = 0$, V = ______
11. $\sqrt[4]{M} = .1$, M = ______
12. $\sqrt[3]{C} = -4.2$, C = ______
13. $\sqrt{x} = \frac{1}{2}$, x = ______
14. $\sqrt{y} = 1\frac{7}{8}$, y = ______
15. $\sqrt[5]{P} = 1$, P = ______
16. $\sqrt[5]{P} = -1$, P = ______
17. $\sqrt[3]{F} = 2.3$, F = ______
18. $\sqrt{x} = \frac{3}{8}$, x = ______
19. $\sqrt[3]{y} = 3\frac{1}{2}$, y = ______
20. $\sqrt[4]{x} = .8$, x = ______

F. SOLVE BY POWERS

Write an equation for each of the following problems, solve for the unknown, and check.

1. The length of a side of a square equals the square root of the area. Given the lengths of the sides of squares, compute the areas.

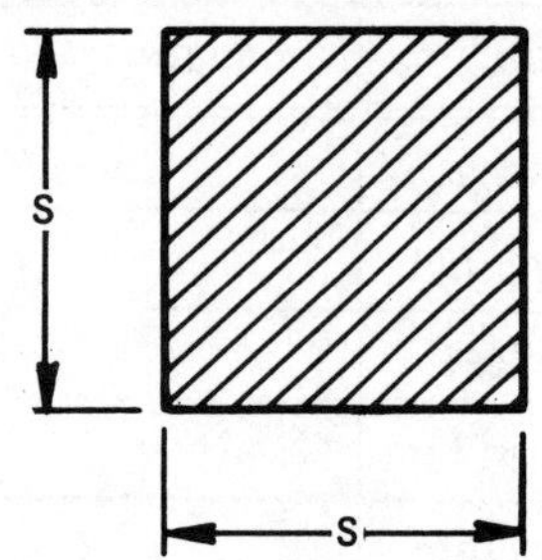

	Length of Side	Equation	Area
a.	2.8″		
b.	.75′		
c.	5/8′		
d.	6¼″		
e.	12.9″		

2. The length of a side of a cube equals the cube root of the volume. Given the lengths of the sides of cubes, compute the volumes.

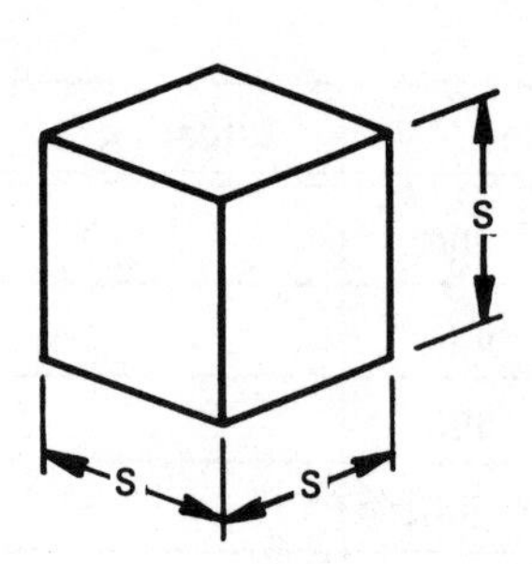

	Length of Side	Equation	Volume
a.	3.3″		
b.	.9′		
c.	3/4′		
d.	2.7″		
e.	4½″		

Unit 27 Solution of Equations Consisting of Combined Operations and Rearrangement of Formulas

OBJECTIVES

After studying this unit, the student should be able to

- Solve equations involving several operations.
- Rearrange formulas in terms of designated unknowns.
- Substitute numerical values for literal values in a formula, rearrange the formula, and solve for the unknown.

The purpose of solving an equation is to find the value of the unknown. In order to solve an equation which requires more than one operation between the unknown and known values, such as $2x - 3 = x + 5$, a definite step-by-step procedure must be followed. Use of the proper procedure results in the unknown standing alone on one side of the equation with its value on the other.

Note: Always solve for a positive unknown. A positive unknown may equal a negative value, but a negative unknown is not permitted. For example, $x = -7$ is correct, but $-x = 7$ is incorrect.

PROCEDURE FOR SOLVING EQUATIONS CONSISTING OF COMBINED OPERATIONS

It is essential that the steps used in solving an equation be taken in the following order. Some or all of these steps may be used depending upon the particular equation.

- Remove parentheses.
- Combine like terms on each side of the equation.
  a. Apply the addition and subtraction principles of equality to get all unknown terms on one side of the equation and all known terms on the other side.
  b. Combine like terms.
- Apply the multiplication and division principles of equality.
- Apply the power and root principles of equality.

Examples:

1. $5x + 7 = 22$

▲ Apply the subtraction principle

$$\begin{array}{rcl} 5x + 7 &=& 22 \\ -7 && -7 \\ \hline 5x &=& 15 \end{array}$$

▲ Apply the division principle.

$$\frac{5x}{5} = \frac{15}{5};\ x = 3$$

Check:
$$\begin{aligned} 5x + 7 &= 22 \\ 5(3) + 7 &= 22 \\ 15 + 7 &= 22 \\ 22 &= 22 \end{aligned}$$

2. $6x + 4x = 3x - 5x + 19 + 5$

▲ Combine like terms on the same side of the equation. $10x = -2x + 24$

▲ Apply the addition principle.

$$\begin{array}{rcl} 10x &=& -2x + 24 \\ +2x && +2x \\ \hline 12x &=& 24 \end{array}$$

▲ Apply the division principle.

$$\frac{12x}{12} = \frac{24}{12};\ x = 2$$

▲ Check:
$$\begin{aligned} 6x + 4x &= 3x - 5x + 19 + 5 \\ 6(2) + 4(2) &= 3(2) - 5(2) + 19 + 5 \\ 12 + 8 &= 6 - 10 + 19 + 5 \\ 20 &= 20 \end{aligned}$$

3. $9x + 7(x + 3) = 25$
   - ▲ Remove parentheses. $9x + 7x + 21 = 25$
   - ▲ Combine like terms on the same side of the equation. $16x + 21 = 25$
   - ▲ Apply the subtraction principle.

$$\begin{array}{rcl} 16x + 21 & = & 25 \\ -21 & & -21 \\ \hline 16x & = & 4 \end{array}$$

   - ▲ Apply the division principle.

$$\frac{16x}{16} = \frac{4}{16};\ x = \frac{1}{4}$$

   - ▲ Check: $9x + 7(x + 3) = 25$

$$9(\tfrac{1}{4}) + 7(\tfrac{1}{4} + 3) = 25$$

$$2\tfrac{1}{4} + 22\tfrac{3}{4} = 25$$

$$25 = 25$$

4. $-x = 14$
   - ▲ Apply the multiplication principle.

$$(-1)(-x) = (-1)(14)$$

$$x = -14$$

   - ▲ Check:

$$-x = 14$$

$$-(-14) = 14$$

$$14 = 14$$

5. $\frac{x^2}{4} - 32 = -23$
   - ▲ Apply the addition principle.

$$\begin{array}{rcl} \frac{x^2}{4} - 32 & = & -23 \\ +32 & & +32 \\ \hline \frac{x^2}{4} & = & 9 \end{array}$$

   - ▲ Apply the multiplication principle.

$$4(\tfrac{x^2}{4}) = 4(9);\ x^2 = 36$$

   - ▲ Apply the root principle.

$$\sqrt{x^2} = \sqrt{36};\ x = 6$$

   - ▲ Check: $\frac{x^2}{4} - 32 = -23$

$$\frac{6^2}{4} - 32 = -23$$

$$\frac{36}{4} - 32 = -23$$

$$9 - 32 = -23$$

$$-23 = -23$$

6. $6\sqrt[3]{x} = 4(\sqrt[3]{x} + 1.5)$
   - ▲ Remove parentheses.

$$6\sqrt[3]{x} = 4\sqrt[3]{x} + 6$$

   - ▲ Apply the subtraction principle.

$$\begin{array}{rcl} 6\sqrt[3]{x} & = & 4\sqrt[3]{x} + 6 \\ -4\sqrt[3]{x} & = & -4\sqrt[3]{x} \\ \hline 2\sqrt[3]{x} & = & 6 \end{array}$$

   - ▲ Apply the division principle.

$$\frac{2\sqrt[3]{x}}{2} = \frac{6}{2}$$

$$\sqrt[3]{x} = 3$$

   - ▲ Apply the power principle.

$$(\sqrt[3]{x})^3 = 3^3$$

$$x = 27$$

   - ▲ Check: $6\sqrt[3]{27} = 4(\sqrt[3]{27} + 1.5)$

$$6(3) = 4(3 + 1.5)$$

$$18 = 4(4.5)$$

$$18 = 18$$

REARRANGING FORMULAS

A formula that is given in terms of a particular value must sometimes be rearranged in terms of another value. The formula $A = 1/2bh$ which is given in terms of A may have to be rearranged in terms of h.

Consider the letter to be solved for as the unknown term and the other letters in the formula as the known values. The formula must be rearranged so that the unknown term is on one side of the equation and all other values are on the other side.

The formula is rearranged using the same procedure that is used for solving equations consisting of combined operations.

Examples: Given the following formulas, rearrange and solve for the designated letter.

1. $A = bh$; solve for h.

▲ Apply the division principle.

$$\frac{A}{b} = \frac{bh}{b} \quad \frac{A}{b} = h$$

2. $\ell = a + b$; solve for a, figure 27-1.

▲ Apply subtraction principle.

$$\begin{array}{l} \ell = a + b \\ \underline{-b \quad -b} \\ \ell - b = a \end{array}$$

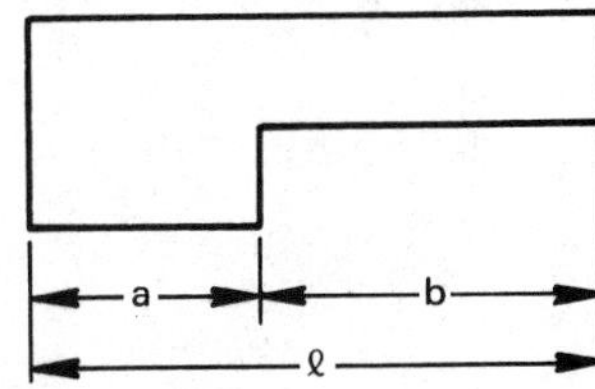

Fig. 27-1

3. $A = \pi(R^2 - r^2)$; solve for R, figure 27-2.

▲ Remove parentheses:

$A = \pi R^2 - \pi r^2$

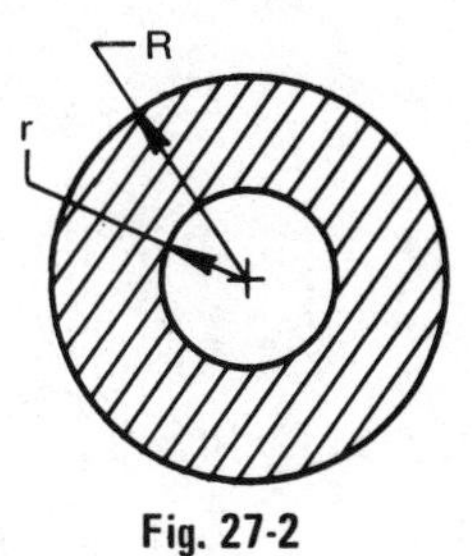

Fig. 27-2

▲ Apply addition principle.

$$\begin{array}{l} A \quad = \pi R^2 - \pi r^2 \\ \underline{+\pi r^2 \qquad\quad + \pi r^2} \\ A \quad + \pi r^2 = \pi R^2 \end{array}$$

▲ Apply division principle.

$$\frac{A + \pi r^2}{\pi} = \frac{\pi R^2}{\pi}; \quad \frac{A + \pi r^2}{\pi} = R^2$$

▲ Apply root principle.

$$\sqrt{\frac{A + \pi r^2}{\pi}} = \sqrt{R^2};$$

$$\sqrt{\frac{A + \pi r^2}{\pi}} = R$$

4. $M = D - 1.5155P + 3W$; solve for W, figure 27-3.

▲ Apply subtraction principle.

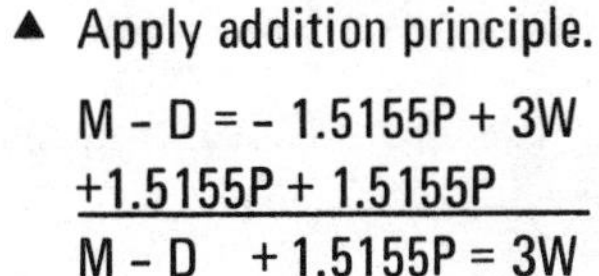

$$\begin{array}{l} M = D \quad - 1.5155P + 3W \\ \underline{-D - D} \\ M - D = - 1.5155P + 3W \end{array}$$

▲ Apply addition principle.

$$\begin{array}{l} M - D = - 1.5155P + 3W \\ \underline{+1.5155P + 1.5155P} \\ M - D \quad + 1.5155P = 3W \end{array}$$

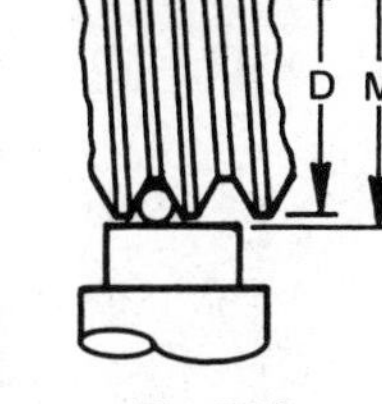

Fig. 27-3

▲ Apply division principle.

$$\frac{M - D + 1.5155P}{3} = \frac{3W}{3}$$

$$\frac{M - D + 1.5155P}{3} = W$$

5. $R - A = \sqrt{R^2 - .25B^2}$; solve for B, figure 27-4.

Note: Observe that both terms, R^2 and $.25B^2$ are enclosed within the radical sign. Therefore, neither term can be removed until the radical sign is eliminated.

▲ Apply the power principle by squaring both sides of the equation to eliminate the radical sign.

$$(R - A)^2 = (\sqrt{R^2 - .25B^2})^2$$

$$(R - A)^2 = R^2 - .25B^2$$

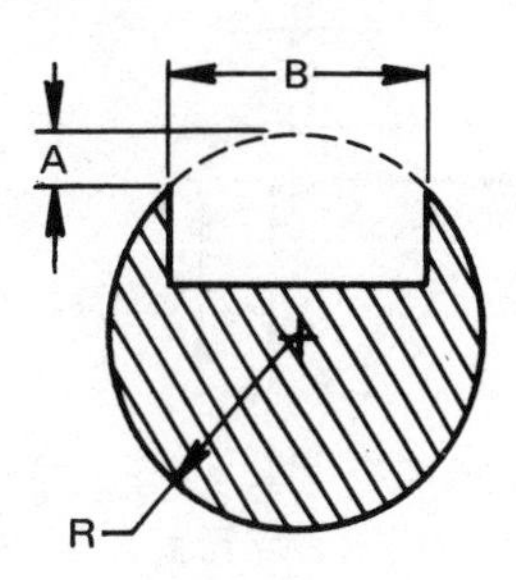

Fig. 27-4

▲ Apply subtraction principle.

$$\begin{array}{l} (R - A)^2 \quad = \quad R^2 - .25B^2 \\ \underline{\qquad - R^2 \qquad\quad -R^2} \\ (R - A)^2 - R^2 \quad = \quad - .25B^2 \end{array}$$

▲ Apply division principle.

$$\frac{(R - A)^2 - R^2}{-.25} = \frac{-.25B^2}{-.25}, \quad \frac{(R - A)^2 - R^2}{-.25} = B^2$$

▲ Apply root principle.

$$\sqrt{\frac{(R - A)^2 - R^2}{-.25}} = \sqrt{B^2}; \quad \sqrt{\frac{(R - A)^2 - R^2}{-.25}} = B$$

APPLICATION

A. EQUATIONS CONSISTING OF COMBINED OPERATIONS

Solve and check the following equations.

1. $8x - 14 = 2x + 28$ $\quad x =$ ____
2. $3x - 9 = 3$ $\quad x =$ ____
3. $5x + 5 + 2x = 40$ $\quad x =$ ____
4. $7y - 14 = 0$ $\quad y =$ ____
5. $3a + 2 = 42 - a$ $\quad a =$ ____
6. $5x + 16 = 30 - 13$ $\quad x =$ ____
7. $12 - (-y + 8) = 18$ $\quad y =$ ____
8. $3x + (2 - x) = 20$ $\quad x =$ ____
9. $-(2 + y) - (4 - 2y) = -6$ $\quad y =$ ____
10. $10(a - 2) = -5(a + 6)$ $\quad a =$ ____
11. $\frac{x}{7} + 8 = 5.9$ $\quad x =$ ____
12. $\frac{1}{8}x - 3(x - 7) = 5\frac{1}{8}x - 3$ $\quad x =$ ____
13. $18.78 - .02d = .58d$ $\quad d =$ ____
14. $8a^2 + 3a = 4a^2 + 3a + 36$ $\quad a =$ ____
15. $(\frac{a}{2})^3 + 34 = 42$ $\quad a =$ ____
16. $(y - 1)(y - 4) = 9 + y^2$ $\quad y =$ ____
17. $\sqrt{m + 2} = 6$ $\quad m =$ ____
18. $\sqrt[3]{y} + 5\sqrt[3]{y} - 3\sqrt[3]{y} = 15$ $\quad y =$ ____
19. $7\sqrt{x} = 3(\sqrt{x} + 8) + 8$ $\quad x =$ ____
20. $-4(x - 3) = 2\sqrt{x} - 4x$ $\quad x =$ ____

B. REARRANGING FORMULAS

The following formulas are used in machine trade calculations. Rearrange the formulas in terms of the designated values.

1. Given: $A = ab$
   a. $a =$ ______
   b. $b =$ ______
   Given: $d = \sqrt{a^2 + b^2}$
   c. $a =$ ______
   d. $b =$ ______

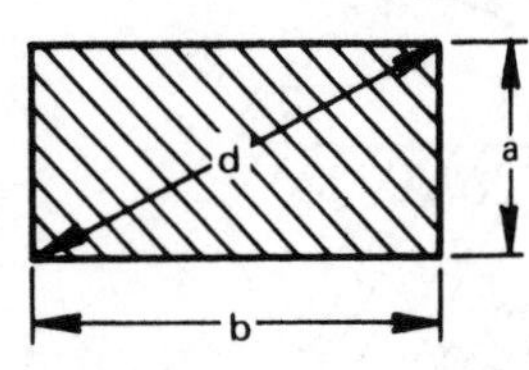

2. Given: $R = 1.155r$
   a. $r =$ ______
   Given: $A = 2.598R^2$
   b. $R =$ ______

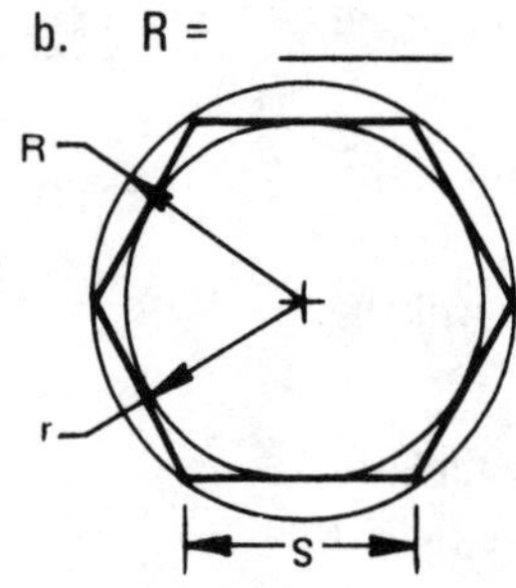

3. Given: $A = \pi(R^2 - r^2)$
   a. $R =$ ______
   b. $r =$ ______

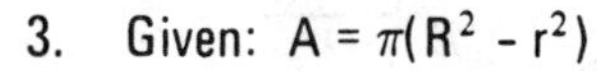

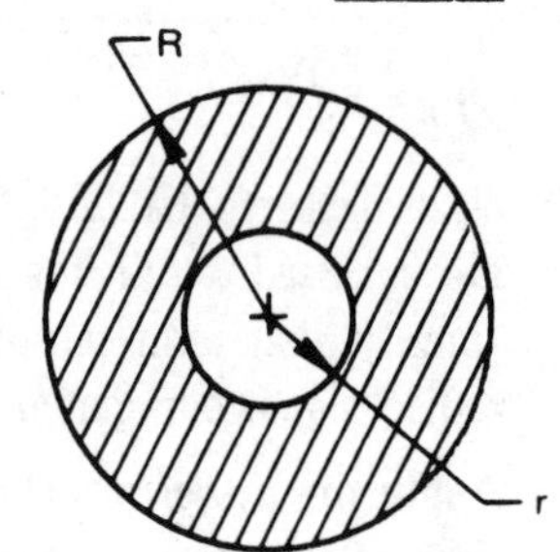

4. Given: $A + B + C = 180$
   a. $A =$ ______
   b. $B =$ ______
   c. $C =$ ______

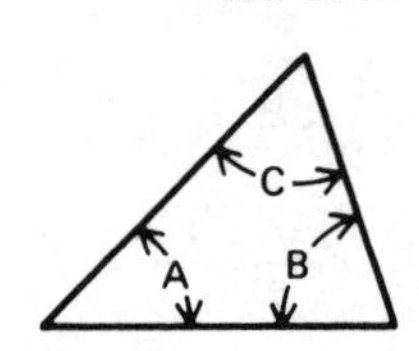

5. Given: $FW = \sqrt{Do^2 - D^2}$
   a. $Do =$ ______
   b. $D =$ ______
   Given: $Do = 2C - d + 2a$
   c. $d =$ ______
   d. $C =$ ______

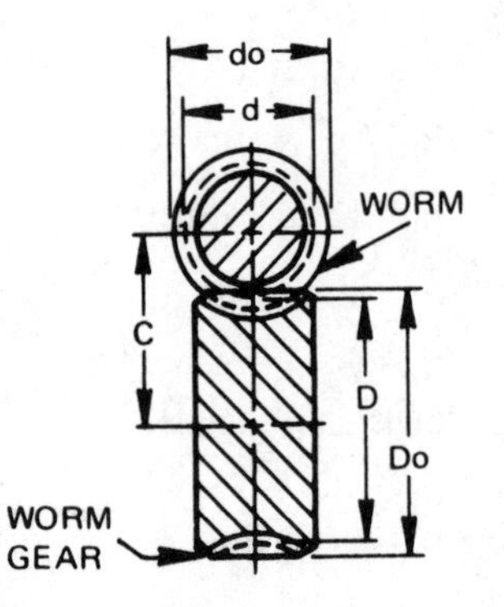

6. Given: $M = D - 1.5155P + 3W$

   a. $D =$ ________

   b. $P =$ ________

   c. $W =$ ________

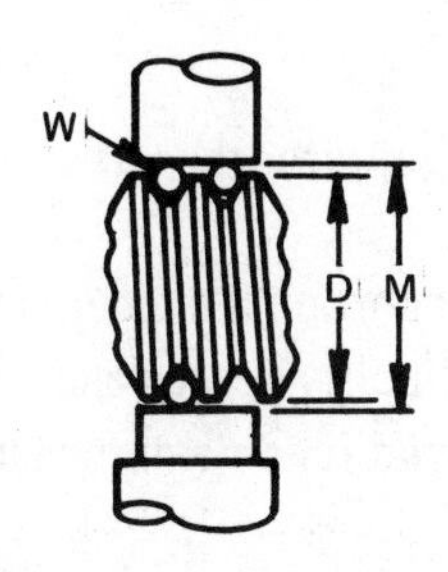

7. Given: $L = 3.14\,(.5D + .5d) + 2x$

   a. $D =$ ________

   b. $d =$ ________

   c. $x =$ ________

PULLEYS—OPEN BELT

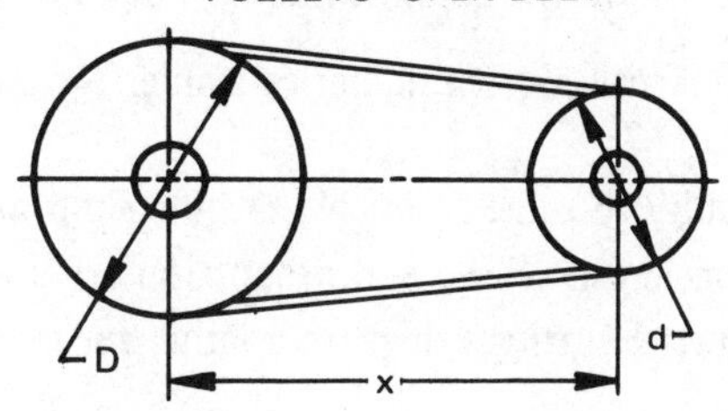

8. Given: $By\,(F - 1) = Cx$

   a. $x =$ ________

   b. $B =$ ________

   c. $F =$ ________

PLANETARY GEARING

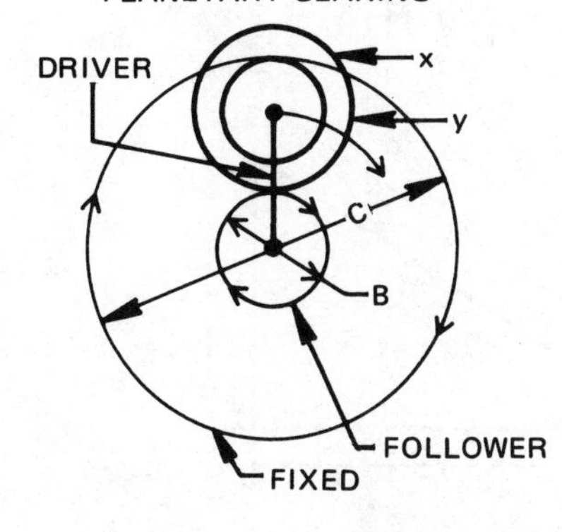

9. Given: $Ca = S\,(C - F)$

   a. $S =$ ________

   b. $F =$ ________

BEVEL GEAR

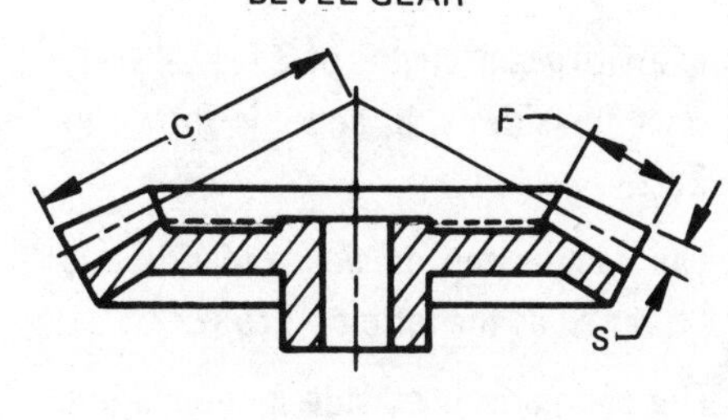

C. The following formulas are used in the machine trades. Substitute the given values in each formula, rearrange the formula, and solve for the unknown. Give answers to 3 decimal places.

1. $F = 2.38P + .25$

   Given: $F = 1.750$ $P =$ ______

2. $H.P. = .000016MN$

   Given: $H.P. = 22$, $N = 50.8$, $M =$ ______

3. $N = .707DPn$

   Given: $N = 24$, $Pn = 8$, $D =$ ______

4. $S = T - 1.732/N$

   Given: $S = .4134$, $N = 20$, $T =$ ______

5. $S = L/\ell\,(1/2\,(D - d))$

   Given: $S = 1/4$, $L = 16$

   $\ell = 4$, $d = 2\frac{1}{2}$, $D =$ ______

6. $W = St(.55d^2 - .25d)$

   Given: $W = 1150$, $d = .750$ $St =$ ______

7. $M = E - .866P + 3W$

   Given: $M = 3.3700$,

   $E = 3.2200$, $P = .125$, $W =$ ______

Unit 28 Ratio and Proportion

OBJECTIVES

After studying this unit, the student should be able to

- Compare quantities by putting them in ratio form.
- Reduce ratios to lowest terms.
- Solve for the unknown term of a proportion.
- Substitute given numerical values for symbols in a proportion and solve for the unknown term.

The ability to solve practical machine shop problems using ratio and proportion is a requirement for the skilled machinist. Ratio and proportion are used for calculating gear and pulley speeds and sizes, for computing thread cutting values on a lathe, for computing taper dimensions, and for determining machine cutting times.

DESCRIPTION OF RATIO

Ratio is the comparison of two like quantities.

Examples of ratios:

1. Figure 28-1 shows two pulleys. The comparison of the diameter of the first pulley to the diameter of the second is expressed as the ratio of 3 to 5.

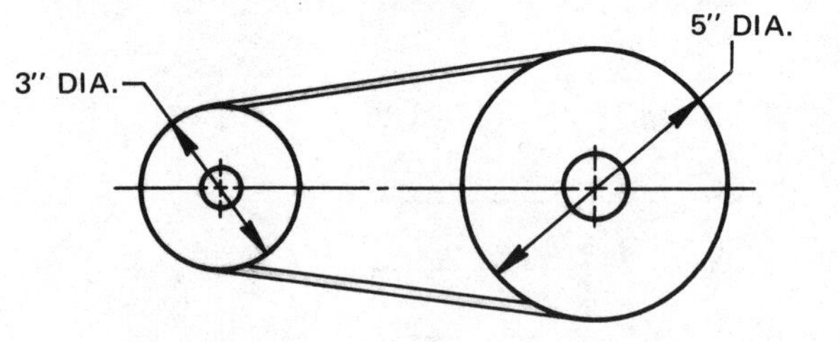

Fig. 28-1

2. A triangle with given lengths of 3 feet, 4 feet, and 5 feet for sides a, b, and c is shown in figure 28-2.
   a. The comparison of side a to side b is expressed as the ratio of 3 to 4.
   b. The comparison of side a to side c is expressed as the ratio of 3 to 5.
   c. The comparison of side b to side c is expressed as the ratio of 4 to 5.

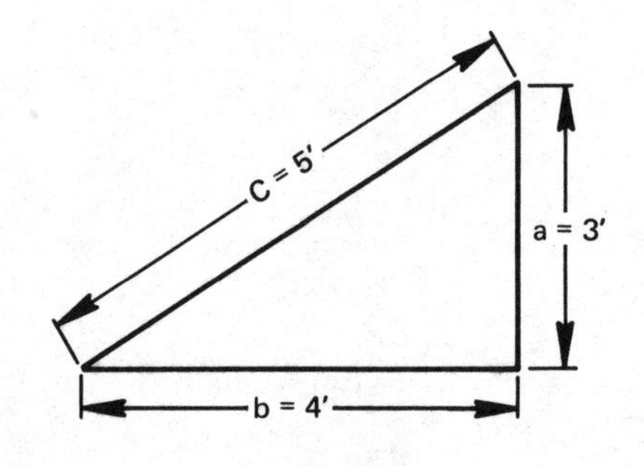

Fig. 28-2

The terms of a ratio are the two numbers that are compared. Both terms of a ratio must be expressed in the same units.

Examples:

1. The two pieces of bar stock shown in figure 28-3 cannot be expressed as a ratio until the 2-foot length is converted to 24 inches. The pieces are in the ratio of 11 to 24.

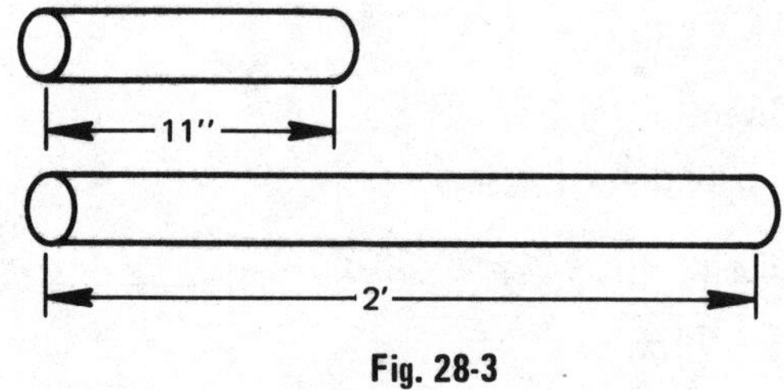

Fig. 28-3

2. It is impossible to express the two quantities shown in figure 28-4 as a ratio. Inches and pounds cannot be compared.

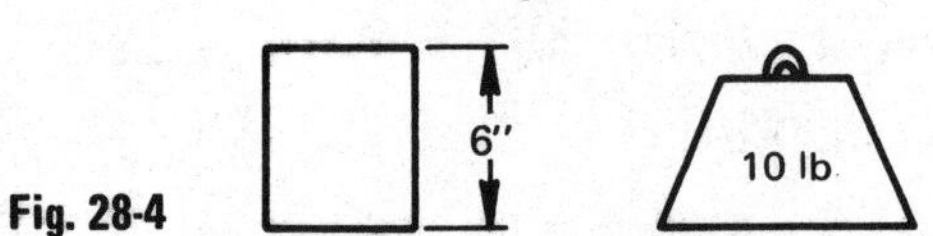

Fig. 28-4

Ratios are expressed in the following ways:

▲ With a colon between the two terms, such as 4 : 7. The colon means "to." Therefore, the ratio 4 : 7 is read "4 to 7."

▲ With a division sign separating the two numbers, such as $4 \div 7$ or as a fraction, $\frac{4}{7}$.

ORDER OF TERMS AND REDUCTION OF RATIOS

The terms of a ratio must be compared in the order in which they are given. The first term is the numerator of a fraction and the second is the denominator.

$$1 \text{ to } 3 = 1 \div 3 = \frac{1}{3} \qquad 3 \text{ to } 1 = 3 \div 1 = \frac{3}{1}$$

$$x : y = x \div y = \frac{x}{y} \qquad y : x = y \div x = \frac{y}{x}$$

Generally, a ratio should be reduced to its lowest terms.

$$3 : 9 = \frac{3}{9} = \frac{1}{3} \qquad 40 \text{ to } 15 = \frac{40}{15} = \frac{8}{3}$$

$$\frac{3}{8} \text{ to } \frac{9}{16} = \frac{3}{8} \div \frac{9}{16} = \frac{3}{8} \times \frac{16}{9} = \frac{2}{3} \qquad 10 : \frac{5}{6} = 10 \div \frac{5}{6} = \frac{10}{1} \times \frac{6}{5} = \frac{12}{1}$$

$$x^3 \text{ to } x^2 = \frac{x^3}{x^2} = \frac{x}{1} \qquad 4ab : 6a = \frac{4ab}{6a} = \frac{2b}{3}$$

DESCRIPTION OF PROPORTIONS

A *proportion* is an expression that states the equality of two ratios.

Proportions are expressed in the following two ways:

- 3 : 4 : : 6 : 8, which reads, "3 is to 4 as 6 is to 8."
- $\frac{3}{4} = \frac{6}{8}$. This equation form is generally the way that proportions are used in the machine trades.

A proportion consists of four terms. The first and the fourth term are called *extremes* and the second and third terms are called *means.*

2 : 3 :: 4 : 6, 2 and 6 are the extremes; 3 and 4 are the means.

$\frac{5}{6} = \frac{10}{12}$, 5 and 12 are the extremes; 6 and 10 are the means.

In a proportion the product of the means equals the product of the extremes. If the terms are cross multiplied, their products are equal.

Examples:

1. $\frac{3}{4} = \frac{6}{8}$. Cross multiply, $\frac{3}{4} \times \frac{6}{8}$; $3 \times 8 = 4 \times 6$; $24 = 24$

2. $\frac{a}{b} = \frac{c}{d}$. Cross multiply, $\frac{a}{b} \times \frac{c}{d}$; $a \times d = b \times c$, $ad = bc$

This method is used in solving proportions which have an unknown term. Since a proportion is an equation, the principles used for solving equations are applied in determining the value of the unknown after the terms have been cross multiplied.

Examples: Solve for the value of x:

1. $\frac{3}{4} = \frac{x}{16}$

▲ Cross multiply:

$4x = 3(16)$

$4x = 48$

▲ Apply the division principle of equality.

$\frac{4x}{4} = \frac{48}{4}$; $x = 12$

▲ Check:

$4x = 3(16)$

$4(12) = 3(16)$

$48 = 48$

2. $\frac{7}{x} = \frac{8}{15}$

▲ $8x = 7(15)$

$8x = 105$

▲ $\frac{8x}{8} = \frac{105}{8}$

$x = 13\frac{1}{8}$

▲ Check:

$8(x) = 7(15)$

$8(13\frac{1}{8}) = 7(15)$

$105 = 105$

3. $\frac{x}{7.5} = \frac{23.4}{20}$

▲ $20x = 7.5(23.4)$

$20x = 175.5$

▲ $\frac{20x}{20} = \frac{175.5}{20}$; $x = 8.775$

▲ Check: $20x = 7.5(23.4)$

$20(8.775) = 7.5(23.4)$

$175.5 = 175.5$

4. $\frac{a}{b} = \frac{c}{x}$

▲ $ax = bc$

▲ $\frac{ax}{a} = \frac{bc}{a}$

$x = \frac{bc}{a}$

▲ Check: $ax = bc$

$a(\frac{bc}{a}) = bc$

$bc = bc$

APPLICATION

A. RATIOS

Express the following ratios in fractional form. Reduce where possible

1.	6:15	____	5.	12′ to 46′	____	9.	$3a^2b$:6ab	____
2.	15:6	____	6.	3 lb. to 18 lb.	____	10.	xy:x^2y	____
3.	2:11	____	7.	17 mi. to 9 mi.	____	11.	2/3 to 1/2	____
4.	7:21	____	8.	156′ : 200′	____	12.	1/2 :2/3	____

B. RATIOS

Refer to the figures which follow and express the following ratios as number fractions. Reduce where possible.

1. Length A : Length B ____

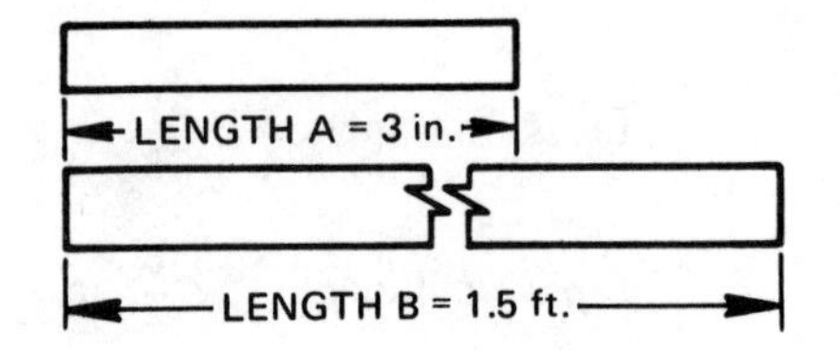

2. The diameters of pulleys E,F,G, and H are given in the accompanying table. Determine the ratios.

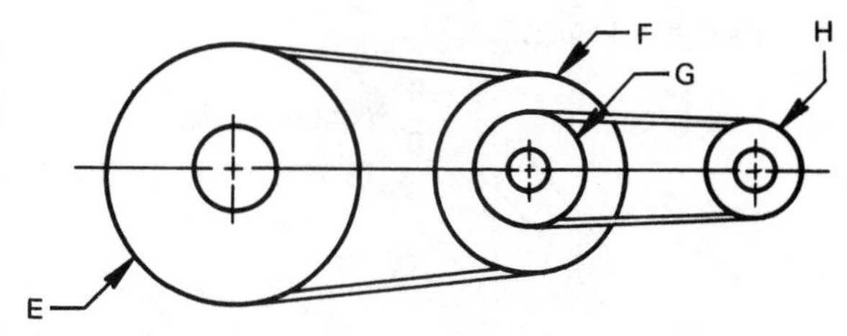

	Diameters (inches)				Ratios							
	E	F	G	H	E:F	E:G	E:H	F:G	F:H	G:H	G:E	H:F
a.	8	6	4	3								
b.	10	6	5	4								
c.	12	9	6	3								
d.	15	12	10	6								

3. Refer to the hole locations given for this plate and determine the ratios.

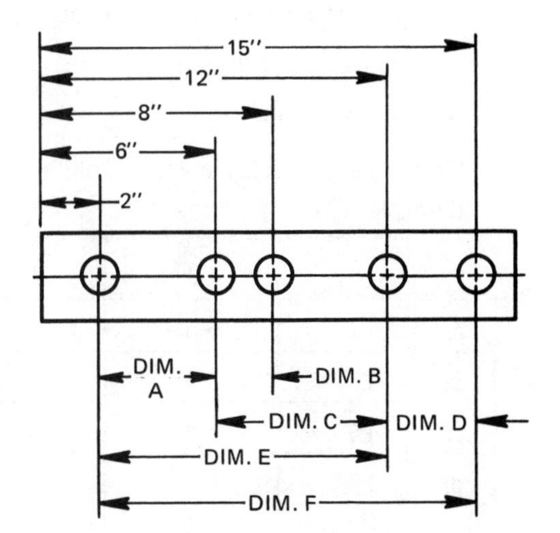

a. Dim. A to Dim. B ____
b. Dim. A to Dim. C ____
c. Dim. C to Dim. D ____
d. Dim. C to Dim. E ____
e. Dim. D to Dim. F ____
f. Dim. F to Dim. B ____
g. Dim. F to Dim. C ____
h. Dim. E to Dim. A ____
i. Dim. D to Dim. B ____
j. Dim. C to Dim. F ____

4. Rev. Gear A : Rev. Gear B ____

C. PROPORTIONS

Solve for x in each of the following proportions. Check.

1. $\frac{2}{3} = \frac{x}{21}$; x = ____

2. $\frac{7}{10} = \frac{x}{50}$; x = ____

3. $\frac{25}{4} = \frac{x}{12}$; x = ____

4. $\frac{19}{x} = \frac{1}{2}$; x = ____

5. $\frac{c^2d^6}{x} = \frac{cd}{m}$; x = ____

6. $\frac{x}{17.76} = \frac{19.3}{37.22}$; x = ____

7. $\frac{x}{.001} = \frac{200}{1}$; x = ____

8. $\frac{1/2}{3} = \frac{x}{5}$; x = ____

9. $\frac{3/8}{7/8} = \frac{1/4}{x}$; x = ____

D. PROPORTIONS

1. The proportion $\frac{A}{B} = \frac{C}{D}$ compares the sides of the two illustrated similar triangles. Determine the missing values in the table.

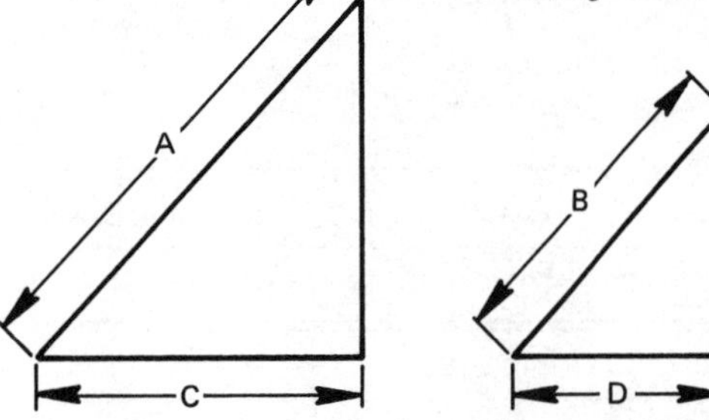

	A	B	C	D
a.	18″	4.5″		4″
b.	6½″	1⅝″	4½″	
c.	7/8″		3/4″	5/8″
d.		25.8″	20.6″	16.4″

2. Where machine parts are doweled in position, it is good practice to extend the pin 1 to 1 1/2 times its diameter into the mating part. The following proportion can be used to determine the length of the dowel extension or the dowel diameter. N/1 = L/D

   N = the number of times the pin extension is greater than the pin diameter

   L = the length of the pin extension

   D = the pin diameter

   Determine the value of each unknown in the table.

	N	D	L
a.	1¼	5/16	
b.	1⅛	1/2	
c.	1½		3/4
d.	1⅜		11/32
e.	1 5/16	5/8	

	N	D	L
f.	1.375		1.032
g.	1.250	.875	
h.	1.500	.3125	
i.	1.125		.281
j.	1.000	.750	

3. It is sometimes impractical to make engineering drawings full size. If the part to be drawn is very large or small, a scale drawing is generally made. The scale which is shown on the drawing compares the lengths of the lines on the drawing to the dimensions on the part.

   Examples:

   a. A scale on a drawing which states 1/4″ = 1″ means the drawing is one-quarter the size of the part. It is expressed as a ratio of 1:4 or 1/4.

   b. A scale drawing which states 2″ = 1″ means that the drawing is double the size of the part. It is expressed as a ratio of 2:1 or 2/1.

   The following proportion can be used to convert drawing lengths to part dimensions or part dimensions to drawing lengths.

$$\frac{\text{numerator of scale ratio}}{\text{denominator of scale ratio}} = \frac{\text{drawing length}}{\text{part dimension}}$$

The actual dimensions of a steel support are given in the figure. Compute the lengths on a drawing for each of the following problems.

	Scale	Drawing Length
a.	1/4" = 1"	B =
b.	4" = 1"	G =
c.	1/2" = 1"	B =
d.	2" = 1"	C =
e.	1½" = 1"	A =
f.	3/4" = 1"	E =
g.	3" = 1"	H =
h.	1/8" = 1"	F =

	Scale	Drawing Length
i.	1/2" = 1"	E =
j.	6" = 1"	G =
k.	3/4" = 1"	F =
l.	1½" = 1"	C =
m.	1/4" = 1"	F =
n.	3" = 1"	G =
o.	1/2" = 1"	B =
p.	2" = 1"	A =

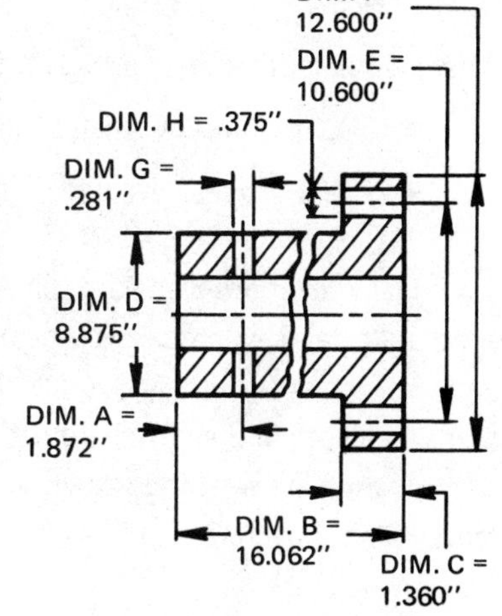

4. The proportion given below is used for lathe thread cutting computations using simple gearing when the fixed stud gear and the spindle gear have the same number of teeth. The figure below shows the relationship of gears in a lathe using a simple gear train.

$$\frac{N_L}{N_C} = \frac{T_S}{T_L}$$

N_L = number of threads per inch on the lead screw

N_C = number of threads per inch to be cut

T_S = number of teeth on stud gear

T_L = number of teeth on lead screw gear

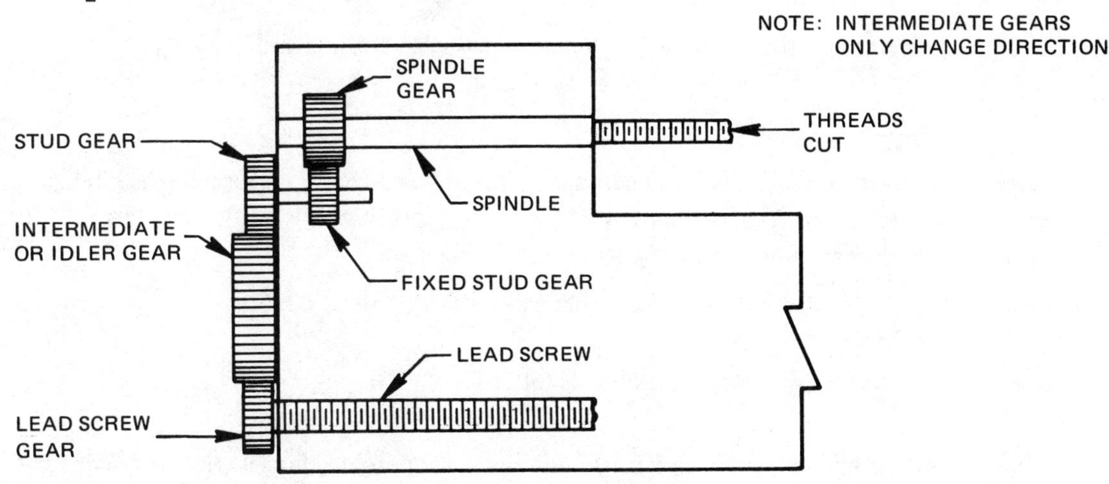

Determine the missing values for each of the following problems:

a. N_L = 4, N_C = 8, T_S = 32, T_L = ________

b. N_L = 7, T_S = 35, N_C = 15, T_L = ________

c. N_C = 10, N_L = 6, T_L = 40, T_S = ________

d. N_L = 8, T_L = 42, T_S = 28, N_C = ________

Unit 29 Direct and Inverse Proportions

OBJECTIVES

After studying this unit, the student should be able to

- Determine by analyzing a problem whether quantities are directly or inversely proportional.
- Set up and solve direct and inverse proportions.

Many shop problems are solved by the use of proportions. A machinist may be required to convert word statements or other given data into proportions. Generally, three of the four terms of a proportion must be known in order to solve the proportion. When setting up a proportion it is important that the terms be placed in their proper positions.

A problem which is to be set up and solved as a proportion must first be analyzed in order to determine where the terms are to be placed. Proportions are either direct or inverse.

DIRECT PROPORTIONS

Two quantities are *directly proportional* if a change in one produces a change in the other in the same direction. If an increase in one produces an increase in the other, or if a decrease in one produces a decrease in the other, the two quantities are directly proportional.

When setting-up a direct proportion in fractional form, the numerator of the first ratio must correspond to the numerator of the second ratio. The denominator of the first ratio must correspond to the denominator of the second ratio.

Examples of Direct Proportions:

1. If a machine produces 120 parts in 2 hours, how many parts does it produce in 3 hours?

▲ Analyze the problem: An increase in time produces an increase in production.
The proportion is direct.

▲ Set up the proportion: Let x represent the number of parts that are produced in 3 hours.

$$\frac{2 \text{ hours}}{3 \text{ hours}} = \frac{120 \text{ parts}}{x \text{ parts}}$$

Note: The numerator of the first ratio corresponds to the numerator of the second ratio: 120 parts are produced in 2 hours. The denominator of the first ratio corresponds to the denominator of the second ratio; x parts are produced in 3 hours.

▲ Solve for x: $\frac{2}{3} = \frac{120}{x}$; $2x = 3(120)$; $2x = 360$; $x = \frac{360}{2}$; $x = 180$

▲ Check: $\frac{2}{3} = \frac{120}{x}$; $\frac{2}{3} = \frac{120}{180}$; $2(180) = 3(120)$: $360 = 360$

2. A tapered shaft is one that varies uniformly in diameter along its length. The shaft shown in figure 29-1 is 15 inches long with a 1.200-inch diameter on the large end. A 9-inch piece is cut from the shaft as shown. Determine the diameter at the large end of the 9-inch piece.

▲ Analyze the problem:

As the length decreases from 15 inches to 9 inches, the diameter also decreases. The proportion is direct.

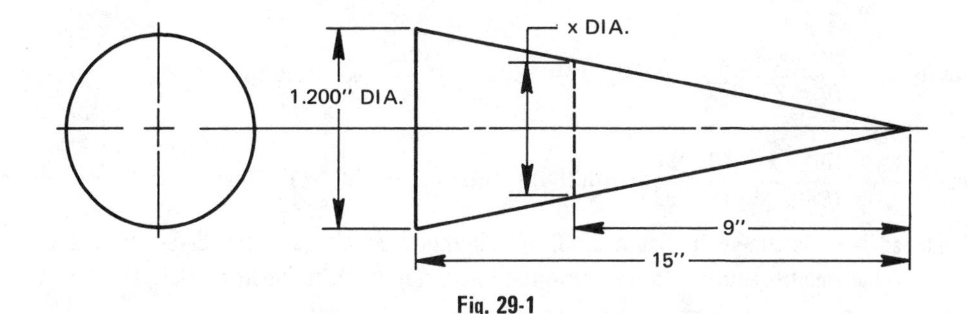

Fig. 29-1

▲ Set up the proportion: Let x represent the diameter at the large end of the 9-inch piece.

$$\frac{15 \text{ inches in length}}{9 \text{ inch in length}} = \frac{1.200 \text{ inches in dia.}}{x \text{ inches in dia.}}$$

Note: The numerator of the first ratio corresponds to the numerator of the second ratio; the 15-inch piece has a 1.200-inch diameter at the large end. The denominator of the first ratio corresponds to the denominator of the second ratio; the 9-inch piece has a diameter of x inches at the large end.

▲ Solve for x: $\frac{\overset{5}{\cancel{15}}}{\underset{3}{\cancel{9}}} = \frac{1.200}{x}$; $5x = 3(1.200)$; $5x = 3.600$; $x = \frac{3.600}{5}$; $x = .720$ inch

▲ Check: $\frac{15}{9} = \frac{1.200}{x}$; $\frac{15}{9} = \frac{1.200}{.720}$; $15(.720) = 9(1.200)$; $10.800 = 10.800$

INVERSE PROPORTIONS

Two quantities are *inversely or indirectly proportional* if a change in one produces a change in the other in the opposite direction. If an increase in one produces a decrease in the other, or if a decrease in one produces an increase in the other, the two quantities are inversely proportional.

When setting up an inverse proportion in fractional form, the numerator of the first ratio must correspond to the denominator of the second ratio. The denominator of the first ratio must correspond to the numerator of the second ratio.

Examples of Inverse Proportions:

1. Two gears in mesh are shown in figure 29-2. The driver gear has 40 teeth and revolves at 360 rpm. Determine the rpm of a driven gear with 16 teeth.

▲ Analyze the problem:

The gear with 16 teeth revolves at greater rpm than the gear with 40 teeth. A decrease in the number of teeth produces an increase in rpm. The proportion is inverse.

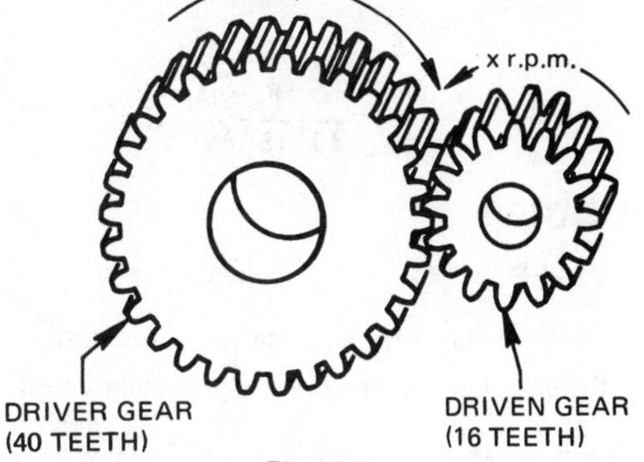

Fig. 29-2

▲ Set up the proportion: Let x represent the rpm of the gear with 16 teeth.

$$\frac{40 \text{ teeth}}{16 \text{ teeth}} = \frac{x \text{ rpm}}{360 \text{ rpm}}$$

Note: The numerator of the first ratio corresponds to the denominator of the second ratio; the gear with 40 teeth revolves 360 rpm. The denominator of the first ratio corresponds to the numerator of the second ratio; the gear with 16 teeth revolves x rpm.

▲ Solve for x: $\frac{\overset{5}{\cancel{40}}}{\underset{2}{\cancel{16}}} = \frac{x}{360}$; $2x = 5(360)$; $2x = 1800$; $x = \frac{1800}{2}$; $x = 900$ rpm

▲ Check: $\frac{40}{16} = \frac{x}{360}$; $\frac{40}{16} = \frac{900}{360}$; $40(360) = 16(900)$; $14{,}400 = 14{,}400$

2. A balanced lever is shown in figure 29-3. A 40-pound weight is placed 6 feet from the fulcrum. Determine the weight required 15 feet from the fulcrum in order to balance the lever.

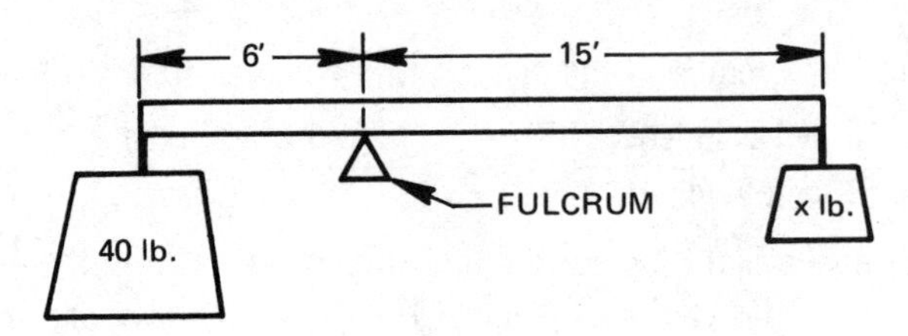

Fig. 29-3

▲ **Analyze the problem:** The 40 pound weight is closer to the fulcrum than the unknown weight. An increase in the distance from the fulcrum produces a decrease in weight required to balance the lever. The proportion is inverse.

▲ Set up the proportion: Let x represent the weight 15 feet from the fulcrum.

$$\frac{6 \text{ feet}}{15 \text{ feet}} = \frac{x \text{ pounds}}{40 \text{ pounds}}$$

Note: The numerator of the first ratio corresponds to the denominator of the second ratio: the 40-pound weight is 6 feet from the fulcrum. The denominator of the first ratio corresponds to the numerator of the second ratio; the unknown weight is 15 feet from the fulcrum.

▲ Solve for x: $\frac{\overset{2}{\cancel{6}}}{\underset{5}{\cancel{15}}} = \frac{x}{40}$; $5x = 2(40)$; $5x = 80$, $x = \frac{80}{5}$; $x = 16$ pounds

▲ Check: $\frac{6}{15} = \frac{x}{40}$; $\frac{6}{15} = \frac{16}{40}$; $6(40) = 16(15)$; $240 = 240$

APPLICATION

A. TAPERS

Taper is the difference between the diameters at each end of a part. Tapers are expressed as the difference in diameters for a particular length along the centerline of a part.

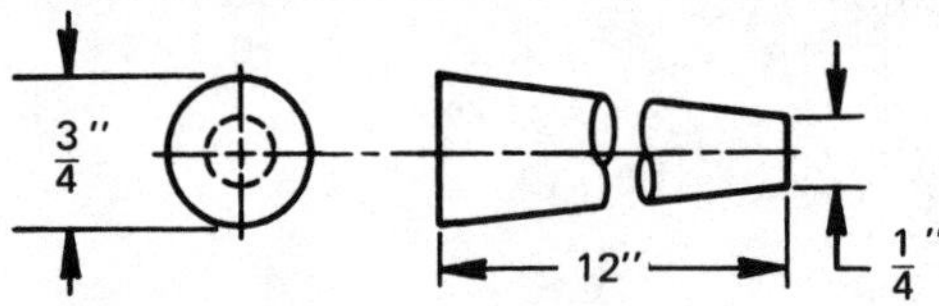

The taper of this shaft is 3/4" - 1/4" = 1/2" taper per foot.

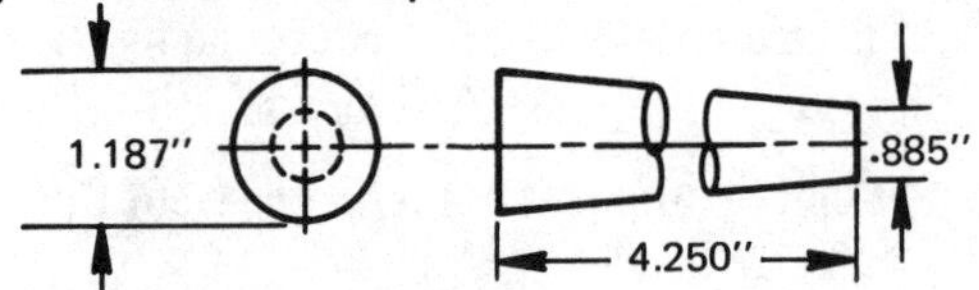

The taper of this shaft is 1.187" - .885" = .302" taper per 4.250".

1. A plug gage tapers .120" along a 1.500" length. Set up a proportion and determine the amount of taper in the workpiece for each of the following problems. Give answers to 3 decimal places.

	Workpiece Thickness	Proportion	Taper in Workpiece
a.	.800"		
b.	1.250"		
c.	9/16"		
d.	1 1/8"		
e.	.937"		

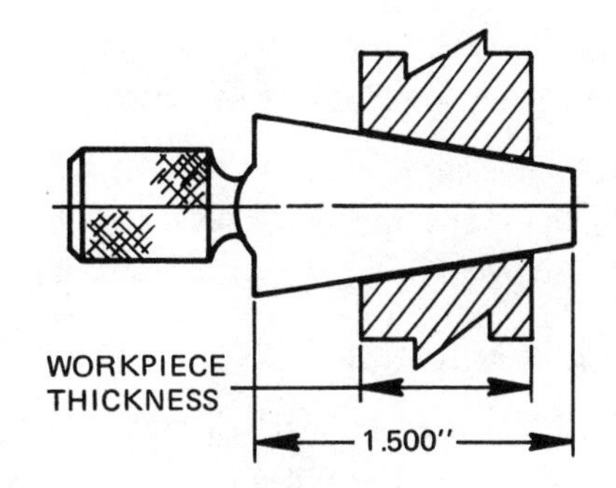

2. A reamer tapers .130" along a 4.250" length. Set up a proportion and determine length A for each of the following problems. Give answers to 3 decimal places.

	Taper in Length A	Proportion	Length A
a.	.030"		
b.	.108"		
c.	.075"		
d.	.008"		
e.	.093"		

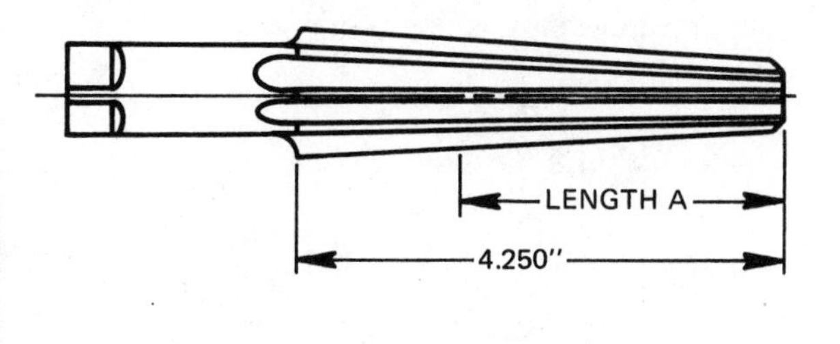

3. A micrometer reading is made at dimension D on a tapered shaft. For each of the problems use the dimensions given in the table, compute the taper, set up a proportion, and determine diameter C to 4 decimal places.

	Length of Shaft	Diameter A	Diameter B	Dimension D	Diameter C
a.	10.000"	1.500"	.700"	6.500"	
b.	8.750"	1.250"	.375"	4.875"	
c.	5 1/2"	1 1/16"	5/8"	3 3/8"	
d.	14 3/4"	2 1/4"	1"	11 1/4"	
e.	9.200"	1.325"	.410"	8.620"	

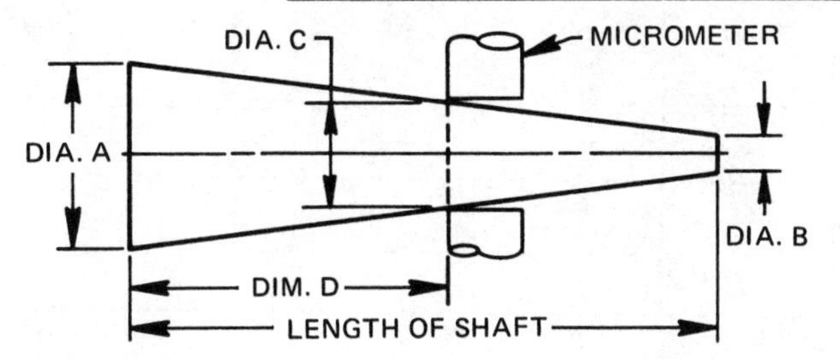

B. Set up proportions for the following problems and solve.

1. A sheet of steel 8¼ feet long weighs 350 pounds. A piece 2½ feet long is sheared from the sheet. Determine the weight of the 2½-foot piece to 2 decimal places. ________

2. A machine produces 1,350 parts in 6.75 hours. How many parts are produced by the machine in 8.25 hours? ________

3. Three machines produce at the same rate. The three machines produce 720 parts in 1.6 hours. How many hours would it take two machines to produce 720 parts? ____________

4. Two forgings are made of the same stainless steel alloy. A forging which weighs 168 pounds contains .85 pound of chromium. How many pounds of chromium does the second forging contain if it weighs 216 pounds? Give answer to 2 decimal places. ____________

C. GEARS AND PULLEYS

1. A belt connects a 10-inch diameter pulley which rotates at 160 rpm with a 6.5-inch diameter pulley. An 8-inch diameter pulley is fixed to the same shaft as the 6.5-inch pulley. A belt connects the 8-inch pulley with a 3.5-inch diameter pulley. Determine the rpm of the 3.5-inch diameter pulley. Give answer to 1 decimal place. ____________

2. Of two gears that mesh, the one which has the greater number of teeth is called the gear, and the one which has the fewer teeth is called the pinion. For each of the problems, set up a proportion, and determine the unknown value.

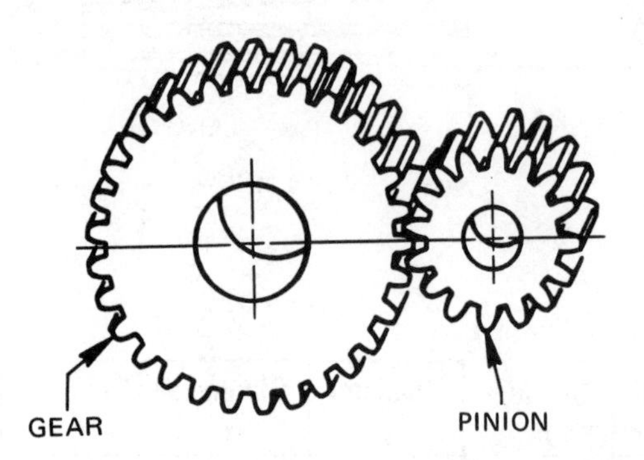

	Number of Teeth on Gear	Number of Teeth on Pinion	Rpm of Gear	Rpm of Pinion
a.	48	20	120	
b.	32	24		210
c.	35		160	200
d.		15	150	250
e.	54	28	80	

3. The figure shows a compound gear train. Gears B and C are keyed to the same shaft; therefore, they turn at the same rpm. Gear A and gear C are driving gears. Gear B and gear D are driven gears. Set up proportions for each problem and determine the unknown values.

	Number of Teeth				RPM			
	Gear A	Gear B	Gear C	Gear D	Gear A	Gear B	Gear C	Gear D
a.	80	30	50	20	120			
b.	60		45		100	300		450
c.		24	60	36	144			280
d.	55	25		15			175	350

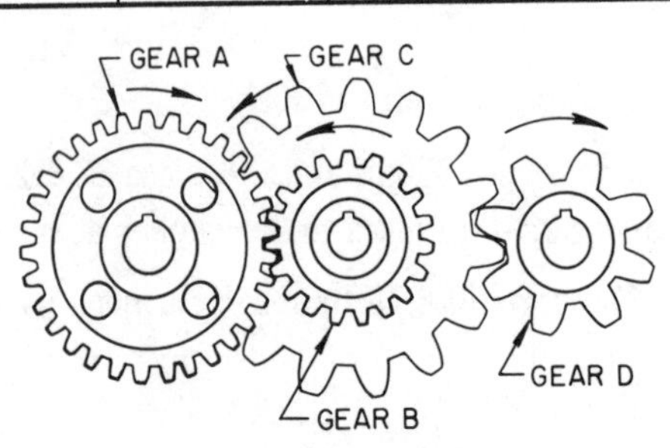

Unit 30 Applications of Formulas to Cutting Speeds, Rpm, and Cutting Time

OBJECTIVES

After studying this unit, the student should be able to

- Solve cutting speed, rpm and cutting time problems by substitution in given formulas.
- Solve production time and cutting feed problems by rearranging and combining formulas.

In order to perform cutting operations efficiently, a machine must be run at the proper cutting speed. Proper cutting speed is largely determined by the type of material that is being cut, the feed and depth of cut, the cutting tool, and the machine characteristics.

The machinist must be able to determine proper cutting speeds by using trade handbook data and formulas.

CUTTING SPEED – LATHE

The *cutting speed* of a lathe is the number of feet that the revolving workpiece travels past the cutting edge of the tool in one minute. The speed of a revolving object equals the product of the circumference times the number of revolutions per minute made by the object.

Circumference = 3.1416 x diameter; cutting speed = 3.1416 x diameter x rpm.

Note: Generally, diameters are expressed in inches. In order to convert speed in inches per minute to feet per minute, it is necessary to divide by 12.

Cutting Speed Formula: $C = \frac{3.1416\,DN}{12}$

C = cutting speed in feet per minute

D = diameter in inches

N = revolutions per minute

Example: A steel shaft 2.5 inches in diameter is turned in a lathe at 184 rpm. Determine the cutting speed to 1 decimal place.

▲ $C = \frac{3.1416\,DN}{12}$

▲ $C = \frac{3.1416(2.5)(184)}{12}$

▲ C = 120.4 feet per minute

REVOLUTIONS PER MINUTE – LATHE

The cutting speed formula is rearranged in terms of N in order to determine the revolutions per minute of a workpiece.

▲ $C = \frac{3.1416\,DN}{12}$

▲ $12C = 3.1416\,DN$

▲ $\frac{12C}{3.1416\,D} = N$

Formula for computing rpm: $N = \frac{12C}{3.1416\ D}$

Example: An aluminum cylinder with a 6" outside diameter is turned in a lathe at a cutting speed of 225 feet per minute. Determine the rpm to 1 decimal place.

▲ $N = \frac{12C}{3.1416\ D}$

▲ $N = \frac{12(225)}{3.1416(6)}$

▲ $N = 143.2$ rpm

CUTTING AND SURFACE SPEED – MILLING MACHINE, DRILL PRESS, AND GRINDER

Cutting speed, surface speeds, and revolutions per minute of drills, milling cutters, and grinding wheels are computed using the same formulas that are used in lathe computations. On the lathe, the workpiece revolves. On drill presses, milling machines, and grinders the tool revolves; speeds and rpm are computed in reference to the tool rather than the workpiece.

The cutting speed or surface speed of a drill press, milling machine, and grinder is the number of feet that a point on the circumference of the tool travels in 1 minute.

Examples:

1. A 10-inch diameter grinding wheel runs at 1910 rpm. Determine the surface speed to the closest whole number

   ▲ $C = \frac{3.1416\ DN}{12}$

   ▲ $C = \frac{3.1416(10)(1910)}{12}$

   ▲ $C = 5000$ feet per minute

2. Determine the cutting speed to the closest whole number of a 3½-inch diameter milling cutter revolving at 120 rpm.

   ▲ $C = \frac{3.1416\ DN}{12}$

   ▲ $C = \frac{3.1416(3.5)(120)}{12}$

   ▲ $C = 110$ feet per minute

REVOLUTIONS PER MINUTE – MILLING MACHINE, DRILL PRESS, AND GRINDER

The formula for determining the number of revolutions per minute is derived by rearranging the cutting speed formula in terms of N.

Examples:

1. A 1/2-inch diameter twist drill has a cutting speed of 60 feet per minute. Determine the rpm to 1 decimal place.

   ▲ $N = \frac{12C}{3.1416\ D}$ ▲ $N = \frac{12(60)}{3.1416(.5)}$ ▲ $N = 458.4$ rpm

2. A 6-inch diameter grinding wheel operates at a cutting speed of 6,000 feet per minute. Determine the rpm to 1 decimal place.

▲ $N = \frac{12C}{3.1416\ D}$

▲ $N = \frac{12(6000)}{3.1416(6)}$

▲ $N = 3819.7$ rpm

CUTTING TIME

The same formula is used to compute cutting times for machines which have a revolving workpiece, such as the lathe, as is used for machines which have a revolving tool, such as the milling machine and drill press.

Cutting time is determined by the length or depth to be cut in inches, the revolutions per minute of the revolving workpiece or revolving tool, and the tool feed in inches for each revolution of the workpiece or tool.

Formula for computing cutting time per cut: $T = \frac{L}{FN}$

T = cutting time per cut in minutes

L = length of cut in inches

F = tool feed in inches per revolution

N = rpm of revolving workpiece or tool

CUTTING TIME – LATHE

Examples:

1. How many minutes are required to take one cut 22 inches in length on a steel shaft when the lathe feed is .050 inch per revolution and the shaft turns 152 rpm?

$$T = \frac{L}{FN};\quad T = \frac{22}{.050(152)};\quad T = 2.9 \text{ minutes}$$

2. What is the total cutting time required to turn a 3.250-inch diameter cast iron sleeve which is 20 inches long to a 2.450-inch diameter? Roughing cuts are each made to a .125 inch depth of cut. One finish cut using a .025 inch depth of cut is made. The feed is .100 inch per revolution for roughing and .030 inch for finishing. Roughing cuts are made at 150 rpm and the finish cut at 200 rpm.

▲ Compute the total number of cuts required: $\frac{3.250 - 2.450}{2} = \frac{.800}{2} = .400$ total depth of cut

Divide the total depth of cut by the roughing depth of cut: $.400 \div .125 = 3$ roughing cuts plus .025 inch for the finish cut.

▲ Time required for roughing: $T = \frac{L}{FN};\ T = \frac{20}{(.100)(150)} =$ 1.33 minute for each cut

Total time for roughing = 3 x 1.33 = 4.0 minutes

▲ Time required for finishing: $T = \frac{L}{FN};\ T = \frac{20}{(.030)(200)} =$ 3.3 minutes

▲ Total cutting time = 4.0 + 3.3 = 7.3 minutes

CUTTING TIME – MILLING MACHINE AND DRILL PRESS

Examples:

1. Determine the cutting time required to drill through a workpiece which is 3.600 inches thick with a drill revolving 300 rpm and a feed of .025 inch per revolution.

$$T = \frac{L}{FN};\ T = \frac{3.600}{.025(300)};\ T = .48 \text{ minutes}$$

2. A milling machine cutter makes 460 rpm with a table feed of .020 inch per revolution. Four cuts are required to mill a slot in an aluminum plate 28 inches long. Compute the total cutting time.

▲ Compute cutting time per cut: $T = \frac{L}{FN};\ T = \frac{28}{.020(460)};\ T = 3.04$ minutes per cut

▲ Compute total cutting time: $4(3.04) = 12.16$ minutes

APPLICATION

A. Given the workpiece or tool diameters and the rpm, determine the cutting speeds in the tables to the closest whole number.

	Workpiece or Tool Diameter	Revolutions per Minute	Cutting Speed(fpm)
1.	.500"	460	
2.	2.750"	50	
3.	4.000"	86	
4.	.875"	175	
5.	1.750"	218	

	Workpiece or Tool Diameter	Revolutions per Minute	Cutting Speeds(fpm)
6.	7.750"	59	
7.	2.125"	764	
8.	.125"	1525	
9.	5.250"	254	
10.	.250"	4584	

B. Given the cutting speed and the tool or workpiece diameter, determine the rpm in the tables to the closest whole number.

	Cutting Speed(fpm)	Workpiece or Tool Diameter	Revolutions per Minute
1.	70	2.375"	
2.	120	.750"	
3.	90	8.000"	
4.	180	8.000"	
5.	225	.375"	

	Cutting Speed(fpm)	Workpiece or Tool Diameter	Revolutions per Minute
6.	400	1.000"	
7.	325	2.625"	
8.	80	.250"	
9.	550	.625"	
10.	450	4.500"	

C. Given the number of cuts, the length of cut, the rpm of the workpiece or tool, and the tool feed, determine the total cutting time in the table to 1 decimal place.

	Number of Cuts	Feed (per revolution)	Length of cut	Revolution per Minute	Total cutting time (minutes)
1.	1	.002"	20"	2100	
2.	1	.0045"	37"	610	
3.	4	.008"	8"	335	

D. CUTTING SPEED AND SURFACE SPEED PROBLEMS

Compute the following problems; give answers to the closest whole number.

1. A 3½-inch diameter high-speed steel cutter, running at 54.5 rpm, is used to rough mill a steel casting. What is the cutting speed? ____________
2. A 2-inch diameter carbon steel drill running at 286 rpm is used to drill an aluminum plate. Find the cutting speed. ____________
3. What is the surface speed of a 16-inch diameter surface grinder wheel running at 1194 rpm? ____________
4. A medium-steel shaft is cut in a lathe using a high-speed steel tool. Determine the cutting speed if the shaft is 2.125 inches in diameter, turning at 262 rpm. ____________
5. A finishing cut is taken on a brass workpiece using a 4-inch diameter carbon steel milling cutter. What is the cutting speed when the cutter is run at 86 rpm? ____________

E. RPM PROBLEMS

Compute the following problems; give answer to the closest whole number.

1. Grooves are cut in a stainless steel plate using a 3.750-inch diameter carbide milling cutter with a cutting speed of 180 feet per minute. Determine the rpm. ____________
2. An annealed cast iron housing is drilled with a cutting speed of 70 feet per minute using a 3/4-inch diameter carbon steel drill. Find the rpm. ____________
3. A grinding operation is performed using a 6-inch diameter wheel with a cutting speed of 5500 feet per minute. Determine the rpm. ____________
4. Determine the rpm of an aluminum alloy rod 1 inch in diameter with a cutting speed of 550 feet per minute. ____________
5. A high-speed steel milling cutter with a 1.750-inch diameter and a cutting speed of 40 feet per minute is used for a roughing operation on an annealed chromium-nickel steel workpiece. Find the rpm. ____________

F. CUTTING TIME PROBLEMS

Compute the following problems; give answer to 1 decimal place.

1. Cast iron 3½ inches in diameter is turned in a lathe at 270 rpm. Each length of cut is 27 inches and five cuts are required. A carbide tool is fed into the work at .015 inch per revolution. What is the total cutting time? ____________
2. A slot 32 inches long is cut into a carbon steel baseplate with a feed of .030 inch per revolution. Find the cutting time using a 3-inch diameter carbide milling cutter running at 640 rpm. ____________
3. Fifteen 1/8-inch diameter holes each 2¼ inches deep are drilled in an aluminum workpiece. The high-speed steel drill runs at 9200 rpm with a feed of .002 inch per revolution. Determine the total cutting time. ____________
4. Thirty 2-inch diameter stainless steel shafts are turned in a lathe at 240 rpm. Two cuts each 14.5 inches long are required using a feed of .020 inch per revolution. Setup and handling time averages 3 minutes per piece. Calculate the total production. ____________

5. Seven brass plates 9 inches wide and 21 inches long are machined with a milling cutter along the length of the plates. The entire top face of each plate is milled. The width of each cut allowing for overlap is 2¼ inches. Using a feed of .020 inch per revolution and 525 rpm determine the total cutting time. __________

G. The solution of the following problems requires more than one formula or the rearrangement of formulas.

1. A 3-inch diameter cylinder is turned for an 11.5-inch length of cut. The cutting speed is 300 feet per minute and the cutting time is 1.02 minutes. Calculate the tool feed in inches per revolution. __________

2. A combination drilling and countersinking operation on bronze round stock is performed on an automatic screw machine. The length of cut per piece is 1¾ inches. The total cutting time for 2300 pieces is 6½ hours running at 1600 rpm. What is the tool feed in inches per revolution? __________

3. Fifteen hundred 1¼-inch diameter steel shafts are turned on an automatic machine. One finishing operation is required for a 16.5-inch length of cut. The tool feed is .015 inch per revolution using a cutting speed of 200 feet per minute. Determine the number of hours of cutting time required for the 1500 parts. __________

4. What is the diameter of a carbide milling cutter used for machining a 22-inch length of cut in stainless steel? The cutting time is 11.95 minutes with a cutting speed of 180 feet per minute and a feed of .010 inch per revolution. __________

5. What is the total number of hours required to produce 850 aluminum baseplates that are 1½ inches thick? Six 1/4-inch diameter holes are drilled in each plate using a feed of .004 inch per revolution and a cutting speed of 300 feet per minute. Setup and handling time is estimated at .5 minutes per piece. __________

Unit 31 Applications of Formulas to Spur Gears

OBJECTIVES

After studying this unit, the student should be able to

- Identify the proper gear formula to use depending on the unknown and the given data.
- Compute gear part dimensions by substituting known values directly into formulas.
- Compute gear part dimensions by rearranging given formulas in terms of the unknowns.
- Compute gear part dimensions by the application of two or more formulas in order to determine an unknown.

Gears have wide application in machine technology. They are basic to the design and operation of machinery. Most machine shops are equipped to cut gears, and some shops specialize in gear design and manufacture.

It is essential that the machinist and draftsman have an understanding of gear parts and the ability to determine gear dimensions by the use of trade handbook formulas.

DESCRIPTION OF GEARS

Gears are used for transmitting power by rotary motion between shafts. Gears are designed to prevent slippage and to insure positive motion while maintaining a high degree of accuracy of the speed ratios between driving and driven gears.

The shape of the gear tooth is of primary importance in providing a smooth transmission of motion. The shape of most gear teeth is called an *involute curve.*

SPUR GEARS

Spur gears are gears that are in mesh between parallel shafts. Of two gears in mesh, the smaller gear is called the pinion and the larger gear is called the gear.

Spur gears and the terms that are applied to these gears are shown in figures 31-1, 31-2, and 31-3. It is essential to study the figures and gear terms before computing gear problems by the use of formulas.

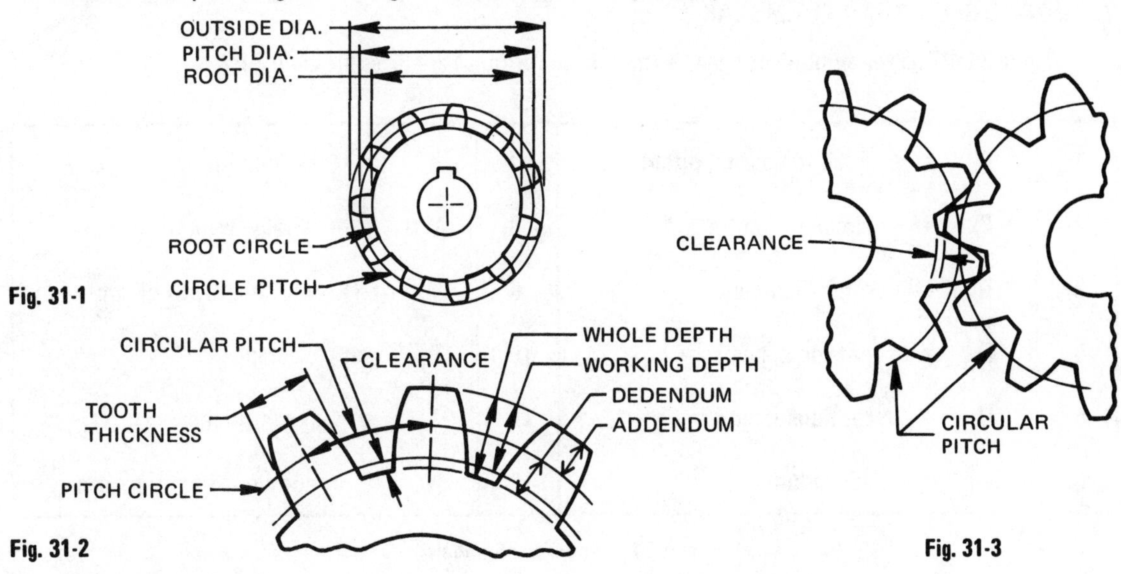

Fig. 31-1

Fig. 31-2

Fig. 31-3

SPUR GEAR DEFINITIONS

- *Pitch Circles* are the imaginary circles of two meshing gears that make contact with each other. The circles are the basis of gear design and gear calculations.
- *Pitch Diameter* is the diameter of the pitch circle.
- *Root Circle* is a circle which coincides with the bottoms of the tooth spaces.
- *Root Diameter* is the diameter of the root circle.
- *Outside Diameter* is the diameter measured to the tops of the gear teeth.
- *Addendum* is the height of the tooth above the pitch circle.
- *Dedendum* is the depth of the tooth space below the pitch circle.
- *Whole Depth* is the total depth of the tooth space. It is equal to the addendum plus the dedendum.
- *Working Depth* is the total depth of mating teeth when two gears are in mesh. It is equal to twice the addendum.
- *Clearance* is the distance between the top of a tooth and the bottom of the mating tooth space of two gears in mesh. It is equal to the whole depth minus the working depth.
- *Tooth Thickness* is the straight line distance from the two points where the pitch circle intersects the edges of a tooth.
- *Circular Pitch* is the length of the arc measured on the pitch circle between the centers of two adjacent teeth. It is equal to the circumference of the pitch circle divided by the number of teeth on the gear.
- *Diametral Pitch (Pitch)* is the ratio of the number of gear teeth to the number of inches of pitch diameter. It is equal to the number of gear teeth for each inch of pitch diameter.

When the pitch of a gear is mentioned, the reference is to diametral pitch, rather than circular pitch. For example, if a gear has 28 teeth and a pitch diameter of 4 inches, it has a pitch (diametral pitch) of 28/4 or 7. It has 7 teeth per inch of pitch diameter, and it is called a 7-pitch gear. It will only mesh with other 7-pitch gears. Gears must have the same pitch in order to mesh.

STANDARD SPUR GEAR FORMULAS

Table 31-1 lists the symbols for gear terms whose formulas are given in table 31-2

1.	P	=	pitch (diametral pitch)	7.	d	=	dedendum
2.	P_C	=	circular pitch	8.	WD	=	whole depth of tooth
3.	D	=	pitch diameter	9.	W_d	=	working depth of tooth
4.	D_O	=	outside diameter	10.	C_ℓ	=	clearance
5.	D_R	=	root diameter	11.	T	=	tooth thickness
6.	a	=	addendum	12.	N	=	number of teeth on gear

Table 31-1 Spur Gear Symbols

	To Find	Formula		To Find	Formula
1.	Pitch	$P = \frac{3.1416}{P_C}$	13.	Dedendum	$d = \frac{1.157}{P}$
2.	Pitch	$P = \frac{N}{D}$	14.	Dedendum	$d = .3683P_C$
3.	Circular Pitch	$P_C = \frac{3.1416D}{N}$	15.	Whole Depth	$WD = \frac{2.157}{P}$
4.	Circular Pitch	$P_C = \frac{3.1416}{P}$	16.	Whole Depth	$WD = .6866P_C$
5.	Pitch Diameter	$D = \frac{N}{P}$	17.	Whole Depth	$WD = a + d$
6.	Pitch Diameter	$D = \frac{NP_C}{3.1416}$	18.	Working Depth	$W_D = \frac{2.000}{P}$
7.	Outside Diameter	$D_O = \frac{N + 2}{P}$	19.	Working Depth	$W_D = .6366P_C$
8.	Outside Diameter	$D_O = \frac{P_c(N + 2)}{3.1416}$	20.	Clearance	$C_\ell = \frac{.157}{P}$
9.	Outside Diameter	$D_O = D + 2a$	21.	Clearance	$C_\ell = .050P_C$
10.	Root Diameter	$D_R = D - 2d$	22.	Tooth Thickness	$T = \frac{1.5708}{P}$
11.	Addendum	$a = \frac{1}{P}$	23.	Number of Teeth	$N = P\,D$
12.	Addendum	$a = .3183P_c$	24.	Number of Teeth	$N = \frac{3.1416D}{P_c}$

Table 31-2 Spur Gear Formulas

GEAR CALCULATIONS

Most gear calculations are made by identifying the proper formula which is given in terms of the unknown, and substituting the known dimensions. It is sometimes necessary to rearrange a formula in terms of a particular unknown. The solution of a problem may require the substitution of values in two or more formulas.

Examples of Gear Calculations: (Refer to tables 31-1 and 31-2.)

1. Determine the pitch diameter of a 5-pitch gear which has 28 teeth.
   - ▲ Identify the formula whose parts consist of pitch diameter, pitch, and number of teeth: $D = N/P$
   - ▲ Solve: $D = \frac{N}{P}$; $D = \frac{28}{5}$; $D = 5.6000$ inches
2. Determine the outside diameter of a gear which has 16 teeth and a circular pitch of .7854 inch.
   - ▲ Identify the formula whose parts consist of outside diameter, number of teeth, and circular pitch:

$$D_o = \frac{P_c(N + 2)}{3.1416}$$

   - ▲ Solve: $D_o = \frac{P_c(N + 2)}{3.1416}$; $D_o = \frac{.7854(16 + 2)}{3.1416}$; $D_o = \frac{.7854(18)}{3.1416}$; $D_o = \frac{14.1372}{3.1416}$; $D_o = 4.5000$ inches

3. Determine the circular pitch of a gear with a whole depth dimension of .3081 inch.
   - ▲ Identify the formula whose parts consist of circular pitch and whole depth: $WD = .6866P_c$.
   - ▲ The formula must be rearranged in terms of circular pitch: $P_c = \frac{WD}{.6866}$
   - ▲ Solve: $P_c = \frac{WD}{.6866}$; $P_c = \frac{.3081}{.6866}$; $P_c = .4487$
4. Determine the addendum of a gear which has an outside diameter of 3.0000 inches and a pitch diameter of 2.7500 inches.
   - ▲ Identify the formula whose parts consist of addendum, outside diameter, and pitch diameter; $D_o = D + 2a$.
   - ▲ The formula must be rearranged in terms of the addendum: $D_o = D + 2a$; $D_o - D = 2a$; $\frac{D_o - D}{2} = a$.
   - ▲ Solve: $a = \frac{D_o - D}{2}$; $a = \frac{3.0000 - 2.7500}{2}$; $a = .1250$ inch
5. Determine the working depth of a gear which has 46 teeth and a pitch diameter of 11.5000 inches.
   - ▲ There is no single formula in table 31-2 that consists of working depth, number of teeth, and pitch diameter. Therefore, it is necessary to substitute in two formulas in order to solve the problem.
   - ▲ Refer to table 31-2 and observe $W_d = \frac{2.0000}{P}$ and $P = \frac{N}{D}$.
   - ▲ Given N and D, find P: $P = \frac{N}{D}$; $P = \frac{46}{11.5000}$; $P = 4$.
   - ▲ Substitute 4 for P and solve for W_d in the formula $W_d = \frac{2.0000}{P}$:

$$W_d = \frac{2.000}{P};\ W_d = \frac{2.000}{4};\ W_d = .5000 \text{ inch}$$

APPLICATION

A. Refer to tables 31-1 and 31-2, choose the proper formula, substitute the given dimensions, and determine the value of the unknown for each problem in the table below.

	Pitch	Circular Pitch	Pitch Diameter	Number of Teeth	Find	
1.		1.5708"			Pitch =	______
2.	12				Circular Pitch =	______
3.			5.2000"	26	Circular Pitch =	______
4.			12.5714"	44	Pitch =	______
5.	7			25	Pitch Diameter =	______
6.		.3142"		12	Pitch Diameter =	______
7.	20				Circular Pitch =	______
8.			.7273"	16	Pitch =	______
9.	12		1.1667"		Number of Teeth =	______
10.		.6283"	8.4000"		Number of Teeth =	______

B. Refer to tables 31-1 and 31-2, choose the proper formula, substitute the given dimensions and determine the value of the unknown for each problem in the table below.

	Number of Teeth	Pitch	Circular Pitch	Pitch Diameter	Addendum	Dedendum	Find
1.	56	7					Outside Diameter = ____
2.		14					Addendum = ____
3.				1.3333"		.0643"	Root Diameter = ____
4.		3.5					Whole Depth = ____
5.			.2856"				Working Depth = ____
6.			1.2566"				Clearance = ____
7.		20					Tooth Thickness = ____
8.				3.5000"	.1818"		Outside Diameter = ____
9.			.0924"				Dedendum = ____
10.			.8976"				Addendum = ____
11.					.1429"	.1653"	Whole Depth = ____
12.		4					Dedendum = ____
13.			.2094"				Whole Depth = ____
14.		17					Clearance = ____
15.		9					Working Depth = ____

C. Refer to tables 31-1 and 31-2. The formula in terms of the unknown is not given in table 31-2. Choose the formula that consists of the given parts, rearrange it in terms of the unknown, and solve.

	Addendum	Outside Diameter	Number of Teeth	Circular Pitch	Working Depth	Pitch Diameter	Find
1.	.0769"						Circular Pitch = ____
2.	.0666"						Pitch = ____
3.	.2000"	4.8000"					Pitch Diameter = ____
4.		2.7144"	17				Pitch = ____
5.		4.3750"		.3927"			Number of Teeth = ____
6.					.0769"		Pitch = ____
7.					.5000"		Circular Pitch = ____
8.		4.7144"				4.4286"	Addendum = ____

D. Refer to tables 31-1 and 31-2. No single formula is given in table 31-2 which consists of the given parts and the unknown. Two or more formulas, some in rearranged form, must be used in solving the problem.

	Number of Teeth	Pitch Diameter	Pitch	Addendum	Find
1.	72	6.0000"			Addendum =
2.	44	3.6667"			Dedendum =
3.	10	2.5000"			Whole Depth =
4.	90	12.8571"			Working Depth =
5.		1.0625"	16		Outside Diameter =
6.		2.9167"	12		Root Diameter =
7.	29	2.0714"			Root Diameter =
8.	75	6.8182"			Clearance =
9.				.1429"	Tooth Thickness =
10.		1.0455"		.0455"	Number of Teeth =

E. Backlash is the amount that a tooth space is greater than the engaging tooth on the pitch circles of two gears. Average backlash = .030/P

Determine the average backlash of

1. a 6-pitch gear ______
2. a 13-pitch gear ______
3. a 3.5-pitch gear ______
4. a gear with a whole depth of .2696" ______
5. a gear with a working depth of .1176" ______
6. a gear with a pitch diameter of 4.800" and 24 teeth ______

PITCH CIRCLE
PITCH CIRCLE
BACKLASH

F. The center distance of a pinion and a gear is the distance between the centers of the pitch circles.

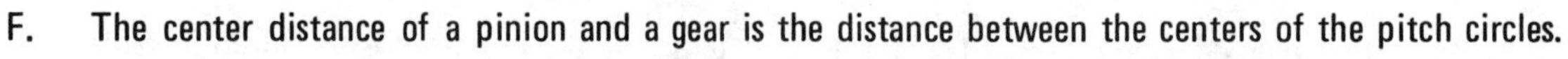

$$\text{Center distance} = \frac{\text{pitch diameter of gear} + \text{pitch diameter of pinion}}{2}$$

Determine the center distance of

1. a pinion with a pitch diameter of 2.8333 inches and a gear with a pitch diameter of 4.1667 inches. ______
2. a pinion with a pitch diameter of 4.8889 inches and a gear with a pitch diameter of 8.6667 inches. ______
3. a 9-pitch pinion and gear; the pinion has 23 teeth and the gear has 38 teeth. ______
4. a 16-pitch pinion and gear; the pinion has 18 teeth and the gear has 44 teeth. ______
5. a gear and pinion with a circular pitch of .1745 inch; the gear has 55 teeth and the pinion has 37 teeth. ______

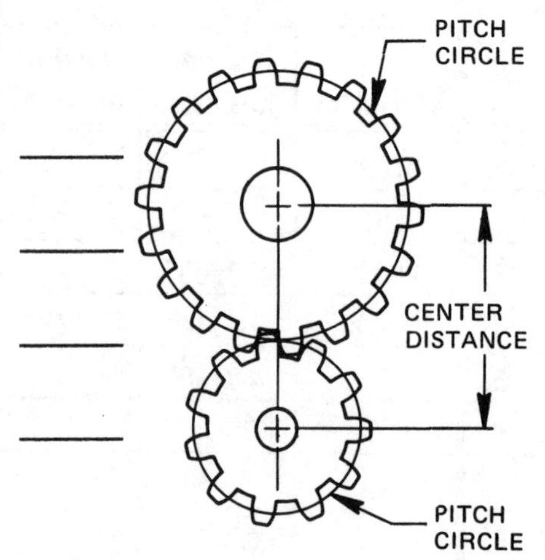

SECTION IV
Fundamentals of Plane Geometry

Unit 32 Introduction to Geometric Figures

OBJECTIVES

After studying this unit, the student should be able to

- Add, subtract, multiply, and divide angles in terms of degrees, minutes, and seconds.
- Convert decimal degrees to degrees, minutes, and seconds.

The fundamental principles of geometry generally applied to machine shop problems are those used to make the calculations required for machining parts from engineering drawings. An engineering drawing is an example of applied geometry.

Since geometry is fundamental to machine technology, it is essential to understand the definitions and terms of geometry. It is equally important to be able to apply the geometric principles in problem solving. The methods and procedures used in problem solving are the same as those required for the planning, making, and checking of machined parts.

Rule: To solve a geometry problem

- Study the figure.
- Relate it to the principle or principles that are needed for the solution.
- Base all conclusions on fact: given information and geometric principles.
- Do not assume that something is true because of its appearance or because of the way it is drawn.

  Note: The same requirements are applied in reading engineering drawings.

PLANE GEOMETRY

Plane geometry is the branch of mathematics that deals with points, lines, and various figures that are made of combinations of points and lines. The figures lie on a flat surface, or *plane*. Examples of plane geometry are the views of a part as shown on an engineering drawing.

AXIOMS

In the study of geometry certain basic statements called *axioms* are assumed to be true.

The following statements are geometric axioms:

- Things equal to the same thing, or to equal things, are equal to each other. Equals may be substituted for equals.
- If equals are added to or subtracted from equals, the sums or remainders are equal.
- If equals are multiplied or divided by equals, the products or quotients are equal.
- The whole is equal to the sum of its parts.
- Only one straight line can be drawn between two given points.
- Through a given point, only one line can be drawn parallel to a given line.
- Two straight lines can intersect at one point only.

LINES

The word *line,* as it is used in this book, always means a straight line. A line other than a straight line, such as a curved line, will be identified.

- *Parallel lines* do not meet regardless of how far they are extended. They are the same distance apart *(equidistant)* at all points. The symbol ∥ means parallel. In figure 32-1 line AB is parallel to line CD; therefore, AB and CD are equidistant (distance x) at all points.
- *Perpendicular lines* meet or intersect at a right, or 90°, angle. The symbol ⊥ means perpendicular. Figure 32-2 shows examples of perpendicular lines.
- *Oblique lines* meet or intersect at an angle other than 90°. Figure 32-3 shows examples of oblique lines.

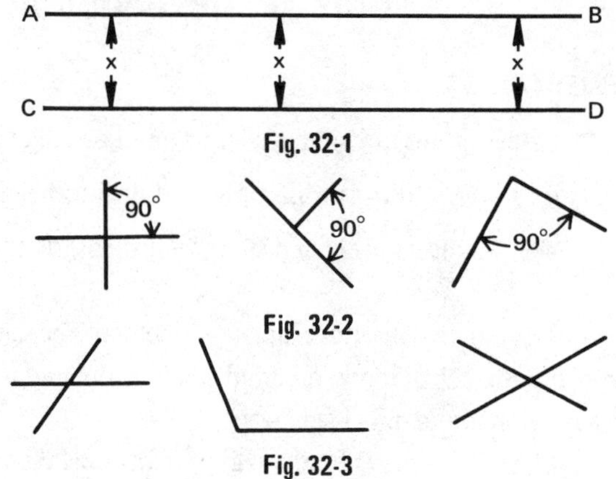

Fig. 32-1

Fig. 32-2

Fig. 32-3

ANGLES

An *angle* is a figure which consists of two lines that meet at a point called the vertex. The symbol ∠ means angle and the symbol ∠'s means angles. The size of an angle is determined by the number of degrees one side of the angle is rotated from the other. The length of the side does not determine the size of the angle. For example, in figure 32-4, angle 1 is equal to angle 2. The rotation of side AC from side AB is equal to the rotation of side DF from side DE although the lengths of the sides are not equal.

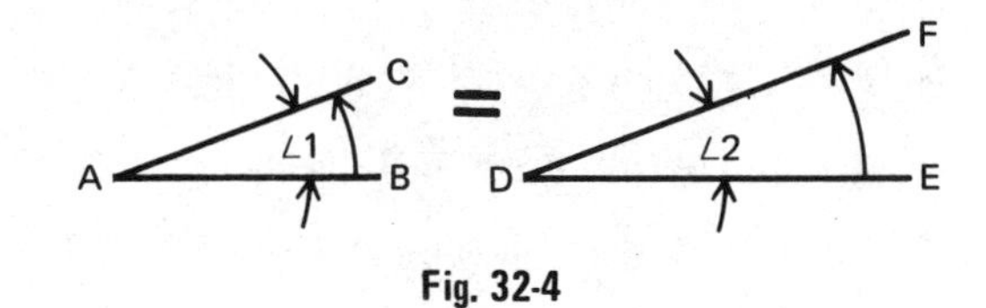

Fig. 32-4

UNITS OF ANGULAR MEASURE

The circumference of a circle can be considered as a complete rotation of the end point of a radius. Therefore, a circumference is equal to 360 degrees.

A machinist must often work to the closest minute of angular measure and to seconds when a very high degree of precision is required.

The units of angular measure are shown in figures 32-5, 32-6, and 32-7.

$$1\text{ degree}\quad (1^\circ) = \frac{1}{360}\text{ of a circumference}$$

$$1\text{ minute}\quad (1') = \frac{1}{60}\text{ of a degree.}$$

$$1\text{ second}\quad (1'') = \frac{1}{60}\text{ of a minute.}$$

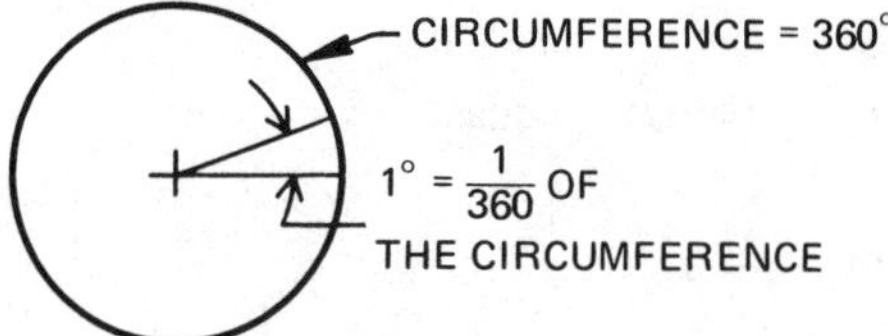

Fig. 32-5

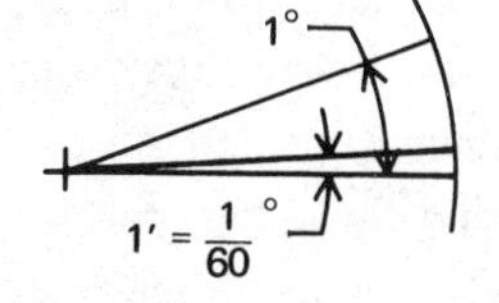

Fig. 32-6

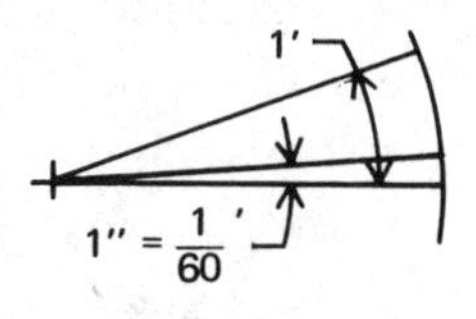

Fig. 32-7

ARITHMETIC OPERATIONS ON ANGULAR MEASURE

The following are examples of addition, subtraction, multiplication, and division of angles. It is sometimes necessary to convert degrees to minutes and minutes to seconds in order to perform arithmetic operations.

Adding Angles:

1. Refer to figure 32-8 and determine $\angle 1$.

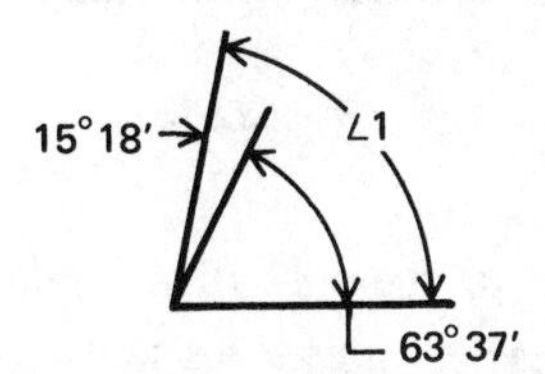

Fig. 32-8

$\angle 1 = 15^\circ\ 18' + 63^\circ\ 37'$

$$\begin{array}{r} 15^\circ\ 18' \\ +\,63^\circ\ 37' \\ \hline 78^\circ\ 55' \end{array}$$

2. Refer to figure 32-9 and determine $\angle 2$.

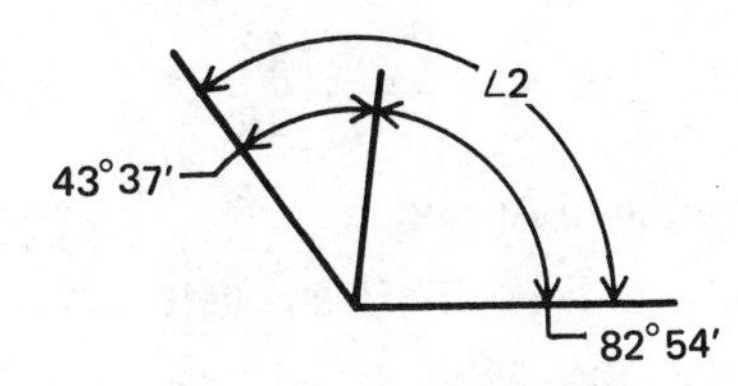

Fig. 32-9

$\angle 2 = 43^\circ\ 37' + 82^\circ\ 54'$

$$\begin{array}{r} 43^\circ\ 37' \\ +\ 82^\circ\ 54' \\ \hline 125^\circ\ 91' \\ +\ \ 1^\circ\ 31' \\ \hline 126^\circ\ 31' \end{array}$$

Note: $91' = 60' + 31' = 1^\circ\ 31'$ therefore, $125^\circ\ 91' = 126^\circ\ 31'$

3. Refer to figure 32-10 and determine $\angle 3$.

$\angle 3 = 78^\circ\ 43'\ 27'' + 29^\circ\ 38'\ 52''$

$$\begin{array}{r} 78^\circ\ 43'\ 27'' \\ +29^\circ\ 38'\ 52'' \\ \hline 107^\circ\ 81'\ 79'' \\ +\qquad 1'\ 19'' \\ \hline 107^\circ\ 82'\ 19'' \\ +\ 1^\circ\ 22'\qquad \\ \hline 108^\circ\ 22'\ 19'' \end{array}$$

Note:

a. $79'' = 60'' + 19'' = 1'\ 19''$ therefore, $107^\circ\ 81'\ 79'' = 107^\circ\ 82'\ 19''$

b. $82' = 60' + 22' = 1^\circ\ 22'$; therefore, $107^\circ\ 82'\ 19'' = 108^\circ\ 22'\ 19''$

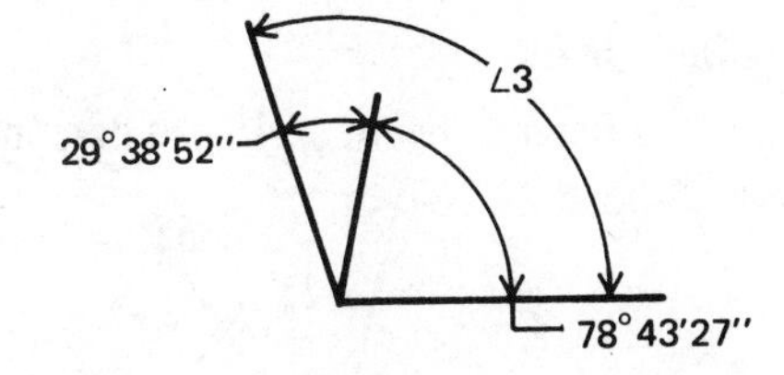

Fig. 32-10

Subtracting Angles:

1. Refer to figure 32-11 and determine $\angle 1$.

$\angle 1 = 123^\circ\ 47'\ 32'' - 86^\circ\ 13'\ 07''$

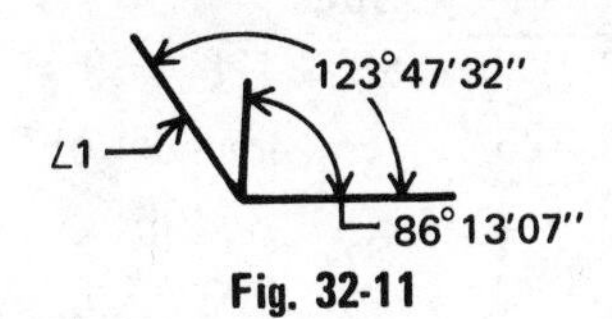

Fig. 32-11

$$\begin{array}{r} 123^\circ\ 47'\ 32'' \\ -86^\circ\ 13'\ 07'' \\ \hline 37^\circ\ 34'\ 25'' \end{array}$$

2. Refer to figure 32-12 and determine $\angle 2$.

$\angle 2 = 97^\circ\ 12' - 45^\circ\ 26'$

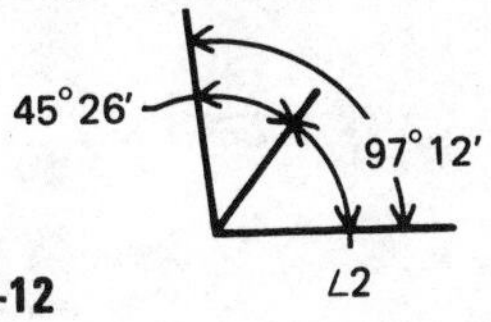

Fig. 32-12

Note: Since 26′ cannot be subtracted from 12′, change $97^\circ\ 12'$ to $96^\circ + 1^\circ + 12' = 96^\circ + 60' + 12' = 96^\circ\ 72'$

$$\begin{array}{r} 96^\circ\ 72' \\ -\ 45^\circ\ 26' \\ \hline 51^\circ\ 46' \end{array}$$

3. Refer to figure 32-13 and determine ∠ 3.
   ∠3 = 57° 13′ 28″ - 44° 19′ 42″

   Note: Since 19′ cannot be subtracted from 13′, and 42″ cannot be subtracted from 28″ ; change 57° 13′ 28″ to 56° + 1° + 13′ + 28″ = 56°73′28″ = 56°72′ + 1′ + 28″ = 56°72′88″.

$$\begin{array}{r} 56^\circ\ 72'\ 88'' \\ -\ 44^\circ\ 19'\ 42'' \\ \hline 12^\circ\ 53'\ 46'' \end{array}$$

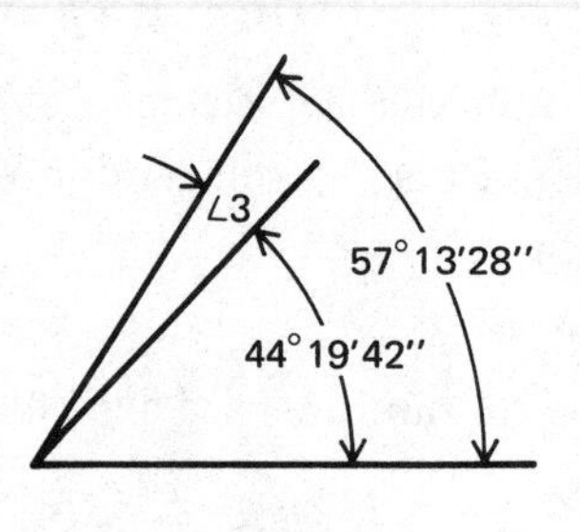

Fig. 32-13

Multiplication of Angles:

1. Refer to figure 32-14 and determine ∠ 1.
   ∠ 1 = 2(64° 29′)

$$\begin{array}{r} 64^\circ\ 29' \\ \times 2 \\ \hline 128^\circ\ 58' \end{array}$$

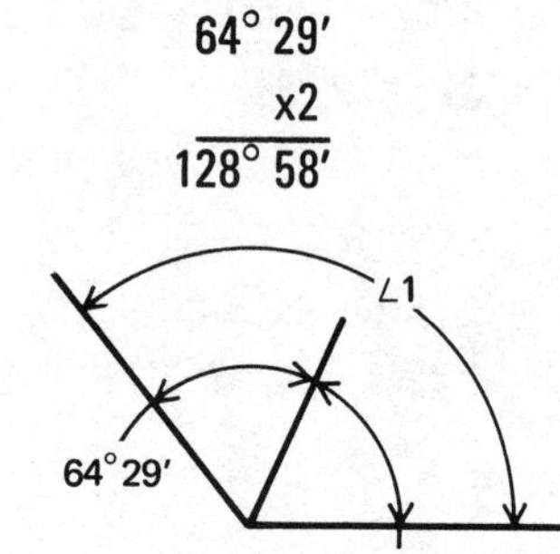

Fig. 32-14

2. Refer to figure 32-15 and determine ∠ 2.
   x = 41° 27′ 42″; ∠ 2 = 5x° = 5(41° 27′ 42″)

$$\begin{array}{r} 41^\circ\ \ 27'\ \ 42'' \\ \times 5 \\ \hline 205^\circ\ 135'\ \not{2}\not{1}\not{0}'' \\ +\quad 3'\ \ 30'' \\ \hline 205^\circ\ \not{1}\not{3}\not{8}'\ \ 30'' \\ +2^\circ\ \ 18'\quad \\ \hline 207^\circ\ \ 18'\ \ 30'' \end{array}$$

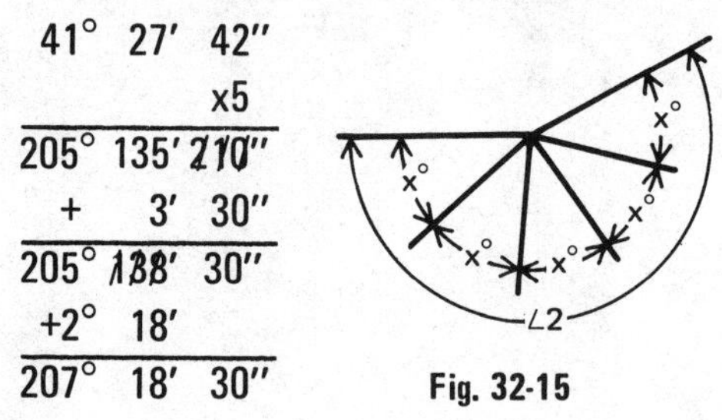

Fig. 32-15

Note:

a. 210″ = 3′ 30″
b. 138′ = 2° 18′

Division of Angles:

1. Refer to figure 32-16 and determine ∠'s 1 and 2.

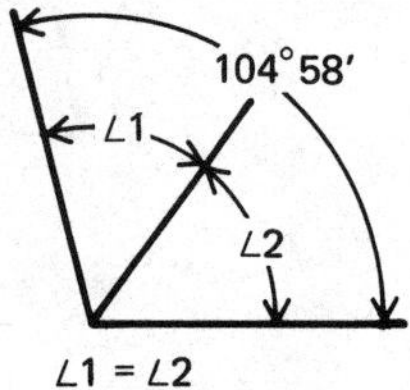

Fig. 32-16

∠ 1 = ∠ 2 = 104° 58′ ÷ 2

$$\begin{array}{r} 52^\circ\ 29' \\ 2\overline{)\,104^\circ\ 58'} \end{array}$$

2. Refer to figure 32-17 and determine ∠'s 1,2, and 3.

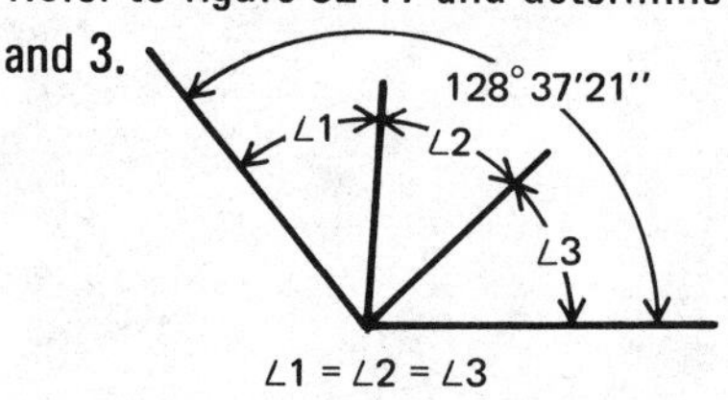

Fig. 32-17

∠ 1 = ∠ 2 = ∠ 3 = 128° 37′ 21″ ÷ 3

$$\begin{array}{r} 42^\circ \\ 3\overline{)\,128^\circ} \\ 126 \\ \hline 2^\circ \end{array} \qquad \begin{array}{r} 52' \\ 3\overline{)\,157'} \\ 156 \\ \hline 1' \end{array} \qquad \begin{array}{r} 27'' \\ 3\overline{)\,81''} \end{array}$$

Note:

a. 128° ÷ 3 = 42° plus a remainder of 2°.
b. Add the 2° to the 37′; 2° = 120′, 120′ + 37′ = 157′.
c. 157′ ÷ 3 = 52′ plus a remainder of 1′.
d. Add the 1′ to the 21″; 1′ = 60″, 60″ + 21″ = 81″.
e. 81″ ÷ 3 = 27″.
f. Combine: 42° 52′ 27″

CONVERSION OF DECIMAL DEGREES

The measure of an angle given in the form of decimal degrees, such as 47.1938°, must usually be converted to degrees, minutes, and seconds.

Rule: To convert decimal degrees to degrees, minutes, and seconds

- Multiply the decimal part of the degrees by 60′ in order to obtain minutes.
- If the number of minutes obtained is not a whole number, multiply the decimal part of the minutes by 60″ in order to obtain seconds.

Example: Express 47.1938° as degrees, minutes, and seconds.

- ▲ Multiply .1938 by 60′, 60′ (.1938) = 11.6280′
- ▲ Multiply .6280 by 60″, 60″ (.6280) = 37.6800″ = 38″
- ▲ Combine: 47.1938° = 47° 11′ 38″

APPLICATION

A. DEFINITIONS AND TERMS

Answer the following questions.

1. Refer to the figure at the right

   a. Line AB and line CD are ____.

   b. Line AB and EF are ____.

   c. Line CD and GH are ____.

   d. In ∠1 point G is called a ____.

2. a. How many degrees are there in a circumference? ____

   b. How many minutes are there in 1 degree? ____

   c. How many seconds are there in 1 minute? ____

   d. How many seconds are there in 1 degree? ____

   e. How many minutes are there in a circumference? ____

3. Give the symbols for the following words:

   a. parallel ________ c. degree ________ e. second ________

   b. perpendicular ________ d. minute ________

B. ADDITION OF ANGLES

1. ∠1 = ____________ 2. ∠2 = ____________ 3. ∠3 = ____________

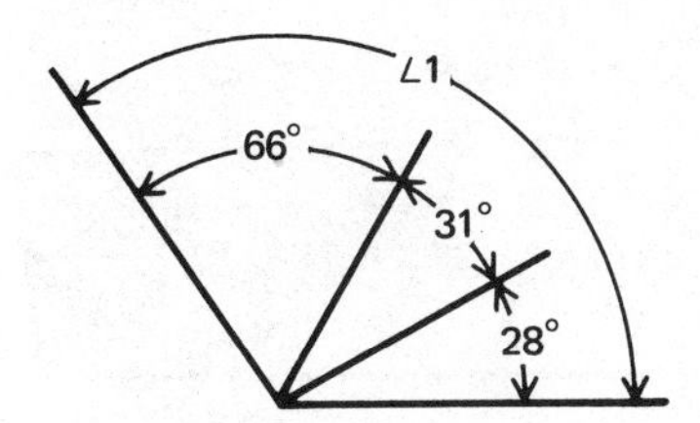

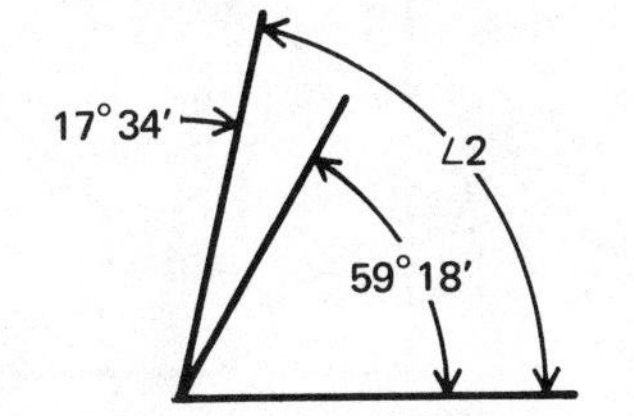

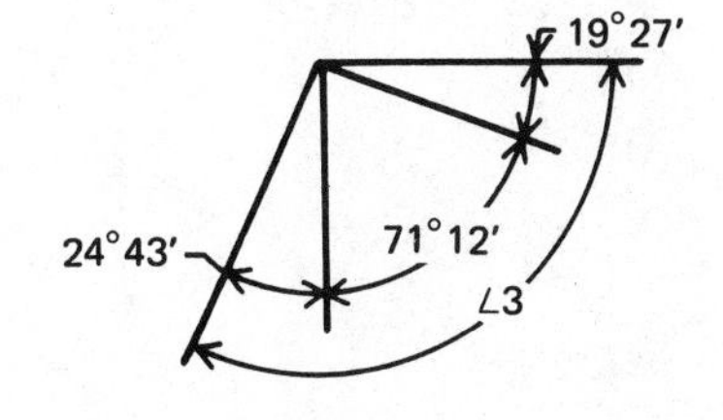

4. ∠1 + ∠2 + ∠3 = ________ 5. ∠5 = ________ 6. ∠6 = ________

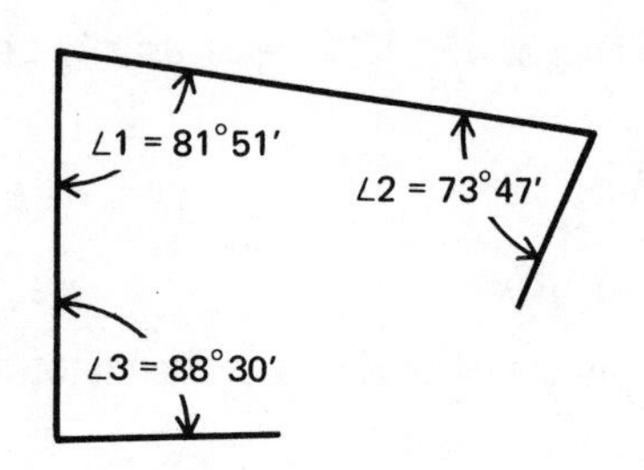

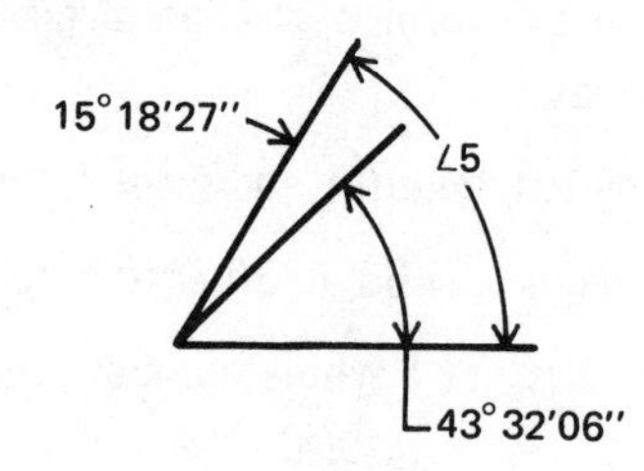

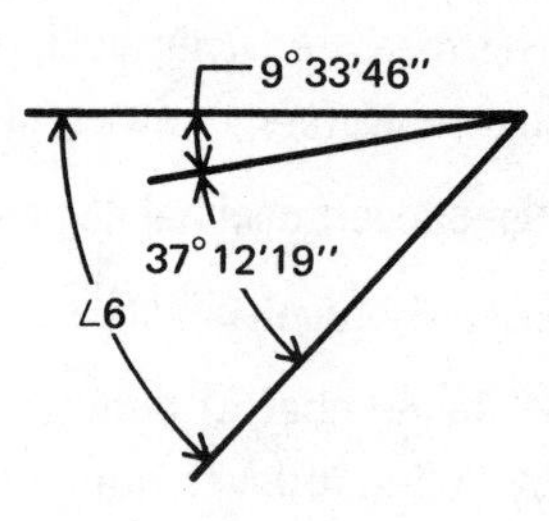

7. ∠7 + ∠8 + ∠9 = ________ 8. ∠1 + ∠2 + ∠3 + ∠4 + ∠5 = ________

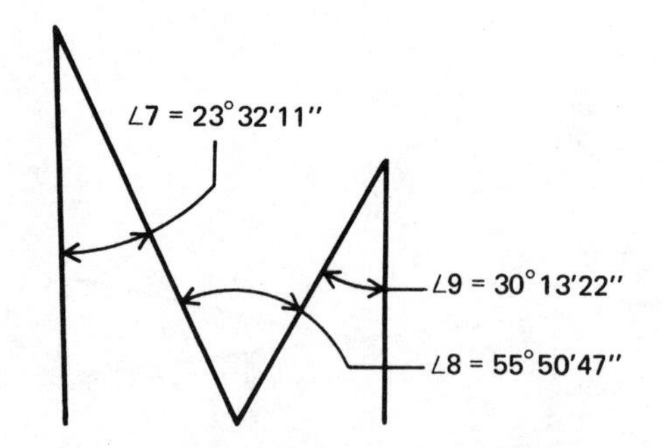

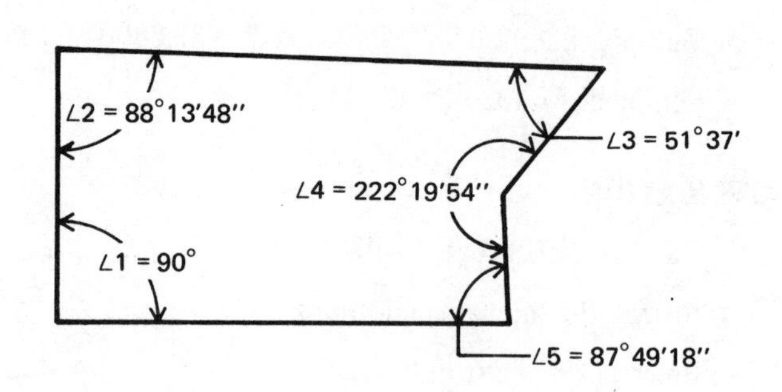

C. SUBTRACTION OF ANGLES

1. 114° - 89° = ____ 2. 92° 35′ - 73° 16′ = ____ 3. 63° 23′ - 32° 58′ = ____
4. 122° 36′ 17″ - 13° 15′ 08″ = ____ 5. 54° 43′ 13″ - 19° 13′ 42″ = ____
6. ∠1 = ________ 7. ∠3 - ∠2 = ________ 8. ∠2 = ________

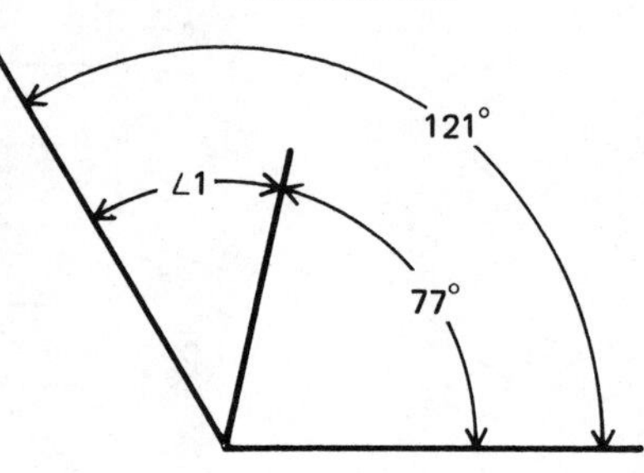

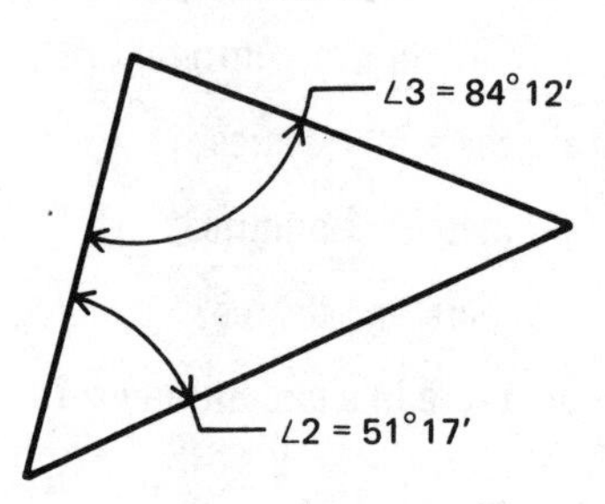

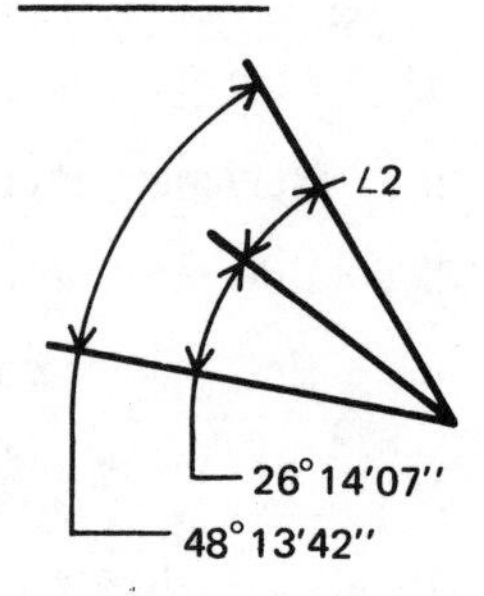

9. ∠1 - ∠2 = ________ 10. ∠6 = 720° - (∠1 + ∠2 + ∠3 + ∠4 + ∠5)

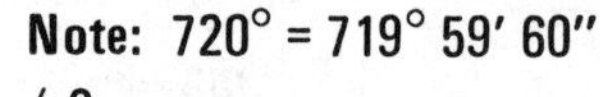
Note: 720° = 719° 59′ 60″

∠6 = ________

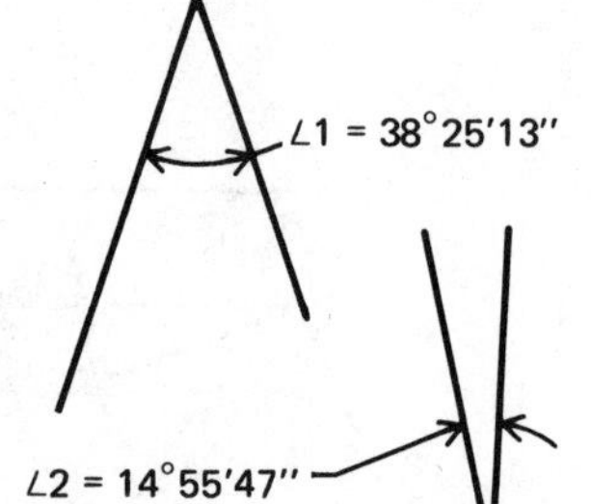

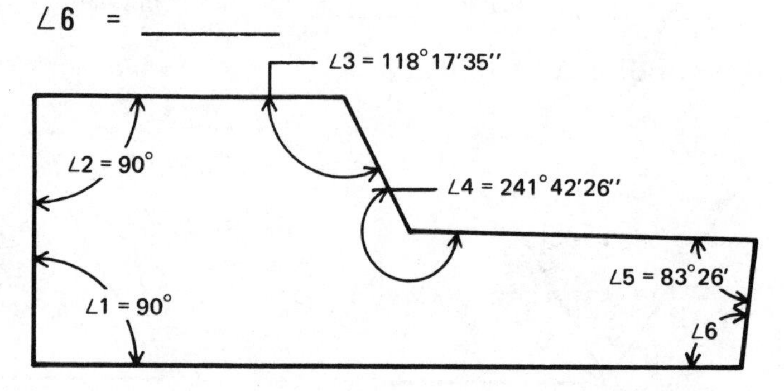

D. MULTIPLICATION OF ANGLES

1. $7(15^\circ) =$ ________
2. $3(27^\circ\ 18') =$ ________
3. $2(43^\circ\ 43') =$ ________
4. $5(22^\circ\ 10'\ 13'') =$ ________
5. $8(34^\circ\ 24'\ 30'') =$ ________
6. $\angle 1 = \angle 2 = 42^\circ$
   $\angle 3 =$ ________
7. $x = 39^\circ\ 14'$
   $\angle 4 =$ ________
8. $\angle 1 = \angle 2 = \angle 3 = \angle 4 = \angle 5 = 53^\circ\ 41'$
   $\angle 6 =$ ________

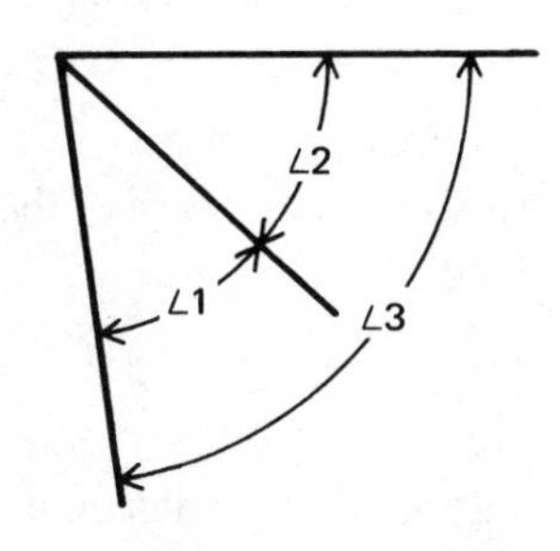

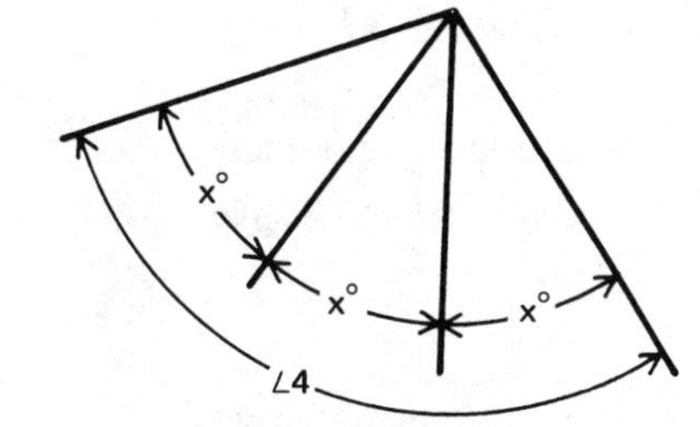

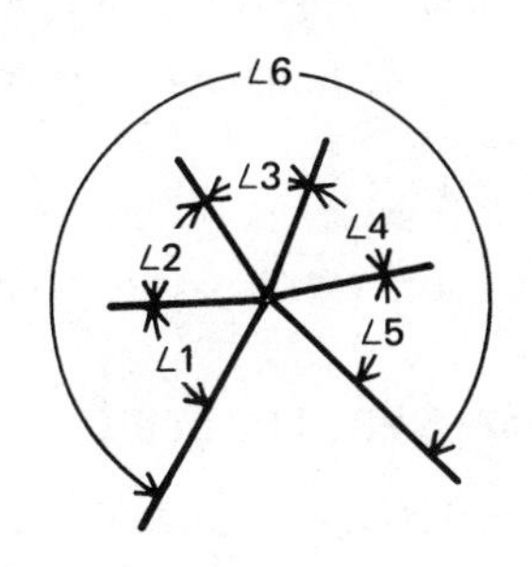

E. DIVISION OF ANGLES

1. $94^\circ \div 2 =$ ________
2. $93^\circ \div 2 =$ ________
3. $105^\circ\ 20' \div 4 =$ ________
4. $\angle 1 = \angle 2, \angle 1 =$ ________
5. $x =$ ________
6. $y =$ ________

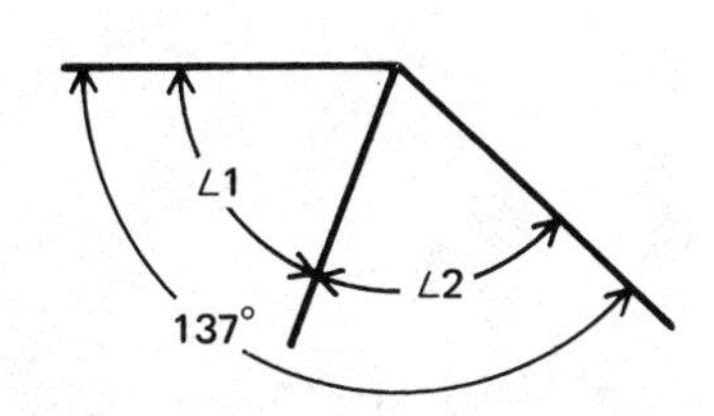

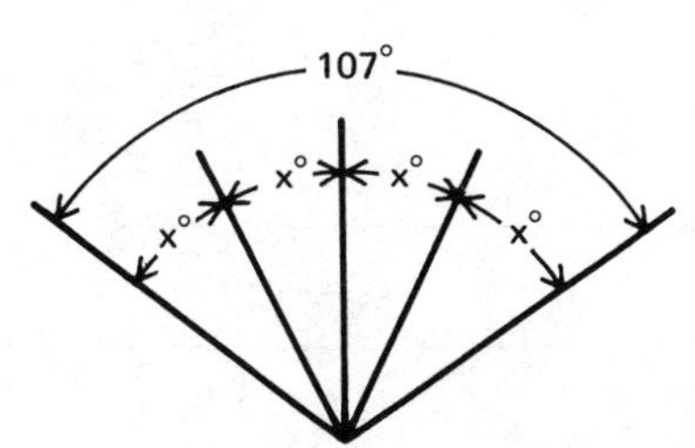

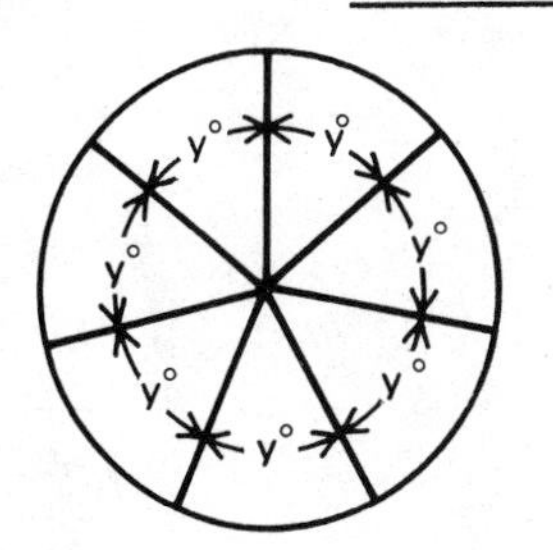

7. $\angle 1 = \angle 2 = \angle 3 = \angle 4$
   $\angle 1 =$ ________

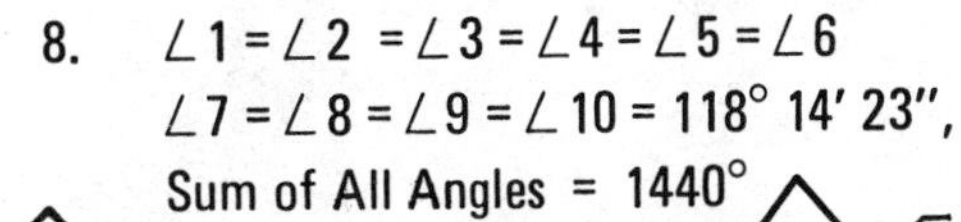

8. $\angle 1 = \angle 2 = \angle 3 = \angle 4 = \angle 5 = \angle 6$
   $\angle 7 = \angle 8 = \angle 9 = \angle 10 = 118^\circ\ 14'\ 23''$,
   Sum of All Angles $= 1440^\circ$
   $\angle 1 =$ ________

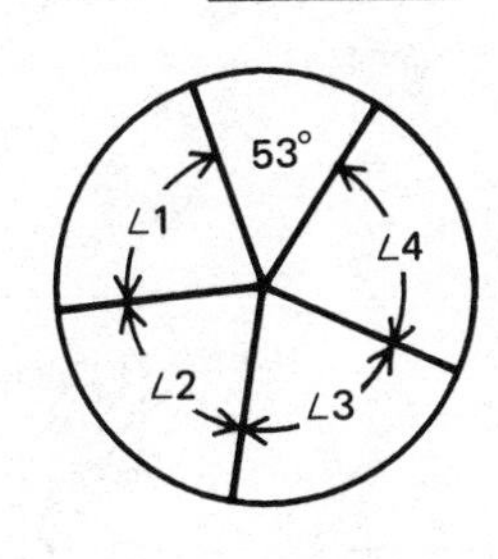

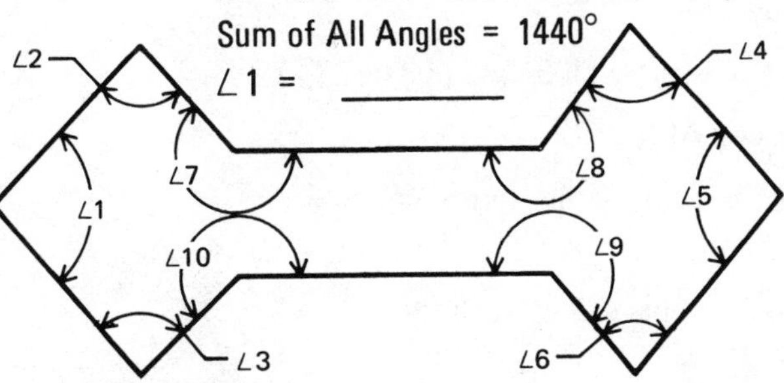

F. CONVERSION OF DECIMAL FRACTIONS

1. Express the following degrees as degrees and minutes.

   a. $13.50^\circ =$ ________
   b. $76.95^\circ =$ ________
   c. $48.10^\circ =$ ________
   d. $117.70^\circ =$ ________
   e. $20.30^\circ =$ ________
   f. $93.15^\circ =$ ________

2. Express the following degrees as degrees, minutes, and seconds.

   a. $52.138^\circ =$ ________
   b. $214.082^\circ =$ ________
   c. $7.925^\circ =$ ________
   d. $44.444^\circ =$ ________
   e. $75.843^\circ =$ ________
   f. $103.009^\circ =$ ________

Unit 33 Protractors – Simple and Caliper

OBJECTIVES

After studying this unit, the student should be able to

- Measure angles with a simple protractor.
- Lay out angles with a simple protractor.
- Read settings on a vernier bevel protractor.
- Compute complements and supplements of angles.

Protractors are used for measuring and laying out angles. Although all protractors are basically the same, different types are available for different uses and degrees of precision required.

SIMPLE PROTRACTOR

A simple protractor has two scales, each graduated from 0° to 180° so that it can be read from either the left or right side. The vertex of the angle to be measured or drawn is located at the middle of the base of the protractor.

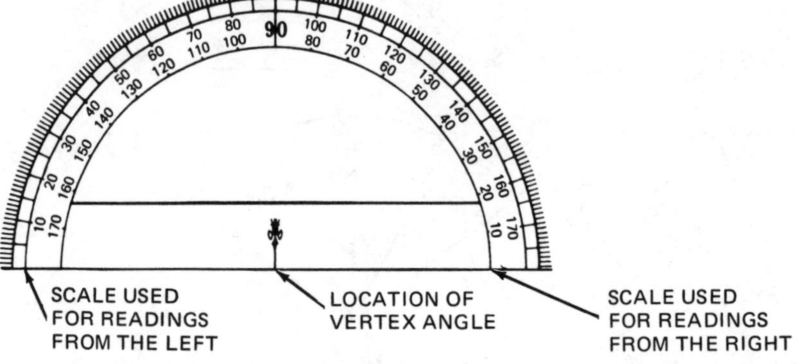

Fig. 33-1

Using a Simple Protractor

Rule: To lay out a given angle (105°, as in figure 33-2)

- Draw a baseline AB.
- On AB mark point O as the vertex of the angle to be drawn.
- Place the protractor base on AB with the center on point O.
- At the scale reading of 105° mark point P.
- Remove the protractor and connect points P and O.

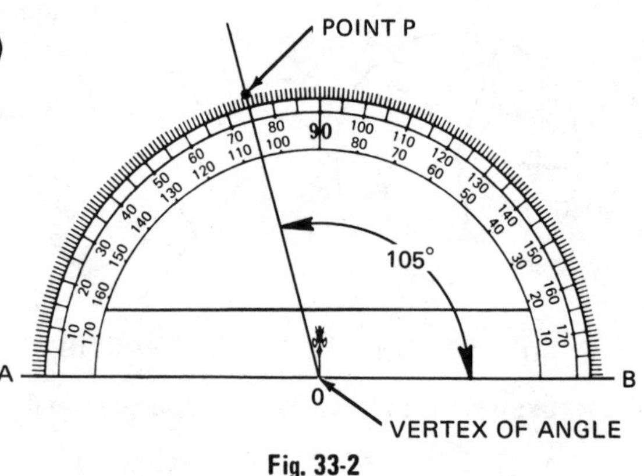

Fig. 33-2

Rule: To measure a given angle (as ∠ 1 (∠ AOB), figure 33-3)

- Extend the sides of the angle.
- Place the protractor base on line OB with the center on point O.
- Read the measurement where the extension of line OA crosses the protractor scale. Angle 1 = 40°.

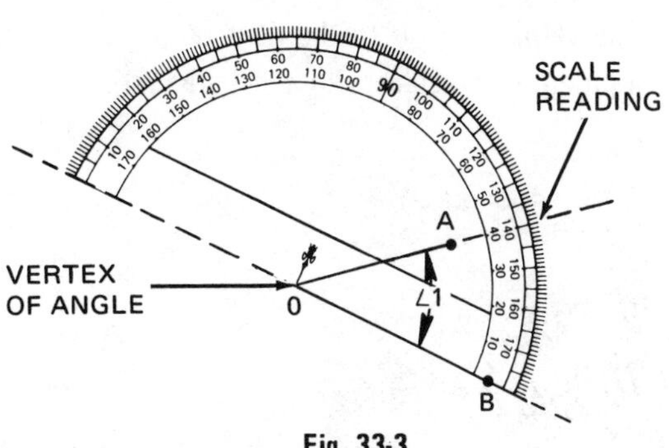

Fig. 33-3

Fig. 33-4

Example: Measure ∠ 2 (DOF) as shown in figure 33-4

- ▲ Extend the sides of the angle.
- ▲ Position the protractor.
- ▲ Read the scale. Angle 2 = 125°

VERNIER PROTRACTOR

The bevel protractor is the most widely used vernier protractor in the machine shop. A bevel protractor is shown in figure 33-5.

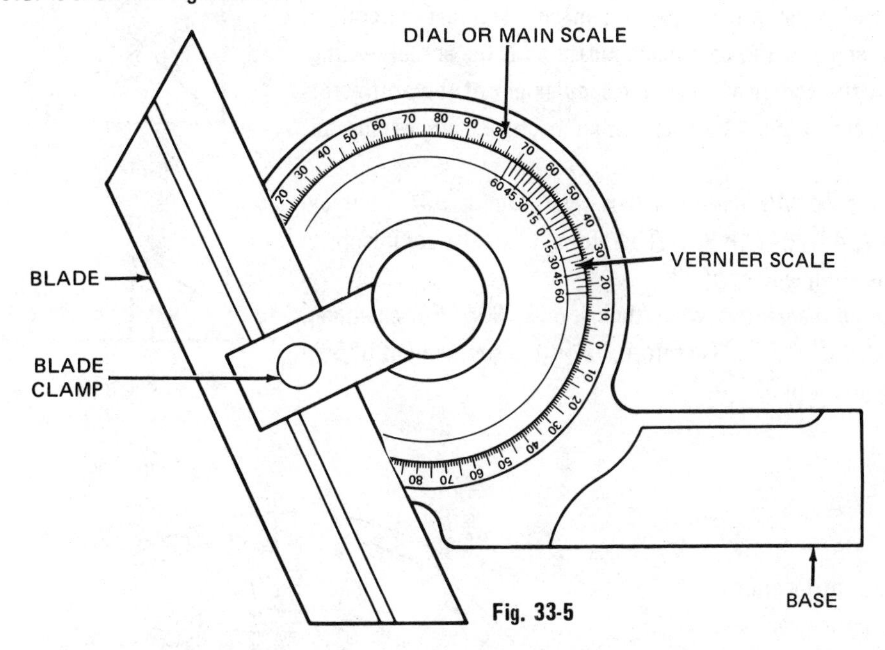

Fig. 33-5

The dial or main scale is fixed and is divided into four sections, each from 0° to 90°. The vernier scale rotates within the main scale. The purpose and use of the vernier scale is similar to that of the vernier caliper and height gage.

The vernier scale is divided into 24 units, with 12 units on each side of zero. The divisions on the vernier scale are in minutes; each division is equal to 5 minutes.

The left vernier scale is used when the vernier zero is to the left of the dial zero; the right vernier scale is used when the vernier zero is to the right of the dial zero.

Figure 33-6 shows a vernier protractor with a reading of 30° 35′. The zero mark on the vernier scale is just to the right of the 30° division on the dial scale. The vernier zero is to the right of the dial zero, therefore, the right vernier scale is read. The 35′ vernier graduation coincides with a dial graduation. The protractor reading is 30° 35′.

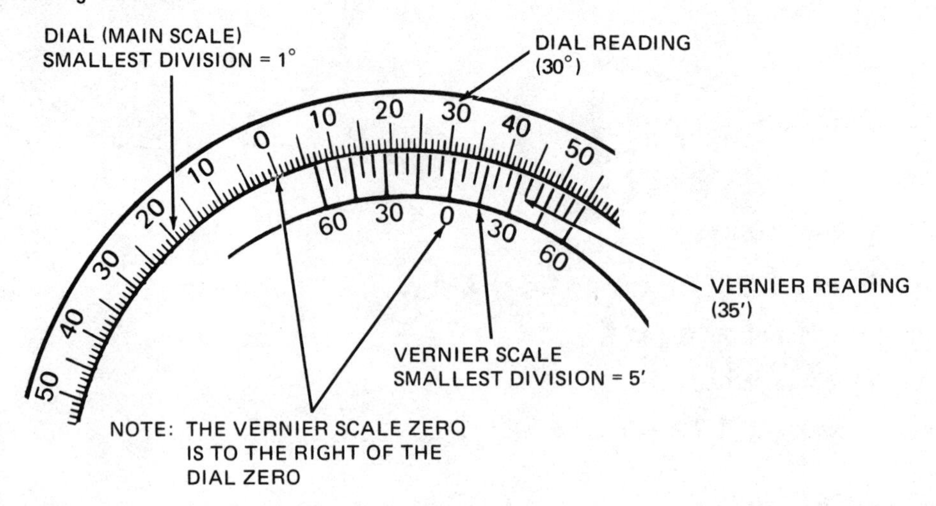

Fig. 33-6

COMPLEMENTS AND SUPPLEMENTS OF SCALE READINGS

When using the bevel protractor, the machinist must determine whether the desired angle of the part being measured is the actual reading on the protractor or the complement or the supplement of the protractor reading. Particular caution must be taken when measuring angles close to 45° and 90°.

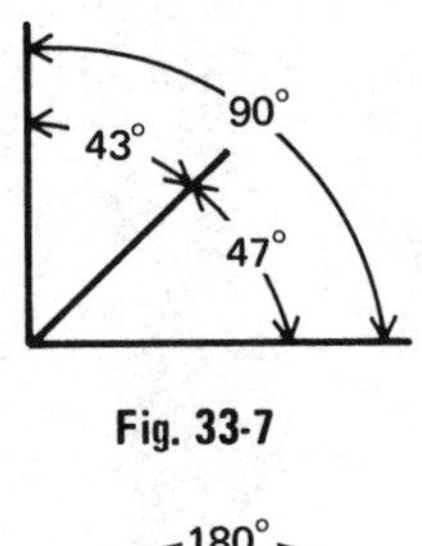

Fig. 33-7

Two angles are *complementary* when their sum is 90°. For example, in figure 33-7, 43° + 47° = 90°. Therefore, 43° is the complement of 47° and 47° is the complement of 43°.

Two angles are *supplementary* when their sum is 180°. For example, in figure 33-8, 92° + 88° = 180°. Therefore, 92° is the supplement of 88° and 88° is the supplement of 92°.

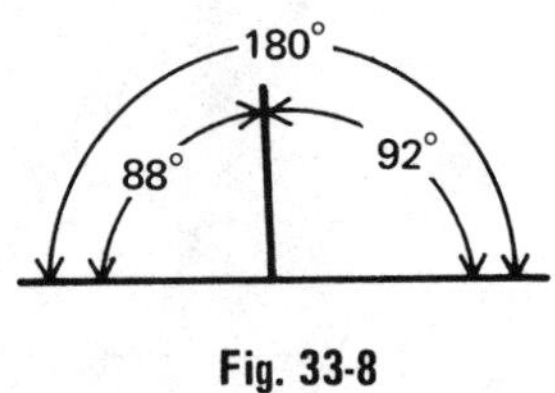

Fig. 33-8

APPLICATION

A. SIMPLE PROTRACTOR

1. Read the values of locations A-J on the protractor scale shown.

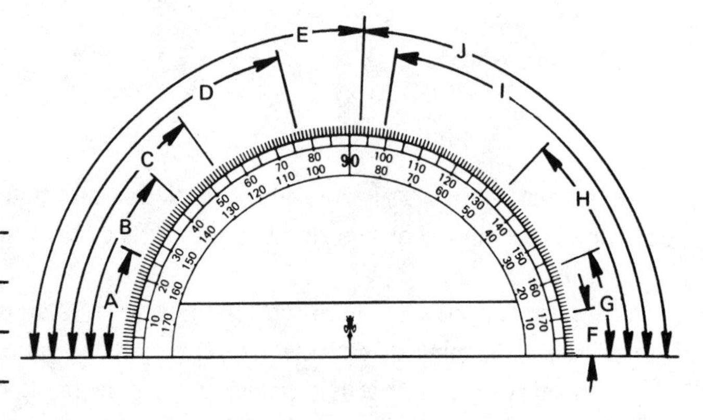

A = ______	F = ______
B = ______	G = ______
C = ______	H = ______
D = ______	I = ______
E = ______	J = ______

2. Measure the following angles to the nearest degree.

   Note: It may be necessary to extend the sides.

   a. ∠1 = ________ b. ∠2 = 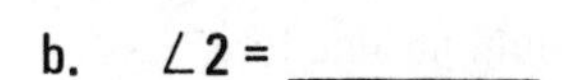 c. ∠3 = ________

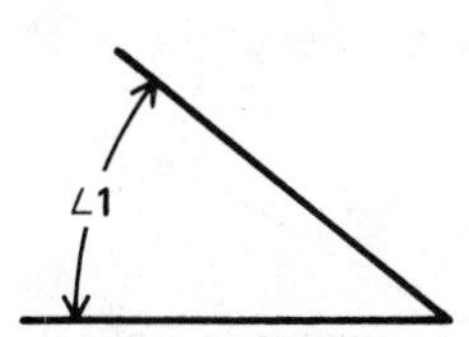

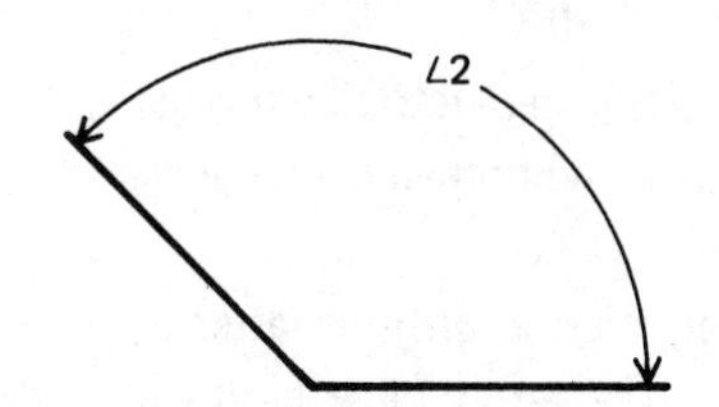

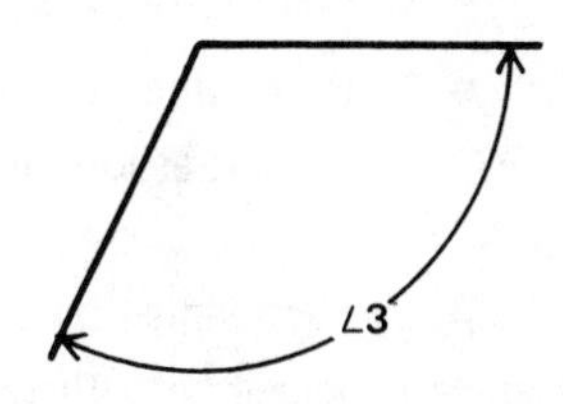

   d. ∠4 = ________ e. ∠5 = ________ f. ∠6 = ____, ∠7 = ____

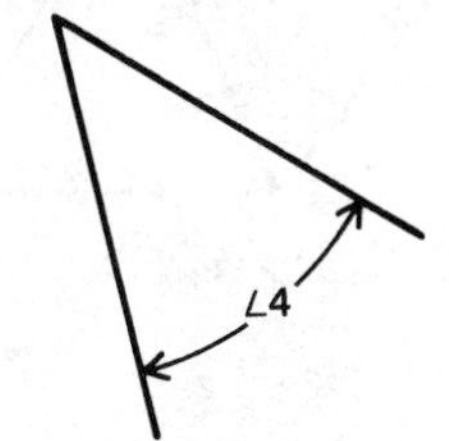

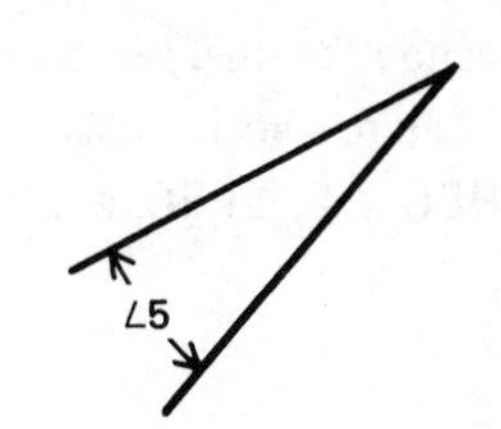

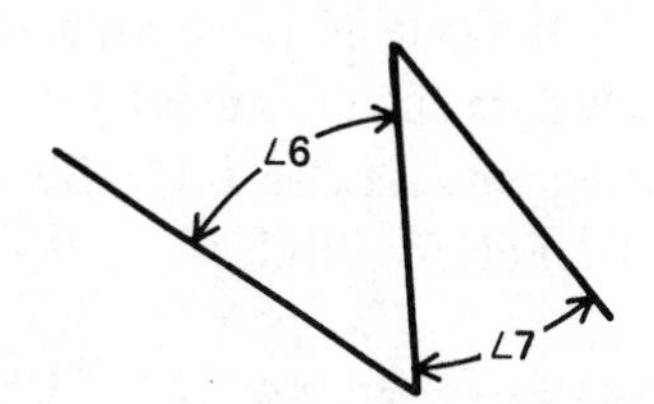

3. Lay out the following angles.

   a. 17° b. 108° c. 93° d. 6° e. 49° f. 174° g. 45° h. 90°

B. VERNIER PROTRACTOR

Read the settings on the vernier protractor scales shown below.

1. ____________ 2. ____________ 3. 

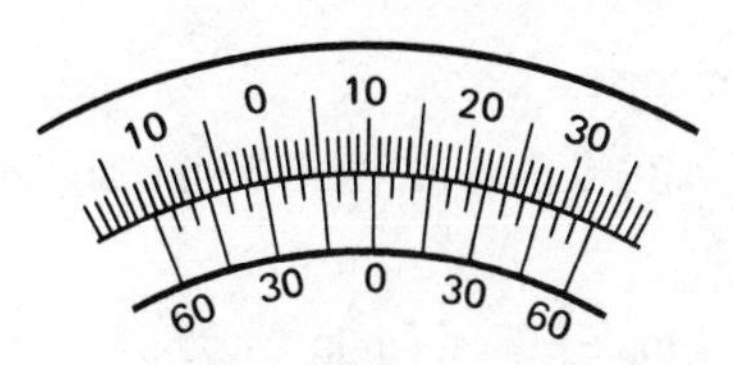

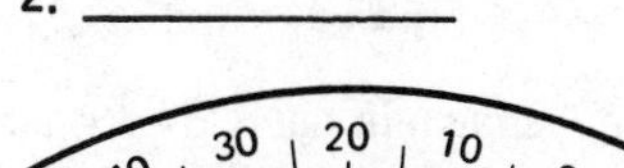
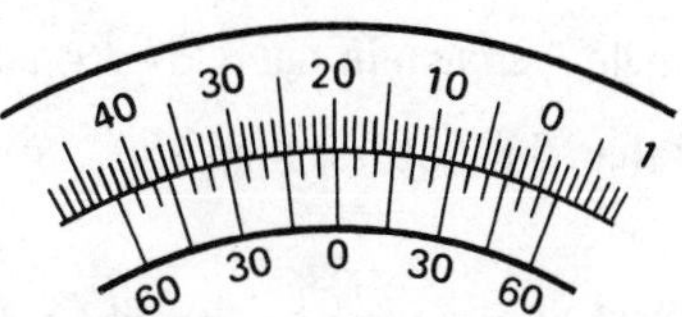

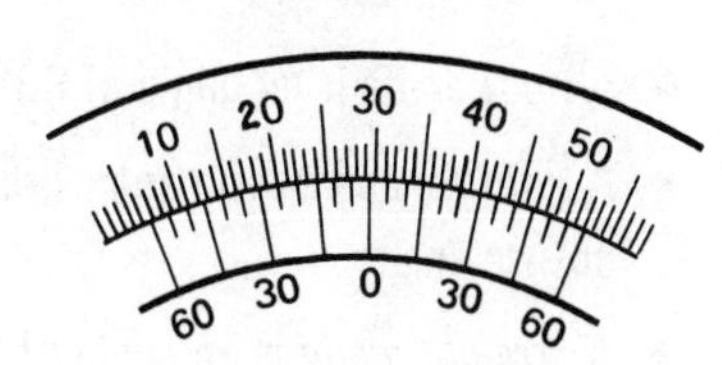

C. COMPLEMENTARY AND SUPPLEMENTARY ANGLES

1. Give the complements of the following angles.

   a. 41° ____ b. 76° ____ c. 17° ____ d. 2° ____

   e. 67° 49′ ____ f. 89° 53′ 07″ ____ g. 44° 59′ 59″ ____

2. Give the supplements of the following angles.

   a. 7° ____ b. 65° ____ c. 91° ____

   d. 179° 59′ ____ e. 67° 18′ 27″ ____ f. 133° 32′ 08″ ____

Unit 34 Angles

OBJECTIVES

After studying this unit, the student should be able to

- Identify different types of angles.
- Determine unknown angles in geometric figures using the principles of opposite, alternate interior, corresponding, parallel, and perpendicular angles.

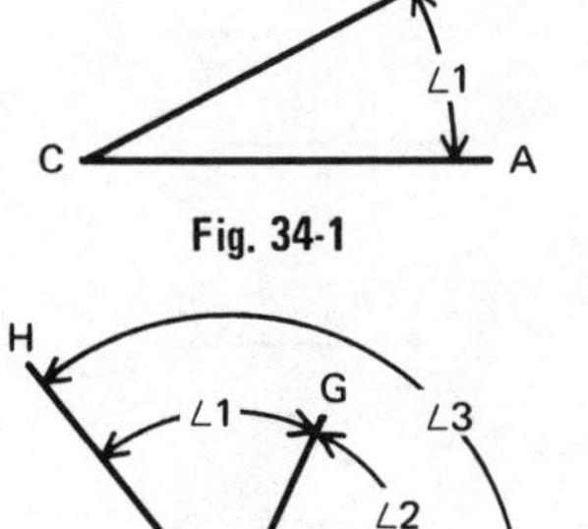

Fig. 34-1

Fig. 34-2

Angles are named by a number, a letter, or three letters. When an angle is named with three letters, the vertex must be the middle letter.

For example, the angle shown in figure 34-1 can be called ∠1, ∠C, ∠ACB, or ∠BCA.

In figure 34-2, the single letter E cannot be used in naming an angle since point E is the vertex of more than one angle. Each of the three angles is called ∠1, ∠GEH, or ∠HEG; ∠2, ∠FEG, or ∠GEF; ∠3, ∠FEH, or ∠HEF.

TYPES OF ANGLES

- An *acute angle* is an angle that is less than 90°. Angle 1 shown in figure 34-3 is acute.

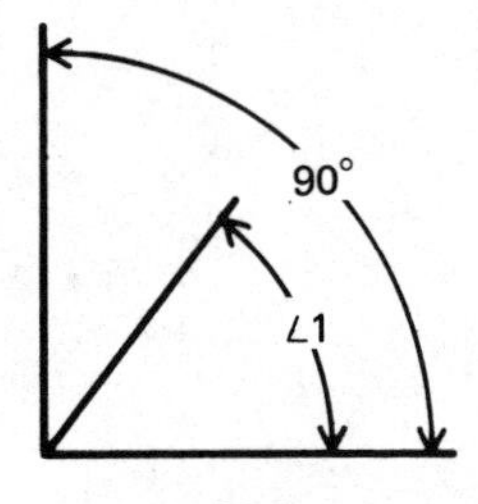

Fig. 34-3

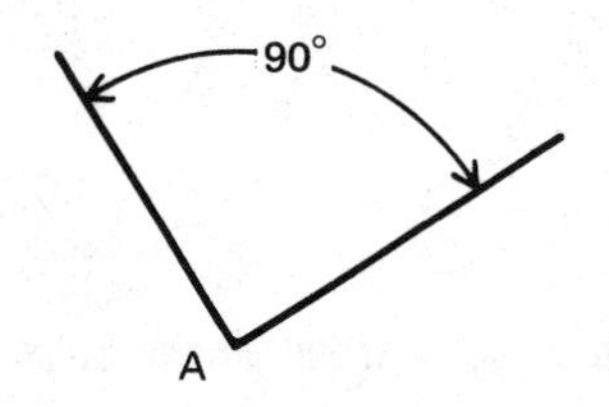

Fig. 34-4

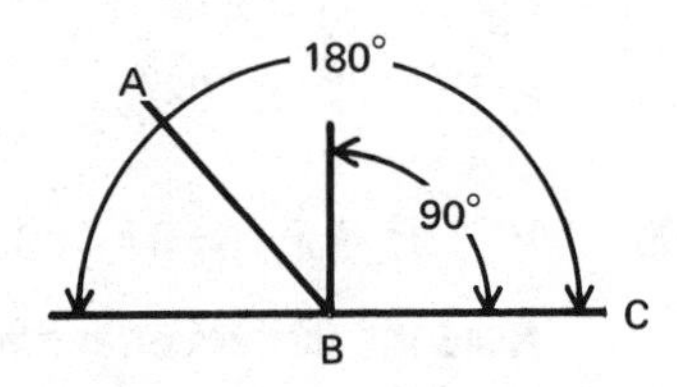

Fig. 34-5

- A *right angle* is an angle of 90°. Angle A shown in figure 34-4 is a right angle.
- An *obtuse angle* is an angle greater than 90° but less than 180°. Angle ABC shown in figure 34-5 is an obtuse angle.
- A *straight angle* is an angle of 180°. A straight line is a straight angle. Line EFG shown in figure 34-6 is a straight angle.

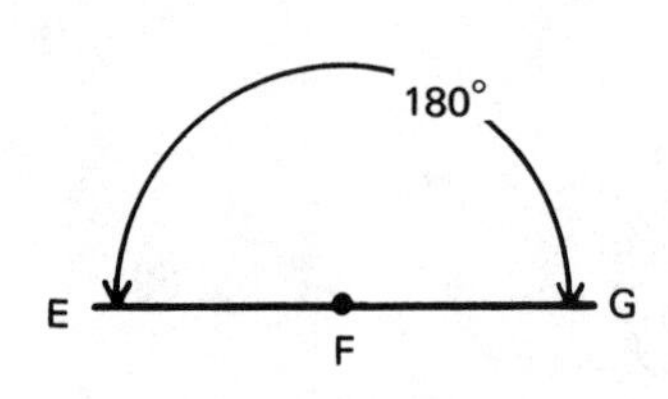

Fig. 34-6

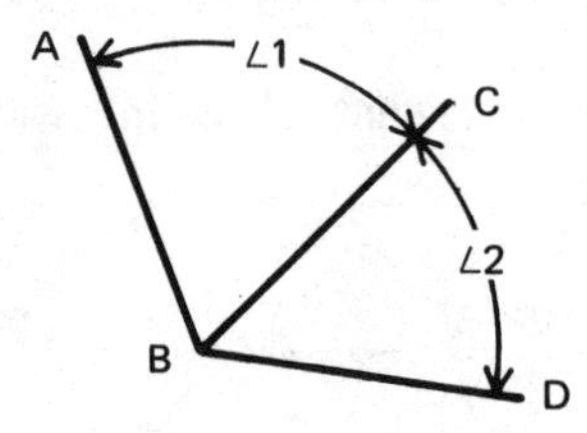

Fig. 34-7

- Two angles are adjacent if they have a common side and a common vertex. Angle 1 and angle 2 shown in figure 34-7 are adjacent since they both contain the common side BC and the common vertex B.

Angles Formed by a Transversal

Note: A *transversal* is a line that cuts two or more lines. Line EF shown in figure 34-8 is a transversal since it cuts lines AB and CD.

- *Alternate interior angles* are pairs of interior angles on opposite sides of the transversal with different vertices. For example, angles 3 and 5 and angles 4 and 6 are alternate interior angles.
- *Corresponding angles* are pairs of angles, one interior and one exterior, with both angles on the same side of the transversal with different vertices. For example, angles 1 and 5, 2 and 6, 3 and 7, and 4 and 8 are corresponding angles.

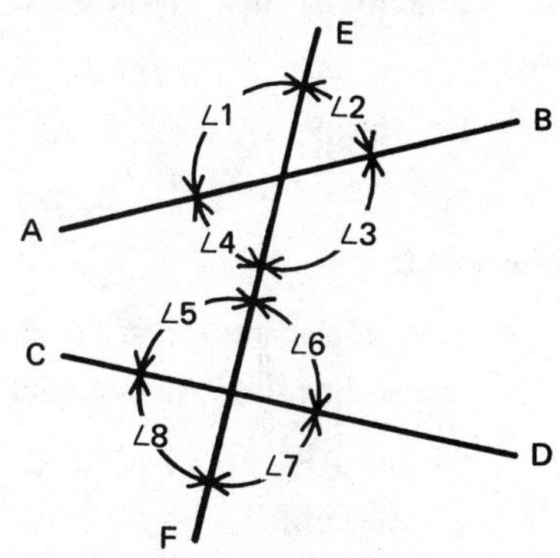

Fig. 34-8

GEOMETRIC PRINCIPLES

In this book, geometric postulates, theorems, and corollaries are grouped together and are called geometric principles. *Geometric principles* are statements of truth which are used as geometric rules.

Note: The principles will not be proved, but will be used as the basis for problem solving.

Principle 1

- **If two lines intersect, the opposite angles are equal.** Figure 34-9 shows angles 1 and 3 and angles 2 and 4 as pairs of vertical angles. Therefore, ∠1 = ∠3 and ∠2 = ∠4.

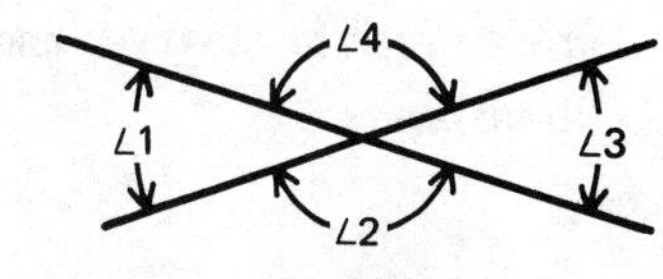

Fig. 34-9

Principle 2

- **If two parallel lines are intersected by a transversal, the alternate interior angles are equal.** In figure 34-10,
  Given: AB ∥ CD
  Conclusion: ∠3 = ∠5 and ∠4 = ∠6.

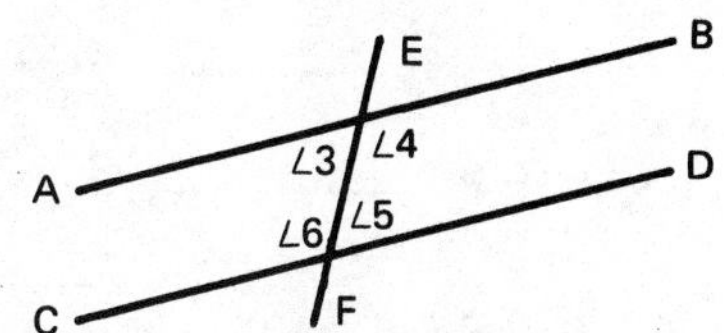

Fig. 34-10

- **If two lines are intersected by a transversal and a pair of alternate interior angles are equal, the lines are parallel.** in figure 34-11,
  Given: ∠1 = ∠2.
  Conclusion: AB ∥ CD.

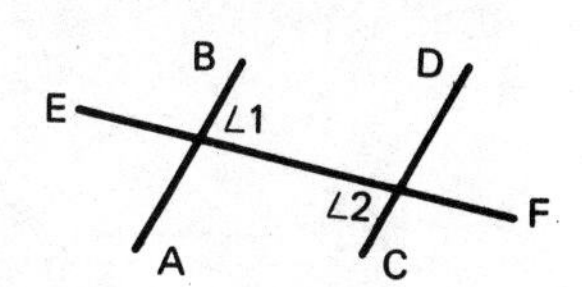

Fig. 34-11

Principle 3

- **If two parallel lines are intersected by a transversal, the corresponding angles are equal.** In figure 34-12,
  Given: AB ∥ CD
  Conclusion: ∠1 = ∠5, ∠2 = ∠6, ∠3 = ∠7 and ∠4 = ∠8.

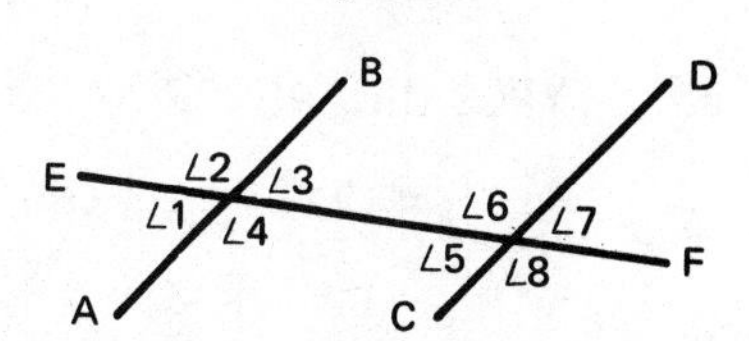

Fig. 34-12

- **If two lines are intersected by a transversal and a pair of corresponding angles are equal, the lines are parallel.** In figure 34-13,
  Given: ∠1 = ∠2
  Conclusion: AB ∥ CD

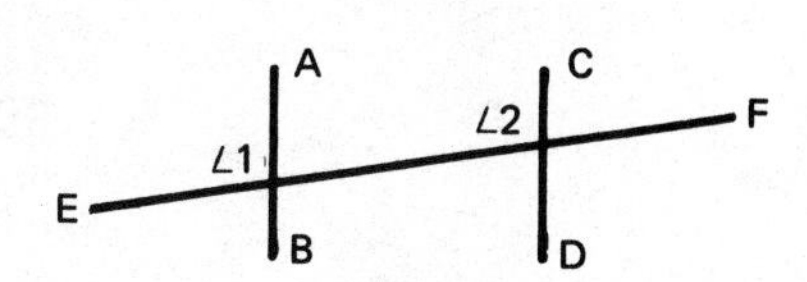

Fig. 34-13

Principle 4

- **Two angles are either equal or supplementary if their corresponding sides are parallel.** In figure 34-14,

  Given: AB ∥ FG and BC ∥ DE
  Conclusion: ∠1 = ∠3 and ∠1 and ∠2 are supplementary. (∠1 + ∠2 = 180°)

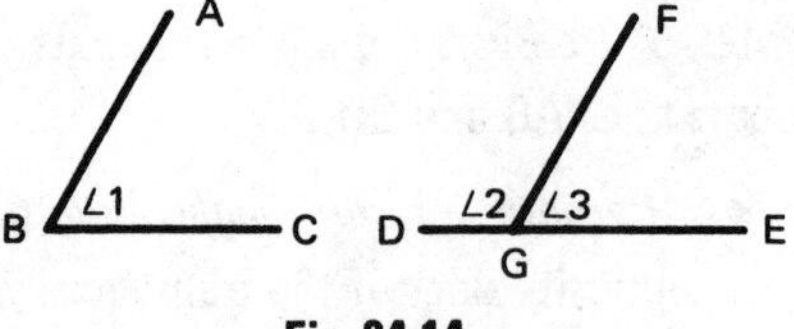

Fig. 34-14

Principle 5

- **Two angles are either equal or supplementary if their corresponding sides are perpendicular.** In figure 34-15,

  Given: AB ⊥ DH and BC ⊥ EF
  Conclusion: ∠1 = ∠2; ∠1 and ∠3 are supplementary. (∠1 + ∠3 = 180°)

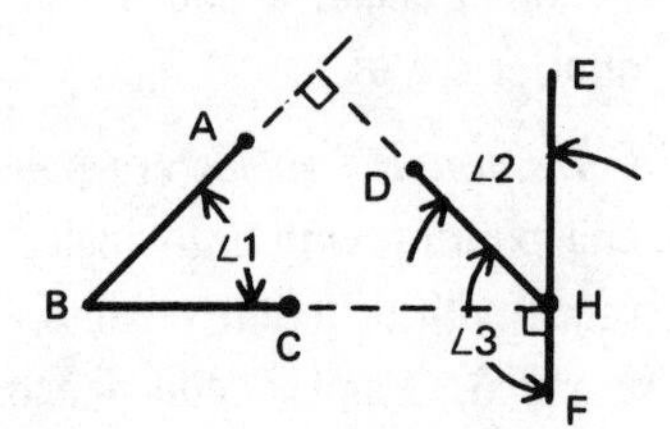

Fig. 34-15

APPLICATION

A. NAMING ANGLES

1. Name each of the given angles in three additional ways.

a. ∠1	________	d. ∠D	________
b. ∠2	________	e. ∠E	________
c. ∠C	________	f. ∠F	________

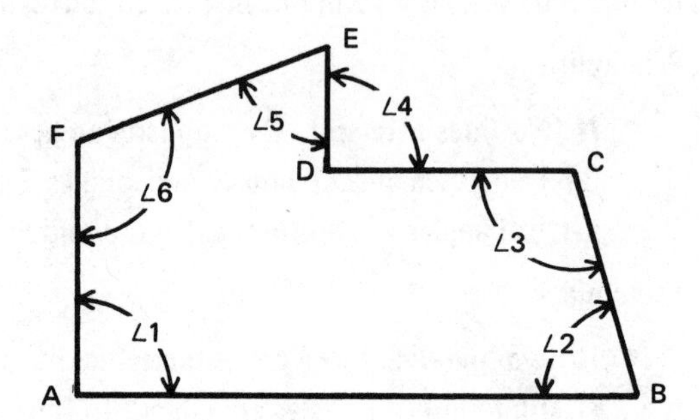

2. Name each of the given angles in two additional ways.

a. ∠1	________	d. ∠ECB	________
b. ∠CBF	________	e. ∠5	________
c. ∠3	________	f. ∠BCD	________

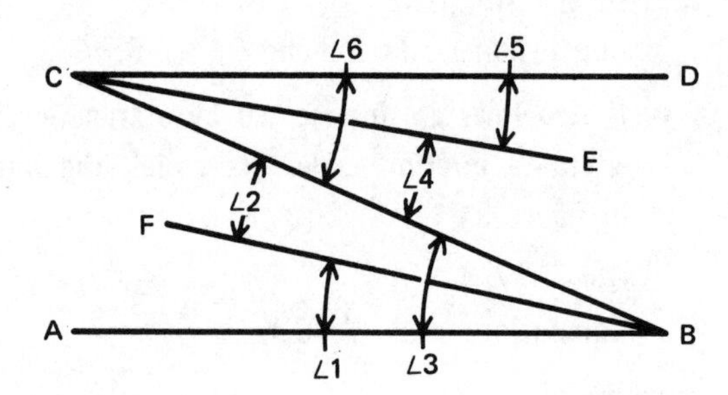

B. TYPES OF ANGLES

1. Identify each of the angles shown as acute, obtuse, right, or straight.

a. ∠A	________	f. ∠BFC	________
b. ∠ABF	________	g. ∠ABC	________
c. ∠CBF	________	h. ∠BCF	________
d. ∠DCE	________	i. ∠AEC	________
e. ∠BFE	________	j. ∠EFC	________

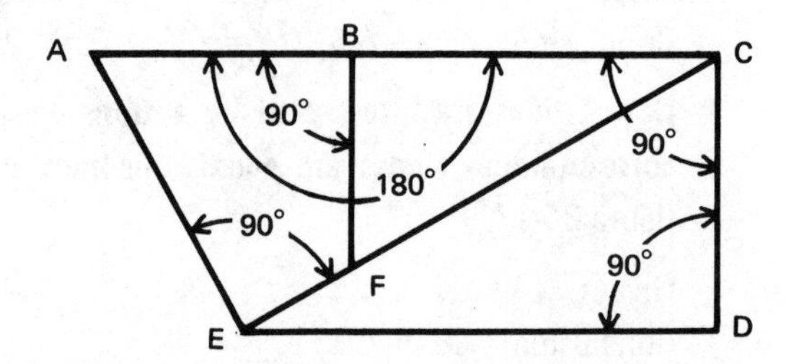

2. Name all pairs of adjacent angles shown in the figure.

   ____

   ____

   ____

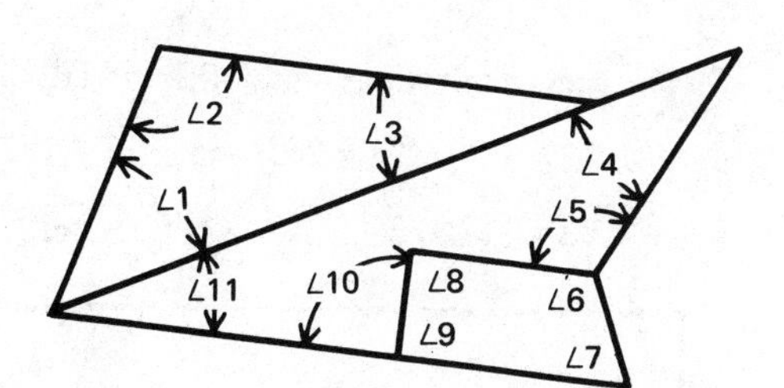

3. Name all pairs of alternate interior angles and all pairs of corresponding angles shown in the figure.

   a. alternate interior angles ____________

   b. corresponding angles ____________

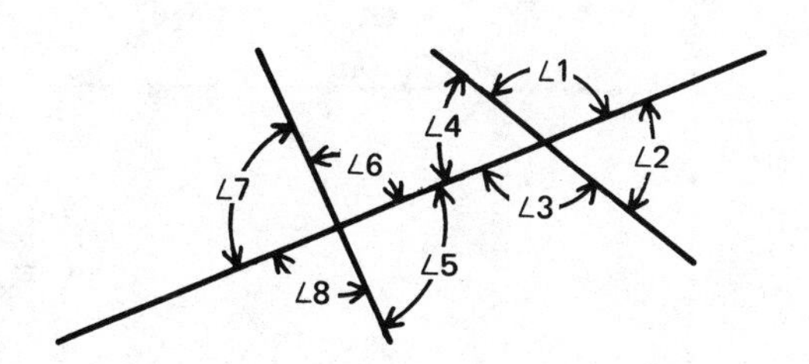

C. APPLICATIONS OF GEOMETRIC PRINCIPLES

Solve the following problems.

1. Determine the values of ∠'s 1-5.

   ____

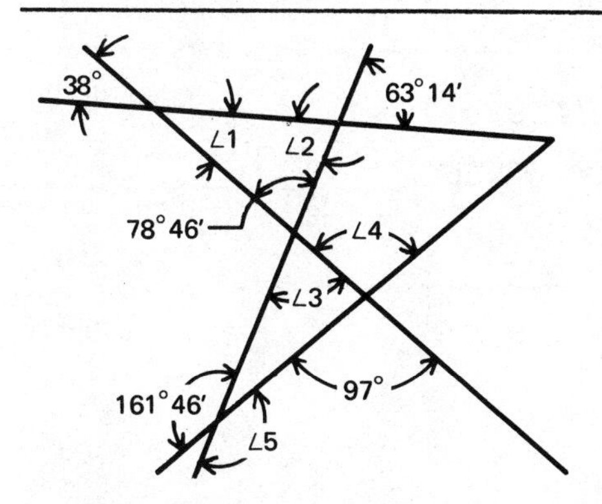

2. Determine the values of ∠'s 2, 3, and 4 when ∠1 =

   a. 28° ____________

   b. 33° 17′ ____________

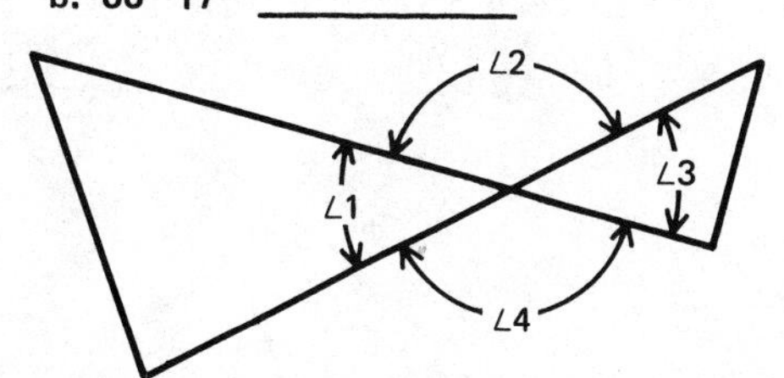

3. Given: AB || CD: Determine the values of ∠'s 2 - 8 when ∠1 =

   a. 68° ____________

   b. 52° 55′ ____________

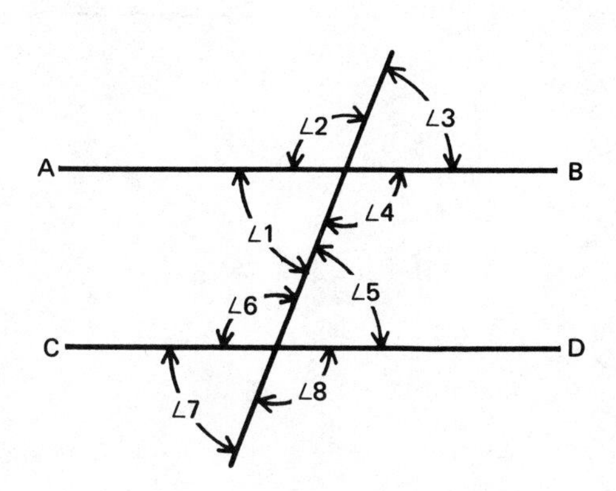

4. Given: Hole centerlines EF || GH and MP || KL. Determine the values of angles 1 - 15 when ∠16 =

   a. 73° ____________

   b. 87° 08′ ____________

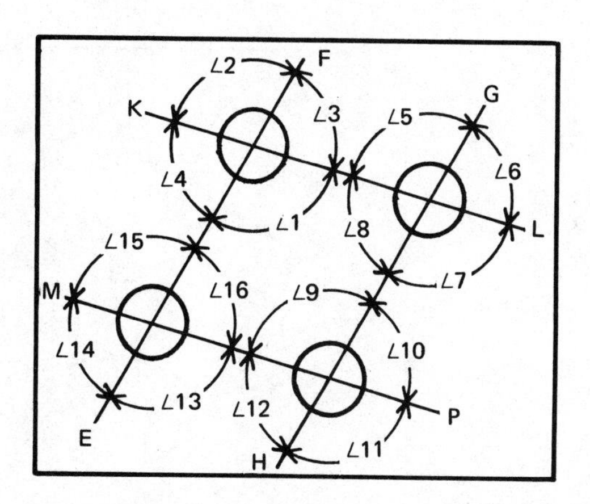

5. Given: Hole centerlines AB ∥ CD and DE ∥ FH. Determine the values of angles 1 – 22 when

a. $\angle 23 = 97°$, $\angle 24 = 34°$, and $\angle 25 = 102°$ ________________

b. $\angle 23 = 112° 23'$, $\angle 24 = 27° 53'$, and $\angle 25 = 95° 18'$ ________________

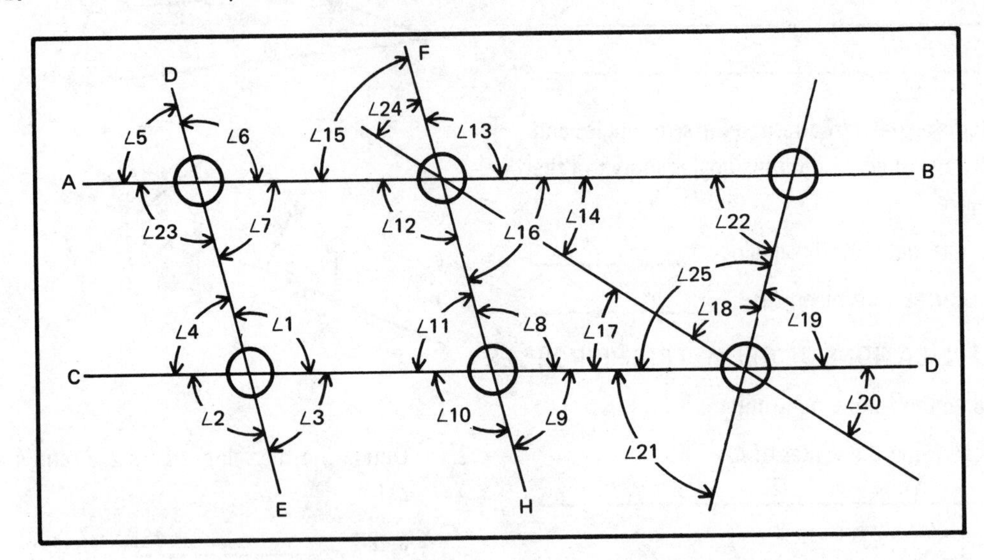

6. Given: AB ∥ CD, AC ∥ ED. Determine the value of ∠'s 2 and 3 when $\angle 1 =$

a. 68° ________________ b. 77° 26' ________________

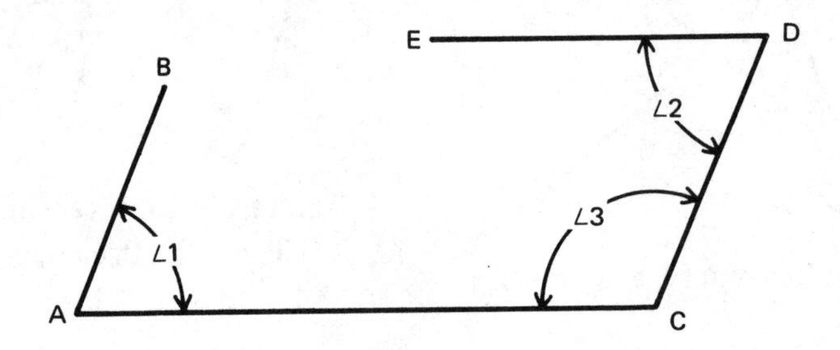

7. Given: FH ∥ GS ∥ KM and FG ∥ HK. Determine the values of ∠'s 1, 2, and 3 when $\angle 4 =$

a. 116° ________________ b. 107° 43' ________________

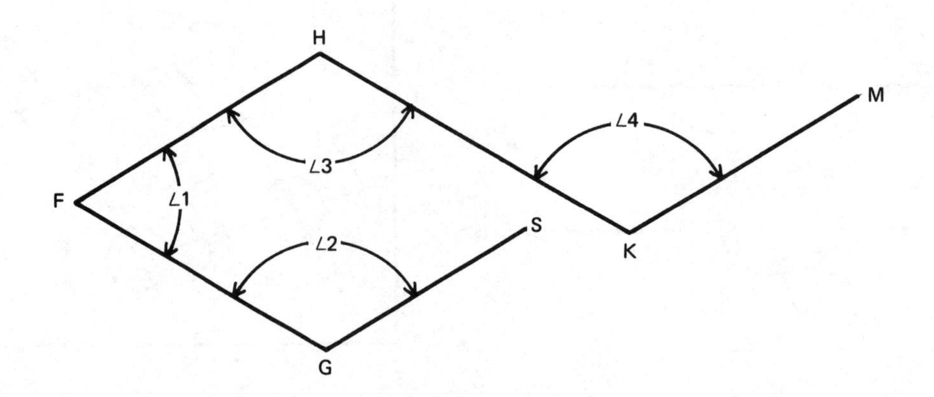

Unit 35 Triangles

OBJECTIVES

After studying this unit, the student should be able to

- Identify different types of triangles.
- Determine unknown angles based on the principle that all triangles contain 180°.
- Determine corresponding parts of triangles.

A *triangle* is a figure consisting of three connected sides. The symbol △ means triangle.

TYPES OF TRIANGLES

- A *scalene triangle* has three unequal sides; therefore, it also has three unequal angles. Triangle ABC shown in figure 35-1 is scalene. Sides AB, AC, and BC are unequal in length, and angles A, B, and C are of unequal size.

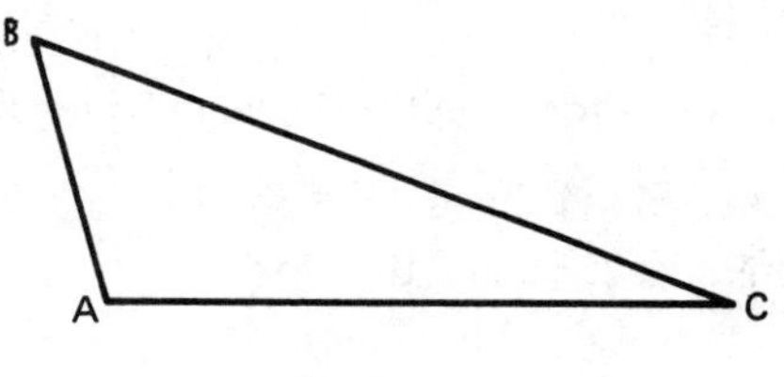

Fig. 35-1

- An *isosceles triangle* has two equal sides. Therefore, it also has two equal base angles. *Base angles* are the angles that are opposite the equal sides. For example, in isosceles triangle ABC shown in figure 35-2, side AC = side BC and ∠A = ∠B.

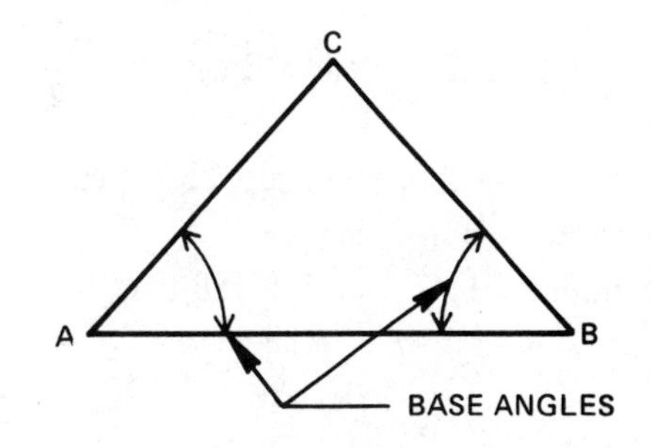

Fig. 35-2

- An *equilateral triangle* has three equal sides. Therefore, it also has three equal angles. In equilateral triangle DEF shown in figure 35-3, sides DE = DF = EF and ∠D = ∠E = ∠F.

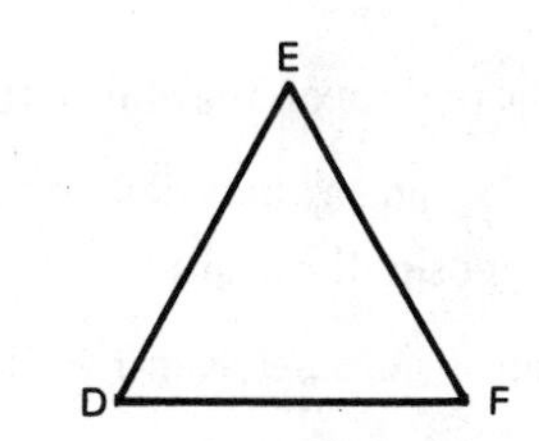

Fig. 35-3

- A right triangle has one right angle. The small arc ⌝ shown at the vertex of an angle means a right angle. The side opposite the right angle is called the *hypotenuse.* Figure 35-4 shows right triangle HJK with ∠H = 90° and hypotenuse JK.

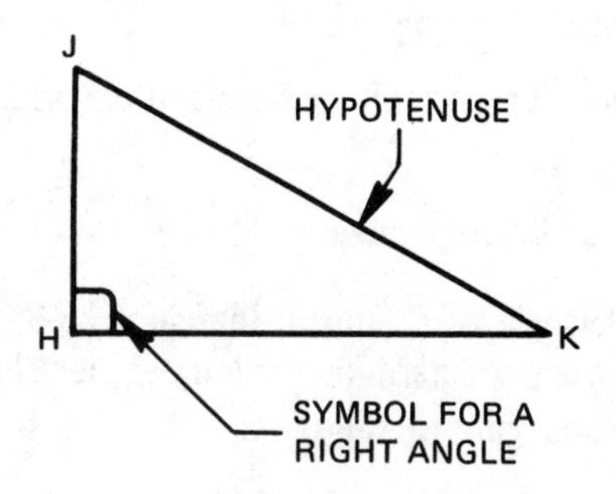

Fig. 35-4

Principle 6

- **The sum of the angles of any triangle is equal to 180°.**

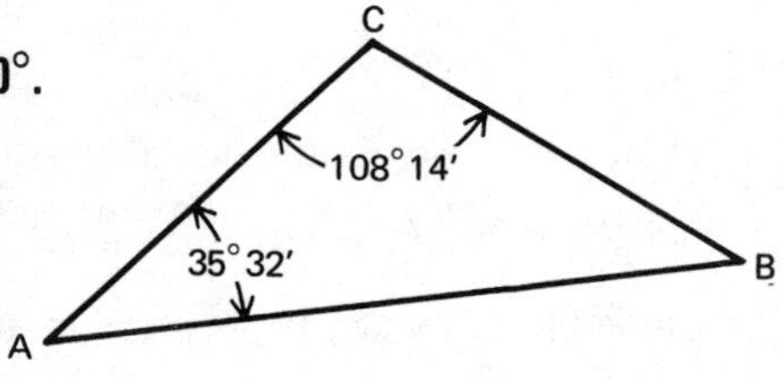

Fig. 35-5

Applications of Principle 6:

1. In triangle ABC shown in figure 35-5,

   Given: $\angle A = 35° \; 32'$, $\angle C = 108° \; 14'$.
   Determine $\angle B$.
   Solution: $\angle A + \angle B + \angle C = 180°$
   $\angle B = 180° - (\angle A + \angle C)$

Note: $180° = 179° \; 60'$, therefore, $\angle B = 179° \; 60' - (35° \; 32' + 108° \; 14')$; $\angle B = 179° \; 60' - 143° \; 46'$; $\angle B = 36° \; 14'$

2. In isosceles triangle EFG shown in figure 35-6,

   Given: $EF = EG$ and $\angle E = 33° \; 18'$.
   Determine $\angle F$ and $\angle G$.
   Solution: $\angle E + \angle F + \angle G = 180°$;
   $180° - \angle E = \angle F + \angle G$; $180° - 33° \; 18' = 146° \; 42'$

   Since $\angle F = \angle G$, $\angle F$ and $\angle G$ each $= \dfrac{146° \; 42'}{2} = 73° \; 21'$

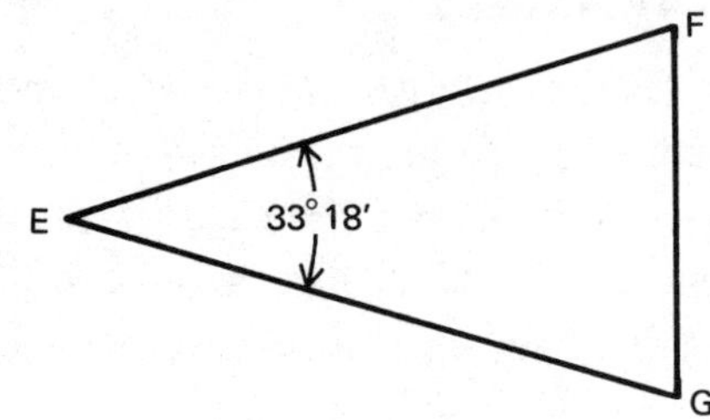

Fig. 35-6

3. Triangle HJK shown in figure 35-7 is equilateral.
   Determine $\angle$'s H, J, and K.
   Solution: $\angle H + \angle J + \angle K = 180°$.
   Since $\angle H = \angle J = \angle K$, each angle $= \dfrac{180°}{3} = 60°$

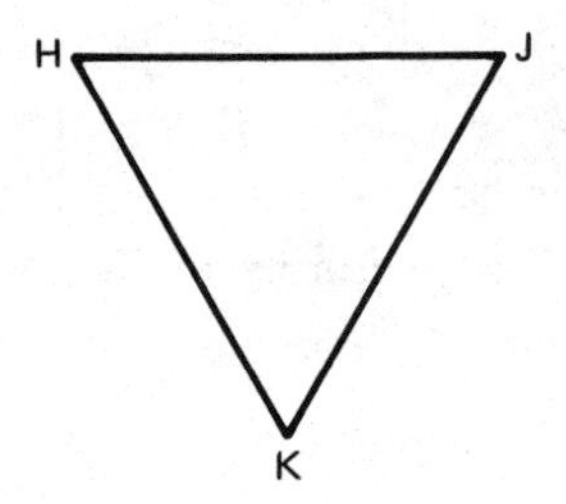

Fig. 35-7

CORRESPONDING PARTS OF TRIANGLES

It is essential to develop the ability to identify corresponding angles and sides of two or more triangles.

- Corresponding angles between two triangles are determined by comparing the lengths of the sides which lie opposite the angles.
- Corresponding sides between two triangles are determined by comparing the sizes of the angles which lie opposite the sides.
- The smallest angle of a triangle lies opposite the shortest side and the largest angle of a triangle lies opposite the longest side.

Note: Corresponding sides and angles between triangles are not determined by the positions of the triangles.

Examples of Correspondence:

1. In triangle ABC, shown in figure 35-8, the given angles determine the longest, next longest, and shortest sides.

▲ The longest side is CB since it lies opposite the largest angle, 107°.

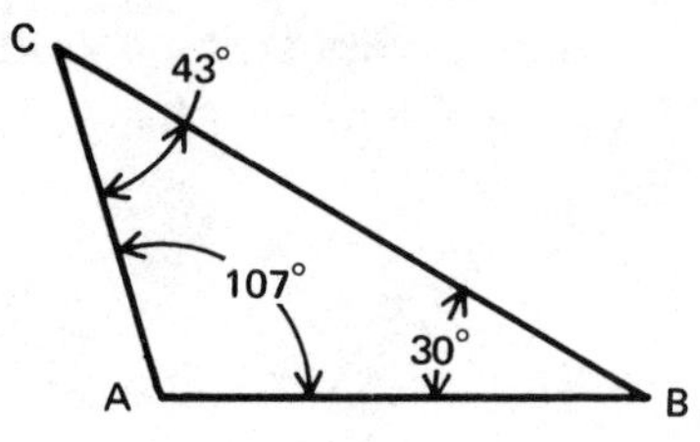

Fig. 35-8

▲ The next longest side is AB since it lies opposite the next largest angle, 43°.

▲ The shortest side is AC since it lies opposite the smallest angle, 30°.

2. In triangle DEF, shown in figure 35-9, the given sides determine the largest, next largest, and smallest angle.

▲ The largest angle is ∠ E since it lies opposite the longest side, 10-inch.

▲ The next largest angle is ∠ D since it lies opposite the next longest side, 7-inch.

▲ The smallest angle is ∠ F since it lies opposite the shortest side, 4-inch.

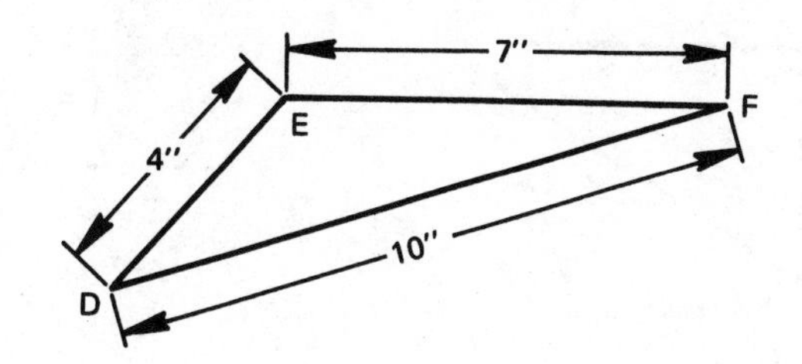

Fig. 35-9

3. In triangles ABC and FED shown in figure 35-10, the given sides determine the pairs of corresponding angles between the two triangles.

▲ Angle C corresponds to ∠ D since each angle lies opposite the longest side of each triangle.

▲ Angle B corresponds to ∠ E since each angle lies opposite the next longest side of each triangle.

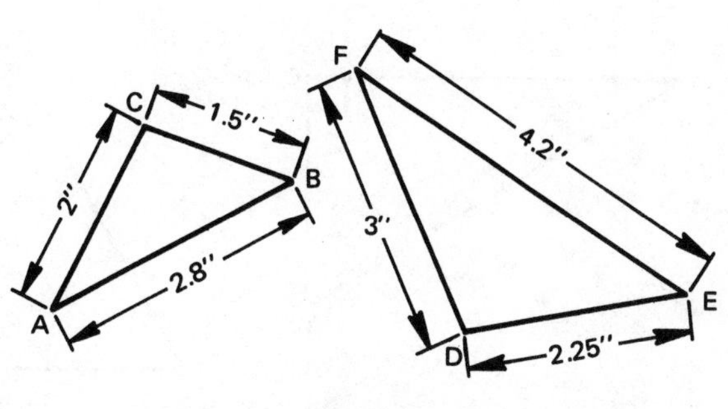

Fig. 35-10

▲ Angle A corresponds to ∠ F since each angle lies opposite the shortest side of each triangle.

APPLICATION

A. Identify each of the triangles 1-8 as scalene, isosceles, equilateral, or right.

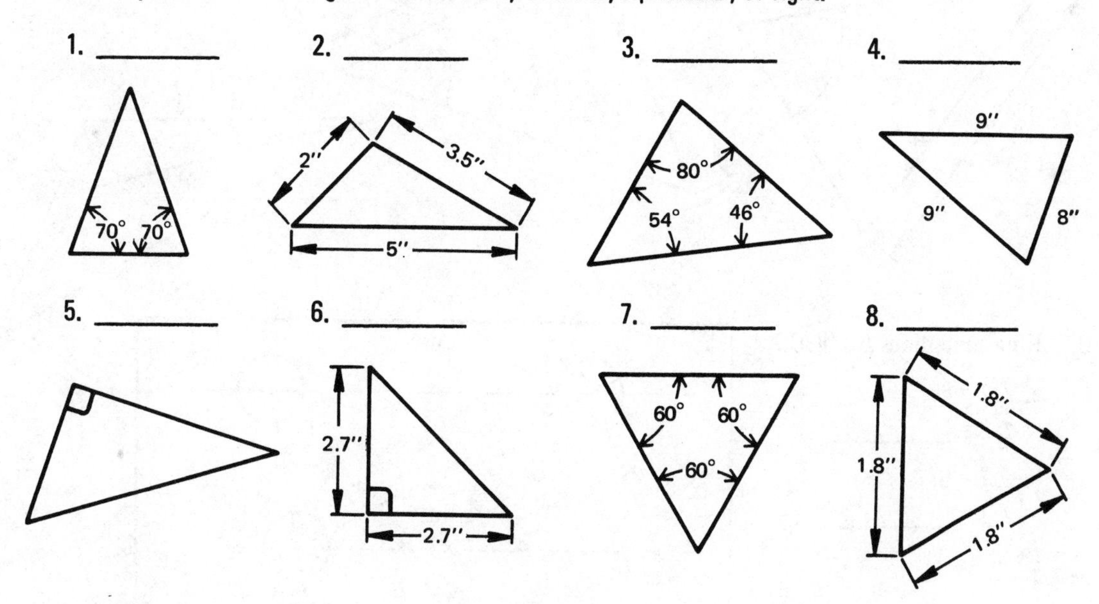

B. Determine the values of the unknown angles for each of the following problems.

1. $\angle A + \angle B + \angle C =$ ____

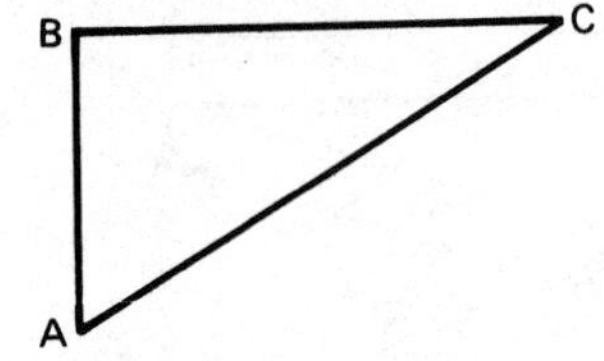

2. a. $\angle 1 = 56°, \angle 2 = 88°$
   $\angle 3 =$ ________
   b. $\angle 2 = 79°, \angle 3 = 46°$
   $\angle 1 =$ ________

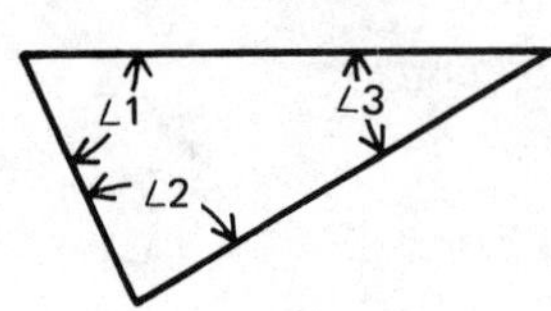

3. a. $\angle 4 = 32° 43', \angle 5 = 119° 17'$
   $\angle 6 =$ ________
   b. $\angle 5 = 123° 17' 13'', \angle 6 = 27°$
   $\angle 4 =$ ________

4. a. $\angle A = 19° 43', \angle B =$ ____
   b. $\angle B = 67° 58', \angle A =$ ____

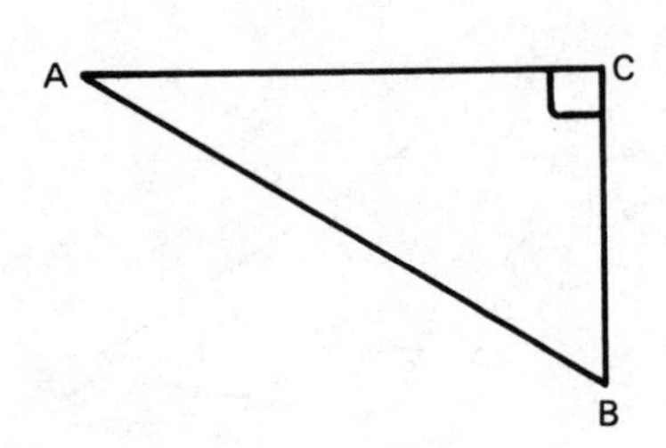

5. BC = 14.2″
   AB = ____ AC = ____

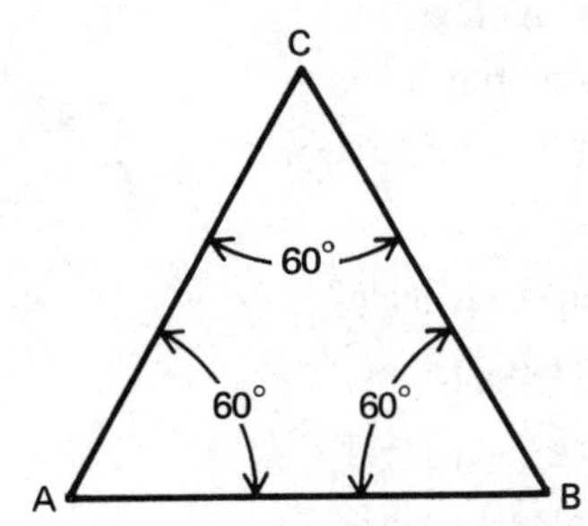

6. a. $\angle E = 78°, \angle G =$ ____
   b. $\angle G = 84° 19', \angle F =$ ____

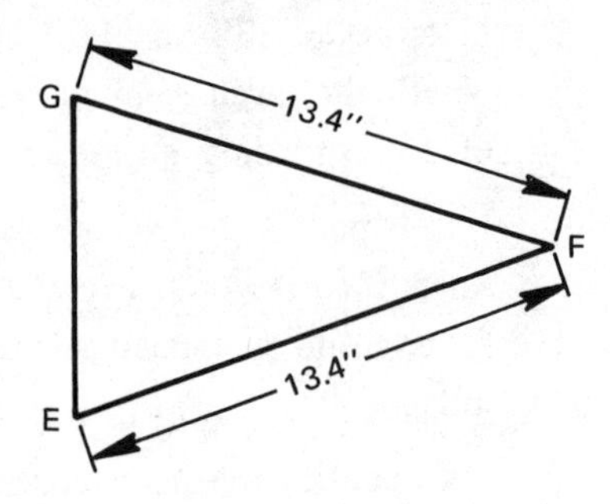

7. a. $\angle 3 = 17°$, $\angle 1 =$ ____
   b. $\angle 3 = 25° 19'$, $\angle 2 =$ ____

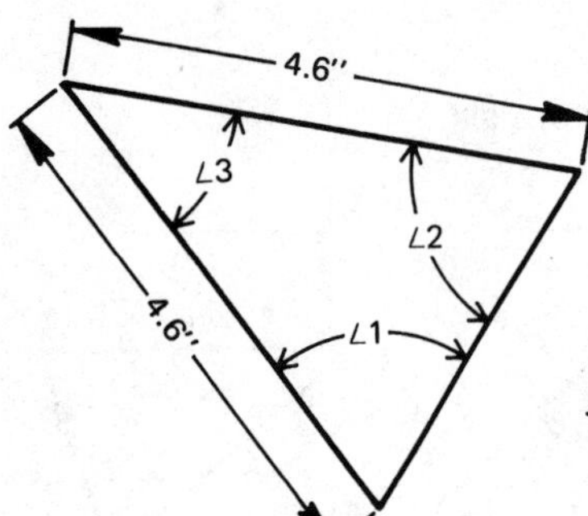

8. a. $\angle 3 =$ ____
   b. $\angle 4 =$ ____

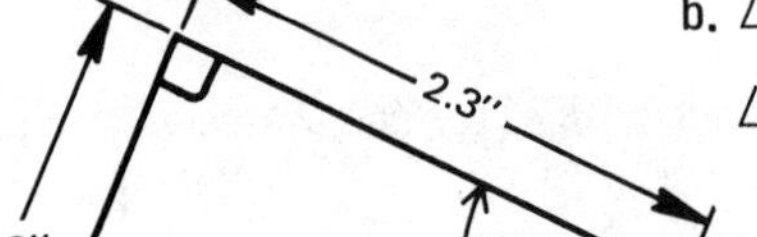
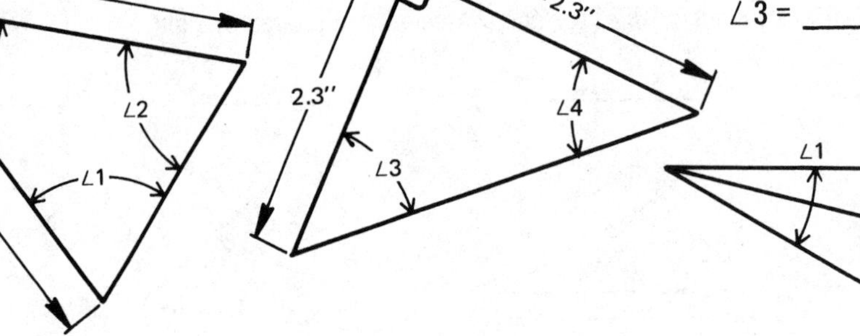

9. a. $\angle 1 = 26°, \angle 3 = 48°$
   $\angle 2 =$ ____
   b. $\angle 1 = 28°, \angle 2 = 15°$
   $\angle 3 =$ ____

10. Hole centerlines AB ∥ CD
    a. $\angle 1 = 86° 15'$
    $\angle 2 =$ ________
    b. $\angle 2 = 67° 17'$,
    $\angle 1 =$ ________

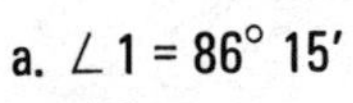

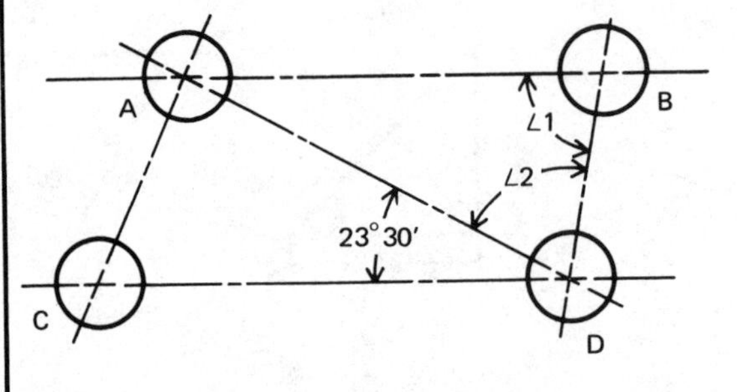

11. a. $\angle 3 =$ ____

b. $\angle 4 =$ ____

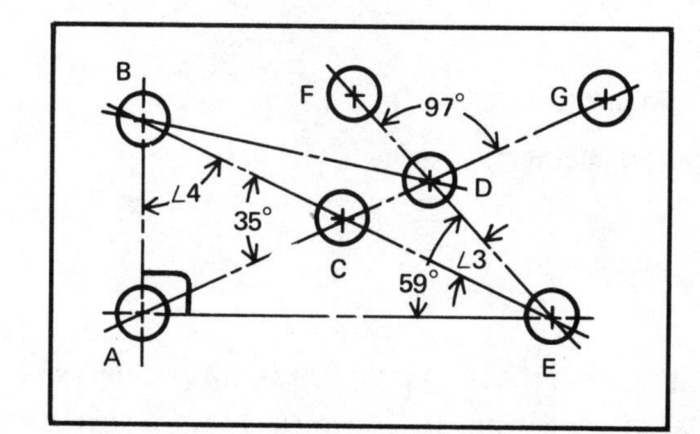

12. AB ∥ DE, BC is an extension of AB.

a. $\angle E = 66° 13'$, $\angle A =$ ________

b. $\angle A = 19° 22'$, $\angle E =$ ________

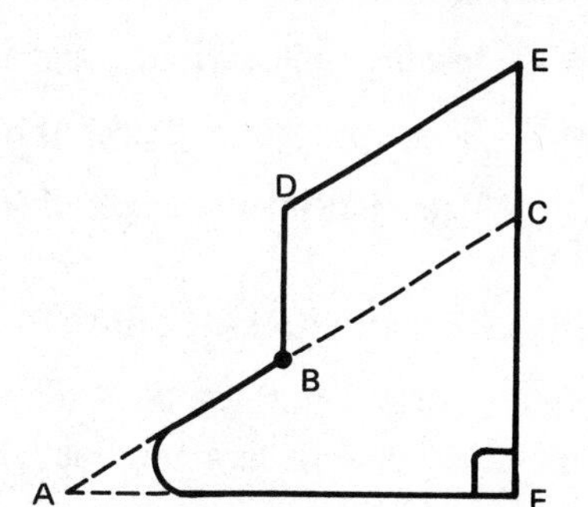

C. Determine the answers to the following problems which are based on corresponding parts.

1. a. The largest angle is ______

   b. The next largest angle is ______

   c. The smallest angle is ______

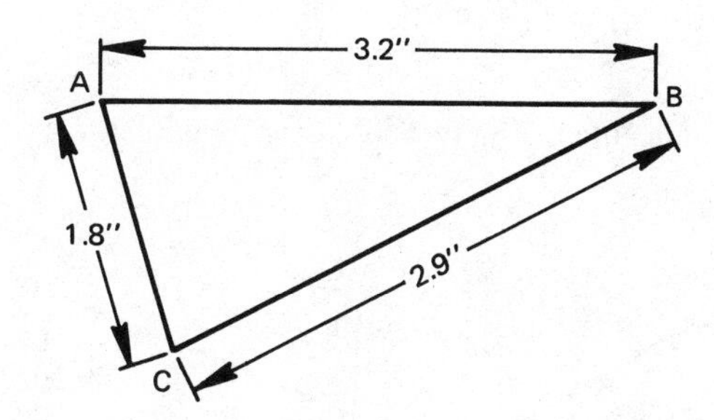

2. a. The shortest side is ______

   b. The next shortest side is ______

   c. The longest side is ______

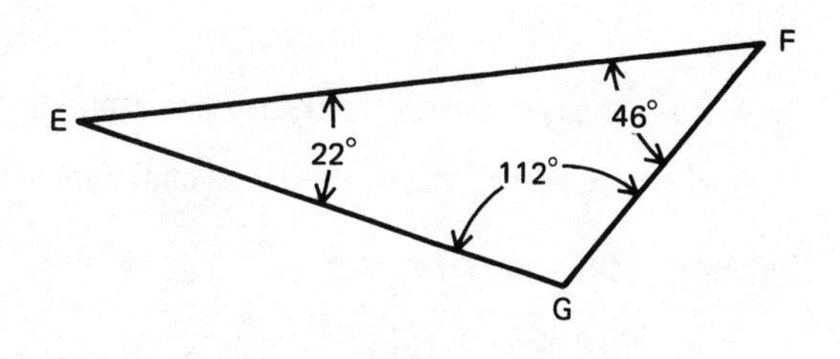

3. a. $\angle A$ corresponds to $\angle$ ______

   b. $\angle B$ corresponds to $\angle$ ______

   c. $\angle 1$ Corresponds to $\angle$ ______

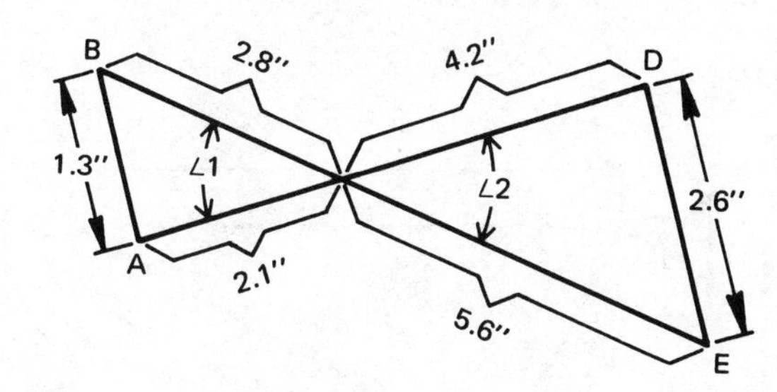

4. a. $\angle F$ corresponds to $\angle$ ______

   b. $\angle G$ corresponds to $\angle$ ______

   c. $\angle H$ corresponds to $\angle$ ______

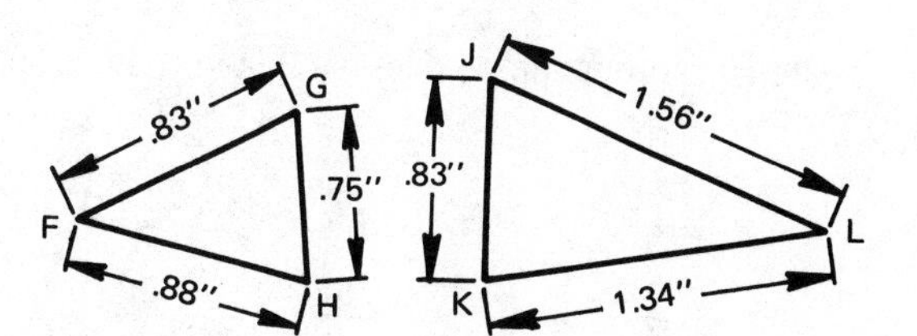

Unit 36 Triangles and Other Common Polygons

OBJECTIVES

After studying this unit, the student should be able to

- Identify similar triangles and compute unknown angles and sides.
- Compute angles and sides of isosceles, equilateral, and right triangles.
- Determine interior angles of any polygon.

CONGRUENT AND SIMILAR TRIANGLES

- Two triangles are *congruent* if they are identical in size and shape. If one triangle is placed on top of the other, they fit together exactly. The symbol ≅ means congruent. Congruent triangles are shown in figure 36-1.
- Corresponding parts of congruent triangles are equal: $\angle A = \angle D, \angle B = \angle E, \angle C = \angle F$; AB = DE, AC = DF, BC = EF.

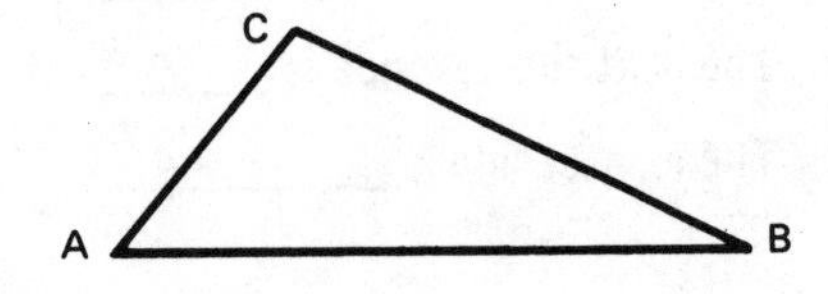

Fig. 36-1

- Two triangles are *similar* if their corresponding angles are equal. The symbol ~ means similar.
- Corresponding sides of similar triangles are proportional.

Examples of Similar Triangles:

1. In figure 36-2,

   ▲ Triangles ABC and DEF have equal corresponding angles.

   ▲ Therefore, triangle ABC is similar to triangle DEF.

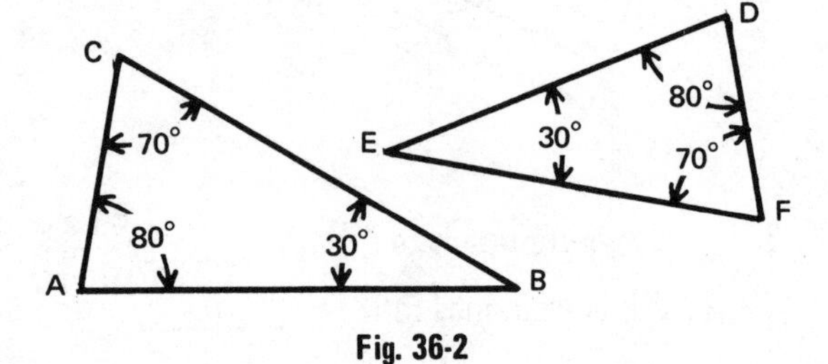

Fig. 36-2

2. In figure 36-3, the lengths of the sides of triangles HJK and LMN are given.

   ▲ The corresponding sides are shown to be proportional:

$$\frac{HJ}{LM} = \frac{JK}{MN} = \frac{HK}{LN}$$

$$\frac{2}{4} = \frac{4}{8} = \frac{5}{10}$$

$$\frac{1}{2} = \frac{1}{2} = \frac{1}{2}$$

   ▲ Therefore, triangle HJK is similar to triangle LMN.

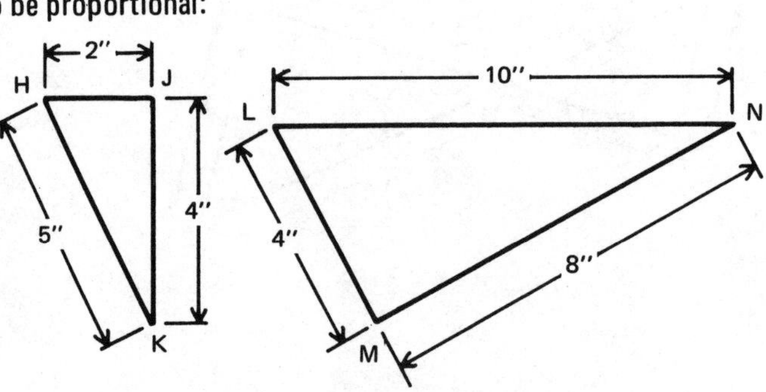

Fig. 36-3

3. In figure 36-4, triangle PRS is similar to triangle TWY.

▲ Determine the lengths of sides PR and TY

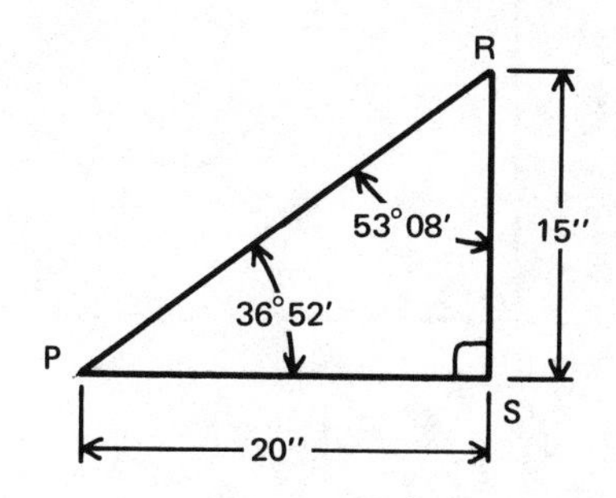

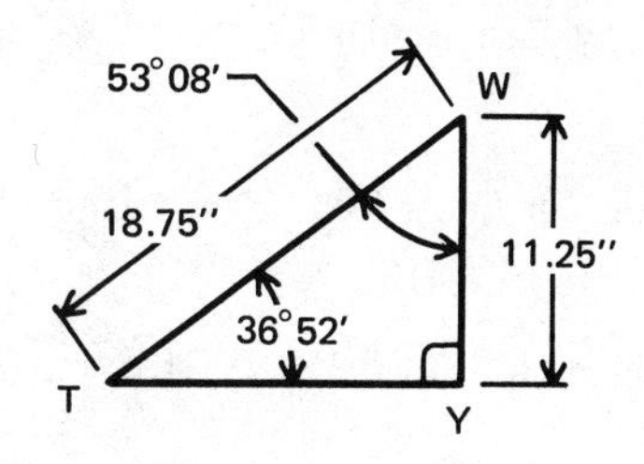

Fig. 36-4

▲ Set up proportions

▲ Solve for the unknown sides

a. $\frac{WY}{RS} = \frac{WT}{PR}$, $\frac{11.25}{15} = \frac{18.75}{PR}$, $PR = \frac{15(18.75)}{11.25}$, PR = 25 inches

b. $\frac{WY}{RS} = \frac{TY}{PS}$, $\frac{11.25}{15} = \frac{TY}{20}$, $TY = \frac{20(11.25)}{15}$, TY = 15 inches

Principle 7

- **Two triangles are similar if their sides are respectively parallel.**

  Example:

  ▲ Given: in figure 36-5, AB || DE, AC || DF, and BC || EF.

  ▲ Conclusion: $\triangle ABC \sim \triangle DEF$.

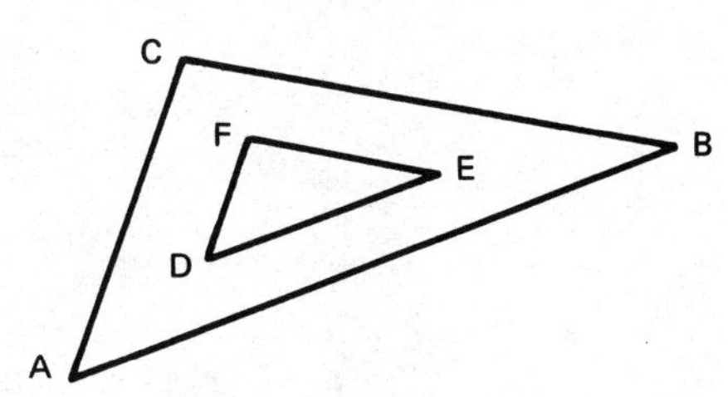

Fig. 36-5

- **Two triangles are similar if their sides are respectively perpendicular.**

  Example:

  ▲ Given: in figure 36-6, $HJ \perp LM$, $HK \perp LP$, and $JK \perp MP$.

  ▲ Conclusion: $\triangle HJK \sim \triangle LMP$.

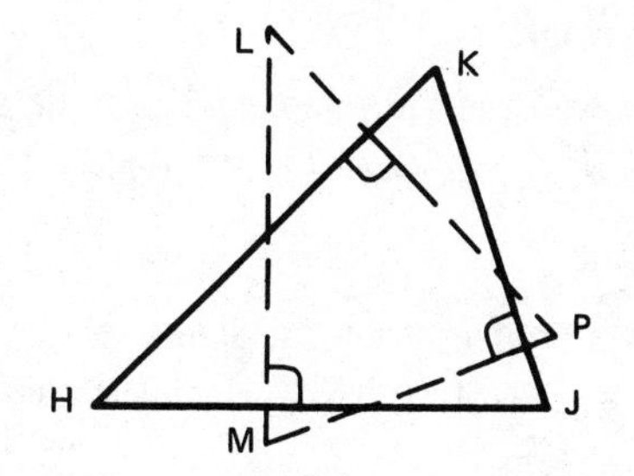

Fig. 36-6

- **Within a triangle, if a line is drawn parallel to one side, the triangle formed is similar to the original triangle.**

  Example:

  ▲ Given: in figure 36-7, line DE is || to side BC.

  ▲ Conclusion: $\triangle ADE \sim \triangle ABC$.

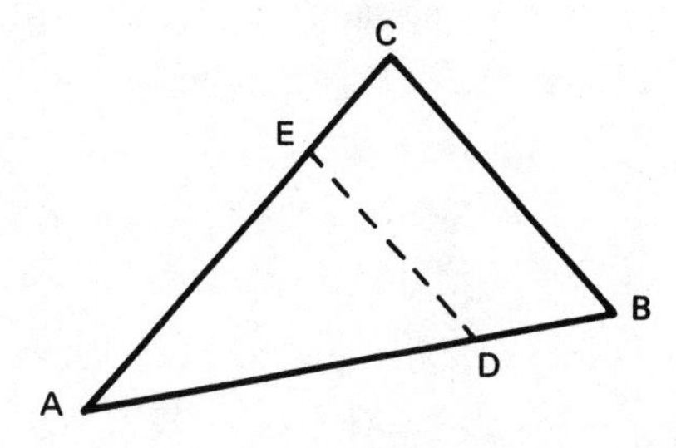

Fig. 36-7

- **In a right triangle, if a line is drawn from the vertex of the right angle perpendicular to the opposite side, the two triangles formed and the original triangle are similar.**

Example:

▲ Given: in figure 36-8, right triangle HFG with line FL ⊥ HG.

▲ Conclusion: △FLH ~ △GLF ~ △GFH.

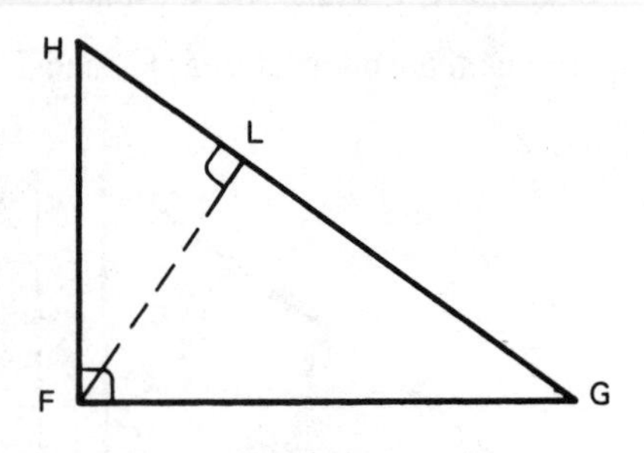

Fig. 36-8

Principle 8

- **In an isosceles triangle, an altitude to the base bisects the base and the vertex angle.**

Note: An *altitude* is a line drawn from a vertex perpendicular to the opposite side.

To *bisect* means to divide into two equal parts.

Example:

▲ Given: in figure 36-9, isosceles triangle ABC with AC = CB and line CD the altitude to base AB.

▲ Conclusion: AD = BD and ∠1 = ∠2.

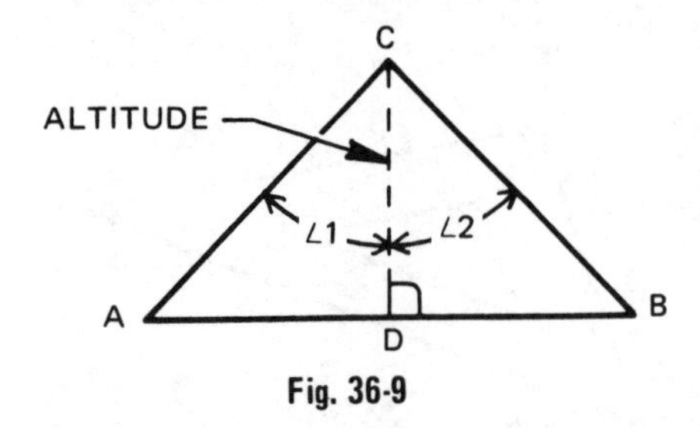

Fig. 36-9

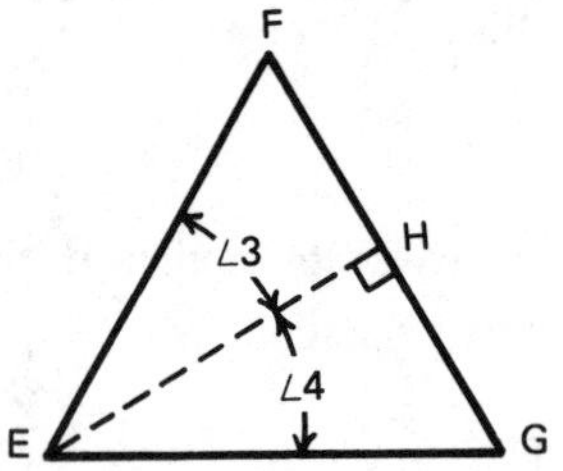

Fig. 36-10

- **In an equilateral triangle, an altitude to any side bisects the side and the vertex angle.**

Example:

▲ Given: in figure 36-10, equilateral triangle EFG with line EH the altitude to side FG.

▲ Conclusion: FH = GH and ∠3 = ∠4.

Principle 9

- **In a right triangle, the square of the hypotenuse is equal to the sum of the squares of the other two sides.** Note: This principle, called the *Pythagorean Theorem*, is often used for solving machine problems.

Examples:

1. Given: in figure 36-11, right triangle ABC with side a = 6 inches and side b = 8 inches. Determine side c.

▲ Side c is the hypotenuse, therefore, $c^2 = a^2 + b^2$.

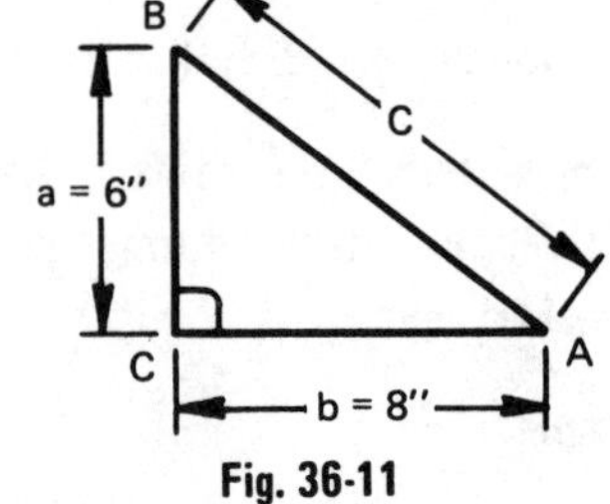

Fig. 36-11

▲ Substitute the given values and solve for side c: $c^2 = a^2 + b^2$, $c^2 = 6^2 + 8^2$, $c^2 = 36 + 64$, $c^2 = 100$, $\sqrt{c^2} = \sqrt{100}$, $c = \sqrt{100}$, c = 10 inches

▲ Check: $10^2 = 6^2 + 8^2$, $100 = 36 + 64$, $100 = 100$.

2. Given: in figure 36-12, right triangle EFG with side f = 5.800 inches and hypotenuse g = 7.200 inches.

   Determine side e:

   ▲ Side g is the hypotenuse; therefore, $g^2 = e^2 + f^2$.

   ▲ Substitute the given values, rearrange the equation, and solve for e:

   $g^2 = e^2 + f^2$; $7.200^2 = e^2 + 5.800^2$; $51.840 = e^2 + 33.640$; $51.840 - 33.640 = e^2$; $18.200 = e^2$; $\sqrt{18.200} = \sqrt{e^2}$; $\sqrt{18.200} = e$; e = 4.266 inches.

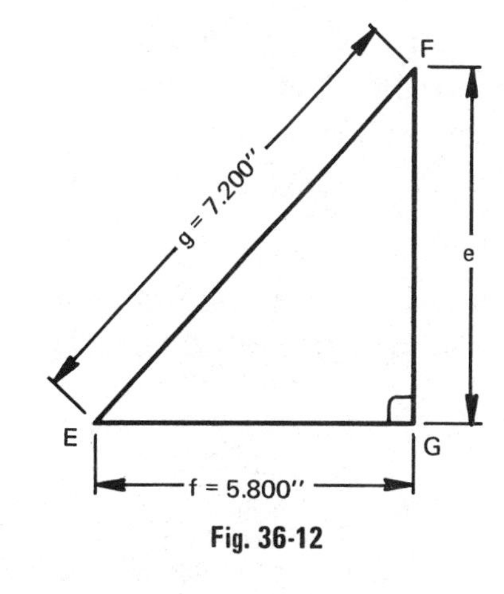

Fig. 36-12

   ▲ Check: $7.200^2 = 4.266^2 + 5.800^2$, 51.840 = 18.200 + 33.640, 51.840 = 51.840

POLYGONS

- A *polygon* is a figure that has three or more connected straight sides.
  The types of polygons most common to machine trade applications in addition to triangles are squares, rectangles, parallelograms, and hexagons.
- A *regular polygon* is one which has equal sides and equal angles.

Examples of Polygons:

▲ A *square* is a regular four-sided polygon. Each angle equals 90°. In the square ABCD shown in figure 36-13, AB = BC = CD = AD and ∠A = ∠B = ∠C = ∠D = 90°.

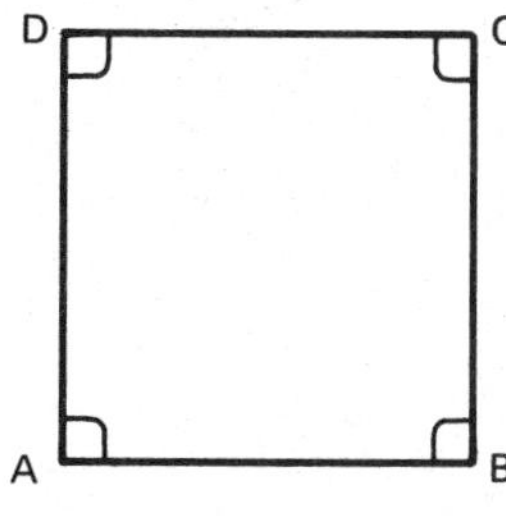

Fig. 36-13

▲ A *rectangle* is a four-sided polygon with opposite sides parallel and equal. Each angle equals 90°. In the rectangle shown in figure 36-14, EF ∥ GH, EH ∥ FG; EF = GH, EH = FG; ∠E = ∠F = ∠G = ∠H = 90°.

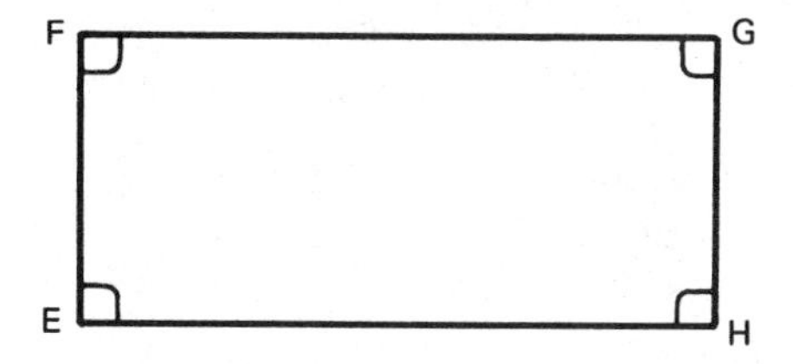

Fig. 36-14

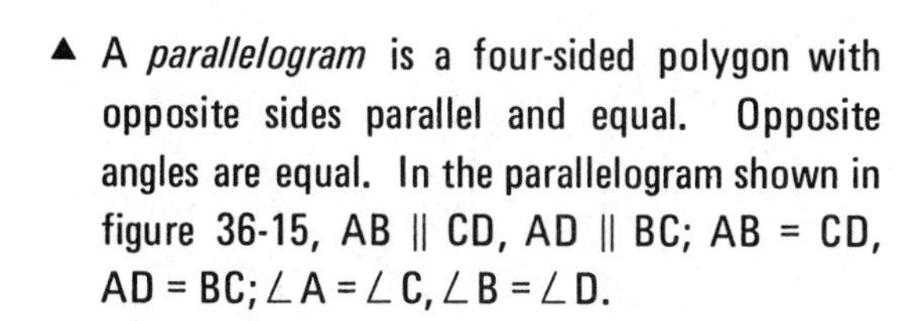

▲ A *parallelogram* is a four-sided polygon with opposite sides parallel and equal. Opposite angles are equal. In the parallelogram shown in figure 36-15, AB ∥ CD, AD ∥ BC; AB = CD, AD = BC; ∠A = ∠C, ∠B = ∠D.

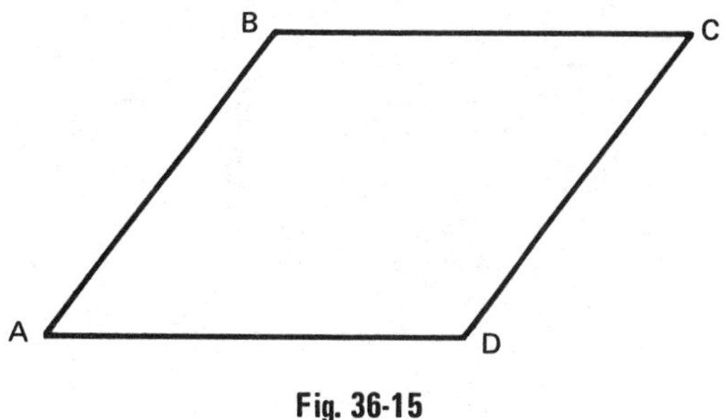

Fig. 36-15

▲ A *regular hexagon* is a six-sided figure with all sides and all angles equal. In the regular hexagon shown in figure 36-16, AB = BC = CD = DE = EF = AF, and ∠A = ∠B = ∠C = ∠D = ∠E = ∠F.

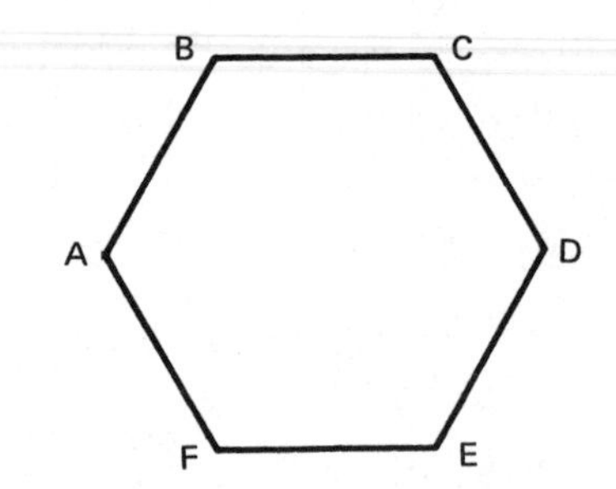

Fig. 36-16

Principle 10

- **The sum of the interior angles of a polygon of N sides is equal to (N - 2) times 180°.**

Examples:

1. Given: in figure 36-17, ∠E = 72°, ∠F = 95°, ∠G = 108°. Determine ∠H.

   ▲ Since EFGH has 4 sides, N = 4.

   ▲ The sum of the 4 angles = (4 - 2)180° = 2(180°) = 360°.

   ▲ Add the 3 given angles and subtract from 360°: ∠H = 360° - (∠E + ∠F + ∠G)
   ∠H = 360° - (72° + 95° + 108°),
   ∠H = 360° - 275°, ∠H = 85°.

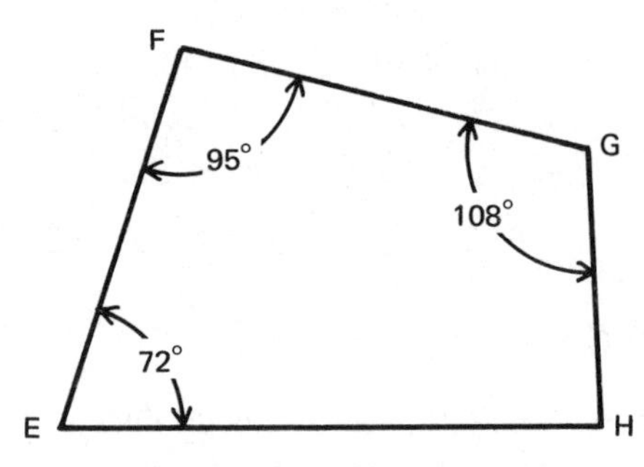

Fig. 36-17

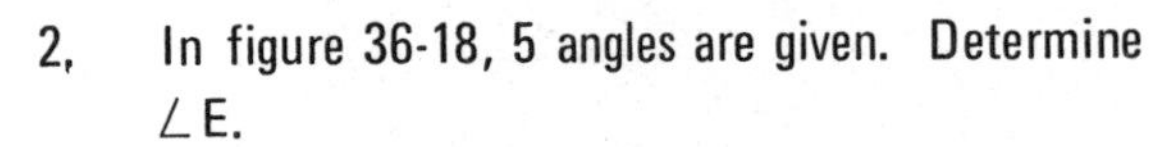

2. In figure 36-18, 5 angles are given. Determine ∠E.

   ▲ Since ABCDEF has 6 sides, N = 6.

   ▲ The sum of the 6 angles = (6 - 2)180° = 4(180°) = 720°.

   ▲ Add the 5 given angles and subtract from 720°. Note: Because ∠B is an exterior angle, it must be subtracted from 360° (a complete revolution) in order to convert it to an interior angle. Angle B = 360° - 114°, ∠B = 246°.

   ∠E = 720° - (∠A + ∠B + ∠C + ∠D + ∠F),
   ∠E = 720° - (57° + 246° + 40° + 77° + 90°), ∠E = 720° - 510°, ∠E = 210°.

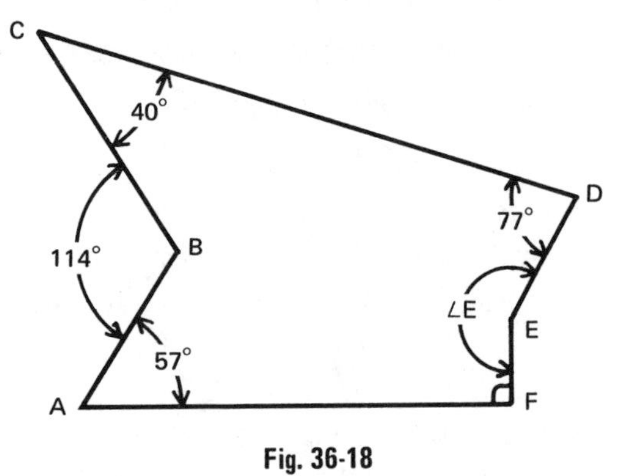

Fig. 36-18

APPLICATION

A. SIMILAR TRIANGLES

Determine which of the following pairs of triangles are similar.

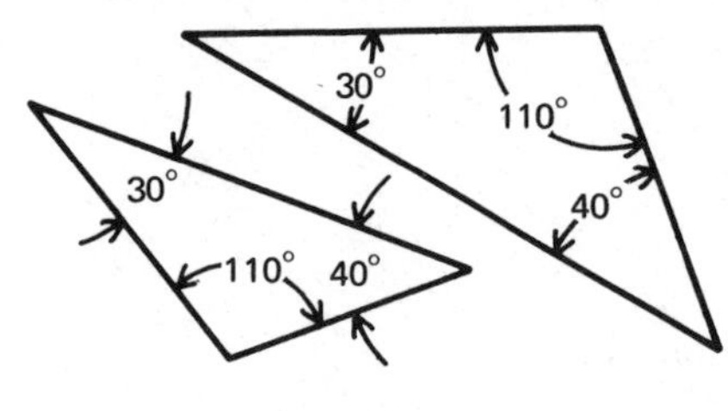

Pair A

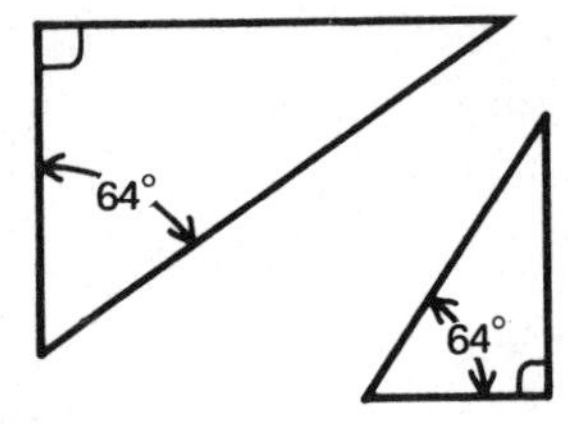

Pair B

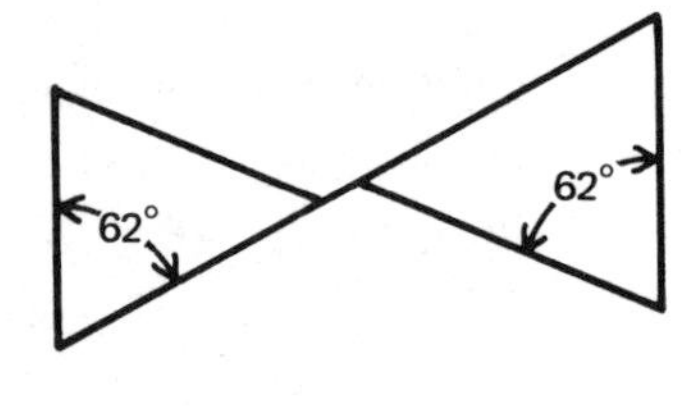

Pair C

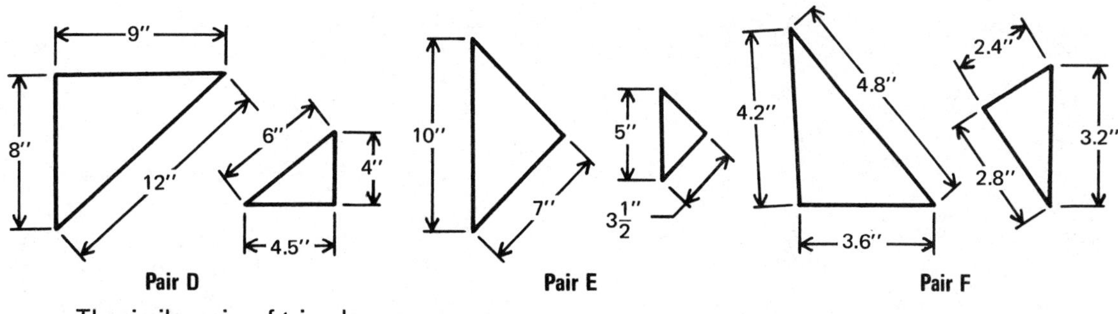

The similar pairs of triangles are ______

B. SIMILAR TRIANGLES

Determine the unknown lengths and angles of the following problems.

1. Given: ∠A = ∠D, ∠B = ∠E, ∠C = ∠F.
   a. Side AC = ______
   b. Side DE = ______

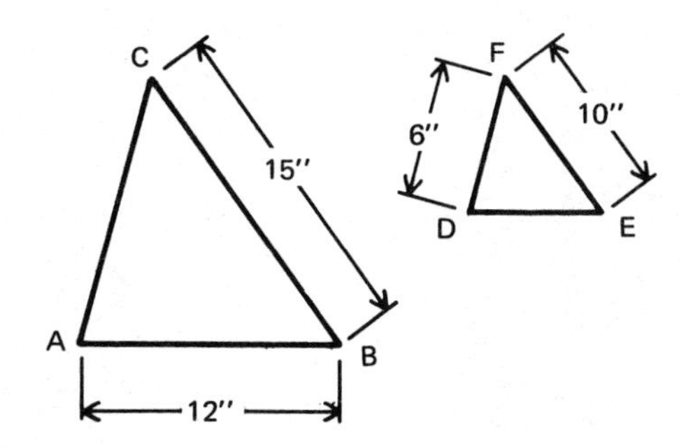

2. Given: ∠H = ∠P, ∠J = ∠M, ∠K = ∠L
   a. Side HK = ______
   b. Side LM = ______

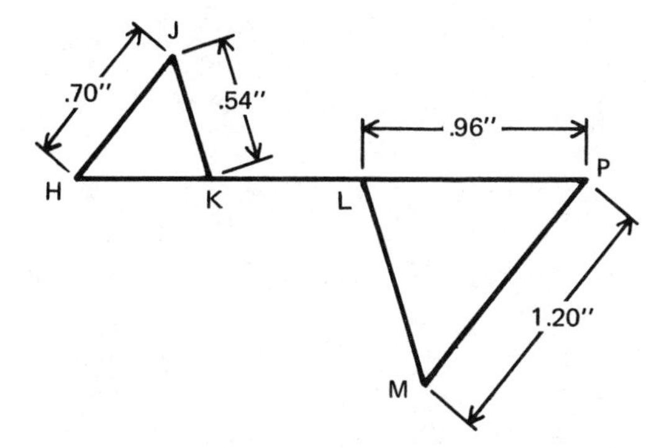

3. Given: AB ∥ DE, AC ∥ DF, BC ∥ EF
   a. ∠A = ______ b. ∠F = ______
   c. ∠B = ______ d. ∠E = ______

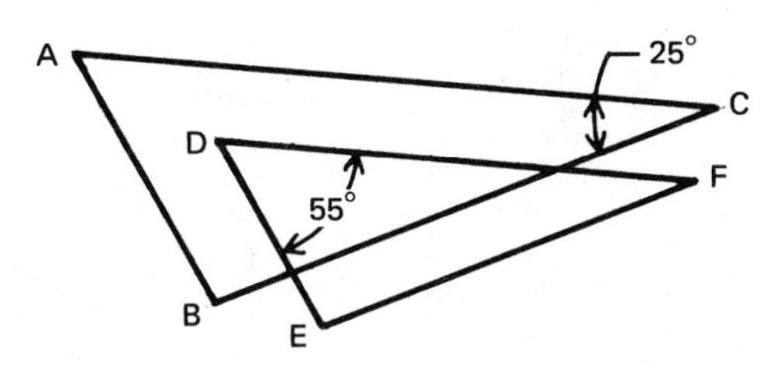

4. a. ∠1 = ______
   b. ∠2 = ______

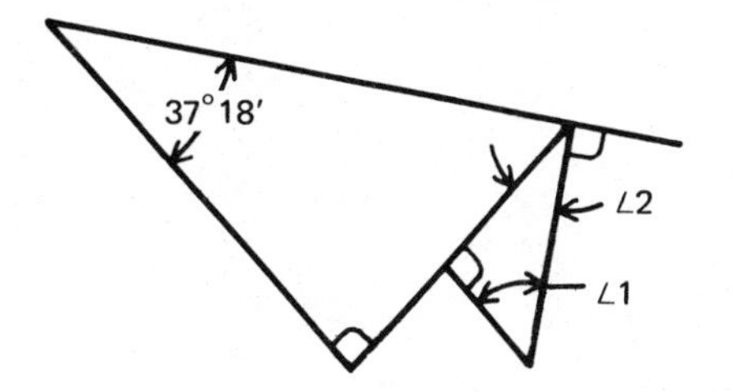

5. Given: PM ∥ JK
   a. ∠HPM = ______ b. ∠PMK = ______

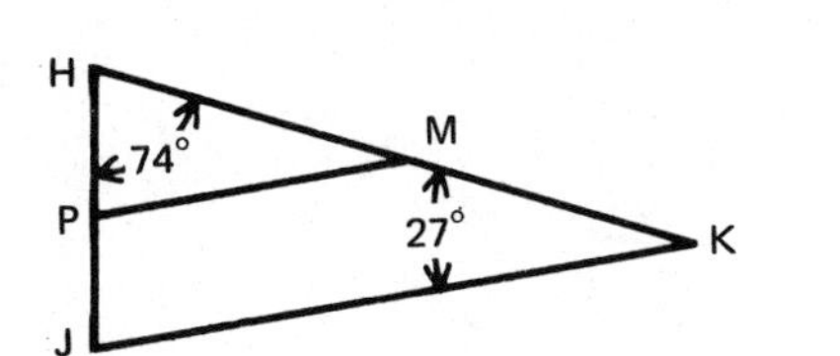

6. a. ∠1 = ______ b. ∠2 = ______
   c. ∠3 = ______

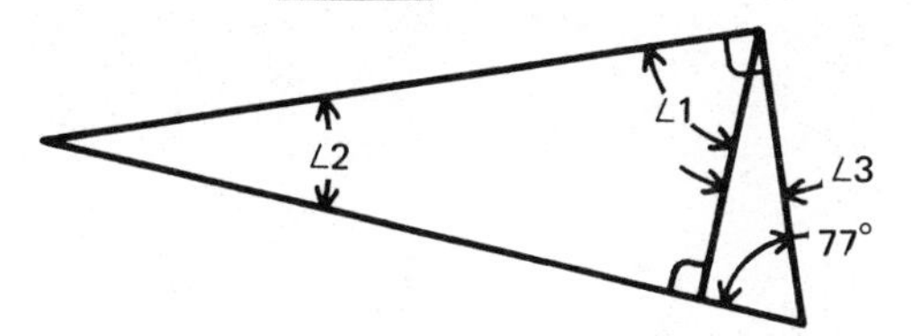

7. a. Dim. A = ____ b. Dim. B = ____

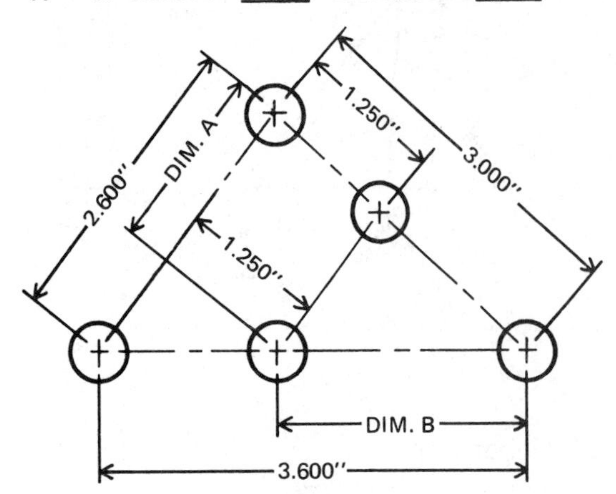

8. Given: AB ∥ DE and CB ∥ EF

a. x = ______ b. y = ______

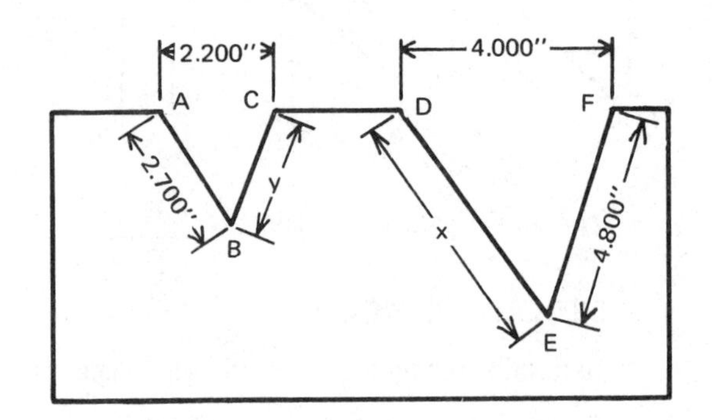

C. ISOSCELES, EQUILATERAL, AND RIGHT TRIANGLES

Determine the unknown values for the following problems.

1. a. x = ______ b. ∠1 = ______

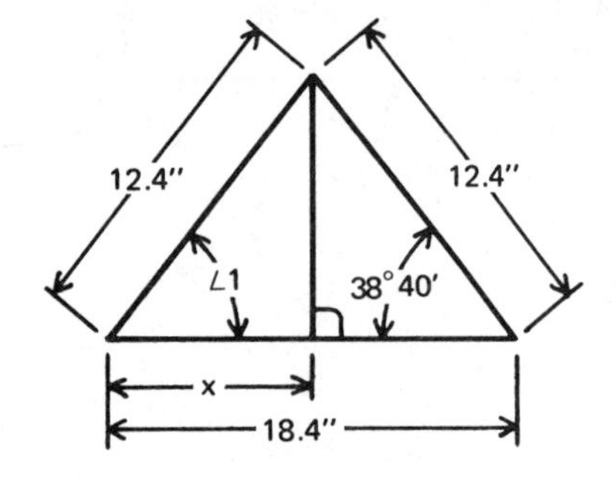

2. a. x = ______ b. y = ______

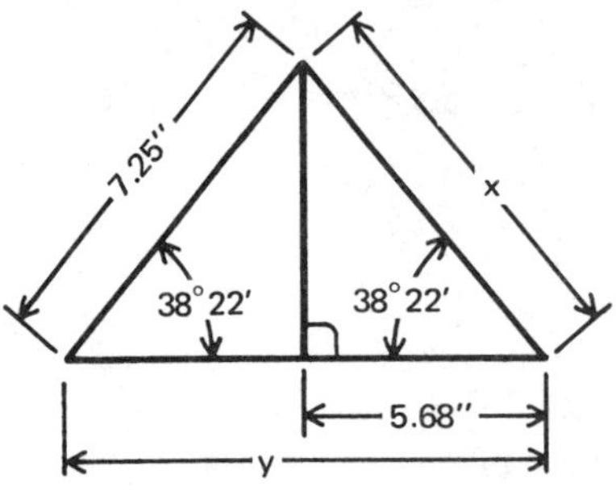

3. a. ∠1 = ______

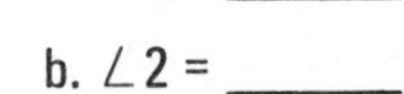

b. ∠2 = ______

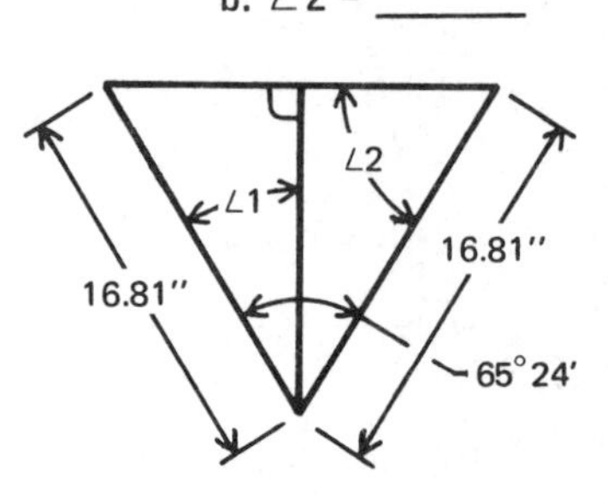

4. a. x = ______ b. y = ______

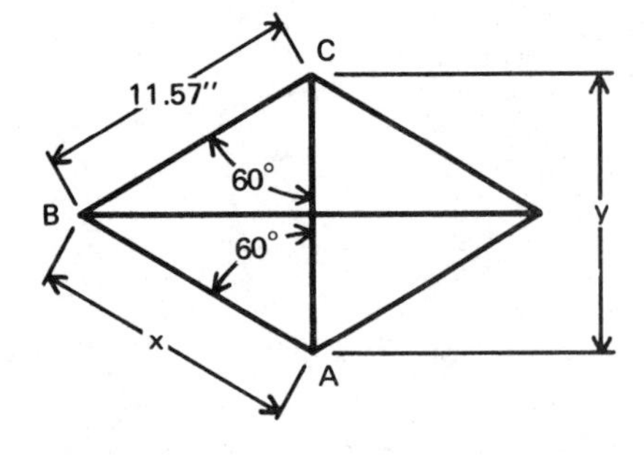

5. a. ∠1 = ______

b. x = ______

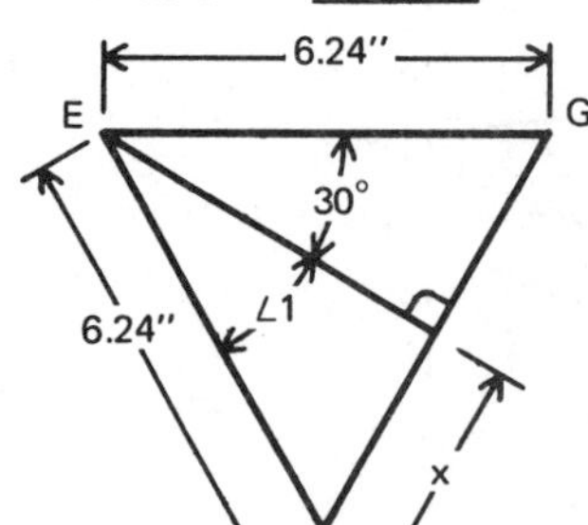

6. a. d = 9", e = 12", x = ______

b. d = 3", e = 4", x = ______

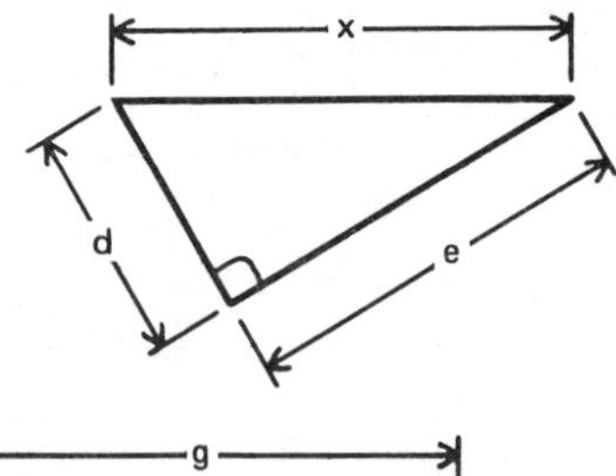

7. a. g = 11.10", m = 12.50", y = ______

b. g = 15.540", m = 17.500", y = ______

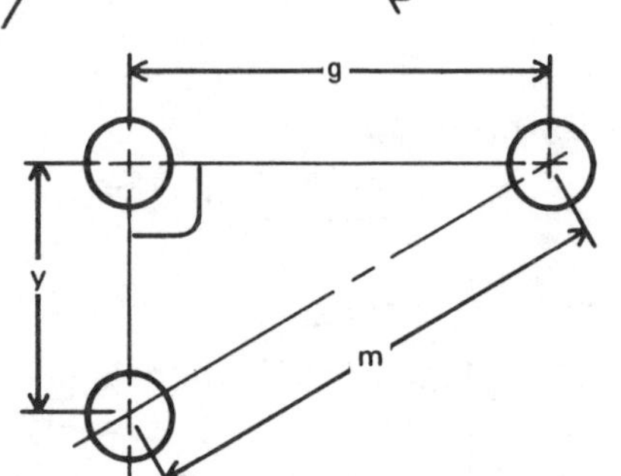

8. a. Radius A = 3.600″, x = 4.800″, y = ______
   b. Radius A = 2.160″, x = 2.880″, y = ______

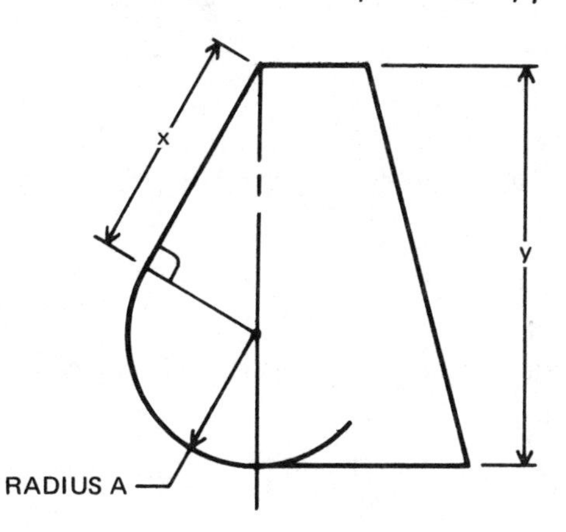

9. a. y = 2.800″, x = ______
   b. y = 3.000″, x = ______

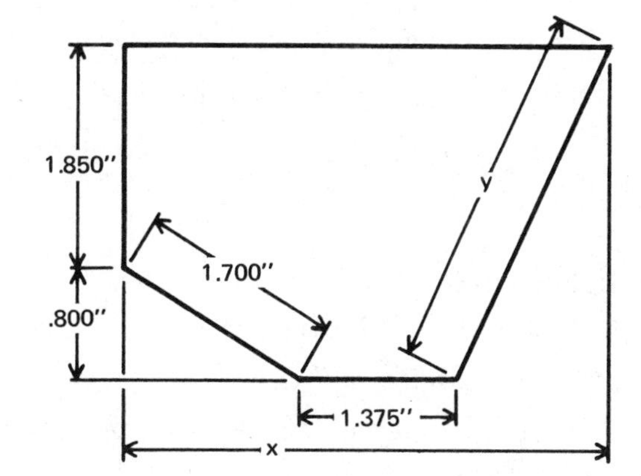

10. a. y = 4.220″, x = ______
    b. y = 4.340″, x = ______

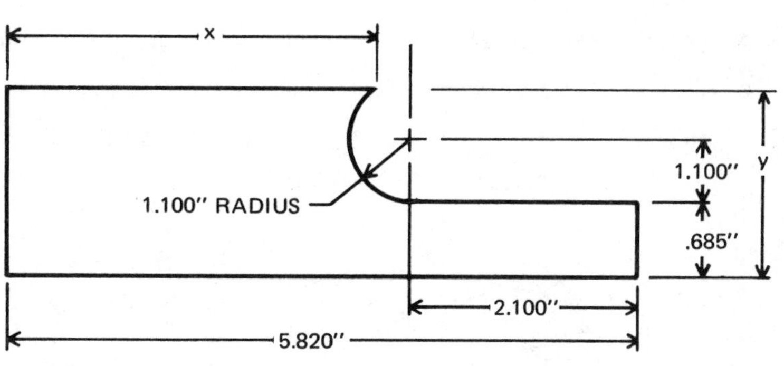

D. POLYGONS

Determine the unknown angles for each of the following problems.

1. a. ∠2 = 87°, ∠1 = ______
   b. ∠1 = 114°, ∠2 = ______

2. a. ∠1 = 116°, ∠2 = ______
   b. ∠2 = 85°, ∠1 = ______

3. a. ∠1 = 37°, ∠2 = ______
   b. ∠1 = 29°, ∠2 = ______

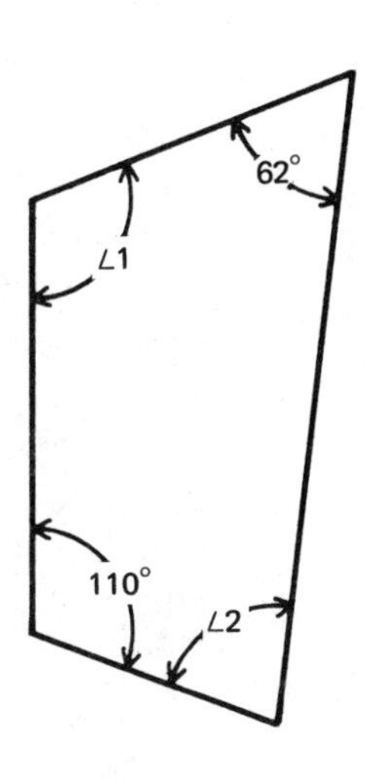

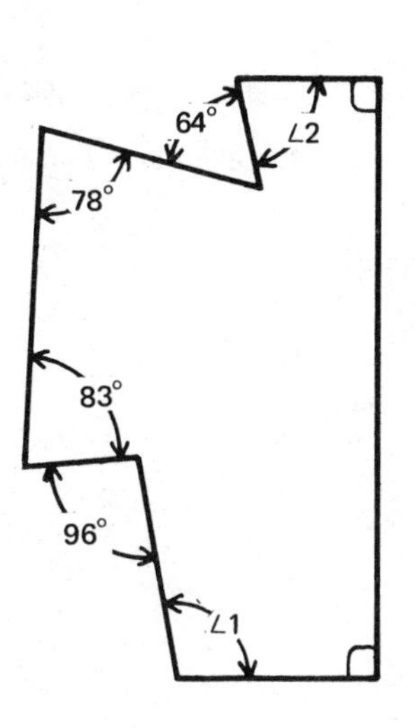

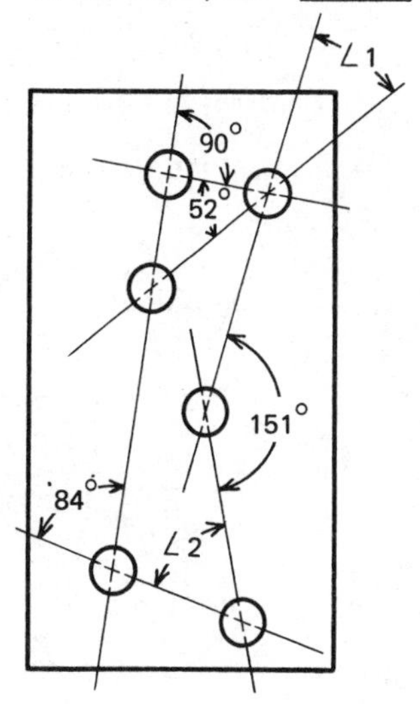

Unit 37 Circles: Part I

OBJECTIVES

After studying this unit, the student should be able to

- Identify parts of a circle.
- Solve problems by using geometric principles which involve chords, arcs, central angles, perpendiculars, and tangents.

DEFINITIONS

- A *circle* is a closed curve of which every point on the curve is equally distant from a fixed point called the center.
- The *circumference* is the length of the curved line which forms the circle, figure 37-1. (Circumference = 2 π radii; circumference = 2(3.1416)r.)
- A *chord* is a straight line that connects two points on the circumference, figure 37-1.
- A *diameter* is a chord that passes through the center of a circle, figure 37-1.
- A *radius* is a straight line that connects the center of the circle with a point on the circumference. (A radius = 1/2 diameter. See figure 37-1.)

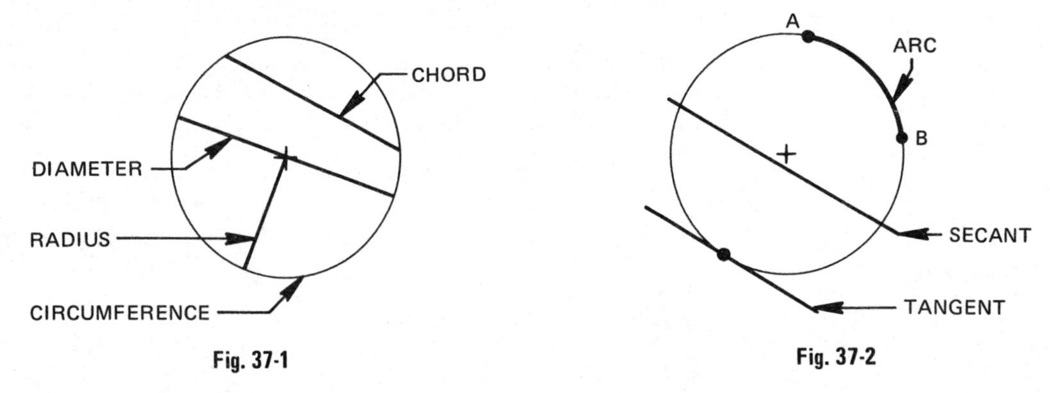

Fig. 37-1

Fig. 37-2

- An *arc* is a curve between two points on a circumference, figure 37-2. The symbol $\frown$ means arc; arc AB is written as $\widehat{AB}$.
- A *tangent* to a circle is a straight line that touches the circle at one point only, figure 37-2.
- A *secant* is a straight line that passes through a circle at two points, figure 37-2.

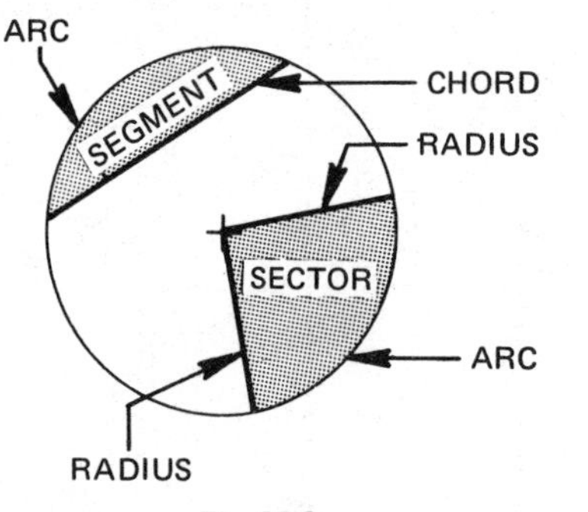

Fig. 37-3

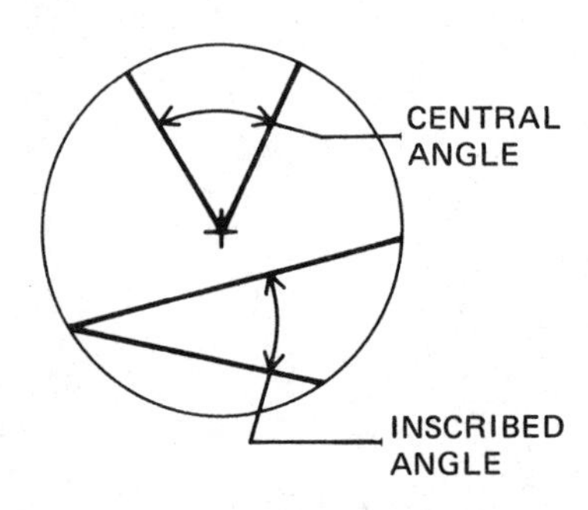

Fig. 37-4

- A *segment* is that part of a circle which is bounded by a chord and its arc, figure 37-3.
- A *sector* is that part of a circle which is bounded by two radii and the arc which the radii intercept, figure 37-3.
- A *central angle* is an angle whose vertex is at the center of a circle and whose sides are radii, figure 37-4.
- An *inscribed angle* is an angle whose vertex is on the circumference of a circle and whose sides are chords, figure 37-4.

Principle 11

- **In the same circle or in equal circles, equal chords cut off equal arcs.**

Example:

▲ Given: In figure 37-5, Circle A = Circle B and chords CD = EF = GH = MS.

▲ Conclusion: $\overset{\frown}{CD} = \overset{\frown}{EF} = \overset{\frown}{GH} = \overset{\frown}{MS}$.

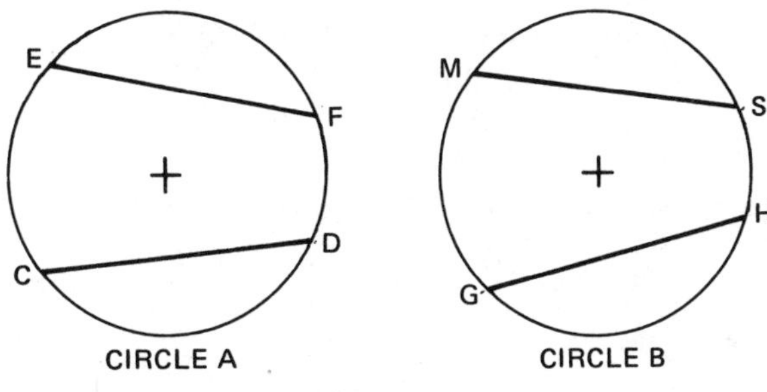

Fig. 37-5

Principle 12

- **In the same circle or in equal circles, equal central angles cut off equal arcs.**

Example:

▲ Given: In figure 37-6, Circle D = Circle E and central angles 1 = 2 = 3 = 4.

▲ Conclusion: $\overset{\frown}{AB} = \overset{\frown}{FG} = \overset{\frown}{HK} = \overset{\frown}{MP}$

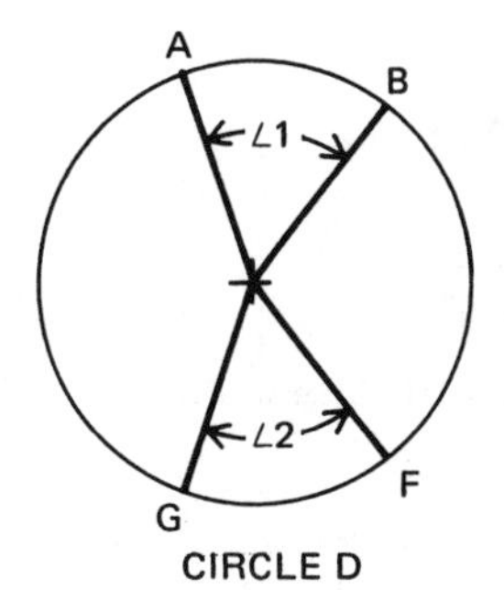

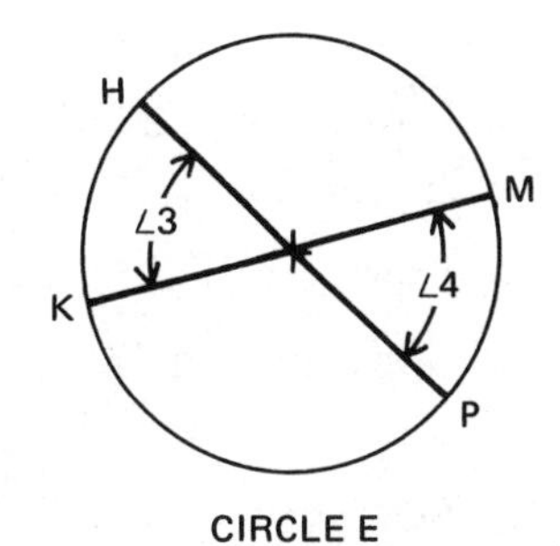

Fig. 37-6

Principle 13

- **In the same circle or in equal circles, two central angles are in the same ratio as the arcs which are cut off by the angles.**

Example:

▲ Given: In figure 37-7, Circle A = Circle B.

$\angle COD = 90^\circ$, $\angle EOF = 50^\circ$, $\overset{\frown}{CD} = 1.400''$, and $\overset{\frown}{GH} = 2.100''$.

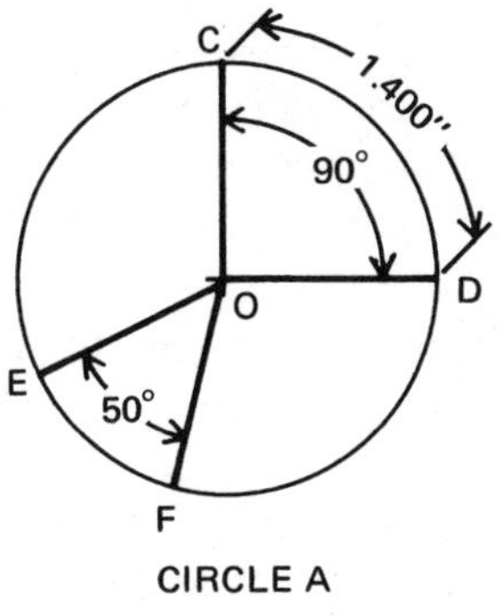

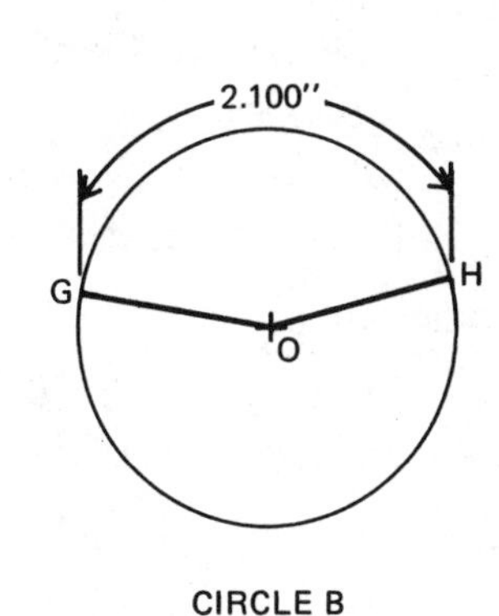

Fig. 37-7

▲ Determine the length of $\overset{\frown}{EF}$ and the size of $\angle GOH$.

a. Set up a proportion between $\overset{\frown}{CD}$ and $\overset{\frown}{EF}$ with their central angles.

$$\frac{\angle COD}{\angle EOF} = \frac{\overset{\frown}{CD}}{\overset{\frown}{EF}},\ \frac{9\not{0}^\circ}{5\not{0}^\circ} = \frac{1.400''}{\overset{\frown}{EF}},\ 9\,\overset{\frown}{EF} = 5(1.400),\quad \overset{\frown}{EF} = \frac{5(1.400)}{9},\quad \overset{\frown}{EF} = .778 \text{ inch}$$

b. Set up a proportion between $\overset{\frown}{CD}$ and $\overset{\frown}{GH}$ with their central angles.

$$\frac{\angle COD}{\angle GOH} = \frac{\overset{\frown}{CD}}{\overset{\frown}{GH}},\ \frac{90^\circ}{\angle GOH} = \frac{1.400''}{2.100''},\quad 1.400\,(\angle GOH) = 90(2.100),$$

$$\angle GOH = \frac{90(2.100)}{1.400},\quad \angle GOH = 135^\circ.$$

Principle 14

- **A line drawn from the center of a circle perpendicular to a chord bisects the chord and the arc cut off by the chord.**

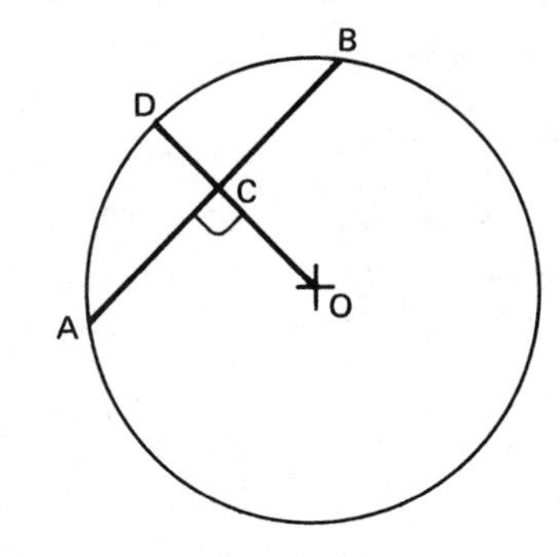

Fig. 37-8

Example:

▲ Given: In figure 37-8, line DO ⊥ chord AB

▲ Conclusion: AC = BC and $\overset{\frown}{AD} = \overset{\frown}{BD}$.

Principle 15

- **A line perpendicular to a radius at its extremity is tangent to the circle, and a tangent is perpendicular to a radius at its tangent point.**

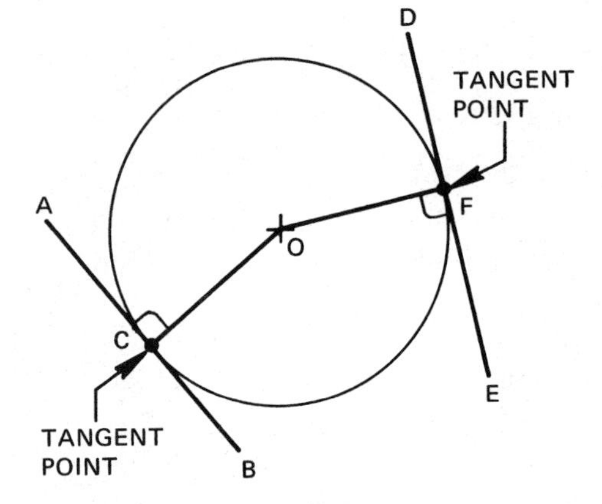

Fig. 37-9

Examples:

1. ▲ Given: In figure 37-9, line AB ⊥ to a radius CO at point C.

   ▲ Conclusion: line AB is a tangent.

2. ▲ Given: In figure 37-9, tangent DE passes through point F of radius FO.

   ▲ Conclusion: tangent DE ⊥ radius FO.

Principle 16

- **Two tangents drawn to a circle from a point outside the circle are equal. The angle at the outside point is bisected by a line drawn from the point to the center of the circle.**

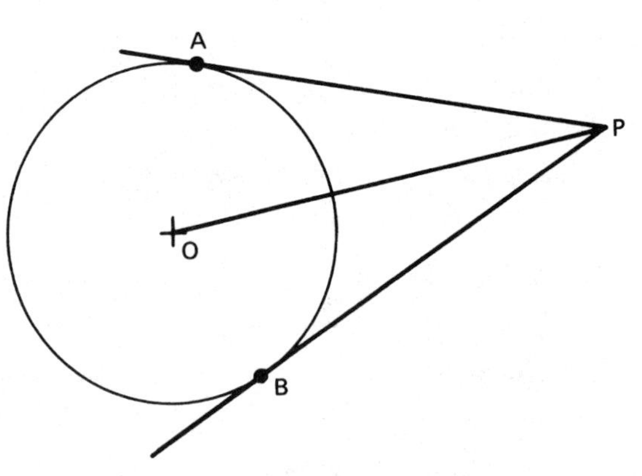

Fig. 37-10

Examples:

1. ▲ Given: In figure 37-10, tangents AP and BP are drawn to the circle from point P.

   ▲ Conclusion: AP = BP.

2. ▲ Given: In figure 37-10, line OP which extends from outside point P to center O.

   ▲ Conclusion: ∠APO = ∠BPO.

Principle 17

- **If two chords intersect inside a circle, the product of the two segments of one chord is equal to the product of the two segments of the other chord.**

Examples:

1. ▲ Given: In figure 37-11 chords AC and DE, which intersect at point B.
   ▲ Conclusion: AB(BC) = BD(BE).
2. ▲ Given: AB = 7.5 inches, BC = 2.8 inches, BD = 2.1 inches.
   ▲ Determine the length of BE: AB(BC) = BD(BE), 7.5(2.8) = 2.10(BE) 21.0 = 2.1BE, BE = 21.0/2.1, BE = 10.0 inches.

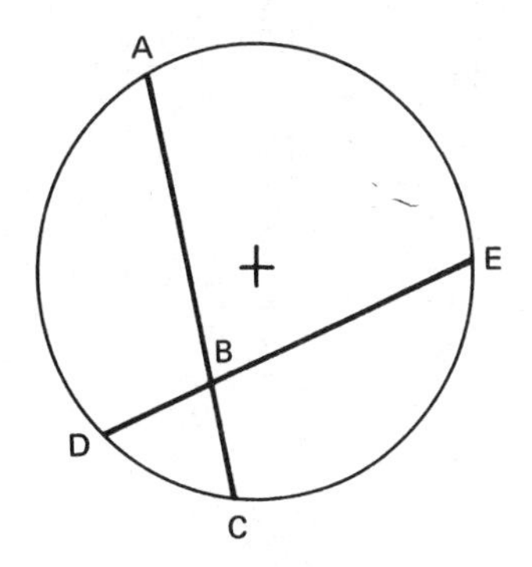

Fig. 37-11

APPLICATION

A. Give the names of the parts of circles for the following problems.

1. a. AB ______
   b. CD ______
   c. EO ______

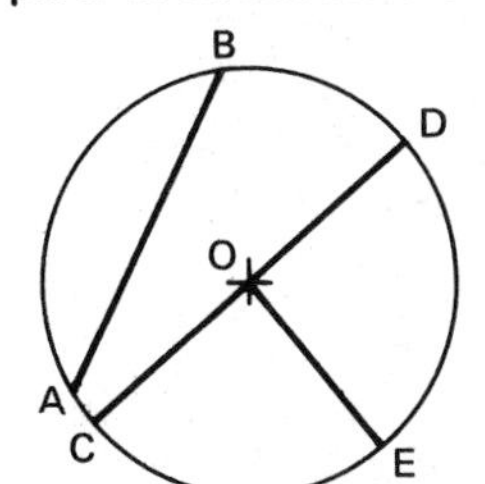

2. a. $\overset{\frown}{GF}$ ______
   b. HK ______
   c. LM ______
   d. GF ______

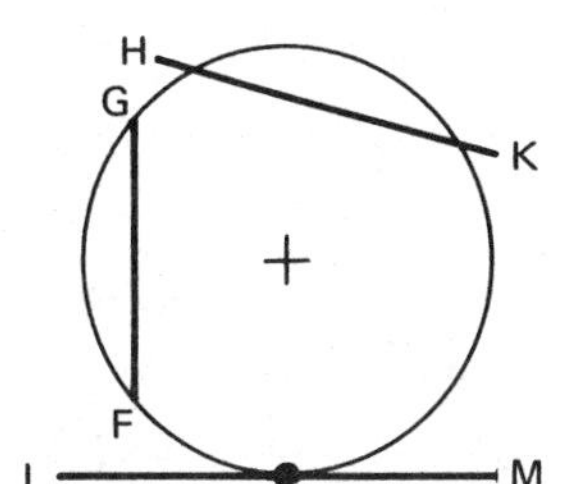

3. a. M ______
   b. P ______
   c. SO ______
   d. TO ______
   e. RW ______

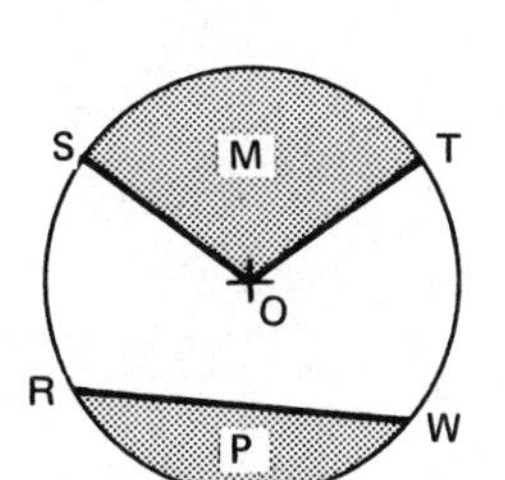

4. a. ∠1 ______
   b. ∠2 ______
   c. AO ______
   d. CD ______
   e. CE ______

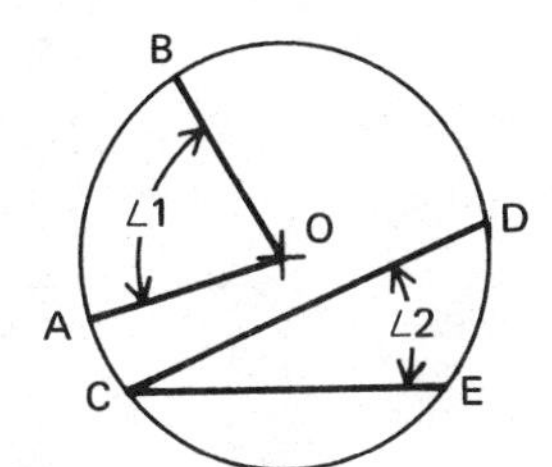

B. Solve the following problems. These problems are based on principles 11-14, although a problem may require the application of two or more of any of the principles.

1. △ABC is equilateral.
   a. $\overset{\frown}{AB}$ = ______
   b. $\overset{\frown}{BC}$ = ______

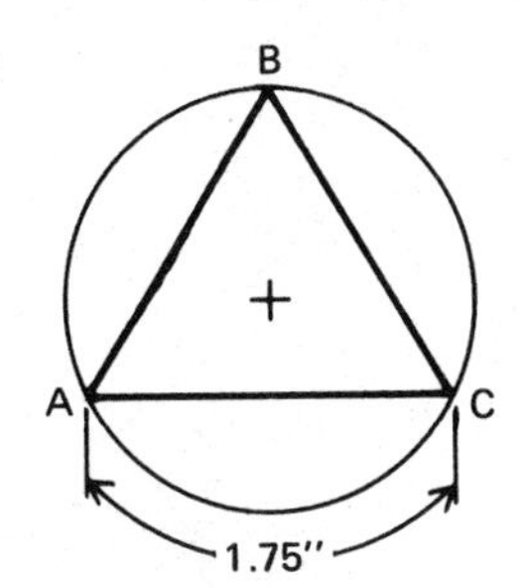

2. a. $\overset{\frown}{AB}$ = ______
   b. $\overset{\frown}{BC}$ = ______

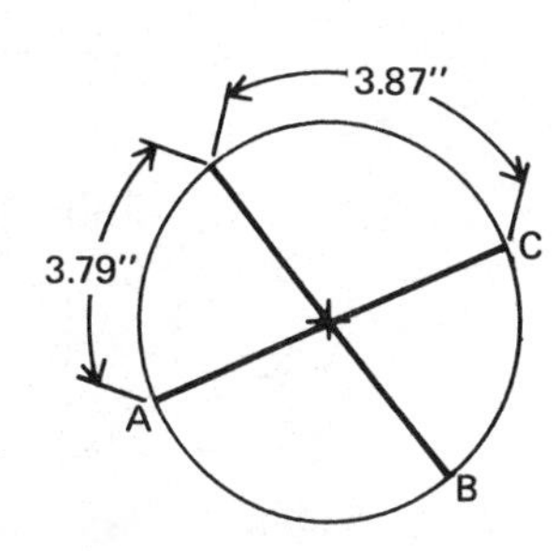

3. a. $\widehat{EF}$ = 1.600″
   $\widehat{HP}$ = ____
   b. $\widehat{HP}$ = 2.840″
   $\widehat{EF}$ = ____

4. a. $\widehat{SW}$ = 3.800″,
   $\widehat{TM}$ = 5.700″
   $\angle 1$ = ____
   b. $\widehat{TM}$ = 4.128″,
   $\widehat{SW}$ = 2.064″
   $\angle 1$ = ____

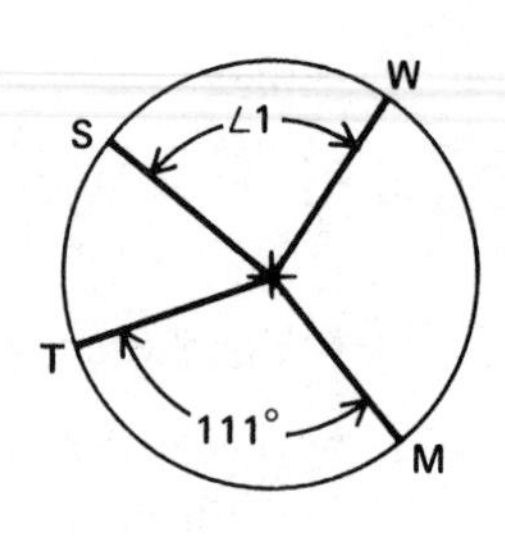

5. a. AB = 5.378″
   $\widehat{AC}$ = 3.782″
   DB = ____
   $\widehat{ACB}$ = ____
   b. DB = 3.017″,
   $\widehat{ACB}$ = 7.308″
   AB = ____
   $\widehat{CB}$ = ____

6. $\widehat{EF}$ = .836″
   $\widehat{HK}$ = ____

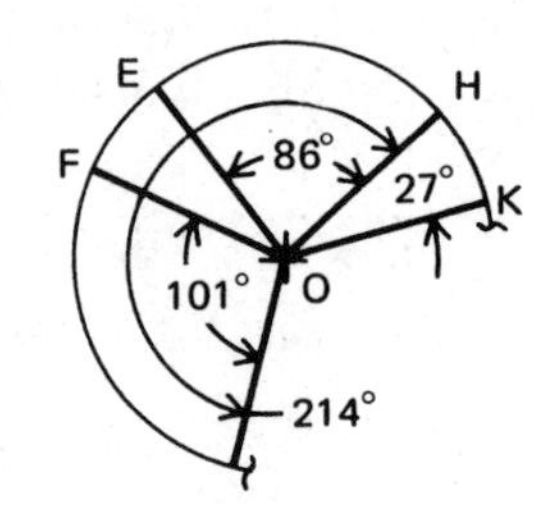

7. a. $\angle 1$ = 240°
   $\widehat{ABC}$ = ____
   b. $\widehat{ABC}$ = 2.500″
   $\angle 1$ = ____

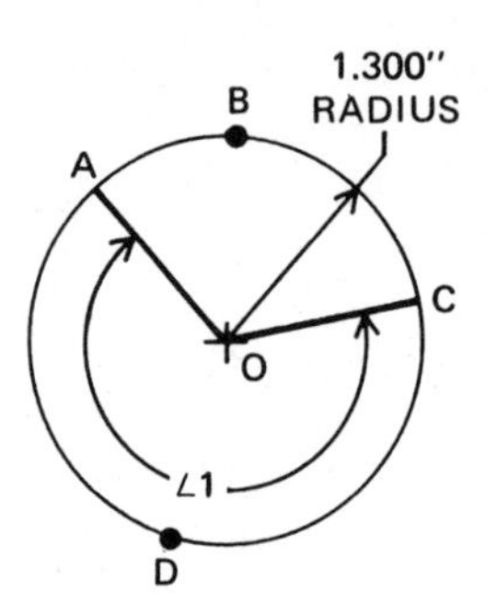

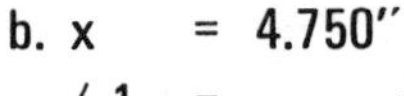

8. a. x = 5.100″
   $\angle 1$ = ____
   b. x = 4.750″
   $\angle 1$ = ____

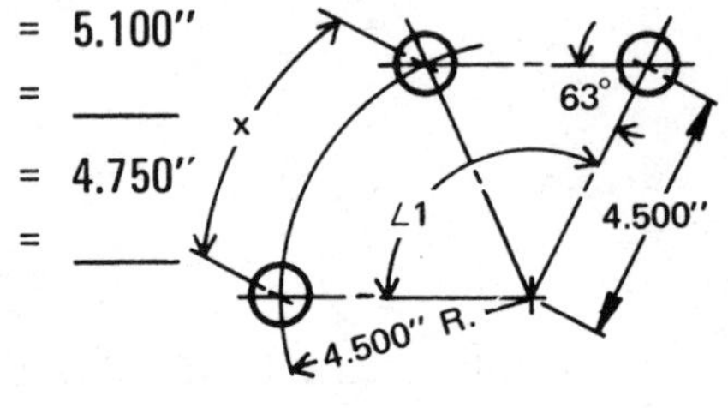

9. a. Radius x = 7.500″,
   Y = 4.500″,
   PM = ____
   b. Radius x = 8.000″,
   y = 4.800″,
   PM = ____

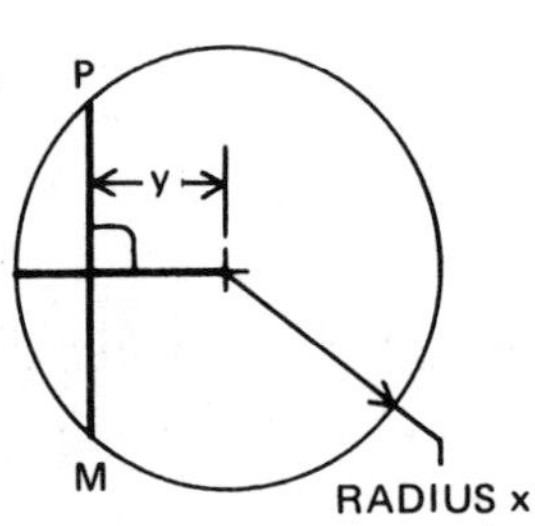

10. Circumference of circle = 15.500″.
   a. x = 3.100″,
   $\angle 1$ = ____
   b. $\angle 1$ = 40°,
   x = ____

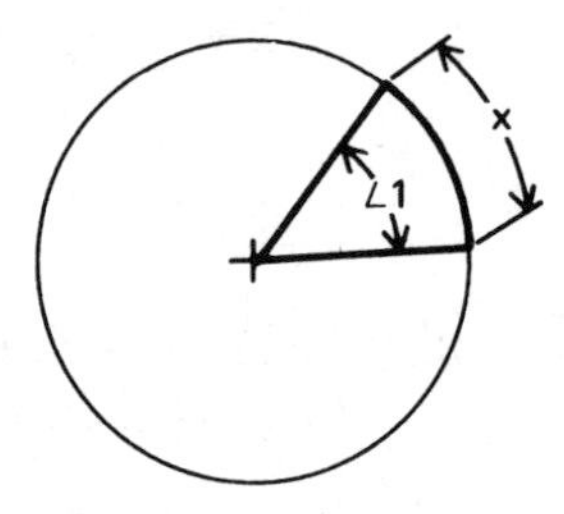

11. Determine the centerline distance between Hole A and Hole B if:
   a. Radius x = 8.000″ and DO = 2.100″

   ____

   b. Radius x = 1.200″ and DO = .700″

   ____

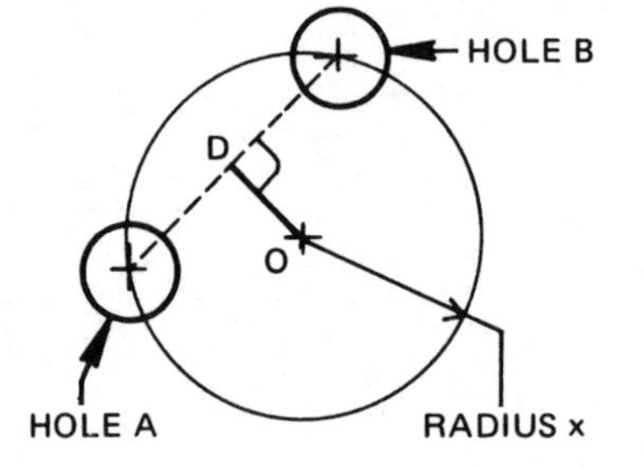

C. Solve the following problems. These problems are based on principles 15-17, although a problem may require the application of two or more of any of the principles.

1. Point P is a tangent point, $\angle 1 = 107^\circ\ 18'$

   a. $\angle 2 = 42^\circ\ 12'$, $\angle E$ = ____, $\angle F$ = ____

   b. $\angle 2 = 49^\circ\ 53'$, $\angle E$ = ____, $\angle F$ = ____

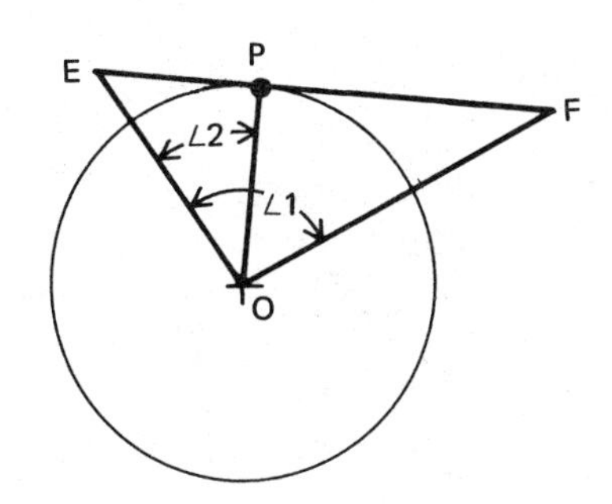

2. AB and CB are tangents.

   a. y = 1.372″, $\angle ABC = 67^\circ$, $\angle 1$ = ____, x = ____

   b. x = 2.077″, $\angle 1 = 33^\circ\ 49'$, $\angle ABC$ = ____, y = ____

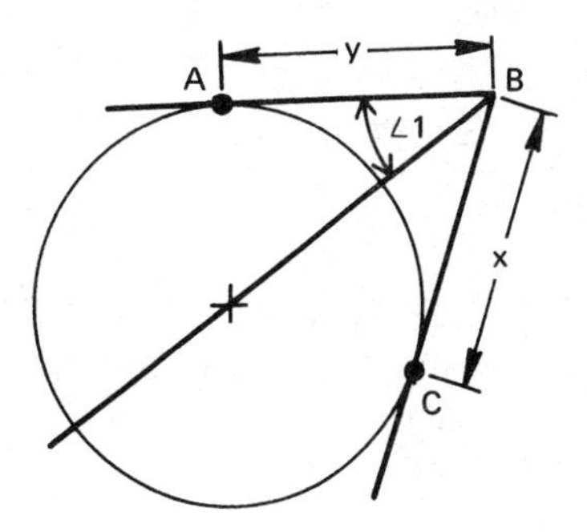

3. Point A is a tangent point.

   a. y = 1.400″, x = ____

   b. y = 1.800″, x = ____

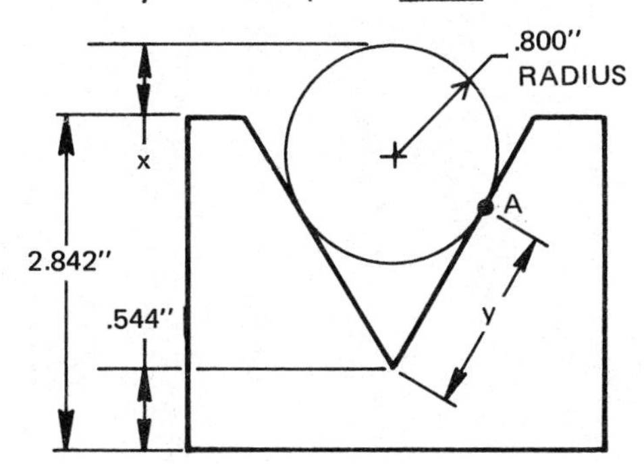

4. Points E, G, and F are tangent points.

   a. $\angle 1 = 117^\circ$, $\angle 2$ = ____

   b. $\angle 1 = 122^\circ\ 15'$, $\angle 2$ = ____

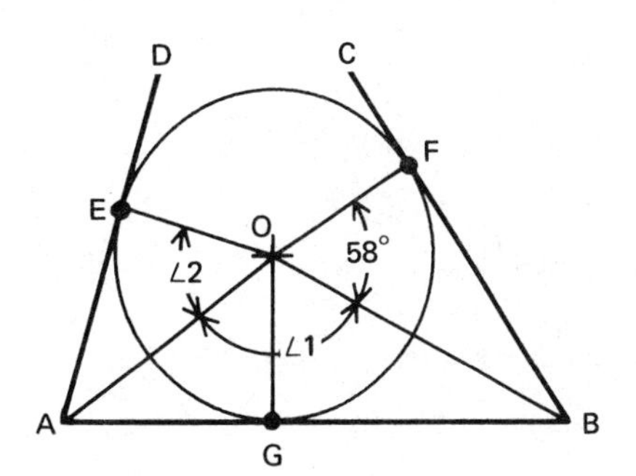

5. a. EK = 6.000″, GK = ____

   b. GK = 4.800″, EK = ____

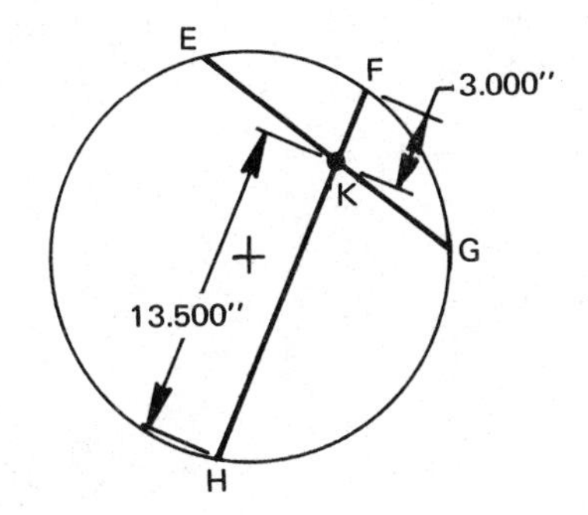

6. a. PT = 1.800″, x = ____

   b. PT = 2.000″, x = ____

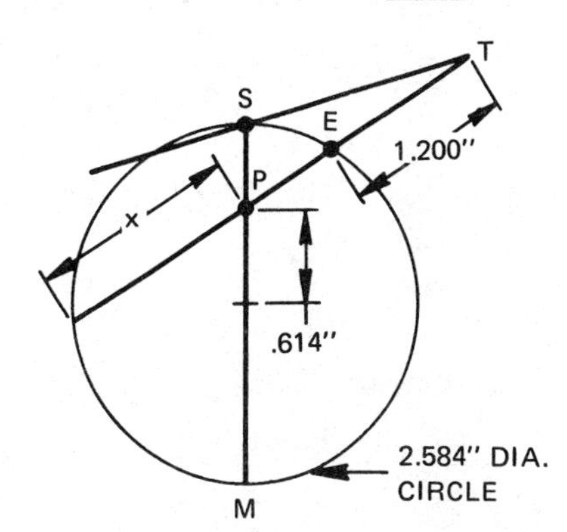

Unit 38 Circles: Part II

OBJECTIVES

After studying this unit, the student should be able to

- Solve problems by using geometric principles which involve angles formed inside and outside a circle.
- Solve problems by using geometric principles which involve internally and externally tangent circles.

ANGLES FORMED INSIDE A CIRCLE

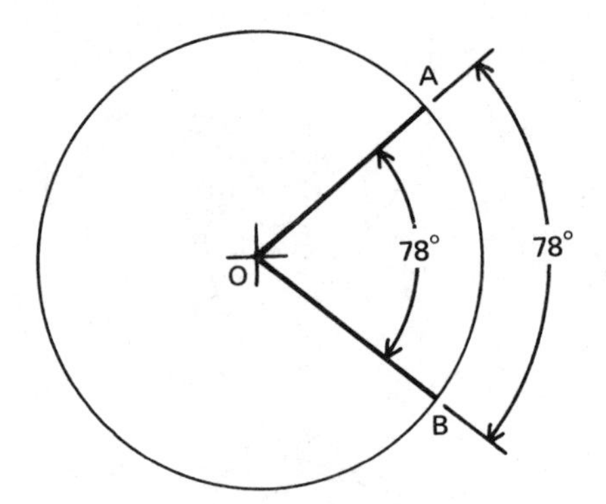

Fig. 38-1

Principle 18

- **A central angle is equal to its intercepted arc.** (An *intercepted arc* is an arc which is cut off by a central angle.)

  Example:

  ▲ Given: In figure 38-1, $\overset{\frown}{AB} = 78°$.

  ▲ Conclusion: $\angle AOB = 78°$.

- **An angle formed by two chords which intersect inside a circle is equal to one-half the sum of its two intercepted arcs.**

  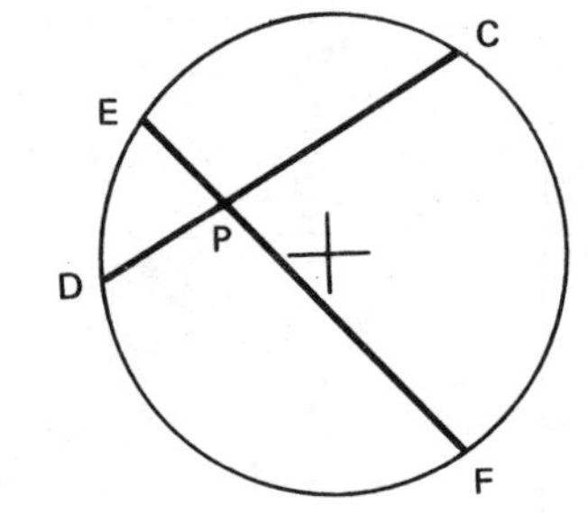

  Fig. 38-2

  Example:

  ▲ Given: In figure 38-2, chords CD and EF intersect at point P.

  ▲ Conclusion: $\angle EPD = 1/2\ (\overset{\frown}{CF} + \overset{\frown}{DE})$

  Examples of use:

  1. ▲ Given: $\overset{\frown}{CF} = 106°$ and $\overset{\frown}{ED} = 42°$

     ▲ Determine $\angle EPD$:
     $\angle EPD = 1/2\ (106° + 42°)$, $\angle EPD = 1/2\ (148°)$, $\angle EPD = 74°$

  2. ▲ Given: $\angle EPD = 64°$ and $\overset{\frown}{CF} = 96°$

     ▲ Determine $\overset{\frown}{DE}$: $\angle EPD = 1/2\ (\overset{\frown}{CF} + \overset{\frown}{DE})$, $64° = 1/2\ (96° + \overset{\frown}{DE})$, $64 = 1/2\ (96) + 1/2\ \overset{\frown}{DE}$, $64 = 48 + 1/2\ \overset{\frown}{DE}$, $64 - 48 = 1/2\ \overset{\frown}{DE}$, $16 = 1/2\ \overset{\frown}{DE}$, $\overset{\frown}{DE} = 16 \div 1/2$, $\overset{\frown}{DE} = 32°$

- **An inscribed angle is equal to one-half of its intercepted arc.**

  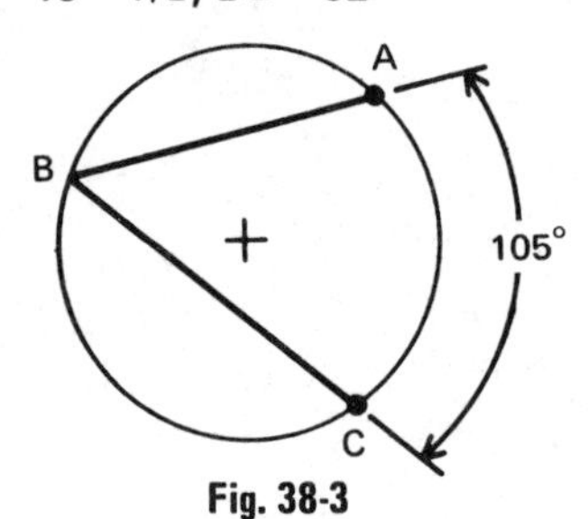

  Fig. 38-3

  Example:

  ▲ Given: In figure 38-3, $\overset{\frown}{AC} = 105°$.

  ▲ Conclusion: $\angle ABC = 1/2\ \overset{\frown}{AC}$, $\angle ABC = 1/2\ (105°)$, $\angle ABC = 52°\ 30'$

ANGLES FORMED ON A CIRCLE

Principle 19

An angle formed by a tangent and a chord at the tangent point is equal to one-half of its intercepted arc.

Example:

▲ Given: In figure 38-4, tangent CD meets chord AB at the point of tangency A, $\overset{\frown}{AB} = 110°$.

▲ Determine $\angle CAB$: $\angle CAB = 1/2\ \overset{\frown}{AB}$, $\angle CAB = 1/2\ (110°)$, $\angle CAB = 55°$.

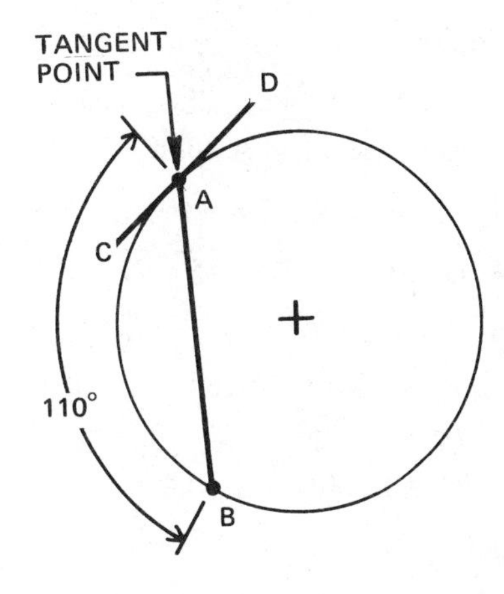

Fig. 38-4

ANGLES FORMED OUTSIDE A CIRCLE

Principle 20

An angle formed at a point outside a circle by two secants, two tangents, or a secant and a tangent is equal to one-half the difference of the intercepted arcs.

Two Secants

Examples:

1. ▲ Given: In Figure 38-5, secants AP and DP meet at point P.

   ▲ Conclusion: $\angle P = 1/2\ (\overset{\frown}{AD} - \overset{\frown}{BC})$

2. ▲ Given: $\overset{\frown}{AD} = 85°\ 40'$, $\overset{\frown}{BC} = 39°\ 17'$

   ▲ Determine $\angle P$: $\angle P = 1/2\ (\overset{\frown}{AD} - \overset{\frown}{BC})$ $\angle P = 1/2\ (85°\ 40' - 39°\ 17')$, $\angle P = 1/2\ (46°\ 23')$, $\angle P = 23°\ 11'\ 30''$.

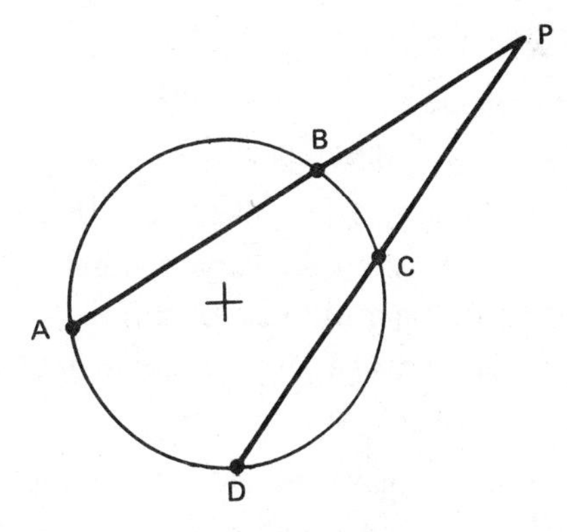

Fig. 38-5

Two Tangents

Examples:

1. ▲ Given: In figure 38-6, tangents DP and EP meet at point P.

   ▲ Conclusion: $\angle P = 1/2\ (\overset{\frown}{DCE} - \overset{\frown}{DE})$

2. ▲ Given: $\overset{\frown}{DCE} = 253°\ 37'$, $\overset{\frown}{DE} = 106°\ 23'$

   ▲ Determine $\angle P$: $\angle P = 1/2\ (\overset{\frown}{DCE} - \overset{\frown}{DE})$, $\angle P = 1/2\ (253°\ 37' - 106°\ 23')$, $\angle P = 1/2\ (147°\ 14')$, $\angle P = 73°\ 37'$

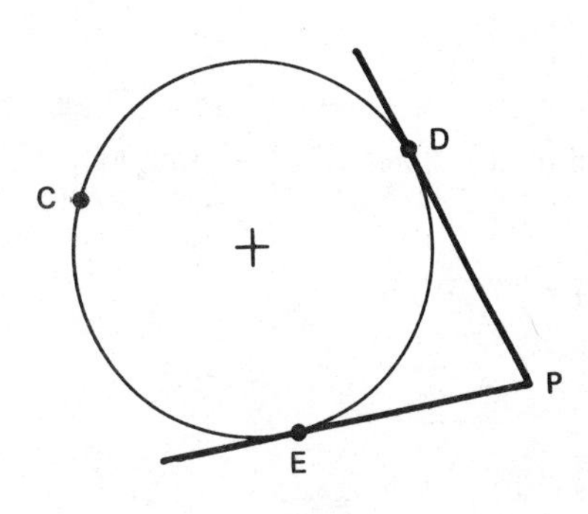

Fig. 38-6

A Tangent and a Secant

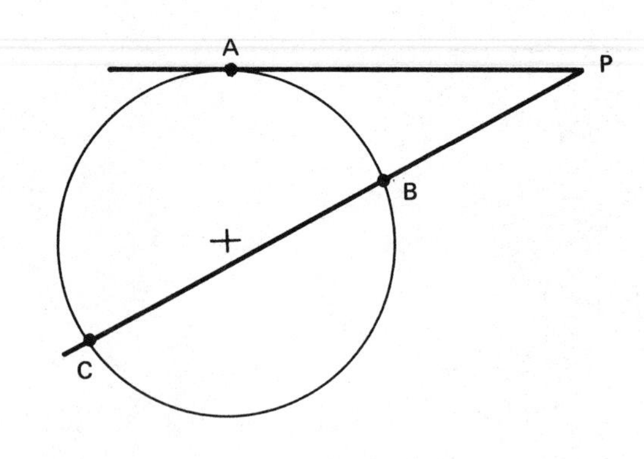

Fig. 38-7

Examples:

1. ▲ Given: In figure 38-7, tangent AP and secant CP meet at point P.

   ▲ Conclusion: $\angle P = 1/2\ (\widehat{AC} - \widehat{AB})$

2. ▲ Given: $\angle P = 28^\circ$, $\widehat{AB} = 72^\circ$

   ▲ Determine $\widehat{AC}$: $\angle P = 1/2\ (\widehat{AC} - \widehat{AB})$, $28^\circ = 1/2\ (\widehat{AC} - 72^\circ)$, $28 = 1/2\ \widehat{AC} - 1/2\ (72)$, $28 = 1/2\ \widehat{AC} - 36$, $28 + 36 = 1/2\ \widehat{AC}$, $64 = 1/2\ \widehat{AC}$, $\widehat{AC} = 64 \div 1/2$, $\widehat{AC} = 128^\circ$

Principle 21

- **If two circles are either internally or externally tangent, a line connecting the centers of the circles passes through the point of tangency and is perpendicular to the tangent line.**

Examples:

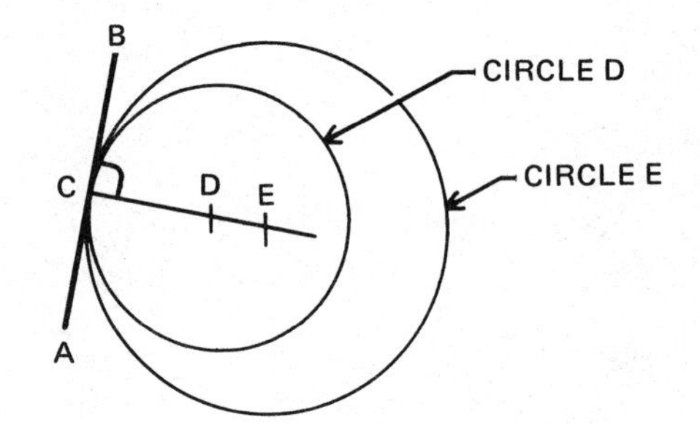

Fig. 38-8

1. ▲ Given: In figure 38-8, Circle D and Circle E are internally tangent at point C; D is the center of Circle D; E is the center of Circle E; line AB is tangent to both circles at point C.

   ▲ Conclusion: An extension of line DE passes through tangent point C and line CDE ⊥ tangent line AB.

2. ▲ Given: In figure 38-9, Circle D and Circle E are externally tangent at point C; D is the center of Circle D; E is the center of Circle E; line AB is tangent to both circles at point C.

   ▲ Conclusion: Line DE passes through tangent point C and ~~line~~ DE ⊥ tangent line AB.

CIRCLE D CIRCLE E

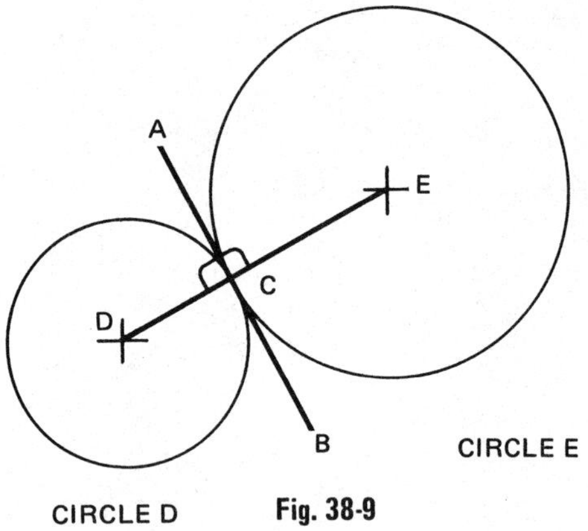

Fig. 38-9

APPLICATION

Solve the following problems. These problems are based on principles 18-21, although a problem may require the application of two or more of any of the principles.

1. a. $\angle 1 = 73^\circ$, $\widehat{DC}$ = ____
   $\angle EOD$ = ____,
   $\widehat{ABC}$ = ____

   b. $\angle 1 = 68^\circ$, $\widehat{DC}$ = ____
   $\angle EOD$ = ____, $\widehat{BCD}$ = ____

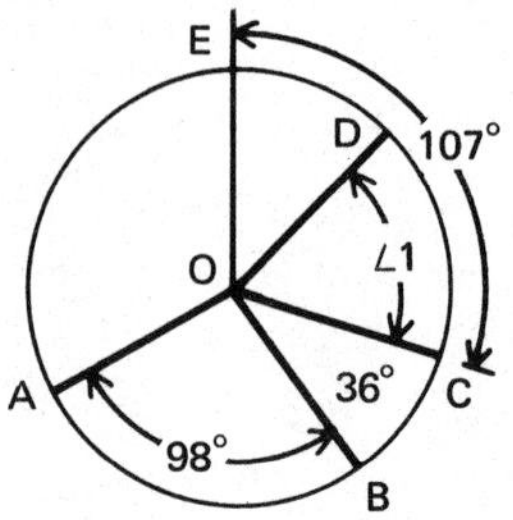

2. a. $\angle 1 = 63°$,
   $\widehat{HK}$ = ____
   $\widehat{HM}$ = ____
   b. $\angle 1 = 59° \, 47'$,
   $\widehat{HK}$ = ____
   $\widehat{HM}$ = ____

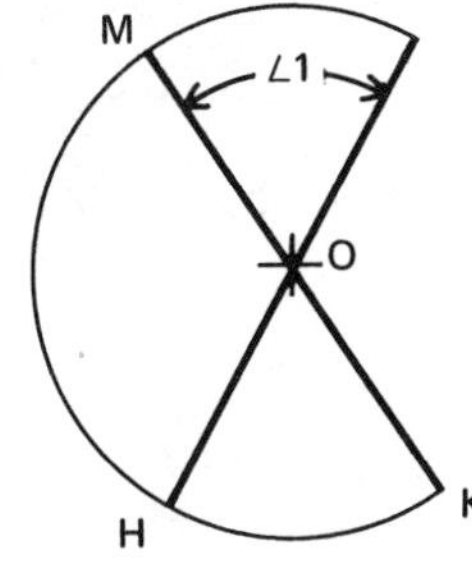

3. a. $\widehat{PS} = 46°$,
   $\angle 1$ = ____
   $\angle 2$ = ____
   b. $\widehat{PS} = 39°$,
   $\angle 1$ = ____
   $\angle 2$ = ____

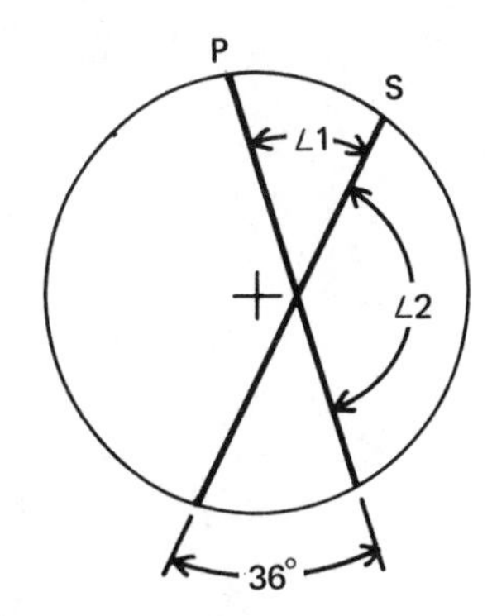

4. a. $\widehat{DC} = 28°$,
   $\widehat{AB}$ = ____
   b. $\widehat{AB} = 131°$,
   $\widehat{DC}$ = ____

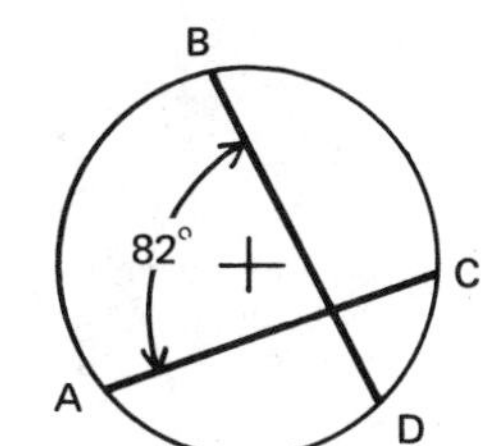

5. a. $\angle 3 = 47°$,
   $\widehat{GH} = 32°$
   $\widehat{EF}$ = ____
   $\angle 4$ = ____
   b. $\angle 4 = 17° \, 53'$,
   $\widehat{EF} = 103°$
   $\angle 3$ = ____
   $\widehat{GH}$ = ____

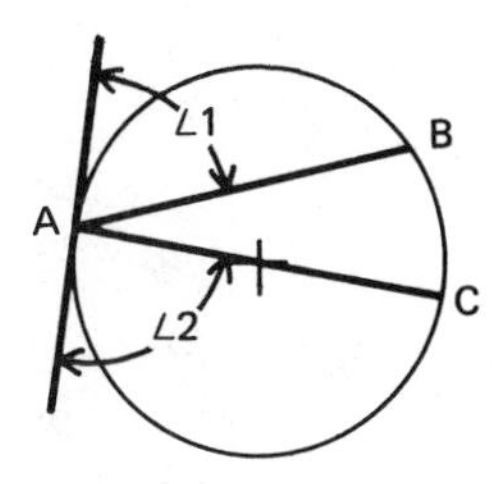

6. a. $\angle 1 = 25°$,
   $\widehat{MPT} = 95°$
   $\widehat{KTP}$ = ____,
   $\widehat{PT}$ = ____
   $\widehat{MP}$ = ____
   b. $\angle 1 = 17° \, 30'$.
   $\widehat{MPT} = 103°$
   $\widehat{KTP}$ = ____,
   $\widehat{PT}$ = ____,
   $\widehat{MP}$ = ____

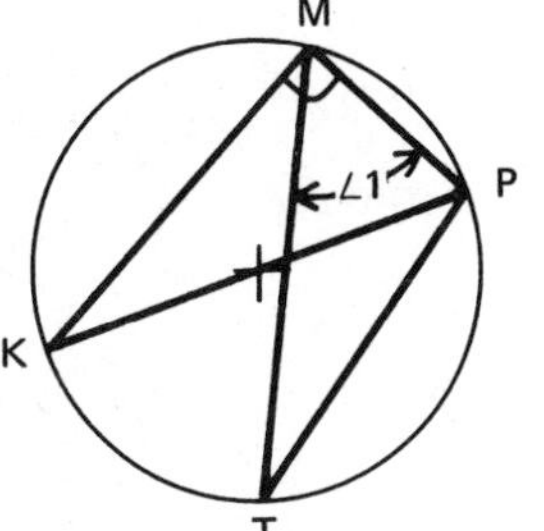
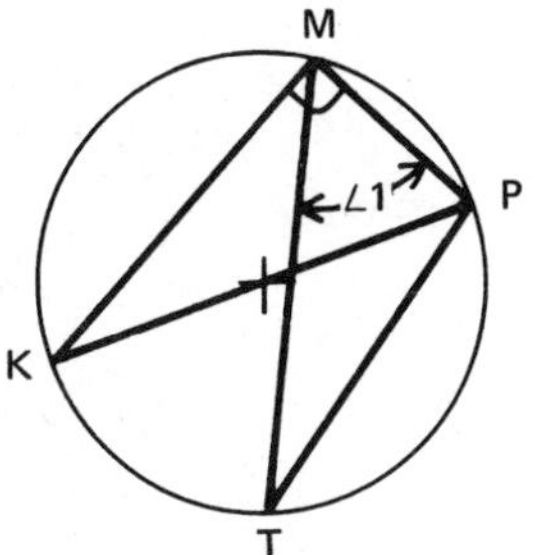

7. a. $\widehat{AB} = 114°$,
   $\angle 2$ = ____
   $\angle 1$ = ____
   b. $\widehat{AB} = 110° \, 42'$,
   $\angle 2$ = ____
   $\angle 1$ = ____

8. a. $\widehat{EF} = 84°$,
   $\angle EFD$ = ____
   $\widehat{HF}$ = ____
   $\angle 1$ = ____
   b. $\widehat{EF} = 79°$,
   $\angle EFD$ = ____
   $\widehat{HF}$ = ____
   $\angle 1$ = ____

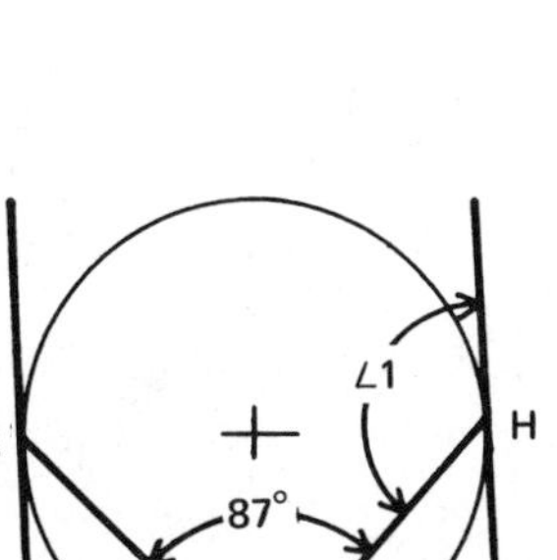
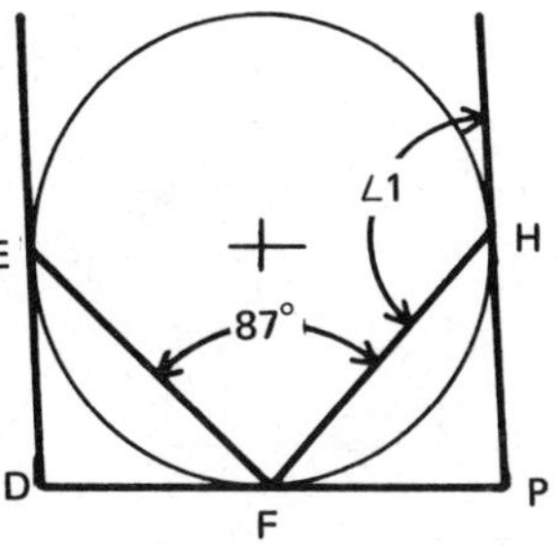

9. a. $\widehat{ST} = 18°$,
   $\widehat{SM} = 40°$
   $\angle 1$ = ____
   $\angle 2$ = ____
   b. $\widehat{ST} = 23°$,
   $\widehat{SM} = 39°$
   $\angle 1$ = ____
   $\angle 2$ = ____

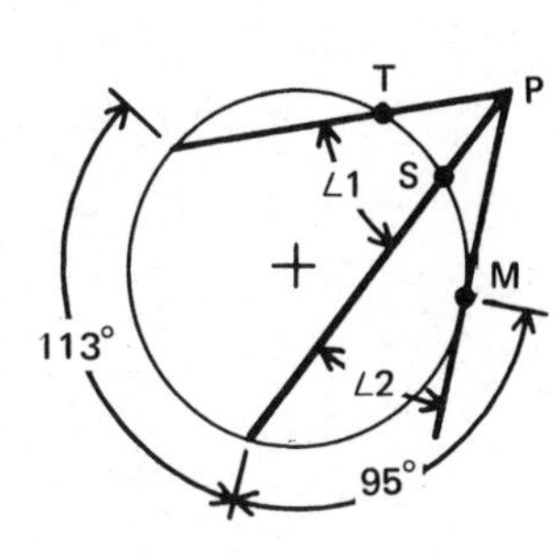

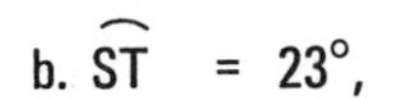

10. a. $\overset{\frown}{AB} = 72°$, $\overset{\frown}{CD} = 50°$

  $\angle 1 =$ ____, $\angle 2 =$ ____, $\angle 3 =$ ____

 b. $\overset{\frown}{CD} = 43°$, $\overset{\frown}{AD} = 106°$

  $\angle 1 =$ ____, $\angle 2 =$ ____, $\angle 3 =$ ____

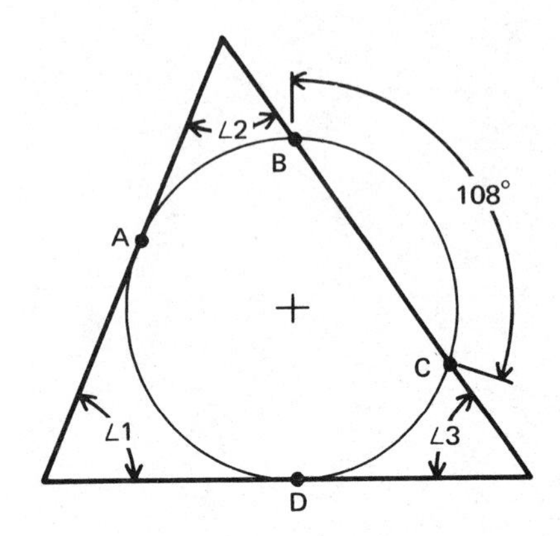

11. a. $\angle 1 = 28°$, $\angle 2 = 62°$

  $\overset{\frown}{DH} =$ ____, $\overset{\frown}{EDH} =$ ____

 b. $\angle 1 = 25°$, $\angle 2 = 67°$

  $\overset{\frown}{DH} =$ ____, $\overset{\frown}{EDH} =$ ____

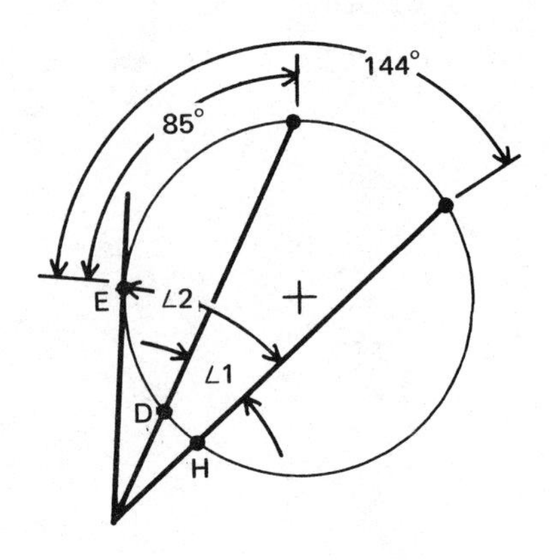

12. a. Dia. A = 3.756″

  Dia. B = 1.622″, x = ____

 b. x = .975″, Dia. B = 1.026″

  Dia. A = ____

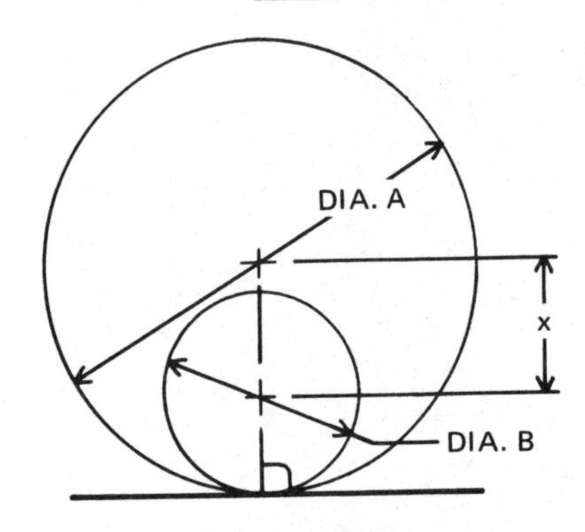

13. a. x = .987, y = 1.175″

  Dia. A = ____

 b. x = 3.133″, y = 4.644″

  Dia. A = ____

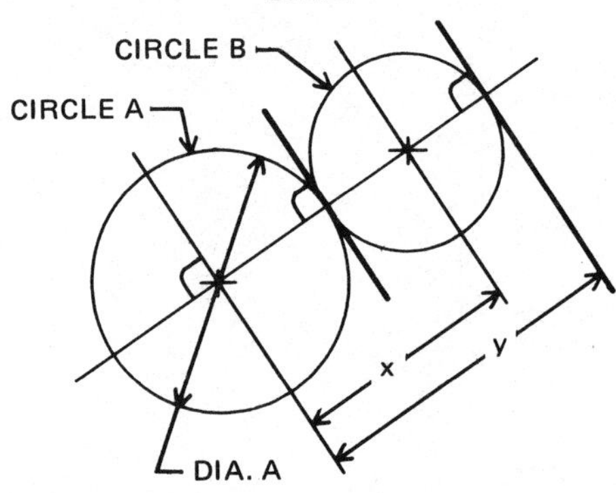

14. a $\angle 1 = 67°$, $\angle 2 = 93°$

  $\overset{\frown}{AB} =$ ____, $\overset{\frown}{DE} =$ ____

 b. $\angle 1 = 75°$, $\angle 2 = 85°$

  $\overset{\frown}{AB} =$ ____, $\overset{\frown}{DE} =$ ____

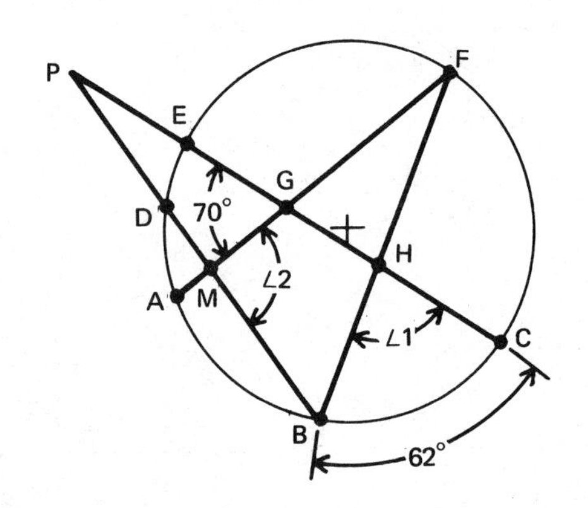

15. a. Dia. A = 1.000", x = _____

 b. Dia. A = .800", x = _____

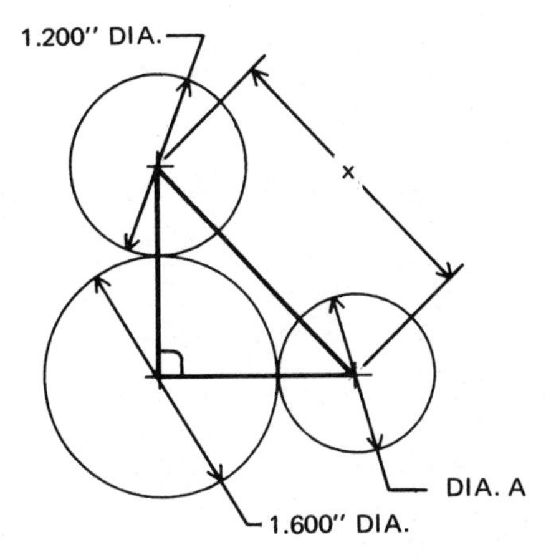

16. a. y = .350", x = _____

 b. y = .410", x = _____

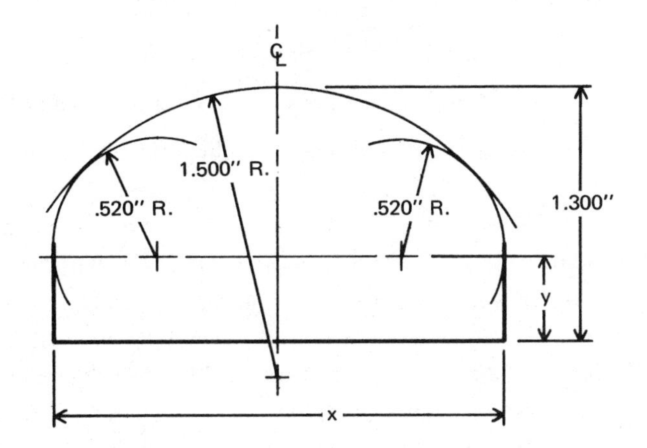

17. AC is a diameter.

 a. ∠2 = 24°, ∠1 = _____

 b. ∠2 = 31° 14′, ∠1 = _____

18. a. Radius R = .600", y = _____

 b. Radius R = .650", y = _____

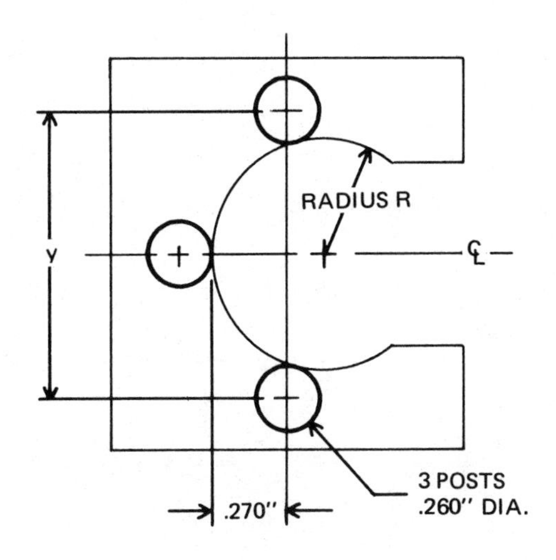

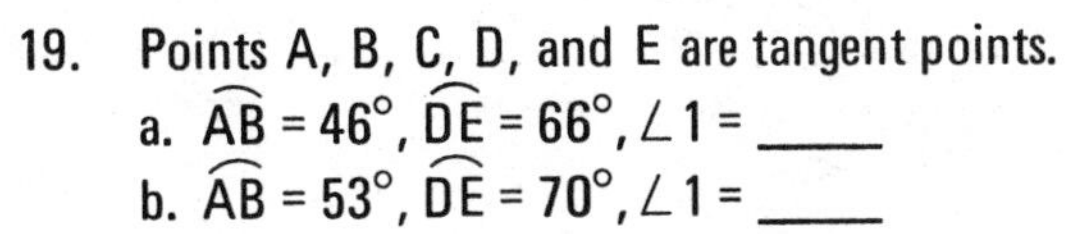

19. Points A, B, C, D, and E are tangent points.

 a. $\widehat{AB}$ = 46°, $\widehat{DE}$ = 66°, ∠1 = _____

 b. $\widehat{AB}$ = 53°, $\widehat{DE}$ = 70°, ∠1 = _____

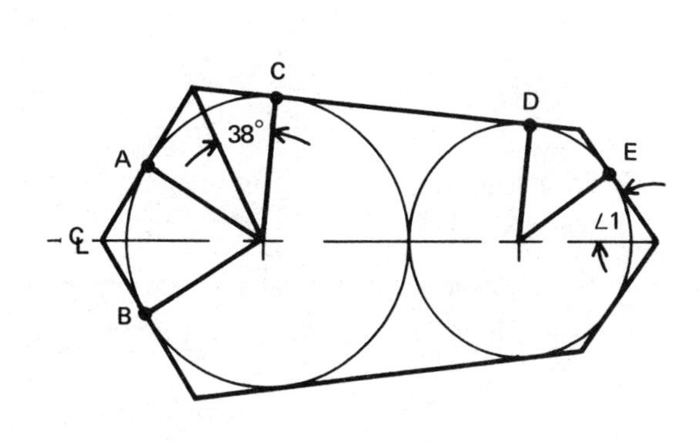

Unit 39 Fundamental Geometric Constructions

OBJECTIVES

After studying this unit, the student should be able to

- Make constructions which are basic to the machine trades.
- Lav out typical machine shop problems using the methods of construction.

A knowledge of basic geometric constructions done with a compass or dividers and a steel rule is required of a machinist in laying out work. The constructions are used in determining stock allowances and reference locations on castings, forgings, and sheet stock.

For certain jobs where wide dimensional tolerances are permissible, the most practical and efficient way of producing a part may be by scribing and centerpunching locations.

Layout dimensions are sometimes used as a reference for machining complex parts which require a high degree of precision. Locations lightly scribed on a part are used as a precaution to insure part or table movement in the proper direction. It is particularly useful in operations which require part rotation or repositioning.

PERPENDICULAR BISECTOR OF A LINE

Rule: To construct a perpendicular bisector to line AB

- With end point A as a center and using a radius equal to more than half AB, draw arcs above and below AB. (See Figure 39-1.)

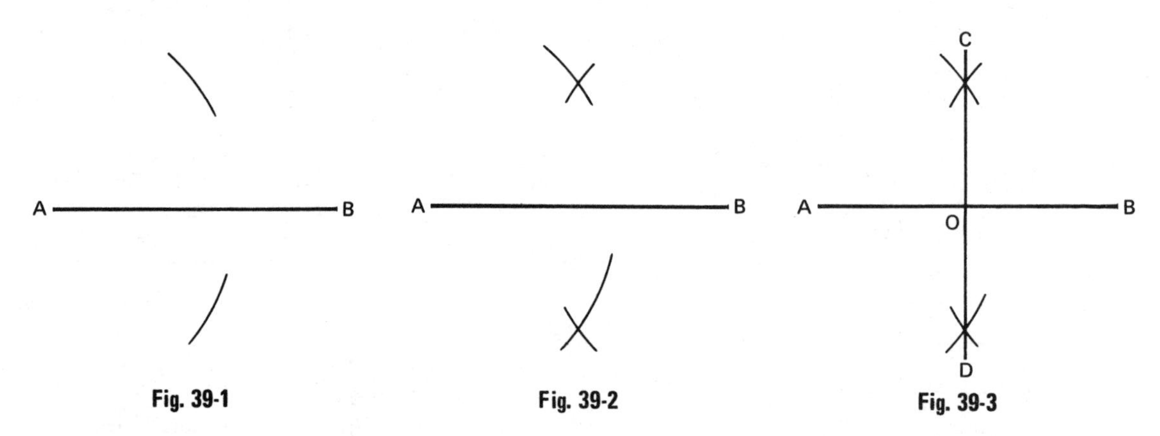

Fig. 39-1 Fig. 39-2 Fig. 39-3

- With end point B as a center and using the same radius, draw arcs above and below AB. (See figure 39-2.)
- Draw a connecting line between the intersection of the arcs above and below AB. Line CD is perpendicular to AB and point O is mid-point of AB. (See figure 39-3.)

Example:

This construction may be used for laying out hole locations, such as the plate shown in figure 39-4. The centerline of holes C and D bisects and is perpendicular to the centerline of holes A and B.

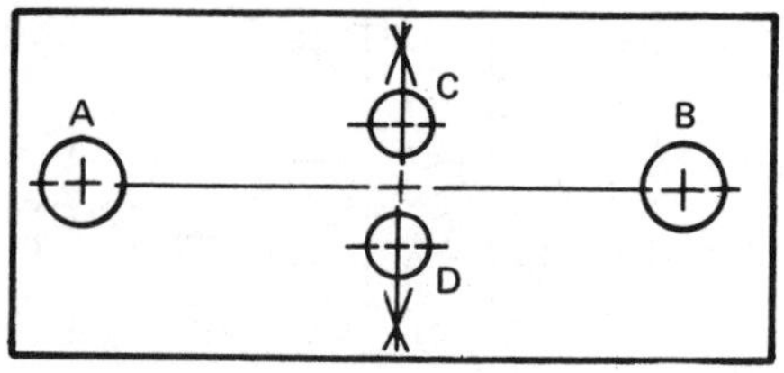

Fig. 39-4

PERPENDICULAR TO A LINE AT A GIVEN POINT ON THE LINE

Rule: To construct a perpendicular to line AB at point O

- With given point O as a center, draw two arcs with the same radius intersecting AB at C and D. (See figure 39-5.)

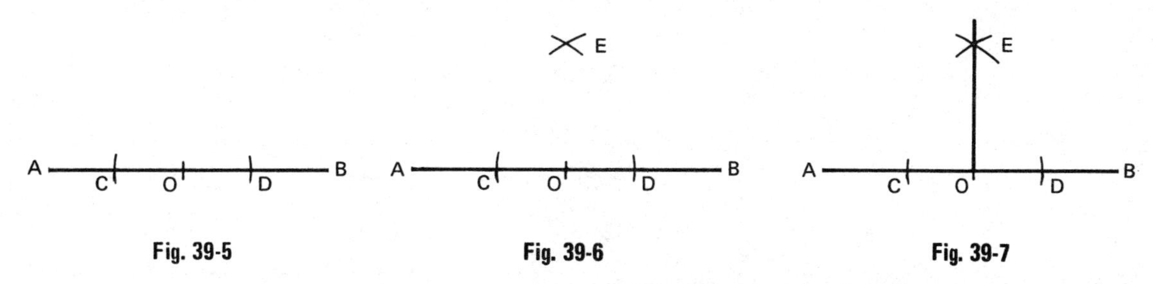

Fig. 39-5 Fig. 39-6 Fig. 39-7

- With points C and D as centers, and with the same radius, draw arcs intersecting at E. (See figure 39-6.)
- Draw a line between point E and point O. Line EO is perpendicular to line AB at point O. (See figure 39-7.)

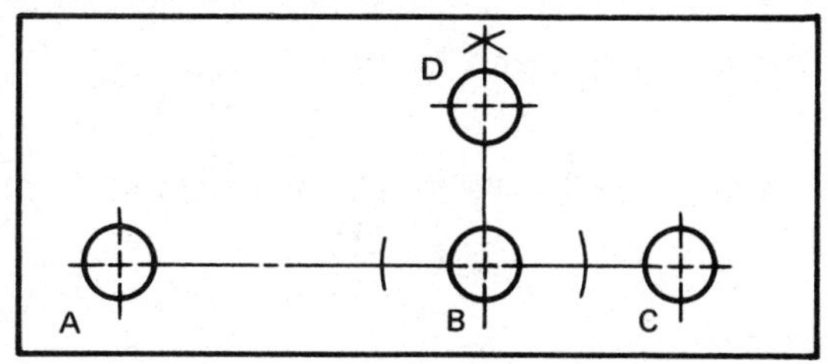

Fig. 39-8

Example:

In figure 39-8, holes A, B, and C are horizontally in line and hole D is vertically above hole B.

PARALLEL TO A LINE

Rule: To construct a line parallel to line AB

- Construct perpendiculars near the ends of line AB with points C and D as centers. (See figure 39-9.)

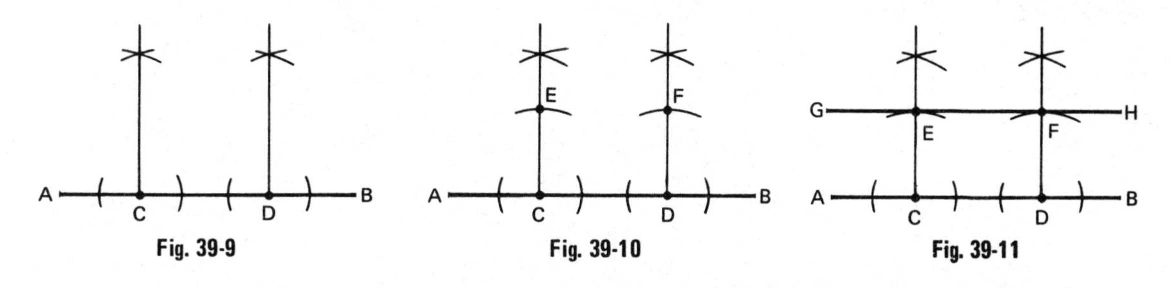

Fig. 39-9 Fig. 39-10 Fig. 39-11

- With points C and D as centers, draw arcs with the same radius intersecting the perpendiculars at points E and F. (See figure 39-10.)
- Connect points E and F. Line GH is parallel to line AB. (See figure 39-11.)

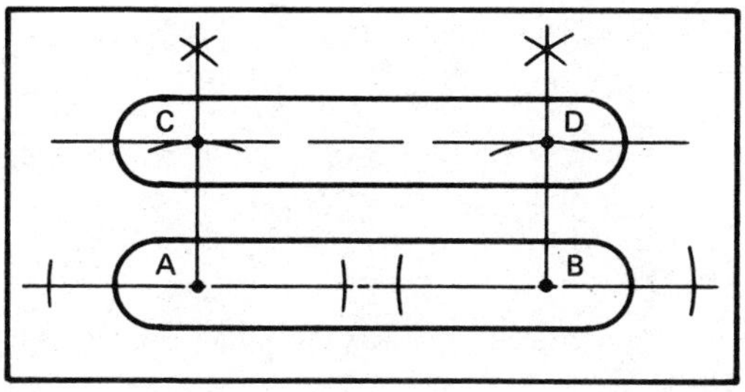

Example:

In figure 39-12, slots AB and CD are parallel.

Fig. 39-12

BISECTING AN ANGLE

Rule: To bisect angle AOB

- With point O as a center, draw an arc intersecting sides OA and OB at points C and D. (See figure 39-13.)

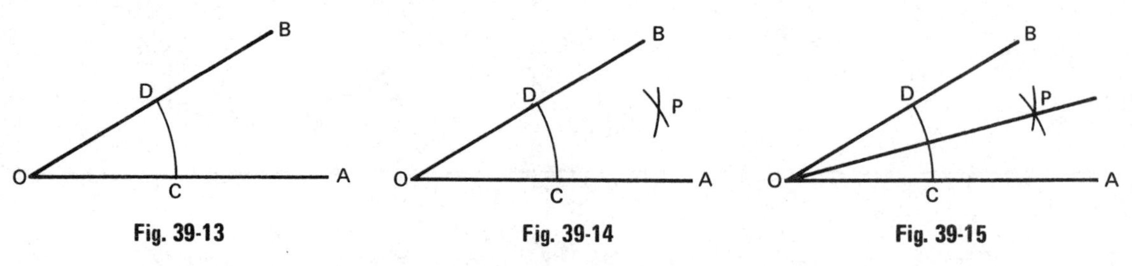

Fig. 39-13 **Fig. 39-14** **Fig. 39-15**

- With points C and D as centers, draw arcs using the same radius intersecting at point P. (See figure 39-14.)
- Connect points P and O. Line OP bisects ∠AOB. (See figure 39-15.)

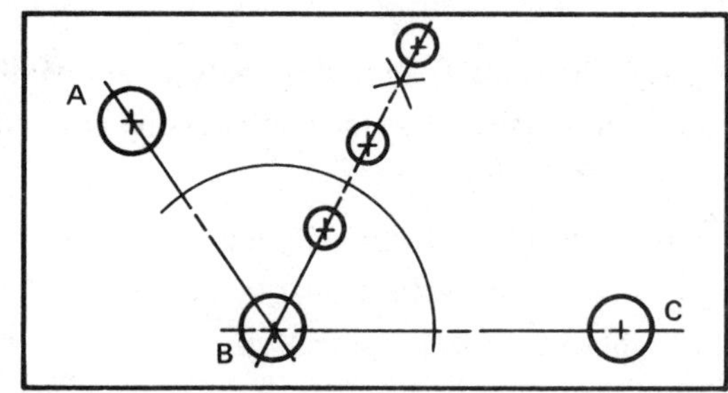

Fig. 39-16

Example:

In figure 39-16 a series of holes are laid out on a centerline which bisects the angle made by connecting the centers of holes A, B, and C.

TANGENTS TO A CIRCLE FROM AN OUTSIDE POINT

Rule: To construct tangents to circle O from point P

- Draw a line connecting center O and point P and bisect OP at point A. (See figure 39-17.)

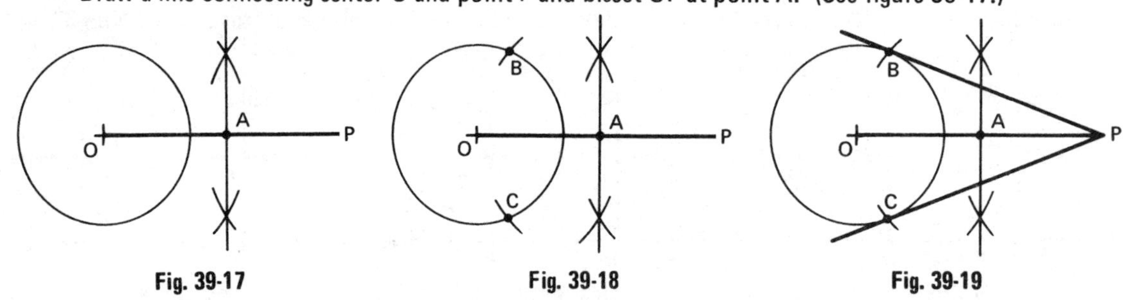

Fig. 39-17 **Fig. 39-18** **Fig. 39-19**

- With point A as a center and AP as a radius, draw arcs intersecting circle O at points B and C. (See figure 39-18.)
- Connect points B and P, and C and P. Lines BP and CP are tangents. (See figure 39-19.)

Example:

The cutout shown in the plate in figure 39-20 is laid out. Lines AB and CD are tangents to arc BD from points A and C.

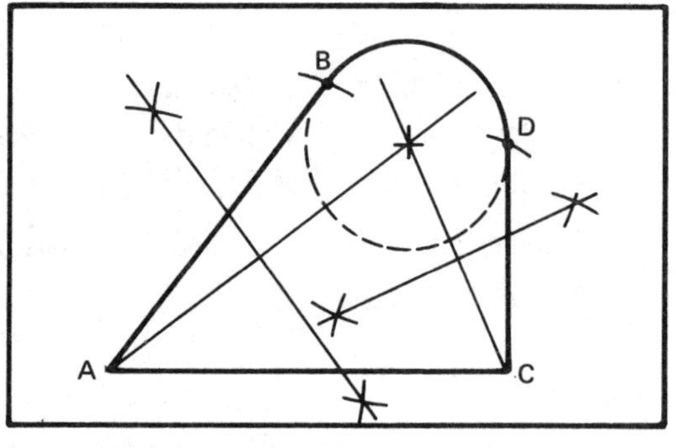

Fig. 39-20

DIVIDING A LINE INTO EQUAL PARTS

Rule: To divide line AB into three equal parts

- Draw line AC at any convenient angle from AB and with a compass lay off three equal segments AD, DE, and EF on AC. (See figure 39-21.)

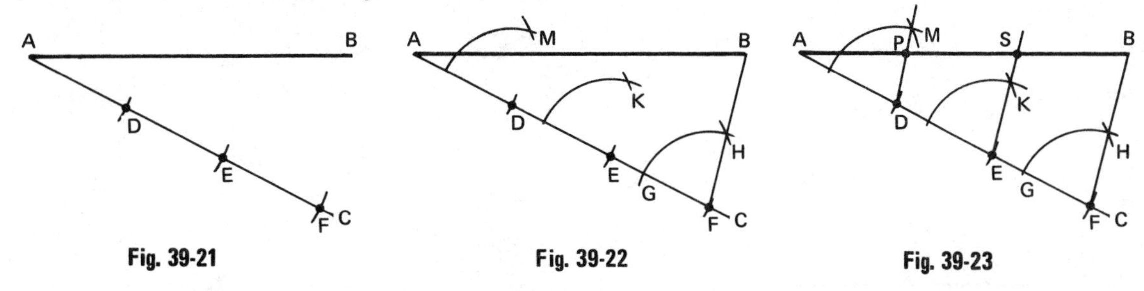

Fig. 39-21 **Fig. 39-22** **Fig. 39-23**

- Connect point F and B. With centers at points D, E, and F, draw arcs using the same radius. The arc with the center at point F intersects AC at point G and BF at point H. Set distance GH on the compass and mark off this distance on the other two arcs; the points of intersection are K and M. (See figure 39-22.)

- Connect points E and K, and D and M. Line AB is divided into three equal parts; AP = PS = SB. (See figure 39-23.)

  Note: Line AB can be divided into any required number of equal parts by laying off the desired number of equal segments on line AC and following the same procedure.

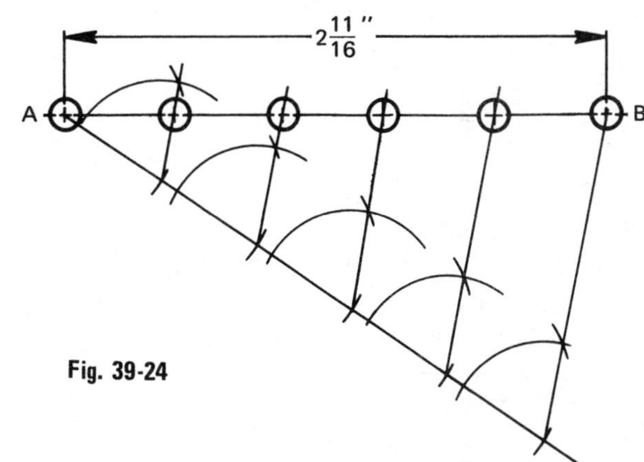

Fig. 39-24

Example:

In figure 39-24, six holes are equally spaced along centerline AB which is 2 11/16 inches long. Since six holes are required, line AB must be divided into five equal parts.

APPLICATION

1. Construct perpendicular bisectors to the given lines.

a.

b.

c.

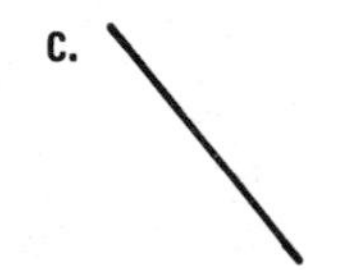

d.

2. Construct perpendiculars to the given lines at the given points.

a.

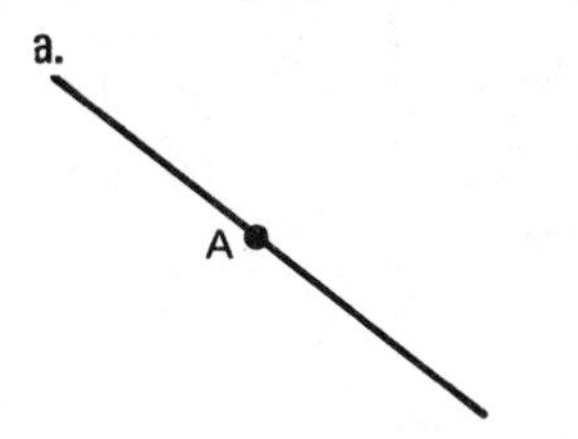

b.

c.

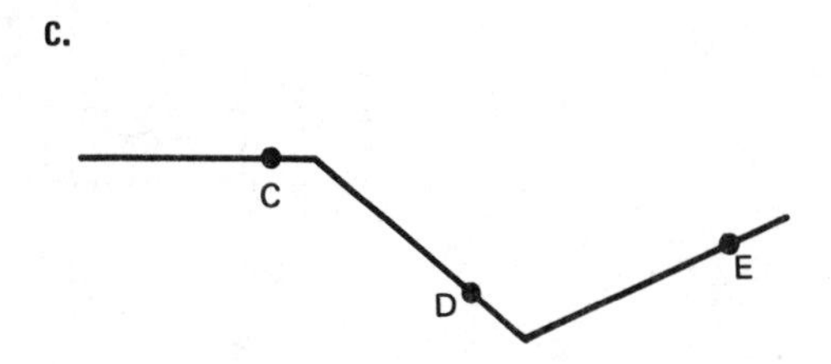

3. Construct a parallel line to each of the given lines.

a. b. c. d.

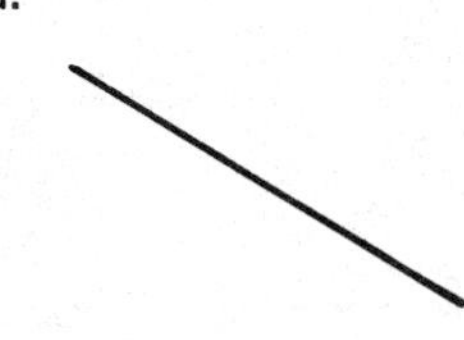

4. Bisect the given angles.

a. b. c.

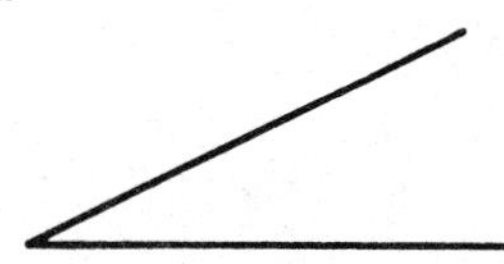
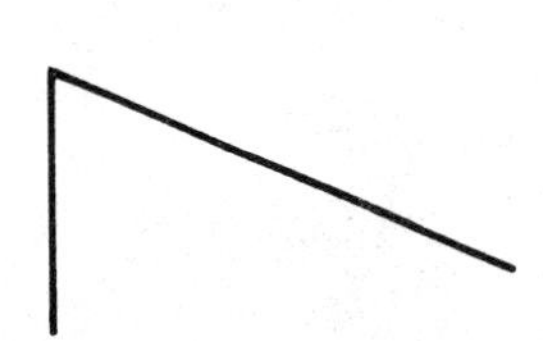
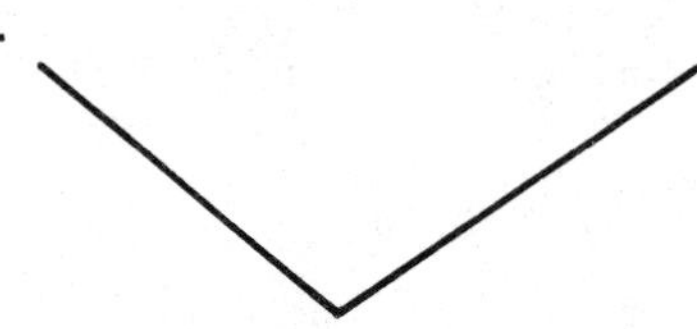

5. Construct tangents to the given circles from the given points.

a. b. c.

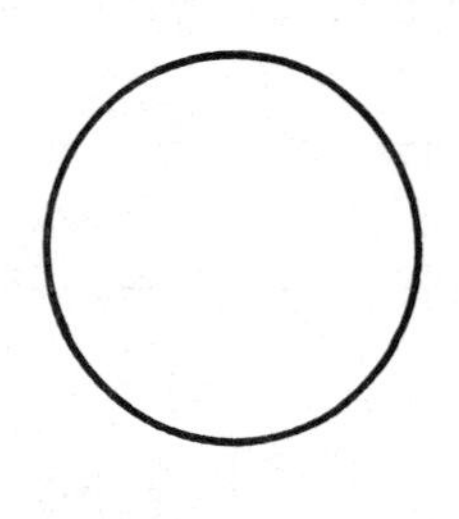

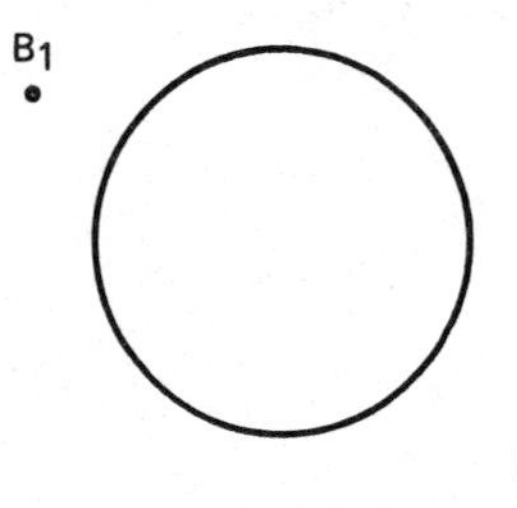

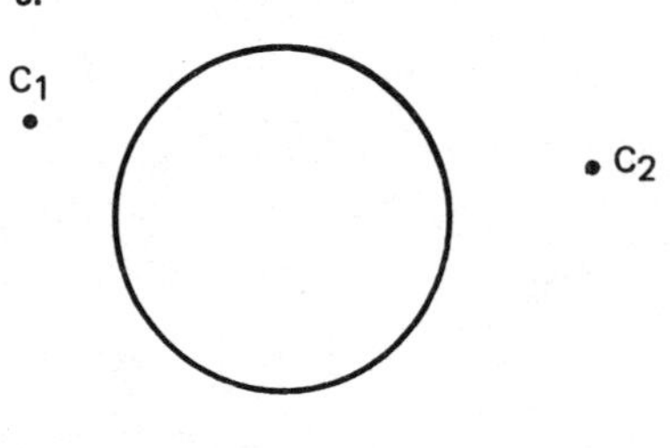

6. Divide the given lines into the designated number of equal parts.

a. b. c.

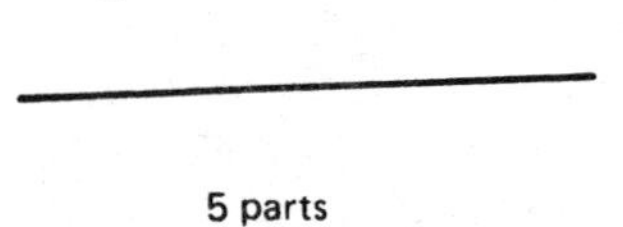

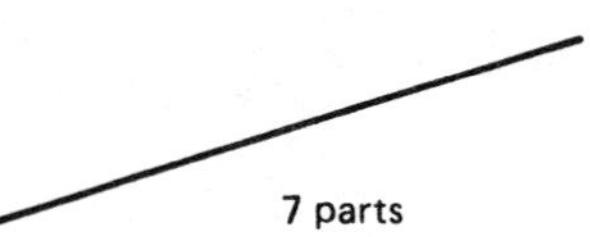

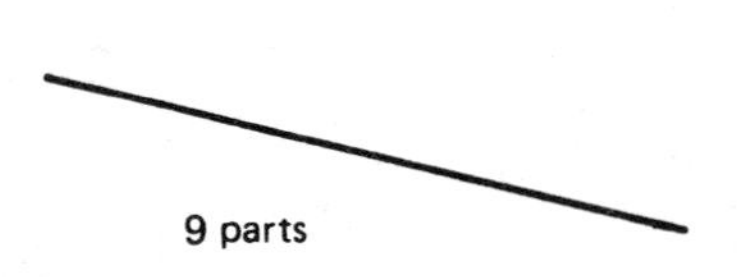

7. Lay out the following angles by construction. (Do not use a protractor.)

a. 45° b. 22° 30′ c. 135°

8. Lay out the illustrated plate. (Do not use a protractor.)

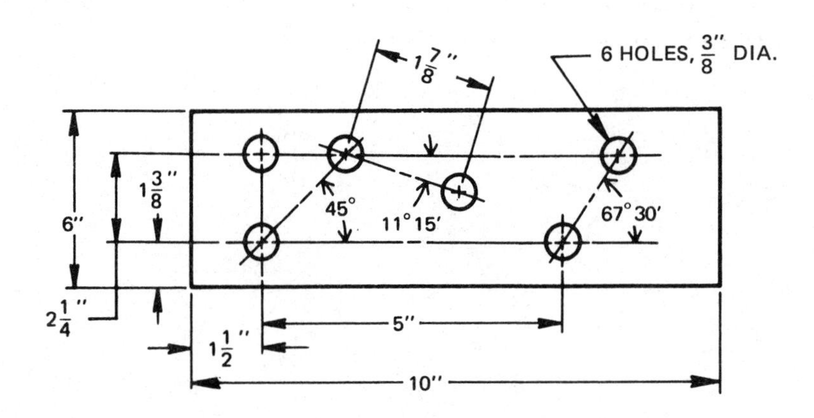

9. Lay out the illustrated part.

10. Lay out the plate shown. The dimensions are in inches. (Do not use a protractor.)

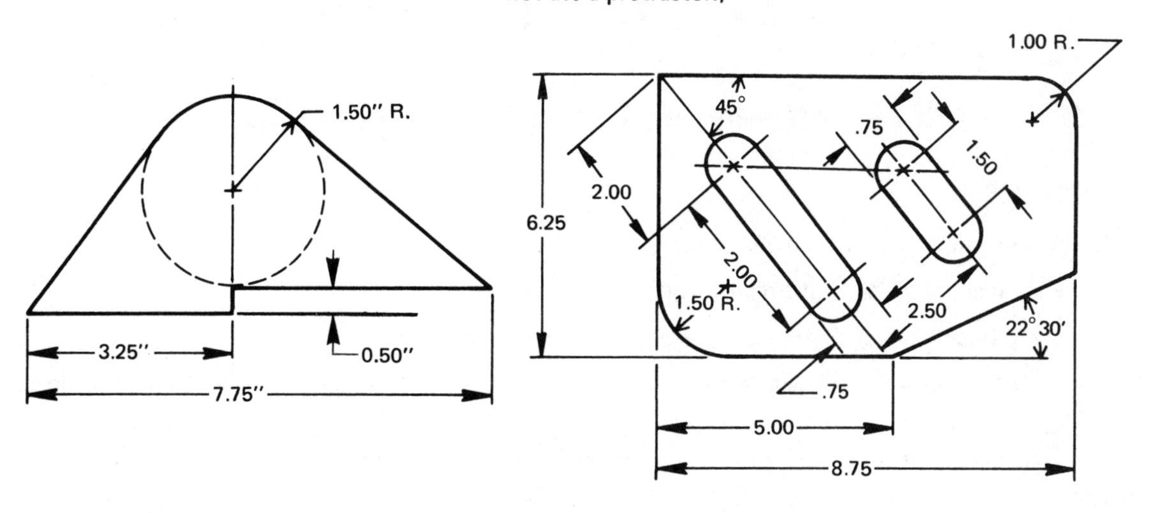

11. Lay out the following sets of holes in this plate following the given directions.

a. Bisect ∠ A and construct 4 equally spaced 3/16" diameter holes. Make the first hole 7/8" from point A and the last hole 2 7/16" from point A.

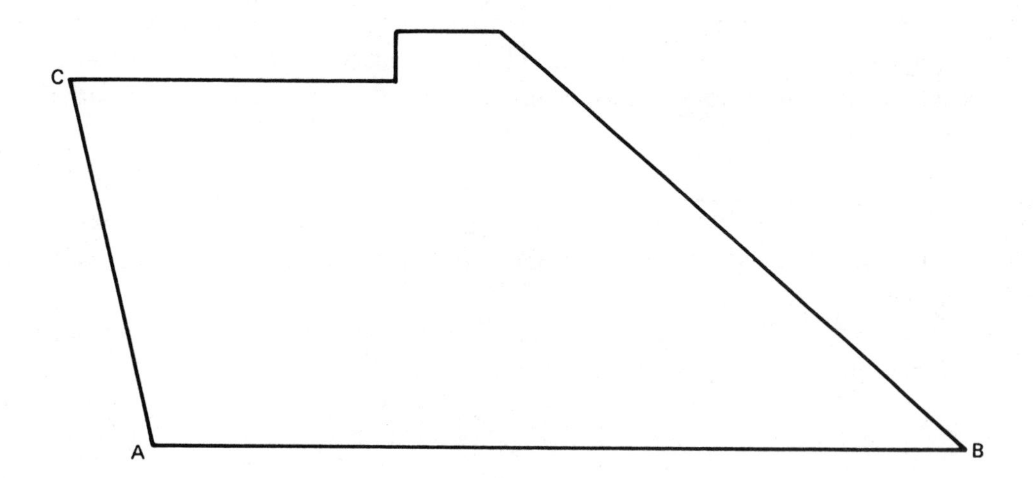

b. Bisect ∠ B and construct 8 equally spaced 1/4" diameter holes. Make the first hole 3/4" from point B and the last hole 3 11/16" from the first hole.

c. Bisect ∠ C and construct 4 equally spaced 3/16" diameter holes. Make the first hole 9/16" from point C and the last hole 2 11/16" from point C.

SECTION V

Trigonometry

Unit 40 Introduction to Right Angle Trigonometry

OBJECTIVES

After studying this unit, the student should be able to

- Identify the sides of a right triangle with reference to any angle.
- Determine functions of given angles using the table of trigonometric functions.
- Determine angles which correspond to given functions using the table of trigonometric functions.

Trigonometry is the branch of mathematics which is used to compute unknown angles and sides of triangles. Many problems that cannot be solved by the use of geometry alone are easily solved by trigonometry.

Practical machine shop problems are often solved by using a combination of elements of algebra, geometry, and trigonometry. Therefore, it is essential to develop the ability to analyze a problem in order to relate and determine the mathematical principles which are involved in its solution. Then the problem must be worked in clear orderly steps, based on mathematical facts.

When solving a problem, it is important to understand the trigonometric operations involved rather than to mechanically "plug in" values. Attempting to solve trigonometry problems without understanding the principles involved will prove to be unsuccessful, particularly in practical shop applications such as those found later in unit 44.

RATIO OF RIGHT TRIANGLE SIDES

In a right triangle, the ratio of two sides of the triangle determine the sizes of the angles, and the angles determine the ratio of two sides.

For example, in figure 40-1, the size of angle A is determined by the ratio of side x to side y. When side x = 2 inches and side y = 1 inch, the ratio of x to y is 2 : 1 or 2/1.

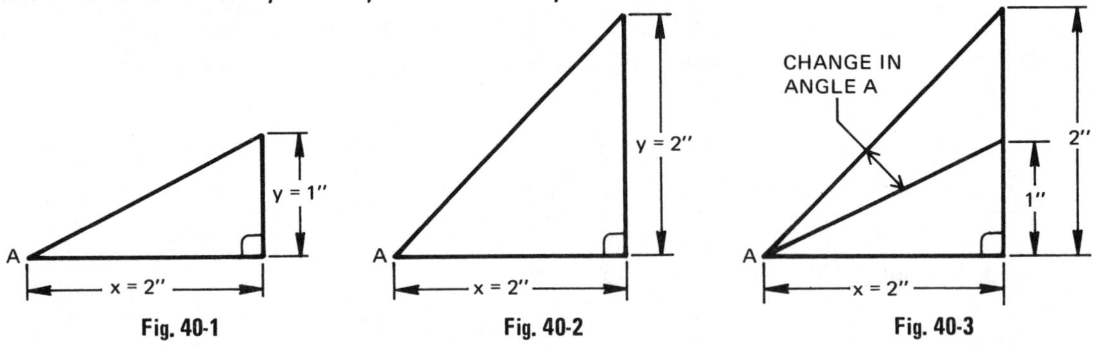

Fig. 40-1 Fig. 40-2 Fig. 40-3

If side x remains 2 inches long and side y is increased to 2 inches, as shown in figure 40-2, the ratio of x to y is 2/2 or 1/1.

Observe the increase in angle A as the ratio changed from 2/1 in figure 40-1, to 1/1 in figure 40-2. Figure 40-3 compares the two ratios and shows the change in angle A. The angles change as the ratio of the sides change.

DESCRIPTION OF SIDES

The sides of a right triangle are named opposite side, adjacent side, and hypotenuse.

The *hypotenuse* is always the side opposite the right angle. The positions of the opposite and adjacent sides depend on the reference angle. The *opposite side* is opposite the reference angle and the *adjacent side* is next to the reference angle.

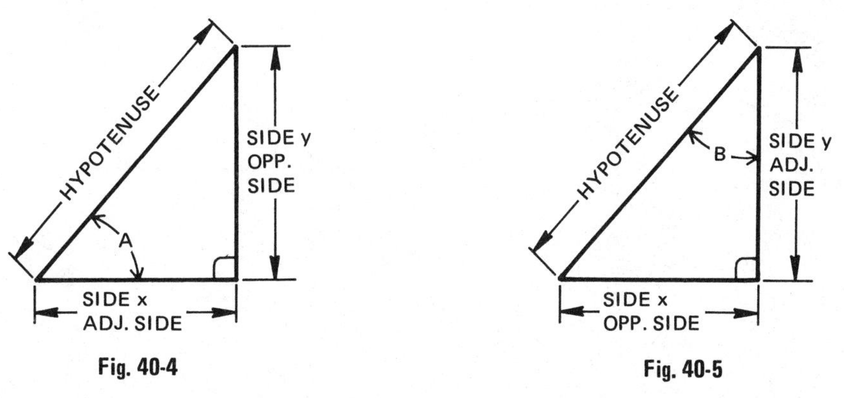

Fig. 40-4 Fig. 40-5

For example, in figure 40-4 in reference to angle A, side x is the adjacent side and side y is the opposite side. In figure 40-5 in reference to angle B, side x is the opposite side and side y is the adjacent side.

It is important to be able to identify the opposite and adjacent sides of right triangles in reference to any angle regardless of the positions of the triangles.

TRIGONOMETRIC FUNCTIONS

Since a triangle has three sides and a ratio is the comparison of any two sides, there are six different ratios. These ratios are the sine, cosine, tangent, cotangent, secant, and cosecant.

The six trigonometric functions, shown in figure 40-6, are defined in table 40-1.

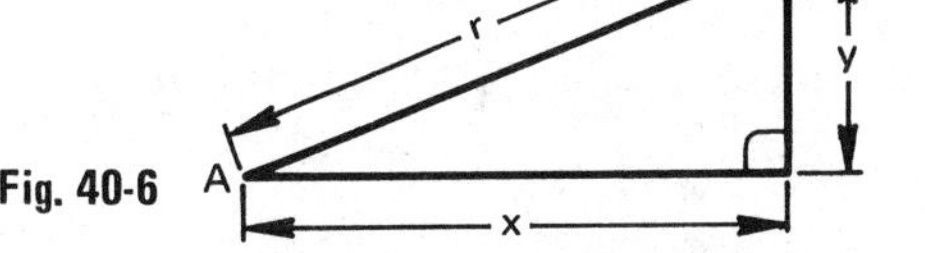

Fig. 40-6

Function	Abbreviation	Description	Formula
sine of an angle	$\sin \angle$	$\frac{\text{opposite side}}{\text{hypotenuse}}$	$\sin \angle A = \frac{y}{r}$
cosine of an angle	$\cos \angle$	$\frac{\text{adjacent side}}{\text{hypotenuse}}$	$\cos \angle A = \frac{x}{r}$
tangent of an angle	$\tan \angle$	$\frac{\text{opposite side}}{\text{adjacent side}}$	$\tan \angle A = \frac{y}{x}$
cotangent of an angle	$\cot \angle$	$\frac{\text{adjacent side}}{\text{opposite side}}$	$\cot \angle A = \frac{x}{y}$
secant of an angle	$\sec \angle$	$\frac{\text{hypotenuse}}{\text{adjacent side}}$	$\sec \angle A = \frac{r}{x}$
cosecant of an angle	$\csc \angle$	$\frac{\text{hypotenuse}}{\text{opposite side}}$	$\csc \angle A = \frac{r}{y}$

Table 40-1

To understand and properly use trigonometric functions, it is essential to realize that the function of an angle depends upon the ratio of the sides and not the size of the triangle. The functions of similar triangles are the same regardless of the sizes of the triangles since the sides of similar triangles are proportional. For example, in the similar triangles shown in figure 40-7, the functions of angle A are the same for the three triangles.

$$\text{In } \triangle ABC, \tan \angle A = \frac{.500}{1.000} = .500$$

$$\text{In } \triangle ADE, \tan \angle A = \frac{.800}{1.600} = .500$$

$$\text{In } \triangle AFG, \tan \angle A = \frac{1.200}{2.400} = .500$$

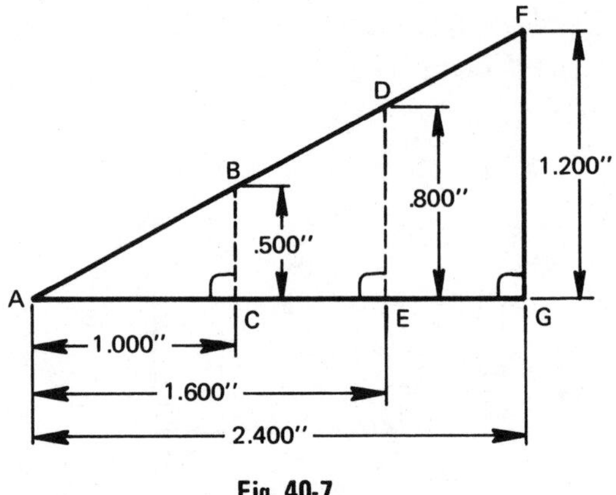

Fig. 40-7

TRIGONOMETRIC FUNCTION TABLE

A table of the six trigonometric functions is given in the appendix of this book. The functions listed in the table are the decimal equivalents of the ratio of two sides of a right triangle. The table lists functions of angles from 0° to 90° in 10′ steps.

Angles from 0° to 45° are located in the left columns and increase in value reading from the top to the bottom of a page. Angles from 45° to 90° are located in the right columns and increase in value from the bottom to the top of the page.

Observe that a column which is labeled sin on the top of a page is labeled cos on the bottom; the same is also true for the other functions. The top function names are used when locating functions of angles from 0° to 45°, the bottom function names are used when locating functions of angles from 45° to 90°.

Examples of Locating Functions of Given Angles

1. Find the sine of 17° 40′.
   - ▲ Locate 17° in the left column and move down to the 40′ row.
   - ▲ Locate the sin function on the top of the page and move down the sin column to the 17° 40′ row.
   - ▲ The sin of 17° 40′ is .30348.
2. Find the tangent of 57° 50′.
   - ▲ Locate 57° in the right column and move up to 50′ row.
   - ▲ Locate the tan function on the bottom of the page and move up the tan column to the 57° 50′ row.
   - ▲ The tangent of 57° 50′ is 1.5900.

Examples of Locating Angles of Given Functions

1. Find the angle whose cosine is .88020.
   - ▲ Locate the cosine column in the table and read down the column until .88020 is located.
   - ▲ Since the top cosine function was used, the corresponding angle is located directly across from .88020 in the left column.
   - ▲ The angle whose cosine is .88020 is 28° 20′.

2. Find the angle whose secant is 2.9957.

▲ Locate the secant column in the table. Observe that 2.9957 cannot be found in the table using the top secant function. Therefore, the angle must be greater than 45° and the bottom secant function is used. Read up the column until 2.9957 is located.

▲ Since the bottom secant function was used, the corresponding angle is located directly across from 2.9957 in the right column.

▲ The angle whose secant is 2.9957 is 70° 30′.

INTERPOLATION

To determine the function of an angle or the angle of a function not listed in the trigonometric function table, a method called interpolation must be used.

Example 1: Determine the tangent of 42° 13′.

10′	3′	tan 42° 10′ = .90568	.00159	.00531
		tan 42° 13′ = .90727		
		tan 42° 20′ = .91099		

▲ The angle 42° 13′ lies between 42° 10′ and 42° 20′. Therefore, the tangent function of 42° 13′ lies between the tangent of 42° 10′ and the tangent of 42° 20′.

▲ The difference between 42° 10′ and 42° 20′ is 10′ and the difference between 42° 10′ and 42° 13′ is 3′. The resulting ratio is 3/10.

▲ The tangent of 42° 10′ is .90568. The tangent of 42° 20′ is .91099. The difference between .91099. and .90658 is .00531.

▲ Multiply: 3/10 x .00531 = .00159.

▲ Add: .90568 + .00159 = .90727.

The tangent of 42° 13′ = .90727.

Example 2: Determine the cosine of 56° 47′.

10′	7′	cos 56° 40′ = .54951	.00170	.00243
		cos 56° 47′ = .54781		
		cos 56° 50′ = .54708		

▲ The cosine function of 56° 47′ lies between the cosine of 56° 40′ and the cosine of 56° 50′.

▲ The difference between 56° 40′ and 56° 50′ is 10′. The difference between 56° 40′ and 56° 47′ is 7′. The resulting ratio is 7/10.

▲ The cosine of 56° 40′ is .54951. The cosine of 56° 50′ is .54708. The difference between .54951 and .54708 is .00243.

▲ Multiply 7/10 x .00243 = .00170.

▲ Subtract .54951 - .00170 = .54781. The cosine of 56° 47′ = .54781.

Example 3: Determine the angle whose sine is .52349.

$$10' \left\{ \begin{array}{l} 4' \left\{ \begin{array}{l} \sin 31^\circ\ 30' = .52250 \\ \sin 31^\circ\ 34' = .52349 \end{array} \right\} .00099 \\ \sin 31^\circ\ 40' = .52498 \end{array} \right\} .00248$$

▲ The sine function .52349 lies between the sine function .52250 whose angle is 31° 30′ and the sine function .52498 whose angle is 31° 40′.

▲ The difference between .52498 and .52250 is .00248, and the difference between .52349 and .52250 is .00099.

The resulting ratio is $\frac{.00099}{.00248}$ or $\frac{99}{248}$.

▲ The difference between 31° 30′ and 31° 40′ is 10′.

▲ Multiply: $\frac{99}{248} \times 10' = \frac{990}{248} = 3\frac{246}{248} = 4'$ The angle whose sine is .52349 is 31° 34′

Note: When interpolating functions from given angles or angles from given functions, do not use the cotangent, secant, or cosecant functions for angles less than 15°. Do not use the tangent, secant, or cosecant functions for angles greater than 75°. The trigonometric function table which is given in steps of 10′ produces changes in these functions which are either too small or too great to obtain accurate interpolated values.

APPLICATION

A. With reference to ∠ 1, name the sides of each of the following triangles as opposite, adjacent, or hypotenuse.

1. Side: r ________
   x ________
   y ________

2. Side: r ________
   x ________
   y ________

3. Side: a ________
   b ________
   c ________

4. Side: a ________
   b ________
   c ________

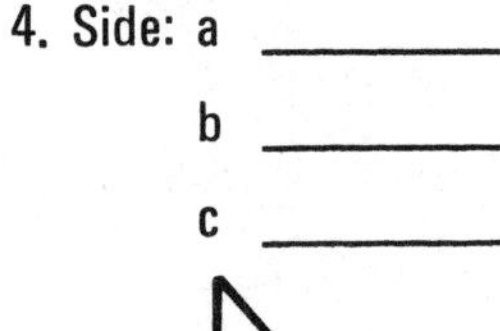

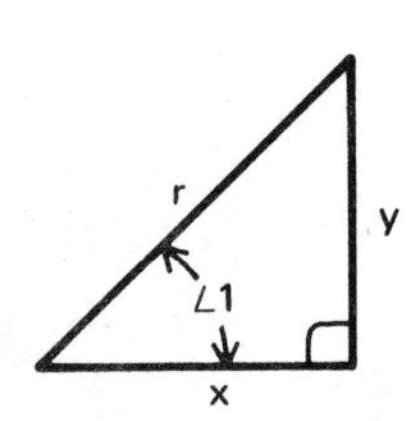

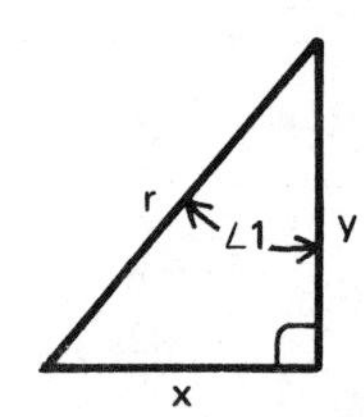

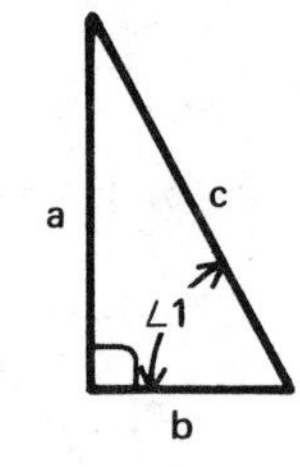

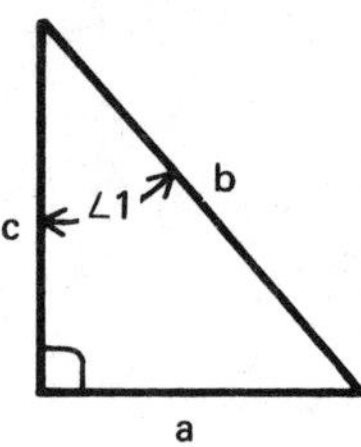

5. Side: a ________
   b ________
   c ________

6. Side: d ________
   m ________
   p ________

7. Side: d ________
   m ________
   p ________

8. Side: e ________
   f ________
   g ________

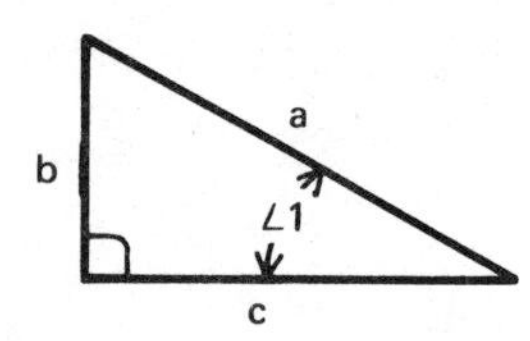

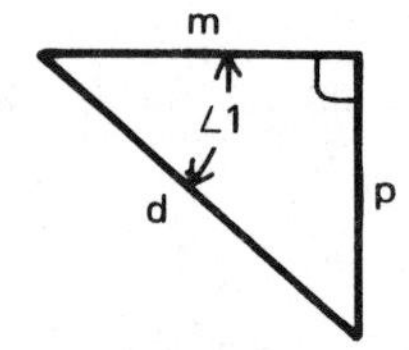

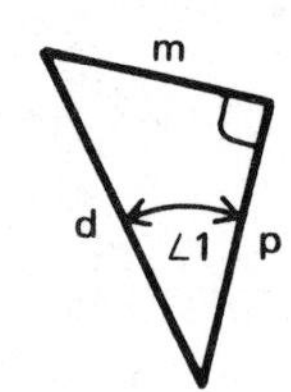

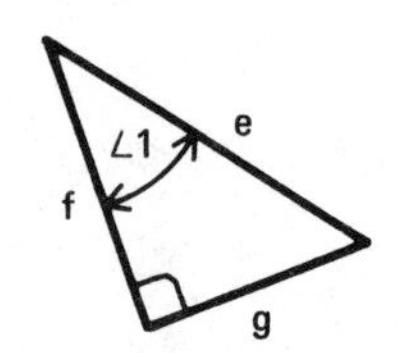

9. Side: h ______
   k ______
   ℓ ______

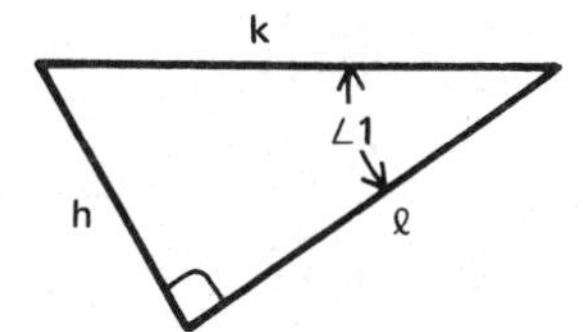

10. Side: h ______
    k ______
    ℓ ______

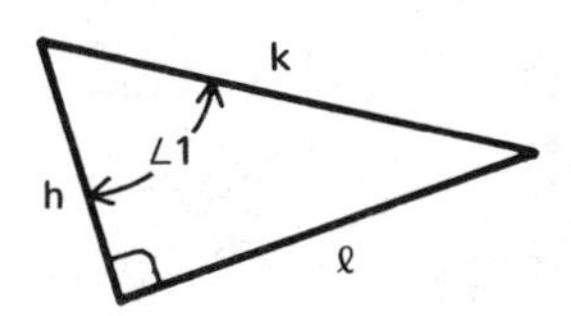

11. Side: m ______
    p ______
    s ______

12. Side: m ______
    p ______
    s ______

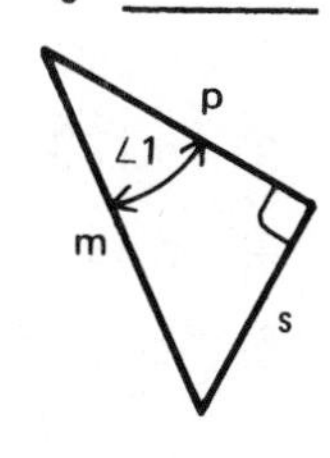

13. Side: m ______
    r ______
    t ______

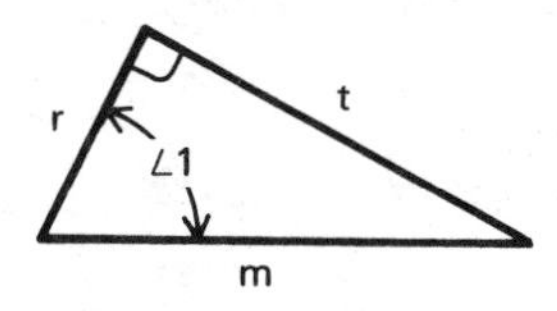

14. Side: m ______
    r ______
    t ______

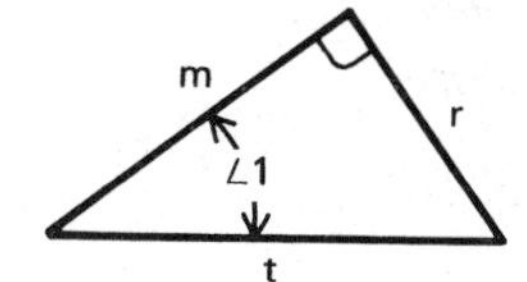

15. Side: f ______
    g ______
    h ______

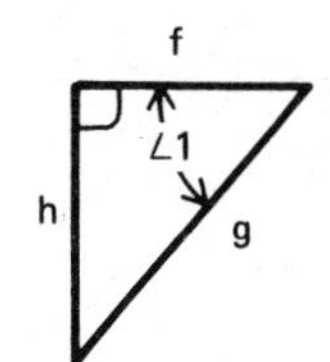

16. Side: f ______
    g ______
    h ______

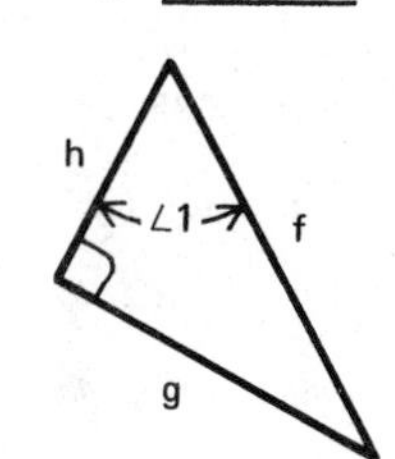

B. Three groups of triangles are given below. Each group consists of four triangles. Within each group name the triangles – a, b, c, or d – in which angles A are equal.

Group 1 ____________

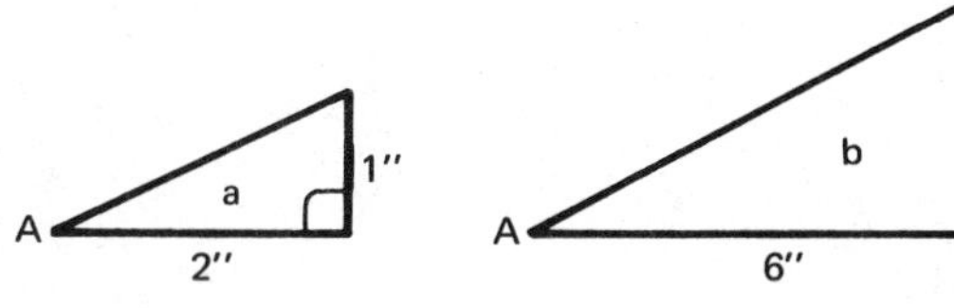

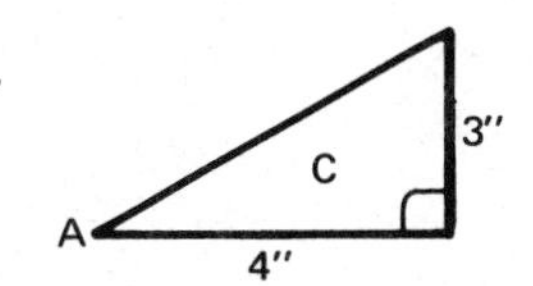

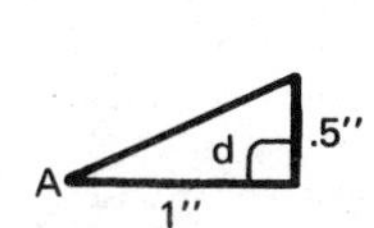

Group 2 ____________

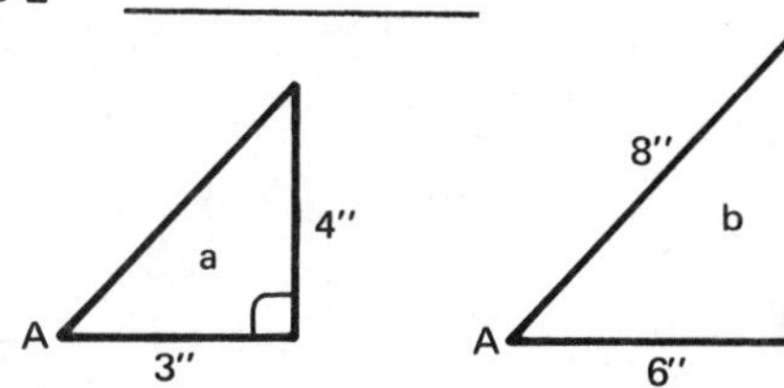

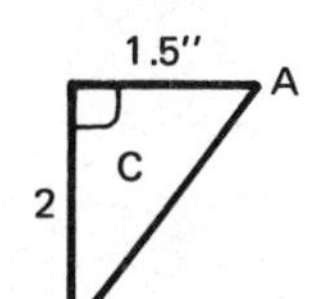

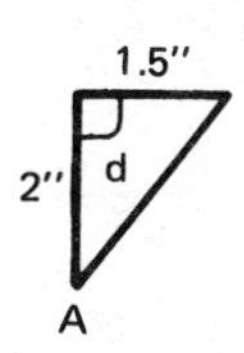

Group 3 ____________

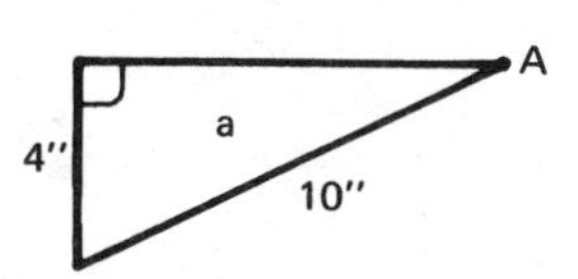

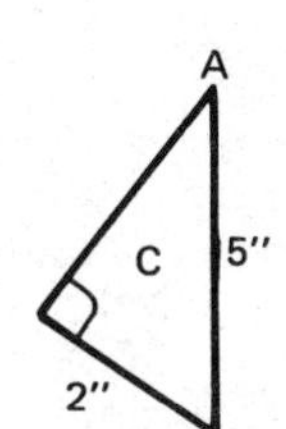

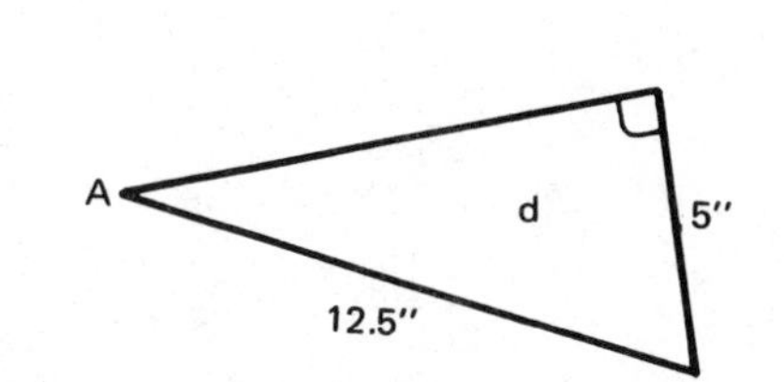

C. Use the table of trigonometric functions to determine the functions of the following angles. Interpolate where necessary.

1. sin 18° ____ 6. sec 47° ____ 11. csc 45° ____ 16. cot 43° 25′ ____

2. cos 23° ____ 7. sin 87° 10′ ____ 12. sec 55° 50′ ____ 17. csc 37° 14′ ____

3. tan 73° ____ 8. cos 7° 50′ ____ 13. sin 10° 12′ ____ 18. sin 0° 58′ ____

4. cot 7° ____ 9. tan 1° 30′ ____ 14. cos 89° 53′ ____ 19. cot 65° 46′ ____

5. csc 39° ____ 10. cot 69° 40′ ____ 15. tan 66° 27′ ____ 20. tan 25° 19′ ____

D. Use the table of trigonometric functions to determine the angles that correspond to the following functions. If an exact value of a function cannot be found in the table, interpolate to obtain angles to the closest minute.

1. sin ∠ = .13917 ____ 7. cot ∠ = .02036 ____ 13. csc ∠ = 3.6243 ____

2. cos ∠ = .78441 ____ 8. sec ∠ = 1.0608 ____ 14. sec ∠ = 1.2322 ____

3. tan ∠ = .90568 ____ 9. sin ∠ = .97502 ____ 15. sin ∠ = .99886 ____

4. cot ∠ = .57735 ____ 10. cos ∠ = .99973 ____ 16. tan ∠ = .96738 ____

5. sec ∠ = 1.7220 ____ 11. tan ∠ = .23455 ____ 17. cos ∠ = .55218 ____

6. csc ∠ = 2.3087 ____ 12. cot ∠ = .35805 ____ 18. cot ∠ = 2.0548 ____

Unit 41 Analysis of Trigonometric Functions, and the Cartesian Coordinate System

OBJECTIVES

After studying this unit, the student should be able to

- Determine the variations of functions as angles change.
- Compute cofunctions of complementary angles.
- Compute the functions of angles greater than 90°.

VARIATION OF FUNCTIONS

As the size of an angle increases from 0° to 90°, the sine, tangent, and secant functions increase while the cofunctions (cosine, cotangent, cosecant) decrease.

Example 1: Variation of an increasing function, the sine function, figure 41-1.

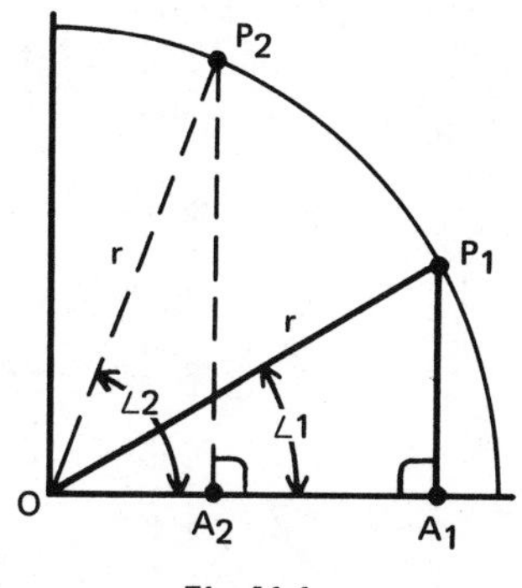

Fig. 41-1

▲ Since OP_1 and OP_2 are radii of the circle, $OP_1 = OP_2 = r$.

▲ The sine of an angle $= \frac{\text{opposite side}}{\text{hypotenuse}}$.

▲ Therefore, $\sin \angle 1 = \frac{A_1P_1}{r}$; $\sin \angle 2 = \frac{A_2P_2}{r}$.

▲ A_2P_2 is greater than A_1P_1; therefore, $\frac{A_2P_2}{r}$ is greater than $\frac{A_1P_1}{r}$.

▲ Conclusion: As the angle is increased from ∠1 to ∠2, the sine of the angle is increased.

Example 2: Variation of a decreasing function, the cosine function, figure 41-1.

▲ The cosine of an angle $= \frac{\text{adjacent side}}{\text{hypotenuse}}$. Therefore, $\cos \angle 1 = \frac{OA_1}{r}$, $\cos \angle 2 = \frac{OA_2}{r}$.

▲ OA_2 is less than OA_1; therefore, $\frac{OA_2}{r}$ is less than $\frac{OA_1}{r}$.

▲ Conclusion: As the angle is increased from ∠1 to ∠2, the cosine of the angle decreases.

A summary of the variations of functions taken from the table of trigonometric functions from 0° to 90° is given.

As an angle increases from 0° to 90°
sin increases from 0 to 1
tan increases from 0 to ∞
sec increases from 1 to ∞
cos decreases from 1 to 0
cot decreases from ∞ to 0
csc decreases from ∞ to 1

The symbol ∞ means an infinitely large quantity and cannot be used for computations at this level of mathematics.

Note: It is advisable to study the table of trigonometric functions in order to observe the changes in functions as angles increase and decrease. It is also helpful to sketch figures similar to figure 41-1 for all functions in order to further develop an understanding of the relationship of angles and their functions. Particular attention should be given to functions of angles close to 0° and 90°.

FUNCTIONS OF COMPLEMENTARY ANGLES

A function of an angle is equal to the cofunction of its complement. For example, in figure 41-2, $\angle A + \angle B = 90°$. Angle A and angle B are complementary angles.

The sine of $\angle A$ (20°) = .34202; the cosine of $\angle B$ (70°) = .34202. Therefore, the sine of 20° = the cosine of 70°.

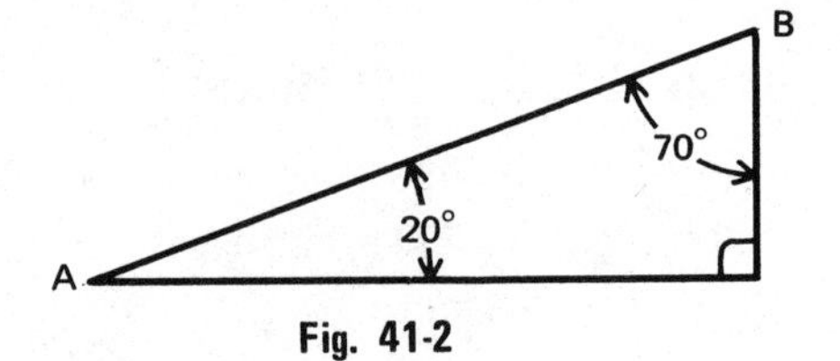

Fig. 41-2

The following is a list of functions and cofunctions of complementary angles:

- $\sin \angle A = \cos (90° - \angle A)$
- $\cos \angle A = \sin (90° - \angle A)$
- $\tan \angle A = \cot (90° - \angle A)$
- $\cot \angle A = \tan (90° - \angle A)$
- $\sec \angle A = \csc (90° - \angle A)$
- $\csc \angle A = \sec (90° - \angle A)$

Examples:

- ▲ $\sin 10° = \cos (90° - 10°) = \cos 80°$
- ▲ $\sec 45° = \csc (90° - 45°) = \csc 45°$
- ▲ $\tan 18° 40' = \cot (90° - 18° 40') = \cot 71° 20'$
- ▲ $\cos 90° = \sin (90° - 90°) = \sin 0°$

CARTESIAN (RECTANGULAR) COORDINATE SYSTEM

It is sometimes necessary to determine functions of angles greater than 90°. In a triangle that is not a right triangle, one of the angles can be greater than 90°. Computations using functions of angles greater than 90° are required in order to solve oblique triangle problems.

Functions of any angles are easily described in reference to the Cartesian Coordinate System, figure 41-3.

A fixed point (*origin*) is located at the intersection of a vertical and horizontal axis. The horizontal axis is the x-axis (*abscissa*) and the vertical axis is the y-axis (*ordinate*).

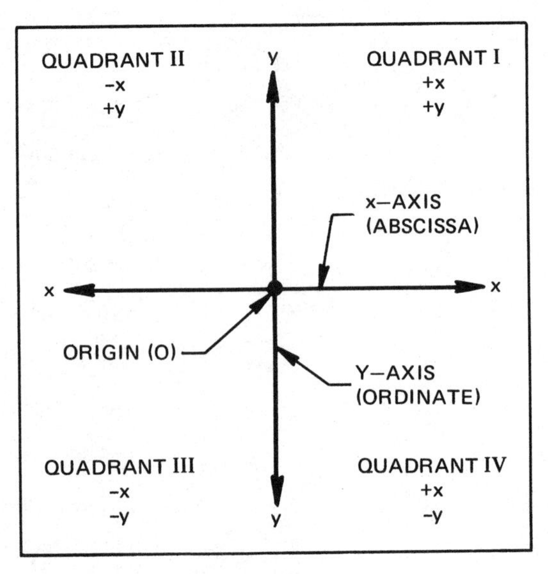

Fig. 41-3

The x and y axes divide a plane into four parts which are called *quadrants.* Quadrant 1 is the upper right section. In a counterclockwise direction from Quadrant 1 are Quadrants II, III, and IV.

All points located to the right of the y-axis have positive x values; all points to the left of the y-axis have negative x values. All points above the x-axis have positive y values; all points below the x-axis have negative y values.

Quadrant I:	+ x + y	Quadrant III:	- x - y
Quadrant II:	- x + y	Quadrant IV:	+ x - y

DETERMINING ANGLES AND FUNCTIONS IN ANY QUADRANT

As an angle is rotated through any of the four quadrants, functions of the angle are determined as follows:

- The angle is rotated in a counterclockwise direction with its vertex at the origin. Zero degrees is on the x-axis in quadrant I.
- From the end point of the rotated side of the angle, a line is projected perpendicular to the x-axis. A right triangle is formed of which the rotated side is the hypotenuse, the projected line the opposite side, and the side on the x-axis the adjacent side. The reference angle is the angle which has its vertex at the origin.
- Functions of reference angles are determined by noting the signs (+ or -) of the sides of the right triangle. The hypotenuse is always positive in all four quadrants.

Examples of Functions of Angles Greater Than 90°:

1. Determine the sin and cos functions of 115° (refer to figure 41-4).

▲ With the vertex at point O, the angle of 115° is rotated counterclockwise.

▲ From the end point of r, side y is projected perpendicular to the x-axis. In the right triangle formed, in reference to angle O, r is the hypotenuse, y is the opposite side, and x is the adjacent side.

Reference $\angle O = 180° - 115° = 65°$

▲ $\text{Sin} \angle O = \frac{\text{opposite side}}{\text{hypotenuse}}$. In Quadrant II, y is positive; r is always positive. Therefore, $\sin \angle O = \frac{+y}{+r} =$ + function.

$\text{Sin } 115° = \sin (180° - 115°) = \sin 65° = .90631.$

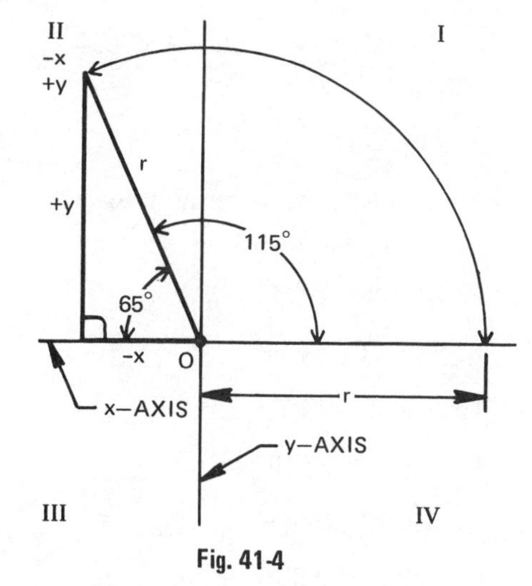

Fig. 41-4

▲ $\text{Cos} \angle O = \frac{\text{adjacent side}}{\text{hypotenuse}}$. Side x is negative, therefore, $\cos \angle O = \frac{-x}{+r} =$ - function. $\text{Cos } 115° = -\cos (180° - 115°) = -\cos 65°.$

▲ Look up cos 65° in the function table and prefix a negative sign. Cos 65° = .42262, - cos 65° = - .42262.

Note: A negative function of an angle does not mean that the angle is negative; it is a negative function of a positive angle. For example, - cos 65° does not mean cos (-65°).

2. Determine the tan and sec functions of 218°, figure 41-5.

▲ Rotate 218° counterclockwise.

▲ Project side y ⊥ to the x-axis.

Reference ∠ O = 218° - 180° = 38°

▲ $\text{Tan} \angle O = \frac{-y}{-x} = +$ function

Tan 218° = tan 38° = .78128

▲ $\text{Sec} \angle O = \frac{+r}{-x} = -$ function

Sec 218° = - sec 38° = - 1.2690

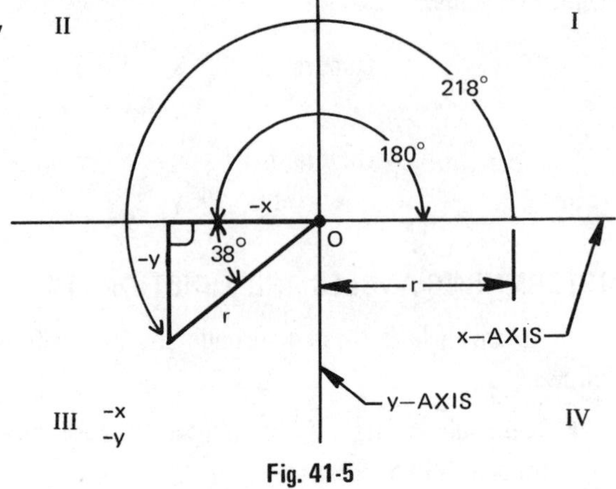

Fig. 41-5

3. Determine the cot and csc functions of 310°, figure 41-6.

▲ Rotate 310° counterclockwise.

▲ Project side y ⊥ to the x-axis.
Reference ∠ O = 360° - 310° = 50°

▲ $\text{Cot} \angle O = \frac{+x}{-y} = -$ function

Cot 310° = - cot 50° = - .83910

▲ $\text{Csc} \angle O = \frac{+r}{-y} = -$ function

Csc 310° = - csc 50° = - 1.3054

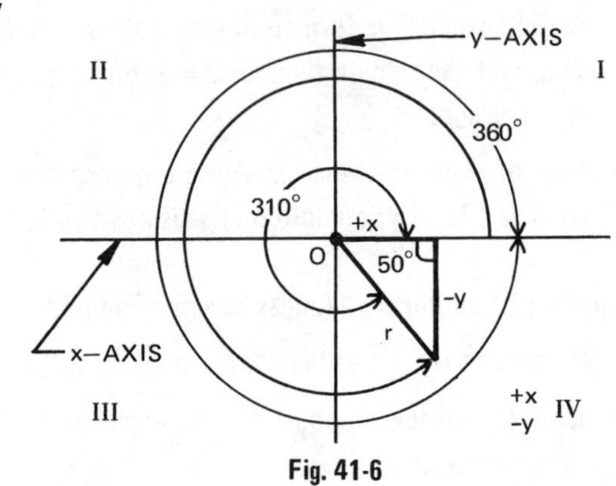

Fig. 41-6

APPLICATION

A. Which of the functions of the two angles given below is the greater? (Do not use the table of trigonometric functions.)

1. sin 43°, sin 39° ____
2. tan 10°, tan 41° ____
3. cos 73°, cos 78° ____
4. sec 37°, sec 19° ____
5. cot 78°, cot 85° ____
6. cot 0°, cot 90° ____
7. sin 45°, sin 46° ____
8. csc 0° 10′, csc 0° 20′ ____
9. cos 66°, cos 90° ____
10. tan 78°, tan 84° ____
11. sin 0°, cos 0° ____
12. tan 0°, cot 0° ____
13. sec 0°, csc 0° ____
14. sin 90°, cos 90° ____
15. tan 90°, cot 90° ____
16. sec 90°, csc 90° ____

B. Give the cofunction of the complement in each of the following.

1. sin 37°	____	6. tan 0°	____	11. sec 45°	____
2. tan 9°	____	7. cos 44°	____	12. sec 0°	____
3. cos 78°	____	8. csc 26°	____	13. tan 89° 50′	____
4. sec 63°	____	9. sin 90°	____	14. sin 12° 20′	____
5. cot 45°	____	10. cot 0° 30′	____	15. cos 90°	____

C. It may be helpful to sketch figures similar to figure 41-1. Refer to the accompanying figure in answering the following questions.

1. When ∠ 1 is almost 90°,
   a. how does side y compare to side r? ____
   b. how does side x compare to side r? ____
2. When ∠ 1 is slightly greater than 0°,
   a. how does side y compare to side r? ____
   b. how does side x compare to side y? ____
3. When side x = side y, what is the value of ∠ 1? ____
4. When side y = side r, what is the size of a. side x ____ b. ∠ 1 ____
5. When side x = side r, what is the size of a. side y ____ b. ∠ 1 ____

D. In each of the following problems, complete the figure by forming a right triangle similar to those in figures 41-4, 41-5, and 41-6. Label the sides + or –. Determine the reference angle and look up the function of the angle in the table. Note: The function of an angle greater than 90° may be a negative value.

1.		2.		3.	
sin 107° =	______	sin 199° 20′ =	______	sin 342° 50′ =	______
cos 107° =	______	cos 199° 20′ =	______	cos 342° 50′ =	______
tan 107° =	______	tan 199° 20′ =	______	tan 342° 50′ =	______
cot 107° =	______	cot 199° 20′ =	______	cot 342° 50′ =	______
sec 107° =	______	sec 199° 20′ =	______	sec 342° 50′ =	______
csc 107° =	______	csc 199° 20′ =	______	csc 342° 50′ =	______

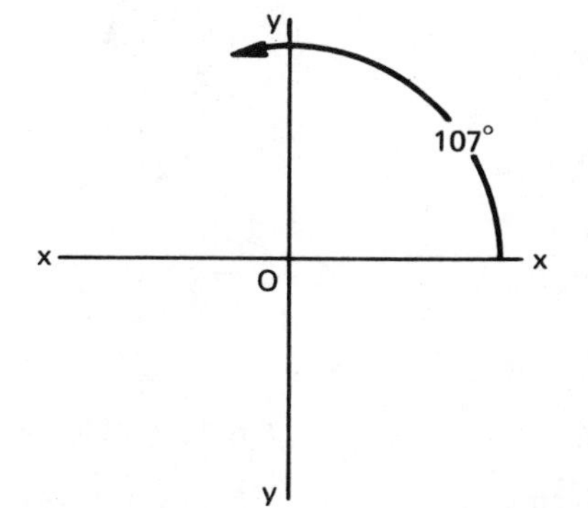

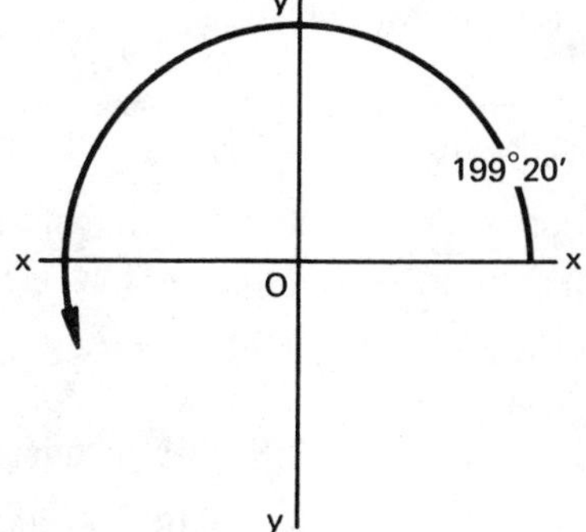

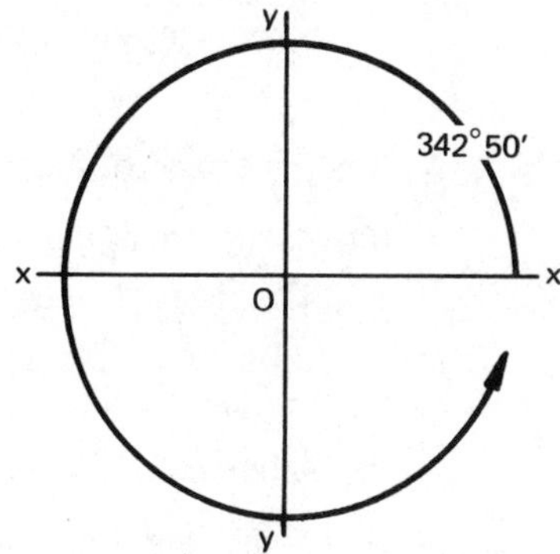

Unit 42 Basic Calculation of Angles and Sides of Right Triangles

OBJECTIVES

After studying this unit, the student should be able to

- Compute an unknown angle of a right triangle when two sides are known.
- Compute an unknown side of a right triangle when an angle and a side are known.

In order to solve for an unknown angle of a right triangle where neither acute angle is known, at least two sides must be known. To solve for an unknown side using trigonometric functions, at least an angle and a side must be known. An understanding of the procedures required for solving for unknown angles and sides is essential to the machinist.

PROCEDURE FOR DETERMINING AN UNKNOWN ANGLE WHEN TWO SIDES ARE GIVEN

- In relation to the desired angle, identify two given sides as adjacent, opposite, or hypotenuse.
- Determine the functions that are ratios of the sides identified in relation to the desired angle.

  Note: Two of the six trigonometric functions are ratios of the two known sides. Either of the two functions can be used. Both produce the same value for the unknown, except for cotangents, secants, and cosecants of angles less than 15° and tangents, secants, and cosecants of angles greater than 75°.
- Choose one of the two functions, substitute the given sides in the ratio, and divide.
- Using the table of trigonometric functions, determine the angle to the nearest minute that corresponds to the quotient obtained. It is often necessary to interpolate.

Example 1: Determine ∠A of the triangle shown in figure 42-1.

▲ In relation to ∠A, the 10.774-inch side is the adjacent side and the 7.500-inch side is the opposite side.

▲ Determine the two functions whose ratios consist of the adjacent and opposite sides. The tan ∠A = $\frac{\text{opposite side}}{\text{adjacent side}}$, and the cot ∠A = $\frac{\text{adjacent side}}{\text{opposite side}}$

Either the tangent or cotangent function can be used.

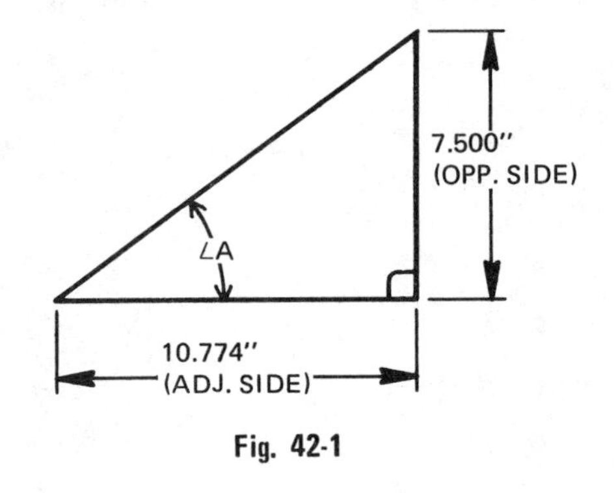

Fig. 42-1

▲ Use the cotangent function: cot ∠A = $\frac{10.774}{7.500}$, cot ∠A = 1.4365

▲ Interpolate from the function table to determine the angle whose cotangent function is nearest 1.4365. The angle interpolated to the nearest minute is 34° 51′, Angle A = 34° 51′.

Example 2: Determine ∠B of the triangle shown in figure 42-2.

▲ In relation to ∠B, the 2.900-inch side is the adjacent side; the 5.750-inch side is the hypotenuse.

▲ Determine the two functions whose ratios consist of the adjacent side and the hypotenuse.

The sec ∠ B = $\frac{\text{hypotenuse}}{\text{adjacent side}}$, and the cos ∠ B = $\frac{\text{adjacent side}}{\text{hypotenuse}}$. Either the secant or cosine function can be used.

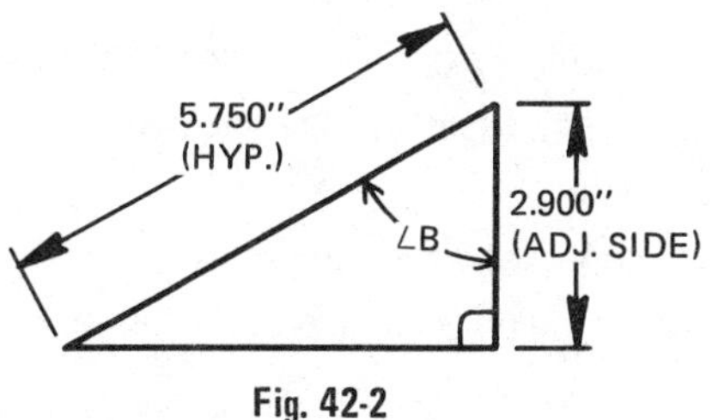

Fig. 42-2

- ▲ Using the secant function: sec ∠ B = $\frac{5.750}{2.900}$, sec ∠ B = 1.9828.
- ▲ Interpolate from the function table to determine the angle whose secant function is nearest 1.9828. The angle interpolated to the nearest minute is 59° 43'. Angle B = 59° 43'.

Example 3: Determine ∠ 1 and ∠ 2 of the triangle shown in figure 42-3.

- ▲ Calculate either ∠ 1 or ∠ 2. Choose any two of the three given sides. In relation to ∠ 1, the 3.420-inch side is the opposite side. The 5.845-inch side is the hypotenuse.
- ▲ Determine the two functions whose ratios consist of the opposite side and the hypotenuse. The sin of ∠ 1 = $\frac{\text{opposite side}}{\text{hypotenuse}}$, and the csc of ∠ 1 = $\frac{\text{hypotenuse}}{\text{opposite side}}$. Either the sine or cosecant function can be used.

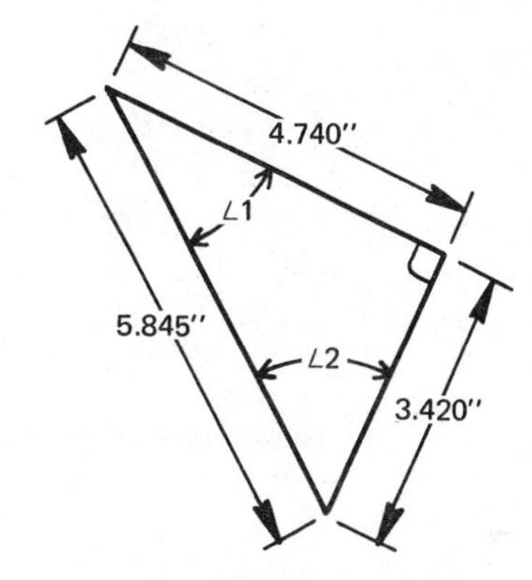

Fig. 42-3

- ▲ Using the sin function: sin ∠ 1 = $\frac{3.420}{5.845}$, sin ∠ 1 = .58512.
- ▲ Interpolate from the function table to determine the angle nearest .58512. The angle interpolated to the nearest minute is 35° 49'. Angle 1 = 35° 49'. Since ∠ 1 + ∠ 2 = 90°, ∠ 2 = 90° - 35° 49', ∠ 2 = 54° 11'.

PROCEDURE FOR DETERMINING AN UNKNOWN SIDE WHEN AN ANGLE AND A SIDE ARE GIVEN

- In relation to the given angle, identify the given side and the unknown side as adjacent, opposite, or hypotenuse.
- Determine the trigonometric functions that are ratios of the sides identified in relation to the given angle.

  Note: Two of the six functions will be found as ratios of the two identified sides. Either of the two functions can be used. Both produce the same value for the unknown, except for cotangents, secants, and cosecants of angles less than 15° and tangents, secants, and cosecants of angles greater than 75°. If the unknown side is made the numerator of the ratio, the problem is solved by multiplication. If the unknown side is made the denominator of the ratio, the problem is solved by division.
- Choose one of the two functions and substitue the given side and given angle.
- Using the trigonometric function table, look up the function of the given angle and substitute this value. If the angle is not given in the table, interpolate the function of the angle.
- Solve as a proportion for the unknown side.

Example 1: Determine side x of the triangle shown in figure 42-4.

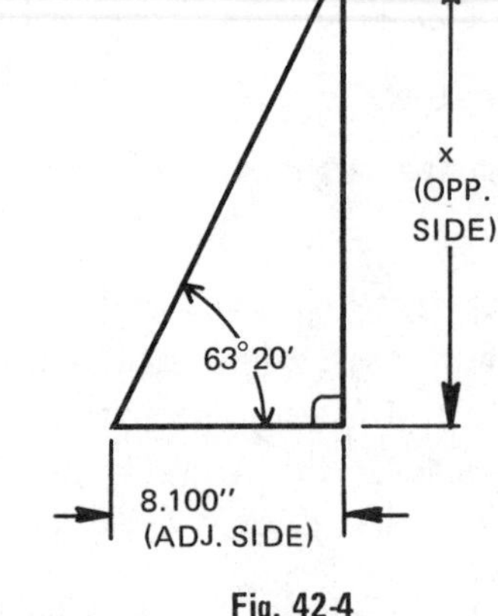

Fig. 42-4

▲ In relation to the 63° 20′ angle, the 8.100-inch side is the adjacent side and side x is the opposite side.

▲ Determine the two functions whose ratios consist of the adjacent and opposite sides. The tan 63° 20′ $= \frac{\text{opposite side}}{\text{adjacent side}}$ and the cot 63° 20′ $= \frac{\text{adjacent side}}{\text{opposite side}}$. Either the tangent or cotangent function can be used.

▲ Using the tangent function: $\tan 63° 20' = \frac{x}{8.100}$. Look up the tangent of 63° 20′ in the function table; tan 63° 20′ = 1.9912. Substitute: $1.9912 = \frac{x}{8.100}$.

▲ Solve as a proportion: $\frac{1.9912}{1} = \frac{x}{8.100}$, x = 1.9912 (8.100), x = 16.129 inches.

Example 2: Determine side x of the triangle shown in figure 42-5.

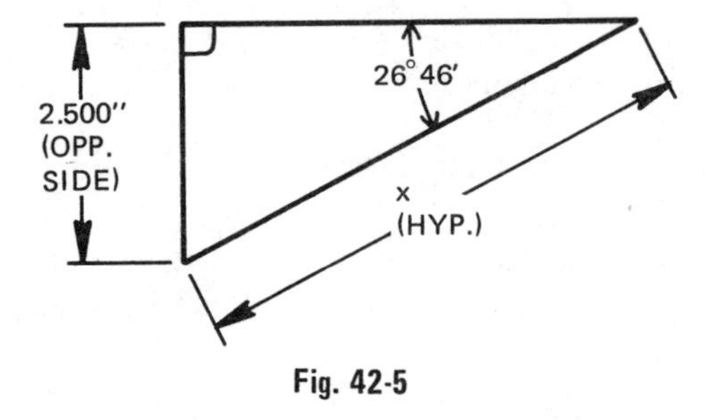

Fig. 42-5

▲ In relation to the 26° 46′ angle, the 2.500-inch side is the opposite side. Side x is the hypotenuse.

▲ Determine the two functions whose ratios consist of the opposite side and the hypotenuse. sin 26° 46′ $= \frac{\text{opposite side}}{\text{hypotenuse}}$ and csc 26° 46′ $= \frac{\text{hypotenuse}}{\text{opposite side}}$. Either the sine or cosecant function can be used.

▲ Using the sine function: $\sin 26° 46' = \frac{2.500}{x}$.

Interpolate the sine of 26° 46′; sin 26° 46′ = .45036.

Substitute: $.45036 = \frac{2.500}{x}$.

▲ Solve as a proportion: $\frac{.45036}{1} = \frac{2.500}{x}$, .45036x = 2.500, $x = \frac{2.500}{.45036}$, x = 5.551 inches.

Example 3: Determine side x, side y, and ∠1 of the triangle shown in figure 42-6.

▲ Calculate either side x or side y. In relation to the 71° 50′ angle, side x is the adjacent side. The 12.250-inch side is the hypotenuse.

▲ Determine the two functions whose ratios consist of the adjacent side and the hypotenuse in order to solve for side x. Either the cosine or secant function can be used.

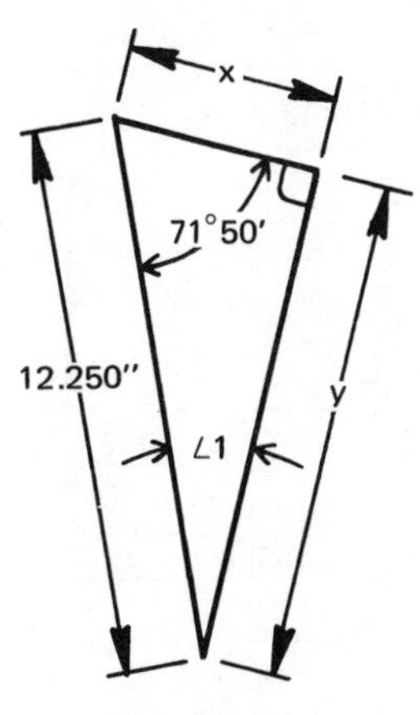

Fig. 42-6

▲ Using the cosine function: $\cos 71° 50' = \frac{x}{12.250}$. Look up the cosine of 71° 50′ in the function table: cos 71° 50′ = .31178.

Substitute: $.31178 = \frac{x}{12.250}$

▲ Solve as a proportion: $\frac{.31178}{1} = \frac{x}{12.250}$, $x = .31178(12.250)$

$x = 3.819$ inches

▲ Solve for side y by using either trigonometric functions or the Pythagorean Theorem. If the Pythagorean Theorem is used to determine y, then $y^2 = 12.250^2 - 3.819^2$. It is more convenient to solve for y using a trigonometric function than with the Pythagorean Theorem.

In relation to the 71° 50′ angle, side y is the opposite side. The 12.250″ side is the hypotenuse.

▲ Determine the two functions whose ratios consist of the opposite side and the hypotenuse. Either the sine or cosecant function can be used.

▲ Using the sine function: $\sin 71° 50' = \frac{y}{12.250}$. Look up the sine of 71° 50′ in the table; sin 71° 50′ = .95015.

Substitute: $.95015 = \frac{y}{12.250}$

▲ Solve as a proportion: $\frac{.95015}{1} = \frac{y}{12.250}$, $y = .95015\,(12.250)$, $y = 11.639$ inches

▲ Determine ∠1: ∠1 = 90° − 71° 50′, ∠1 = 18° 10′

APPLICATION

A. GIVEN TWO SIDES, DETERMINE THE UNKNOWN ANGLE

1. ∠x = ________

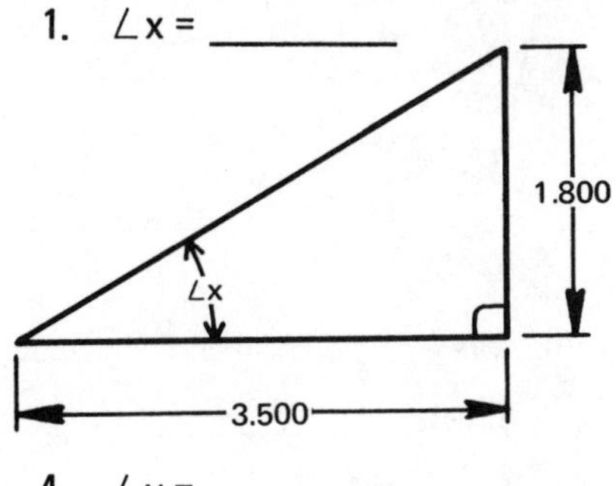

2. ∠1 = ________

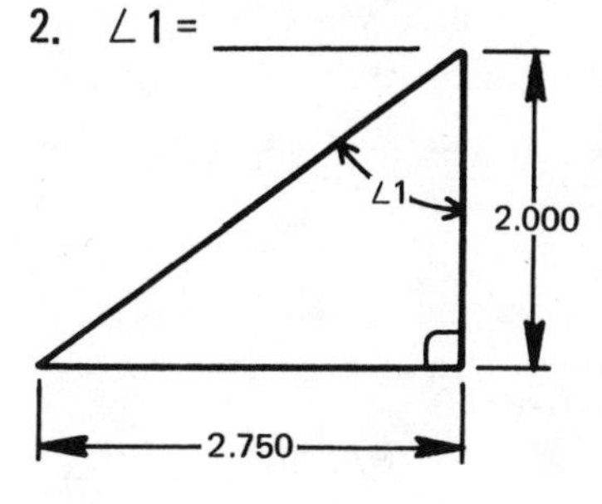

3. ∠A = ________

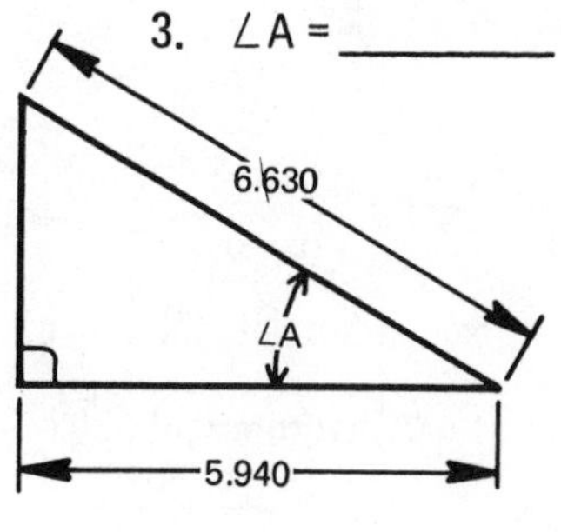

4. ∠y = ________

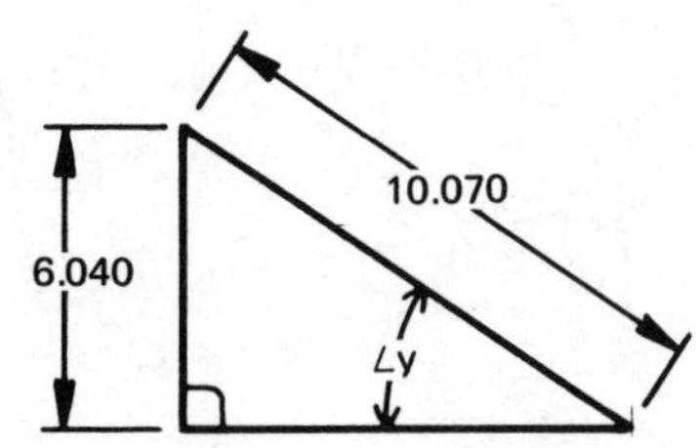

5. ∠y = ________

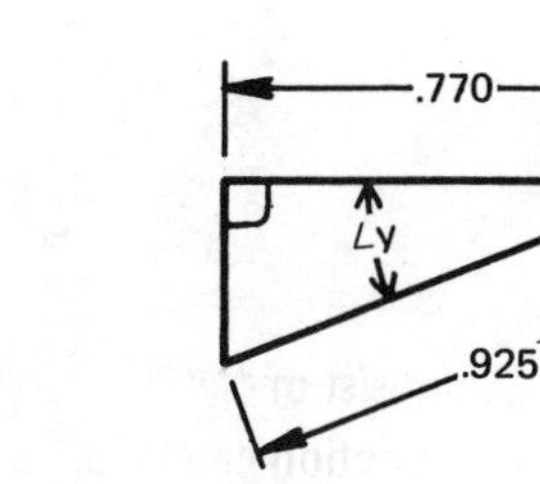

6. ∠B = ________

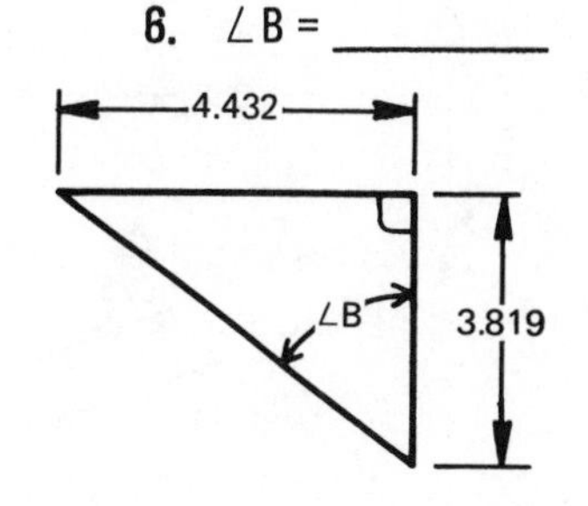

B. GIVEN A SIDE AND AN ANGLE, DETERMINE THE UNKNOWN SIDE

1. Side y = ________

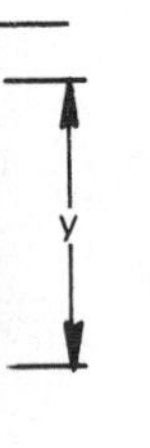

2. Side r = ________

3. Side x = ________

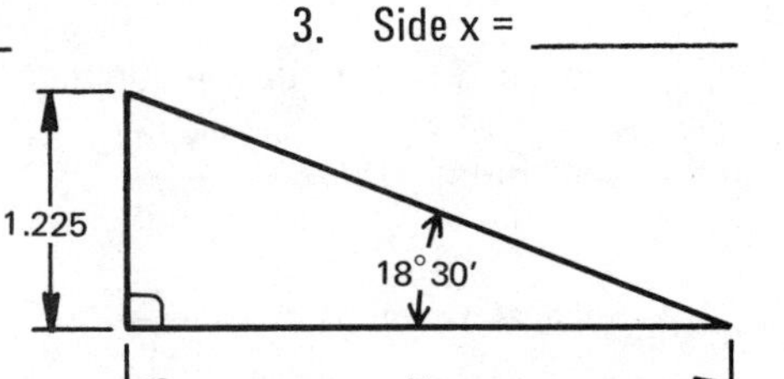

4. Side b = ________

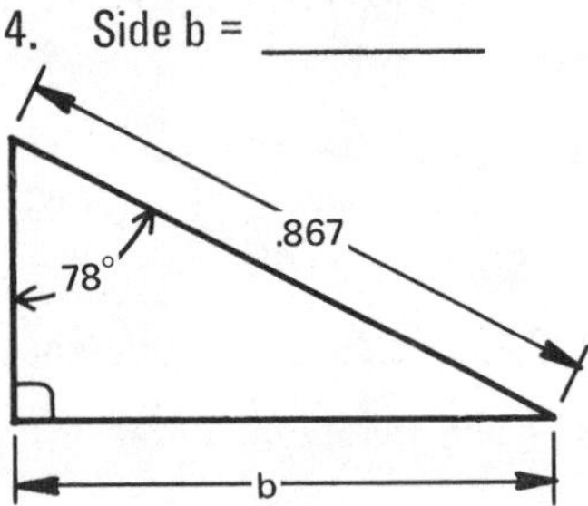

5. Side a = ________

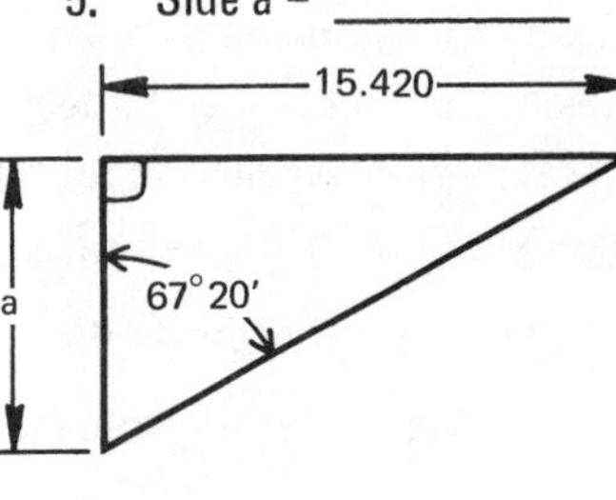

6. Side c = ________

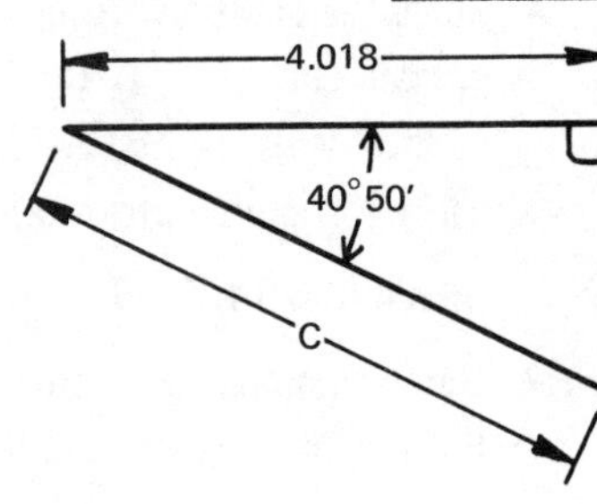

7. Side m = ________

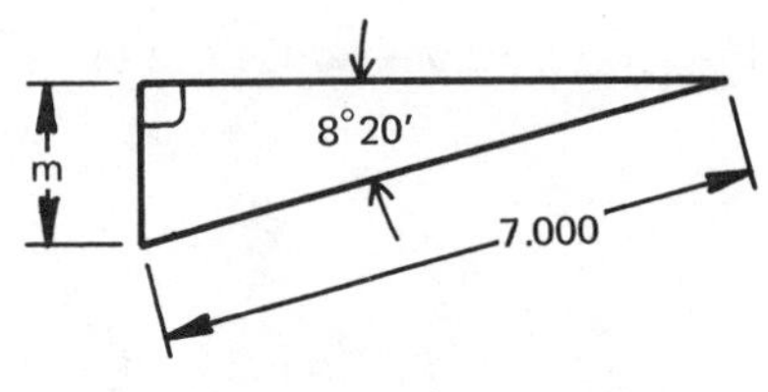

C. DETERMINE THE UNKNOWN SIDES AND ANGLES

1. ∠1 = _____, ∠2 = _____
   Side x = ________

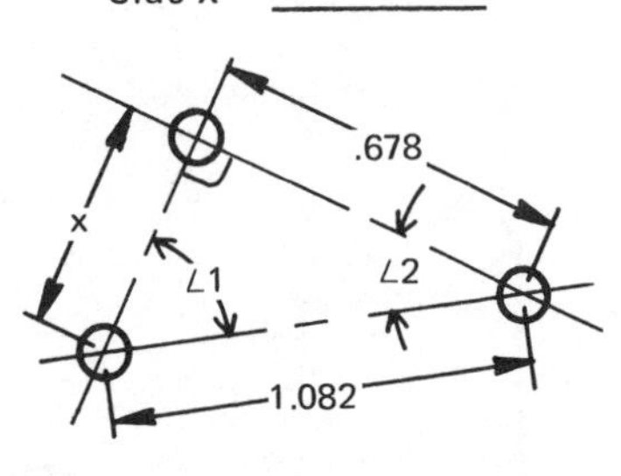

2. ∠2 = _____, Side r = _____
   Side x = ________

3. ∠A = _____
   ∠B = ________

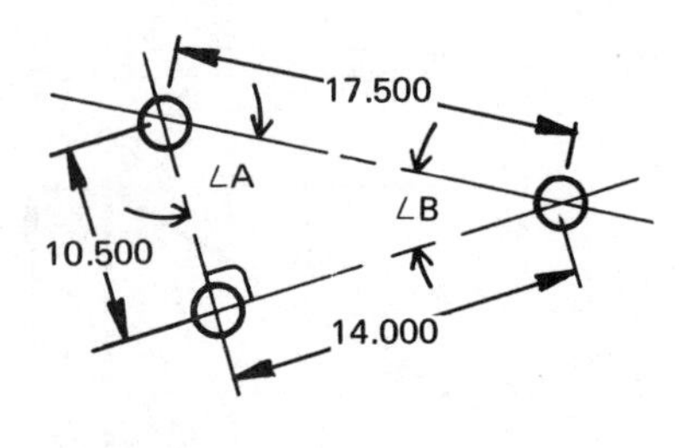

4. ∠B = _____, Side a = _____
   Side C = ________

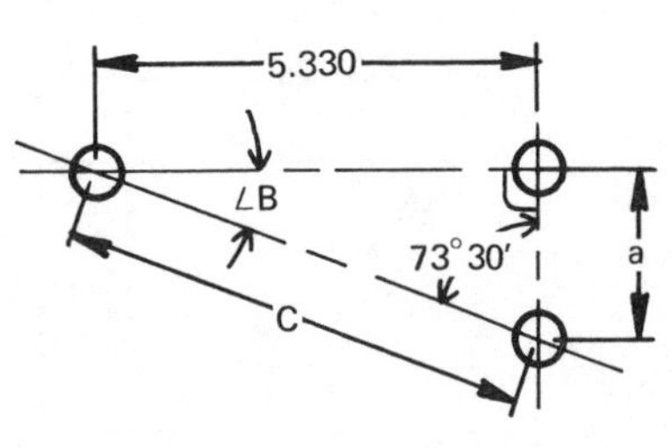

5. ∠D = _____, ∠E = _____
   Side m = ________

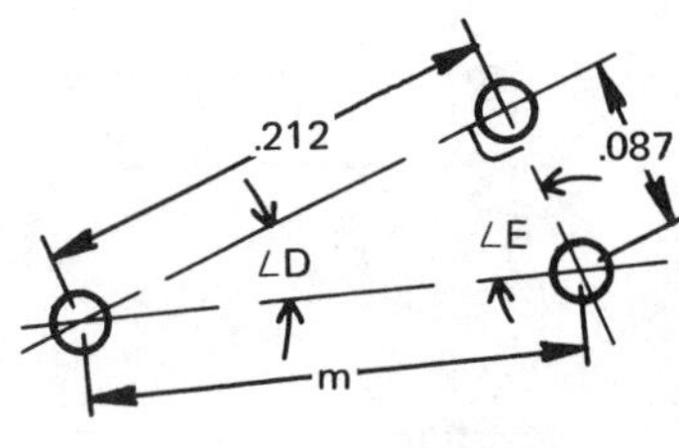

6. ∠1 = _____, Side g = _____
   Side h = ________

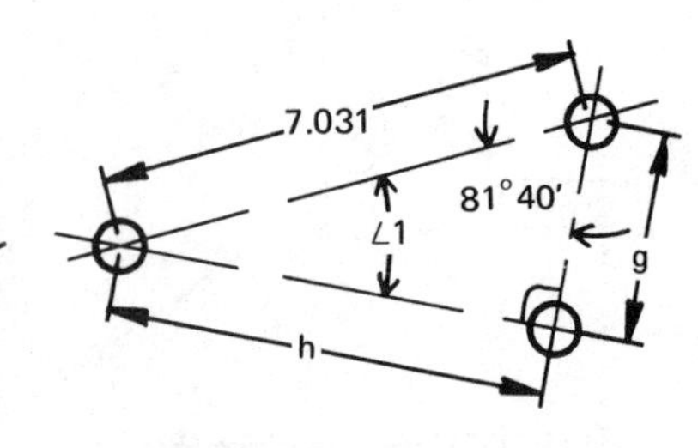

Unit 43 Simple Practical Machine Applications

OBJECTIVE

After studying this unit, the student should be able to

- Solve simple machine shop problems which require the projection of auxiliary lines and the use of geometric principles and trigonometric functions.

METHOD OF SOLUTION

The examples discussed in this unit are simple practical shop applications of right angle trigonometry, although they may not be given directly in the form of right triangles. To solve most of the examples, it is necessary to project auxiliary lines to produce a right triangle. The unknown, or a dimension required to compute the unknown, is part of the triangle. The auxiliary lines may be projected between given points, or from given points, they may be projected parallel or perpendicular to centerlines, tangents, or other reference lines.

It is important to study carefully the procedures and the use of auxiliary lines as they are applied to the examples which follow. The same basic method is used in solving many similar machine shop problems.

A knowledge of both geometric principles and trigonometric functions and the ability to relate and apply them to specific situations are also essential to the work of the machinist.

SINE BAR AND SINE PLATE

Sine bars and sine plates are used to measure angles which have been cut in parts and to position parts which are to be cut at specified angles. One end of the sine bar or plate is raised with gage blocks in order to set a desired angle. The most common sizes of bars and plates are 5 inches and 10 inches between rolls. In setting angles, the sine bar is the hypotenuse of a right triangle, and the gage blocks are the opposite side in reference to the desired angle. The top plate of the sine plate is the hypotenuse.

Example 1: Determine the gage block height x which is required to set an angle of 24° 20′ with a 5-inch sine bar as shown in figure 43-1.

▲ $\text{Sin } 24^\circ\, 20' = \dfrac{\text{gage block height } x}{\text{sine bar length}}$,

$\sin 24^\circ\, 20' = \dfrac{x}{5}$

▲ $\text{Sin } 24^\circ\, 20' = .41204.$

Substitute: $.41204 = \dfrac{x}{5}$

$x = 5\,(.41204) = 2.0602$ inches.

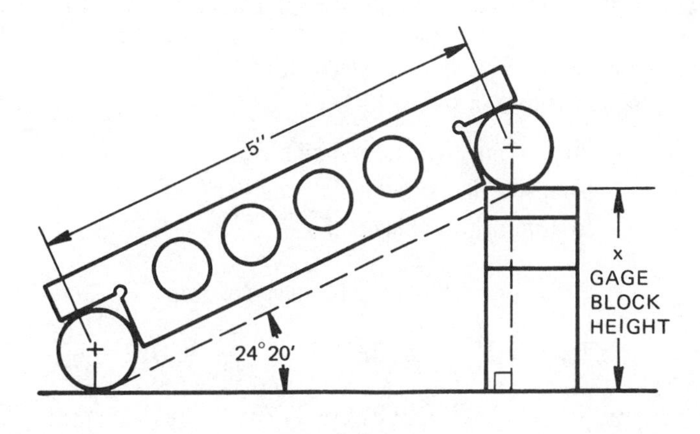

Fig. 43-1

Note: In order to determine gage block heights, look up the sine of the angle in the function table and multiply it by the length of the sine bar or sine plate. When using a 10-inch bar or plate, look up the sine of the angle and move the decimal point one place to the right (the same as multiplying by 10).

Example 2: Determine the angle set on a 10-inch sine plate using a gage block height of 3.0625 inches, figure 43-2.

- Sin $\angle x = \frac{3.0625''}{10''}$, x = .30625.
- Look up the angle that corresponds to sin function, .30625. Angle x = 17° 50′

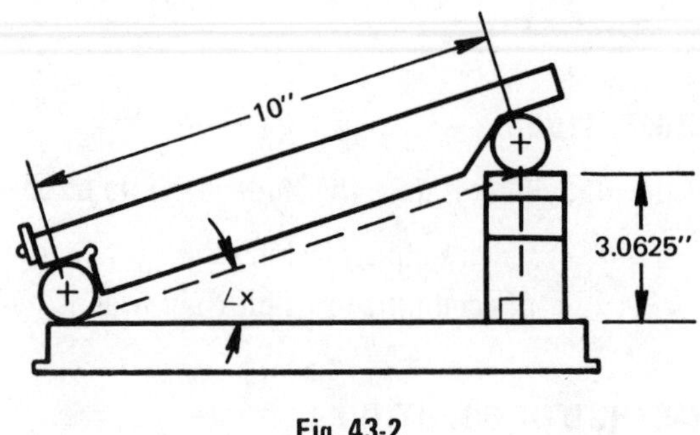

Fig. 43-2

TAPERS AND BEVELS

Example 1: Determine the included taper angle of the shaft shown in figure 43-3.

- The problem must be solved by using a figure in the form of a right triangle. Therefore, project line AB from point A parallel to the centerline. Right △ ABC is formed in which angle A is one-half the included taper angle.
- Using sides AB and BC, solve for ∠A. Side AB = 10.500″, side BC $= \frac{1.800'' - .700''}{2} = .550''$.

  $\text{Tan} \angle A = \frac{BC}{AB}$, $\text{Tan} \angle A = \frac{.550}{10.500} = .05238, \angle A = 3° \, 0'$

- Included taper ∠ = 2(3° 0′) = 6° 0′

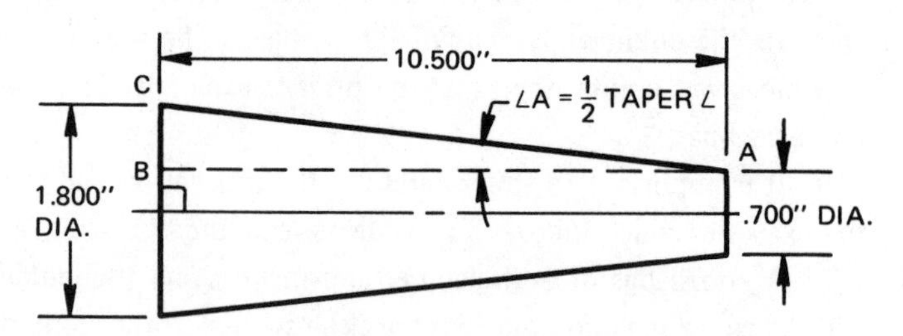

Fig. 43-3

Example 2: Determine diameter x of the part shown in figure 43-4.

- Project line DE from point D parallel to the centerline, in order to form right triangle DEF.

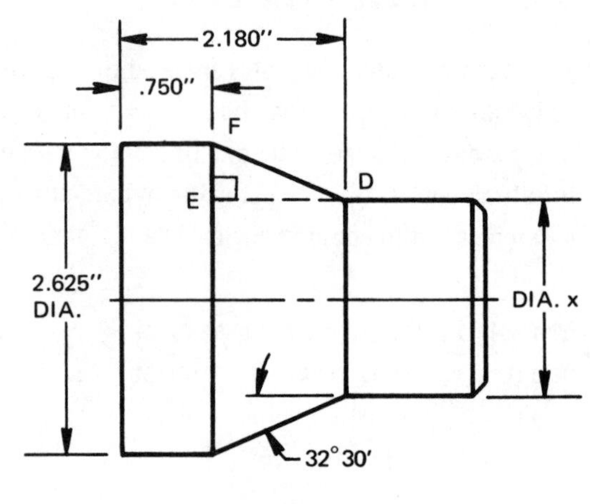

Fig. 43-4

- Using side DE and ∠ D, solve for side EF. Side DE = 2.180″ - .750″ = 1.430″, ∠D = 32° 30′.

  $\text{Tan} \angle D = \frac{EF}{DE}$, $\tan 32° \, 30' = \frac{EF}{1.430}$, $.63707 = \frac{EF}{1.430}$, EF = .63707 (1.430) = .911″

- Dia. x = 2.625 - 2 (.911) = 2.625 - 1.822, Dia. x = .803 inch

ISOSCELES TRIANGLE APPLICATIONS: DISTANCE BETWEEN HOLES, V-SLOTS

Example 1: In figure 43-5, five holes are equally spaced on a 5.200-inch diameter circle. Determine the straight line distance between two consecutive holes.

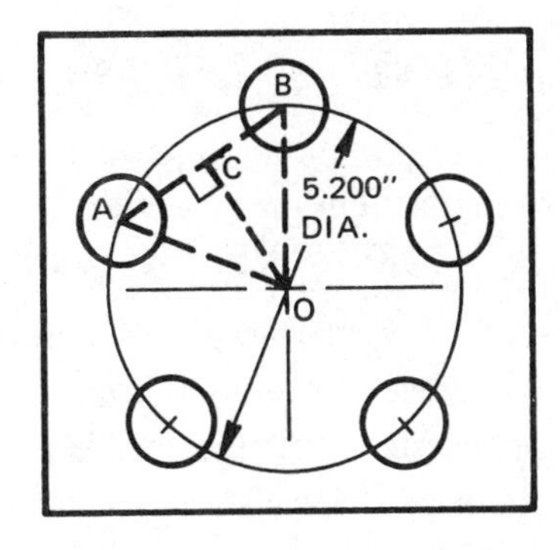

Fig. 43-5

- Project radii from center O to hole centers A and B. Project a line from A to B. Central angle $AOB = \frac{360^\circ}{5} = 72^\circ$.

  Since OA = OB, $\triangle$AOB is isosceles.
- Project line OC $\perp$ to AB from angle O. Line OC bisects $\angle$O and side AB.
- In right $\triangle$AOC, $\angle AOC = \frac{72^\circ}{2} = 36^\circ$ and $AO = \frac{5.200''}{2} = 2.600$ inches
- Solve for side AC: $\sin \angle AOC = \frac{AC}{AO}$, $\sin 36^\circ = \frac{AC}{2.600}$, $.58778 = \frac{AC}{2.600}$, AC = .58778 (2.600) = 1.528 inches
- AB = 2(1.528) = 3.056 inches

Example 2: Determine the depth of cut x required to machine the V-slot shown in figure 43-6.

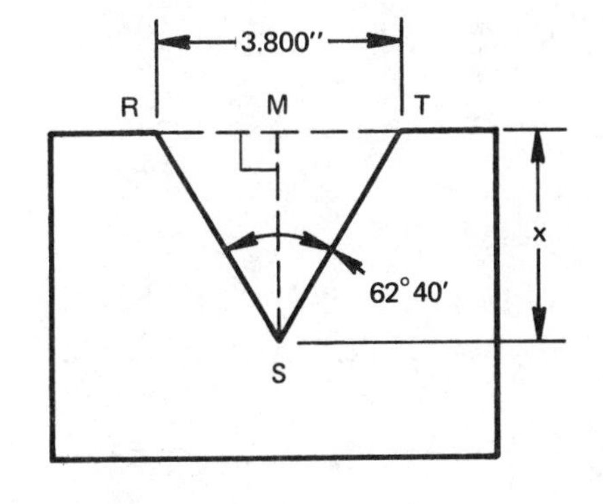

Fig. 43-6

- Connect a line between points R and T. Sides RS = TS; therefore, $\triangle$RST is isosceles.
- Project line SM from point S $\perp$ to RT. Angle S and side RT are bisected.
- In right $\triangle$RMS, $\angle RSM = \frac{62^\circ 40'}{2} = 31^\circ 20'$,

  $RM = \frac{3.800}{2} = 1.900''$. Solve for depth of cut

  MS: $\cot \angle RSM = \frac{MS}{RM}$, $\cot 31^\circ 20' = \frac{MS}{1.900}$, $1.6425 = \frac{MS}{1.900}$, MS = 1.6425 (1.900),

  MS = 3.121. MS = x, x = 3.121 inches.

TANGENTS TO CIRCLE APPLICATIONS: V-BLOCKS, THREAD WIRE CHECKING DIMENSIONS, DOVETAILS, AND ANGLE CUTS

Example 1: A .750-inch diameter pin is used to inspect the groove machined in the block shown in figure 43-7. Determine dimension x.

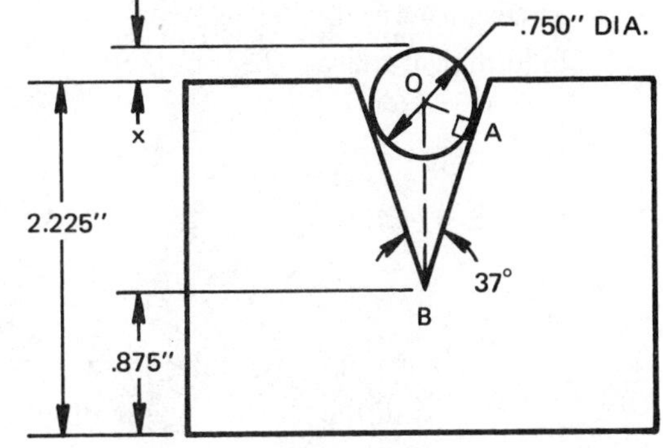

Fig. 43-7

- Project a line from center O to point B. Project radius AO from center O to tangent point A. Since a radius is $\perp$ to a tangent line at the point of tangency, $\triangle$AOB is a right triangle.
- In right triangle AOB, $OA = \frac{.750}{2} = .375''$.

  Since the angle formed by two tangents to a circle from an outside point is bisected by a line from the point to the center of the circle, $\angle ABO = \frac{37^\circ}{2} = 18^\circ 30'$.

▲ Solve for side OB: $\csc \angle ABO = \frac{OB}{OA}$, $\csc 18^\circ 30' = \frac{OB}{.375}$, $3.1515 = \frac{OB}{.375}$, OB = 3.1515 (.375) = 1.182 inches.

▲ The height from the base of the block to the top of the .750" dia. pin = .875" + OB + radius of pin = .875" + 1.182" + .375" = 2.432 inches. Dimension x = 2.432" - 2.225" = .207 inch.

Example 2: An internal dovetail is shown in figure 43-8. Two pins or balls are used to check the dovetail for both location and angular accuracy. Calculate check dimension x.

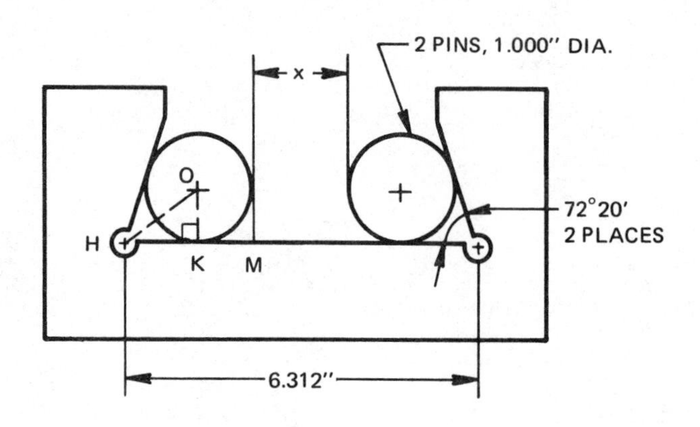

Fig. 43-8

▲ Project line HO from point H to the pin center O; HO bisects the 72° 20' angle. Project a radius from point O to the point of tangency K; $\angle HKO$ is a right angle since a radius is perpendicular to a tangent at the point of tangency.

▲ In right $\triangle HOK$, $\angle KHO = \frac{72^\circ 20'}{2} = 36^\circ 10'$ and $KO = \frac{1.000}{2} = .500$ inch.

▲ Solve for side HK: $\cot \angle KHO = \frac{HK}{KO}$, $\cot 36^\circ 10' = \frac{HK}{.500}$, $1.3680 = \frac{HK}{.500}$, HK = .500 (1.3680) = .684 inch.

▲ Distance KM = pin radius = .500; HM = HK + KM = .684 + .500 = 1.184. Check dimension x = 6.312 - 2 HM = 6.312 - 2 (1.184) = 6.312 - 2.368 = 3.944 inches.

APPLICATION

A. 1. Determine the height of gage blocks required to set the following angles on a 10" sine plate.

a. 25° ____	d. 7° 20' ____	g. 0° 40' ____
b. 13° 10' ____	e. 28° 30' ____	h. 2° 20' ____
c. 36° 50' ____	f. 44° 20' ____	i. 19° 50' ____

2. Determine the height of gage blocks required to set the following angles with a 5" sine bar.

a. 40° 40' ____	d. 0° 30' ____	g. 37° 20' ____
b. 5° ____	e. 21° 50' ____	h. 44° 50' ____
c. 12° 10' ____	f. 9° ____	i. 8° 10' ____

B. Determine the unknown value in each of the following problems. Calculate angles to the closest 1' and lengths to the closest thousandths of an inch.

1. Included taper ∠ x = ____

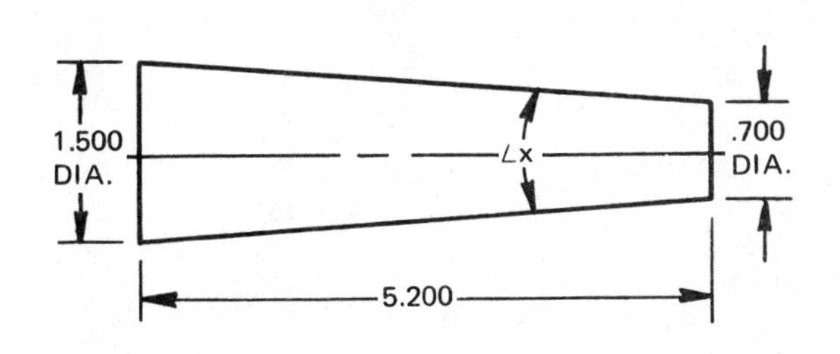

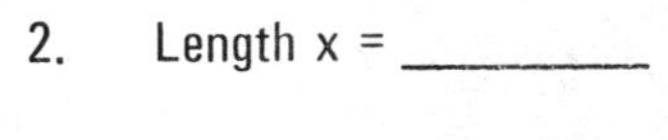

2. Length x = ________

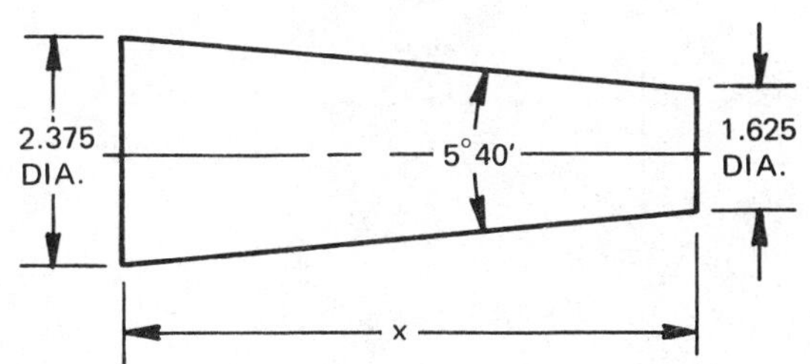

3. Diameter y = ____

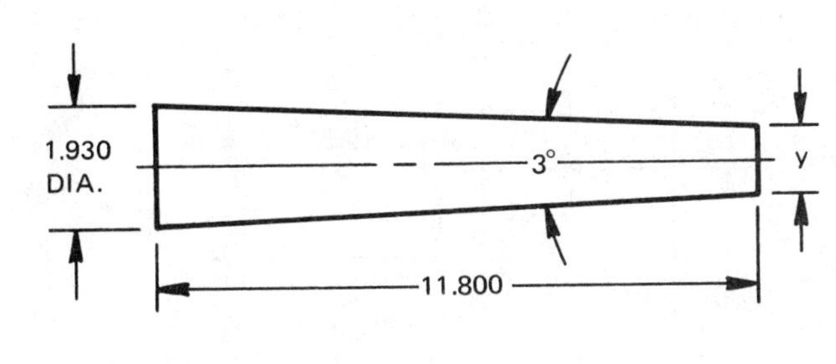

4. Diameter x = ____

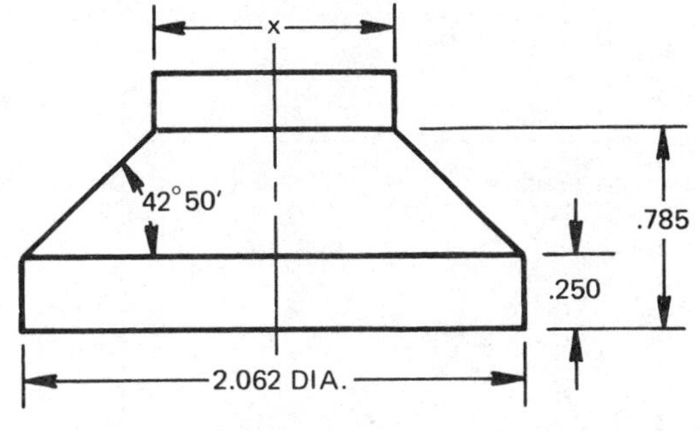

5. ∠ x = ____

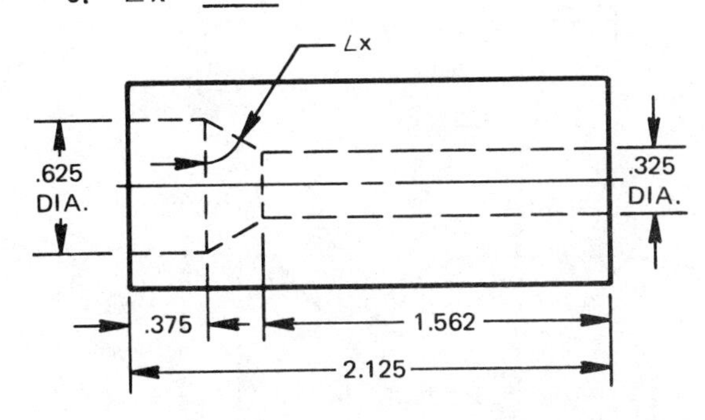

6. Dimension y = ____

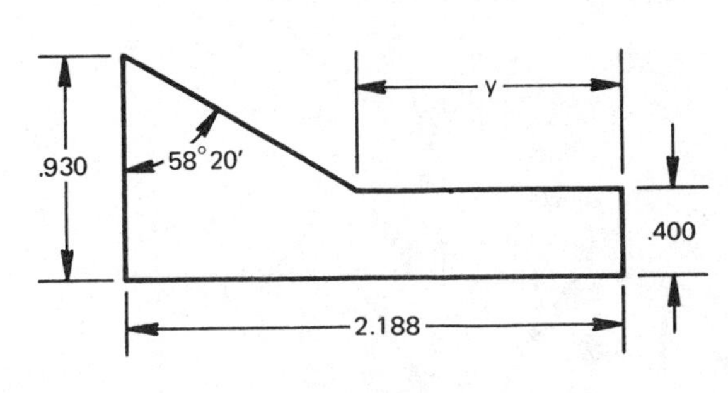

7. Center distance y = ____

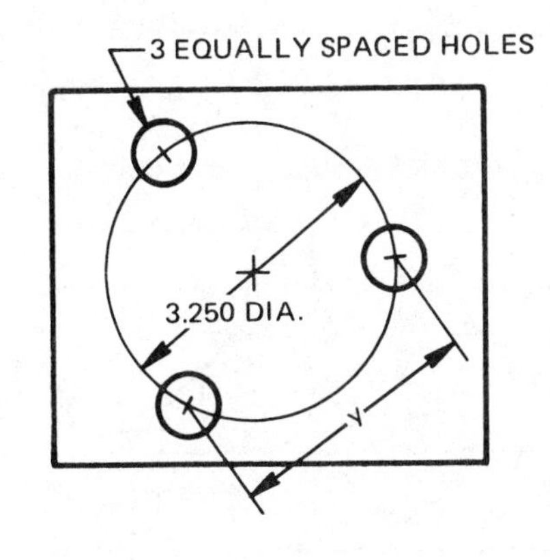

8. Inside caliper dimension x = ____

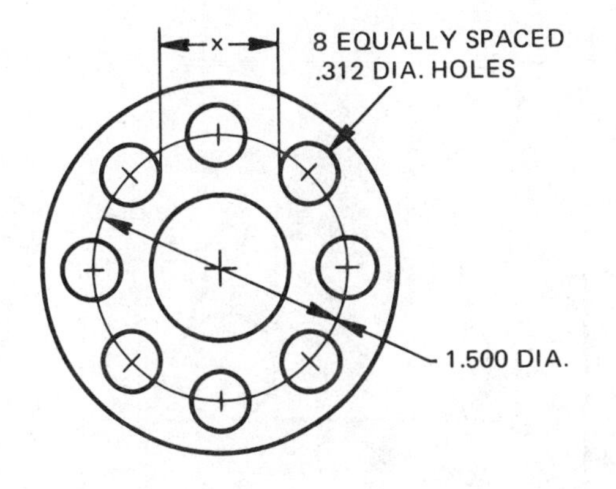

9. Radius r = ____

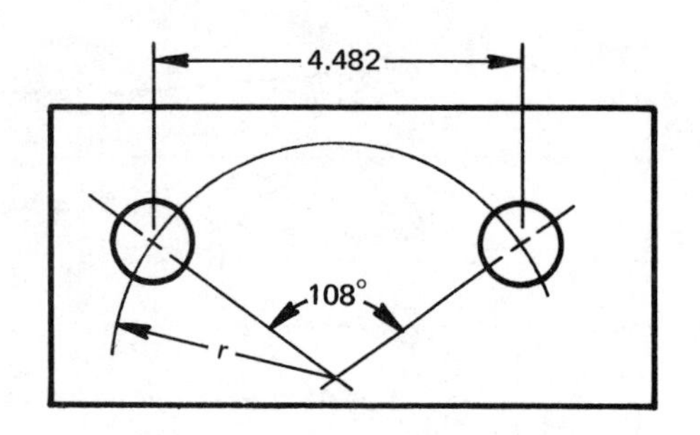

10. Arc dimension x = ____

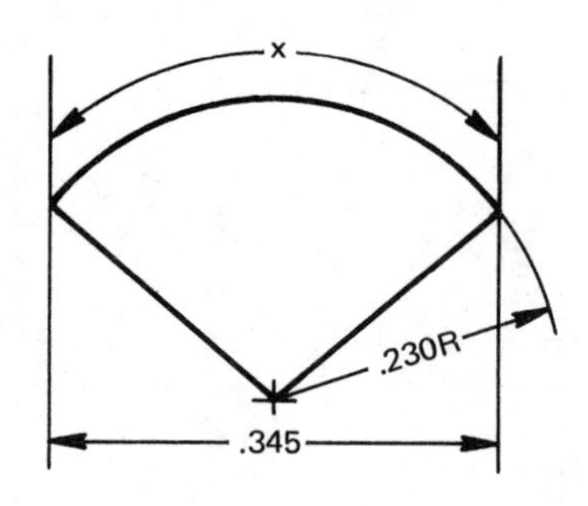

11. Depth of cut x = ____

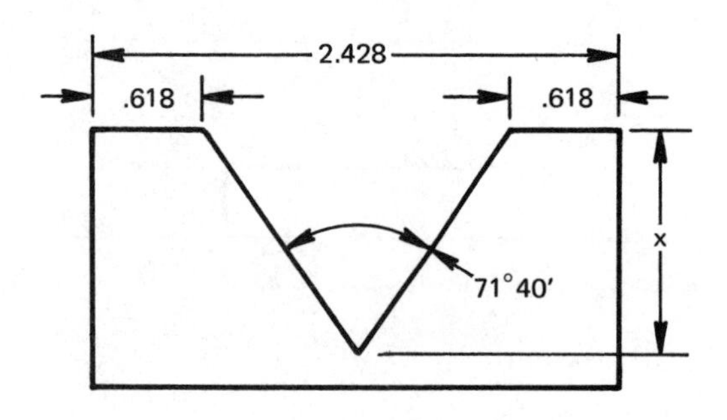

12. ∠x = ____

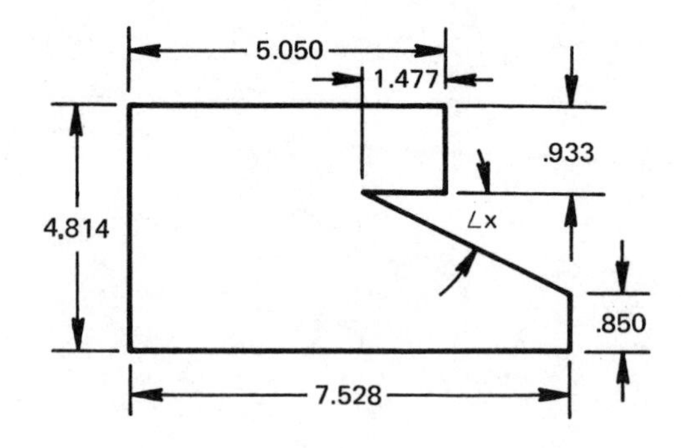

13. Gage dimension y = ____

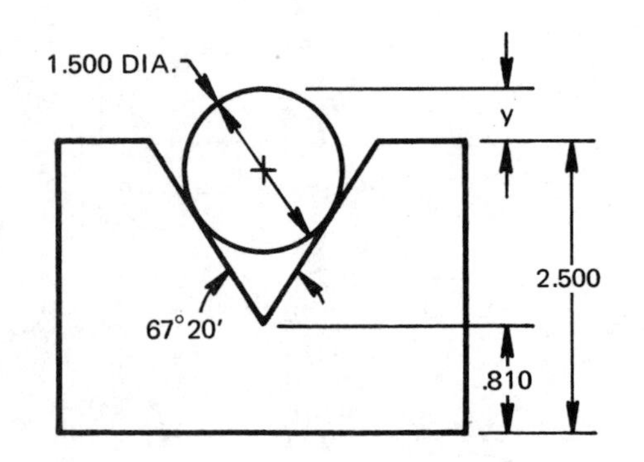

14. ∠y = ____

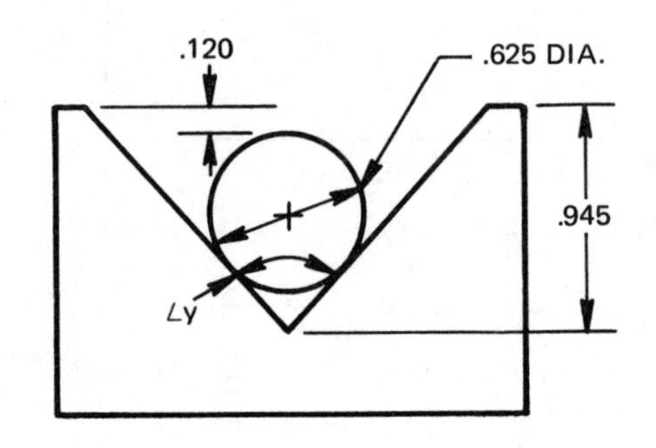

15. Gage dimension x = ____

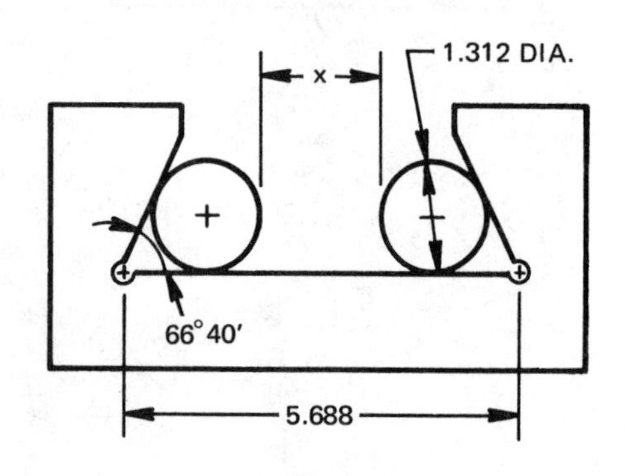

16. ∠x = ____

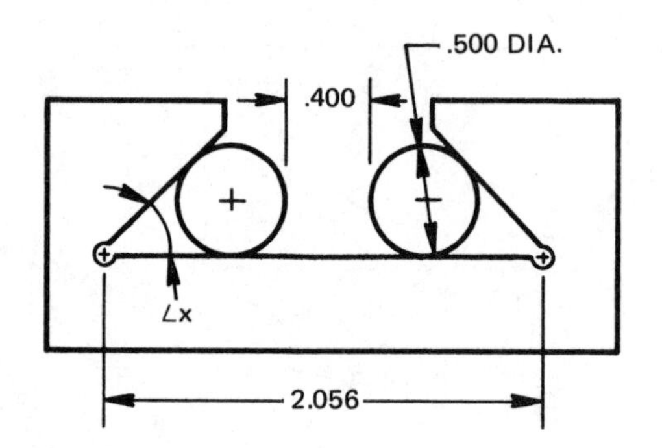

17. Dimension y = ____

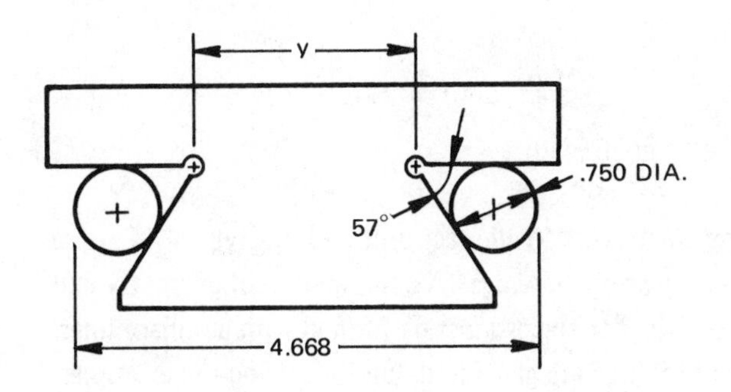

18. ∠y = ____

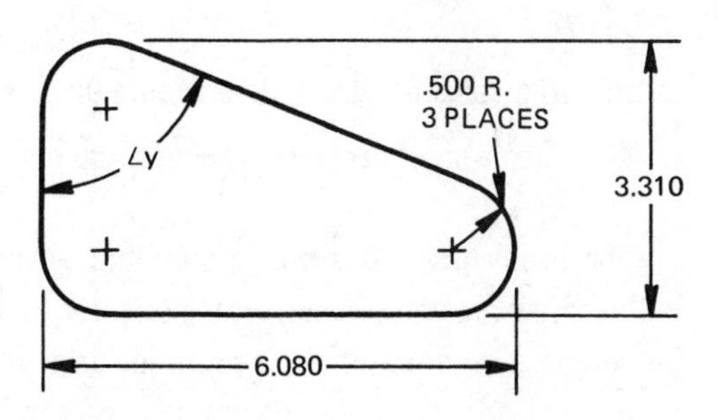

19. Dimension x = ____

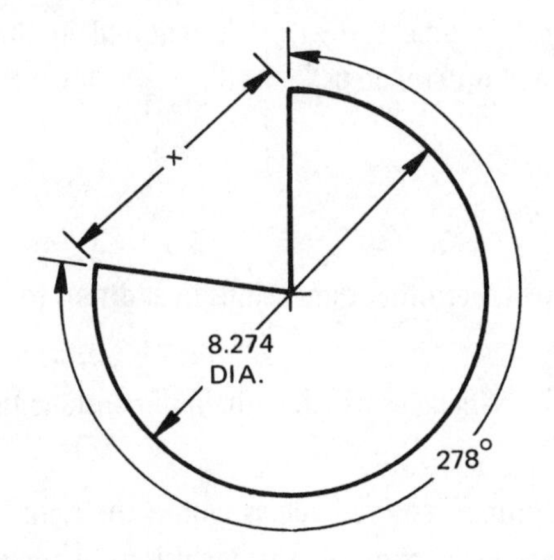

20. ∠x = ____

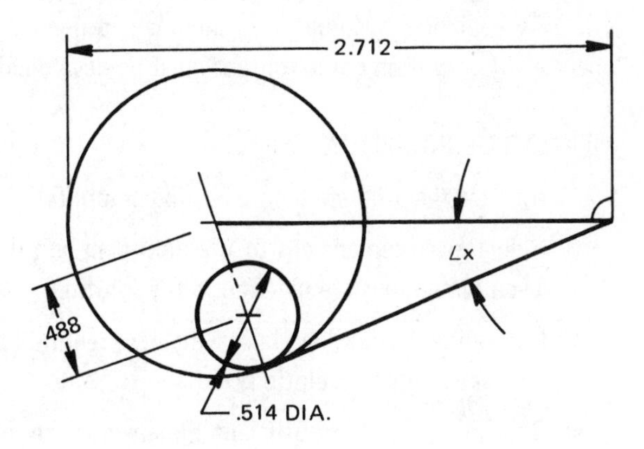

21. Distance y = ____

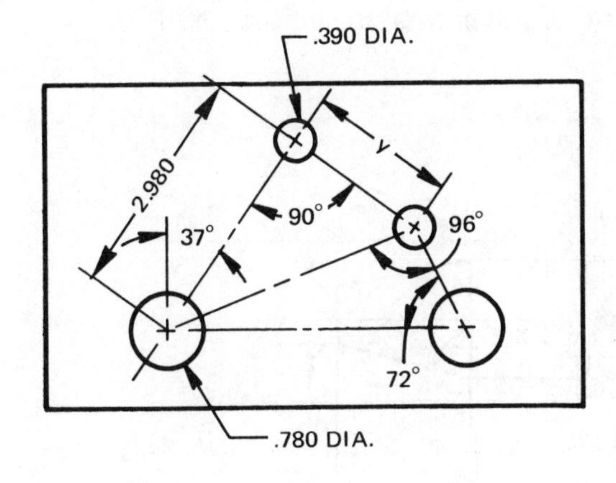

22. Dimension y = ____

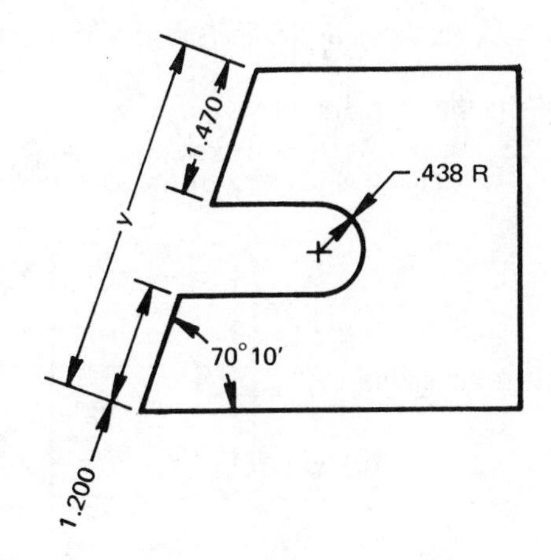

Unit 44 More Complex Practical Machine Applications

OBJECTIVE

After studying this unit, the student should be able to

- Solve practical machine shop problems of a complex nature.

The problems in this unit are more challenging than those in the last unit and are typical of those found in actual practice when working directly from engineering drawings. A combination of geometry and trigonometry is required for their solution. Two or more right triangles must be formed with auxiliary lines.

It is essential to study and restudy the procedures which are given in detail for solving the examples. There is a common tendency to begin writing computations before the complete solution to a problem has been "thought through." This tendency must be avoided.

As problems become more complex, a greater proportion of time and effort is required in their analyses. The written computations must be developed in clear and orderly steps.

METHOD OF SOLUTION

1. Analyze the problem before writing computations.
   - Relate given dimensions to the unknown and determine whether other dimensions in addition to the given dimensions are required in the solution.
   - Determine the auxiliary lines which are required to form right triangles which contain dimensions that are needed for the solution.
   - Determine whether sufficient dimensions are known to obtain required values within the right triangles. If enough information is not available for solving a triangle, continue the analysis until enough information is obtained.
   - Check each step in the analysis to verify that there are no gaps or false assumptions.
2. Write the computations.

Example 1: Determine length x of the part shown in figure 44-1.

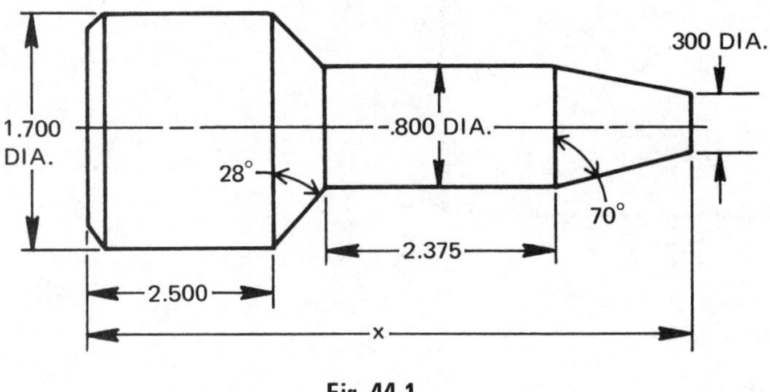

Fig. 44-1

▲ Analyze the problem:

▲ Refer to figure 44-2. If distances AB and CD can be determined, the problem can be solved: x = 2.500 + AB + 2.375 + CD.

▲ Determine whether enough information is given to solve for AB: In right △ABD, ∠A = 90° - 28° = 62° (complementary ∠'s;)

$$BD = \frac{1.700 - .800}{2} = \frac{.900}{2} = .450.$$

There is enough information to determine AB.

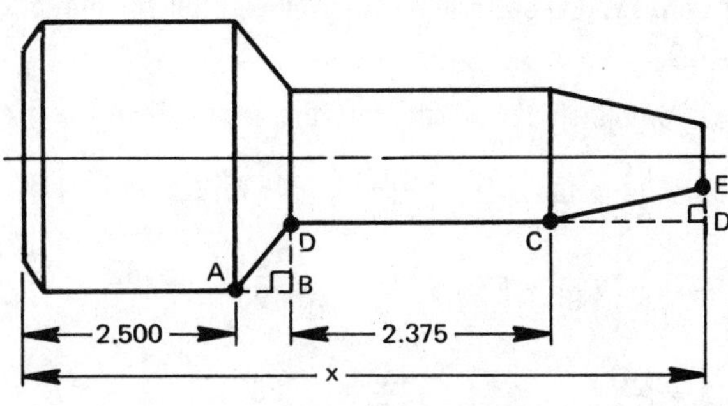

Fig. 44-2

▲ Determine whether enough information is given to solve for CD. In right △CDE, ∠C = 90° - 70° = 20° (complementary ∠'s); $ED = \frac{.800 - .300}{2} = \frac{.500}{2} = .250$. There is enough information to determine CD.

▲ The problem has been completely analyzed, therefore, proceed with the written computations:

▲ Solve for AB: $\cot \angle A = \frac{AB}{BD}$, $\cot 62° = \frac{AB}{.450}$, $.53171 = \frac{AB}{.450}$, AB = .450 (.53171) = .2393

▲ Solve for CD: $\cot \angle C = \frac{CD}{DE}$, $\cot 20° = \frac{CD}{.250}$, $2.7475 = \frac{CD}{.250}$, CD = .250 (2.7475) = .6869

▲ Substitute values in original: x = 2.500 + AB + 2.375 + CD, x = 2.500 + .2393 + 2.375 + .6869, x = 5.801 inches

Example 2: Determine ∠x of the plate shown in figure 44-3.

Note: Generally, when solving problems which involve an arc which is tangent to one or more lines, it is necessary to project the radius of the arc to the tangent point and to project a line from the vertex of the unknown angle to the center of the arc.

▲ Analyze the problem:

▲ See figure 44-4. Project auxiliary lines between the points A and O, from point O to the tangent point B, and from point O to point C. Right triangles ACO and ABO are formed.

▲ If ∠1 and ∠2 can be computed, ∠x can be determined since ∠x = 90° - (∠1 + ∠2).

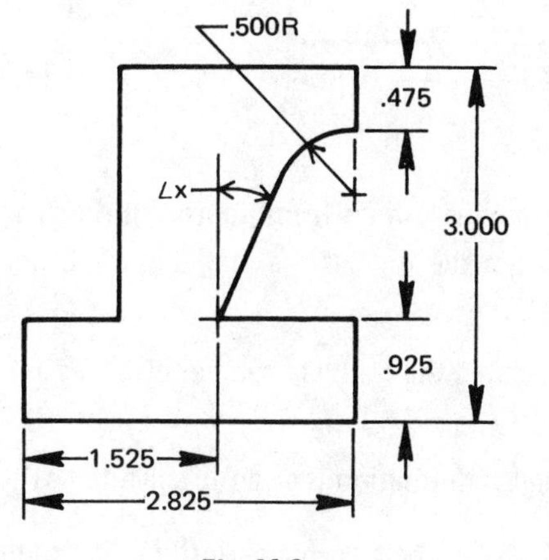

Fig. 44-3

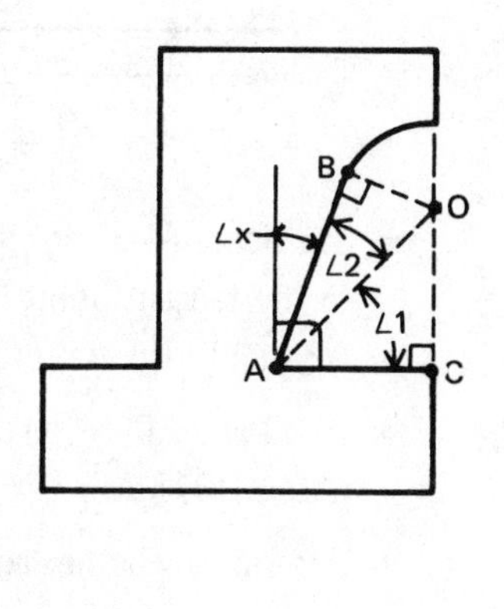

Fig. 44-4

▲ Determine whether enough information is given to solve for $\angle 1$. In right triangle ACO, AC = 2.825 - 1.525 = 1.300; CO = 3.000 - (.925 + .500 + .475) = 3.000 - 1.900 = 1.100. There is enough information to determine $\angle 1$.

▲ Determine if enough information is given to solve for $\angle 2$. In the right triangle ABO, BO = .500. Side AO is also a side of right $\triangle$ACO and can be determined after solving for $\angle 1$. There is enough information to determine $\angle 2$.

▲ The problem has been completely analyzed; therefore, proceed with written computations:

▲ Solve for $\angle 1$: $\tan \angle 1 = \frac{CO}{AC}$, $\tan \angle 1 = \frac{1.100}{1.300} = .84615$, $\angle 1 = 40° \ 14'$.

▲ Solve for side AO: $\csc \angle 1 = \frac{AO}{CO}$, $\csc 40° \ 14' = \frac{AO}{1.100}$. $1.5482 = \frac{AO}{1.100}$, AO = 1.5482 (1.100) = 1.7030.

▲ Solve for $\angle 2$: $\sin \angle 2 = \frac{BO}{AO}$, $\sin \angle 2 = \frac{.500}{1.7030} = .29360$, $\angle 2 = 17° \ 4'$.

▲ Substitute values in original: Angle x = $90° - (\angle 1 + \angle 2) = 90° - (40° \ 14' + 17° \ 4') = 90° - 57° \ 18'$, $\angle x = 32° \ 42'$

Example 3: Determine dimension x which is used to check the V-groove shown in figure 44-5.

▲ Analyze the problem:

▲ Dimension x is determined by the pin size, the points of tangency where the pin touches the groove, the angle of the V-groove, and the depth of the groove. Therefore these dimensions and locations must be part of the calculations.

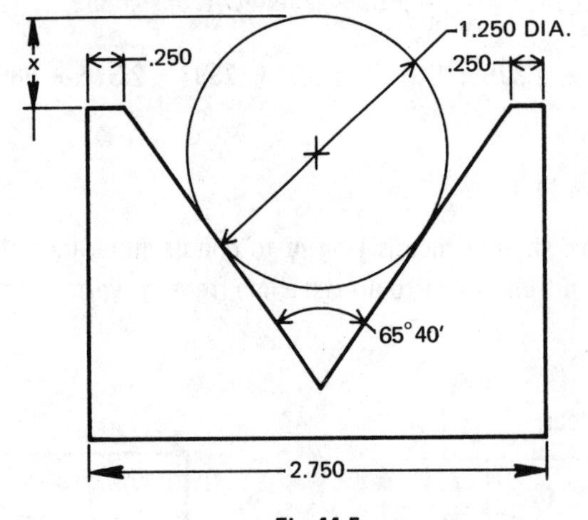

Fig. 44-5

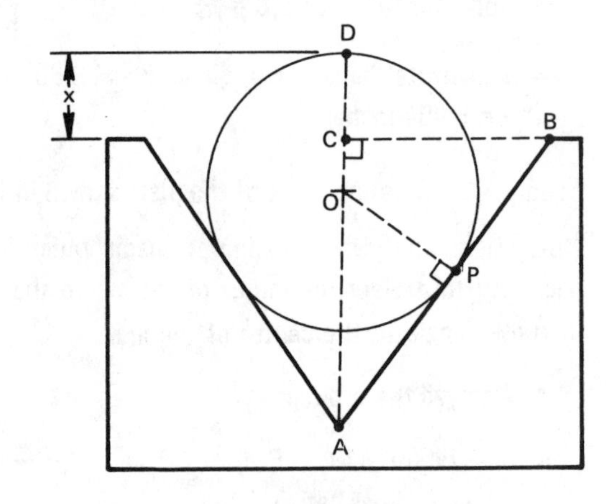

Fig. 44-6

▲ See figure 44-6. Project auxiliary lines from point A through the center of the pin O, from point O to the tangent point P, from the top surface of the block B horizontally to point C. Right triangles APO and ACB are formed.

▲ If AO and AC can be calculated the problem can be solved. DO = radius of pin = .625. Dimension x = (AO + DO) - AC.

▲ Determine whether enough information is given to solve for AO. In right triangle APO, PO = $\frac{1.250}{2}$ = .625. $\angle A = \frac{65° \ 40'}{2} = 32° \ 50'$. There is enough information to determine AO.

▲ Determine if enough information is given to solve for AC. In right triangle ACB, BC = $\frac{2.750}{2}$ - .250 = 1.375 - .250 = 1.125. $\angle$A = 32° 50′. There is enough information to determine AC.

▲ The problem has been completely analyzed, therefore, proceed with the written computations.

▲ Solve for AO: $\csc \angle A = \frac{AO}{PO}$, $\csc 32° 50' = \frac{AO}{.625}$, $1.8443 = \frac{AO}{.625}$, AO = 1.8443 (.625) = 1.1527

▲ Solve for AC: $\cot \angle A = \frac{AC}{BC}$, $\cot 32° 50' = \frac{AC}{1.125}$, $1.5497 = \frac{AC}{1.125}$, AC = 1.5497 (1.125) = 1.7434

▲ Substitute in the original: Dimension x = (AO + DO) - AC, x = (1.1527 + .625) - 1.7434 = 1.7777 - 1.7434, x = .034 inch.

Example 4: Determine $\angle$x in the series of holes shown in the plate in figure 44-7.

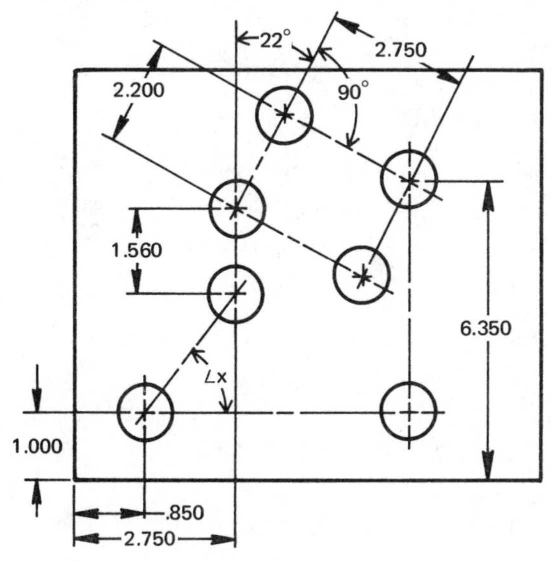

Fig. 44-7

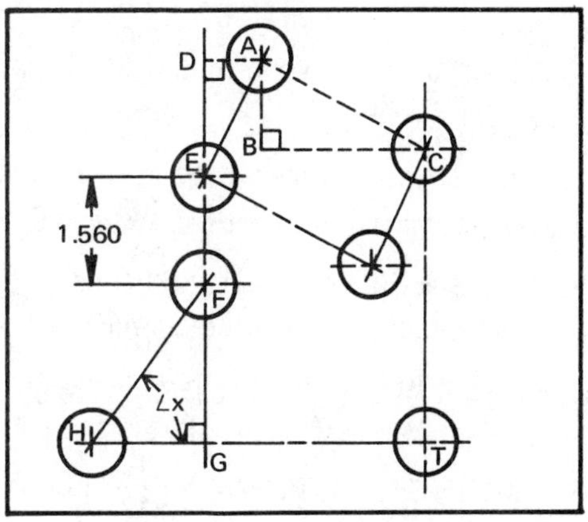

Fig. 44-8

▲ Refer to figure 44-8.

▲ Project auxiliary lines AD, AB, BC. Right triangles ABC, ADE, and FGH are formed.

▲ In right triangle FGH; HG = 2.750 - .850 = 1.900, therefore, FG must be determined to solve for $\angle$x. Side FG = (TC + AB) - (DE + 1.560),

▲ FG = [(6.350 - 1.000) + AB] - (DE + 1.560).

▲ Determine AB. In right triangle ABC, AC = 2.750, $\angle$c = 22° (2 angles whose corresponding sides are $\perp$ are =). $\sin 22° = \frac{AB}{2.750}$, $.37461 = \frac{AB}{2.750}$, AB = .37461 (2.750) = 1.0302.

▲ Determine DE: In right triangle ADE, $\angle$E = 22°, AE = 2.200. $\cos 22° = \frac{DE}{2.200}$; $.92718 = \frac{DE}{2.200}$; DE = .92718 (2.200) = 2.0398.

▲ Determine FG: Side FG = [(6.350 - 1.000) + AB] - (DE + 1.560), FG = (5.350 + 1.0302) - (2.0398 + 1.560) = 6.3802 - 3.5998 = 2.7804.

▲ Substitute in original. Determine $\angle$x: Side FG = 2.7804, HG = 1.900. $\tan \angle x = \frac{FG}{HG}$, $\tan \angle x = \frac{2.7804}{1.900}$ = 1.4634, x = 55° 39′.

Example 5: Determine dimension x of the template shown in figure 44-9.

Procedure: Refer to figure 44-10.

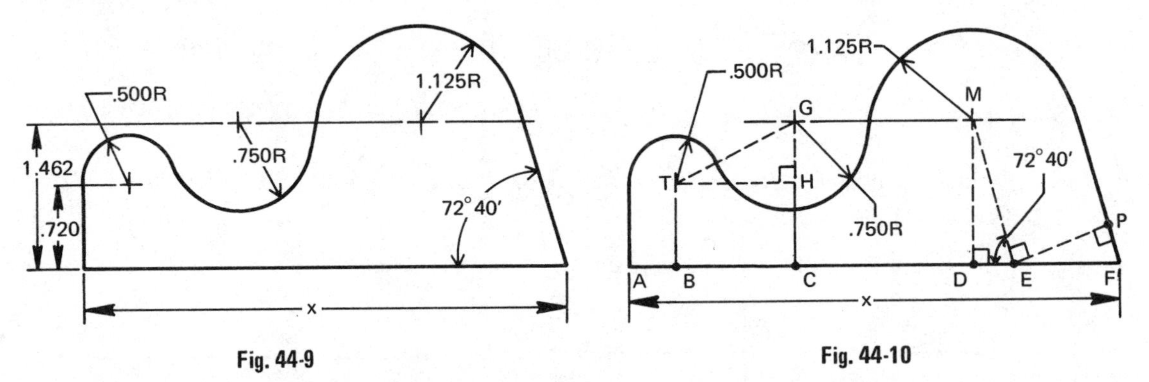

Fig. 44-9

Fig. 44-10

▲ Project auxiliary lines to form right triangles GHT, DEM, and EFP.

▲ Dimension x = AB + BC + CD + DE + EF. Dimensions AB = .500, CD = GM = .750 + 1.125 = 1.875 (a line connecting the centers of two externally tangent circles passes through the point of tangency.) It is necessary to solve for BC, DE, and EF.

▲ Determine BC: BC = TH. In right GHT, GH = 1.462 - .720 = .742, GT = .500 + .750 = 1.250. (GT passes through the point of tangency.)

▲ $\text{Sin} \angle T = \frac{GH}{GT} = \frac{.742}{1.250} = .59360; \angle T = 36° \, 25'$.

▲ $\text{Cot} \angle T = \frac{TH}{GH}$, $\cot 36° \, 25' = \frac{TH}{.742}$, $1.3555 = \frac{TH}{.742}$, TH = .742 (1.3555) = 1.0058 = BC.

▲ Determine DE: In right triangle DEM, $\angle E = 72° \, 40'$, DM = 1.462. $\text{Cot} \angle E = \frac{DE}{DM}$, $\cot 72° \, 40' = \frac{DE}{1.462}$, $.31210 = \frac{DE}{1.462}$, DE = .31210 (1.462) = .4563.

▲ Determine EF: In right triangle EFP, $\angle F = 72° \, 40'$, EP = 1.125 (1.125 radius is ⊥ to tangent line at the point of tangency).

$\text{Csc } 72° \, 40' = \frac{EF}{EP}$, $1.0476 = \frac{EF}{1.125}$, EF = 1.0476 (1.125) = 1.1786

▲ Dimension x = AB + BC + CD + DE + EF = .500 + 1.0058 + 1.875 + .4563 + 1.1786, x = 5.016 inches.

APPLICATION

Solve the following problems. Determine angular values to the closest 1' and length dimensions to closest thousandths of an inch.

1. Length x = ______

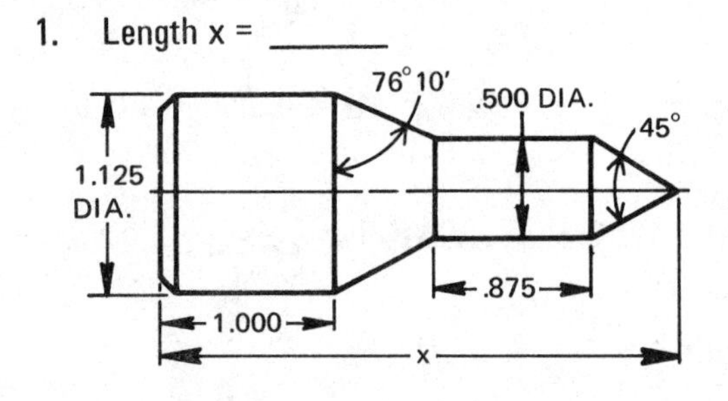

2. ∠x = ______

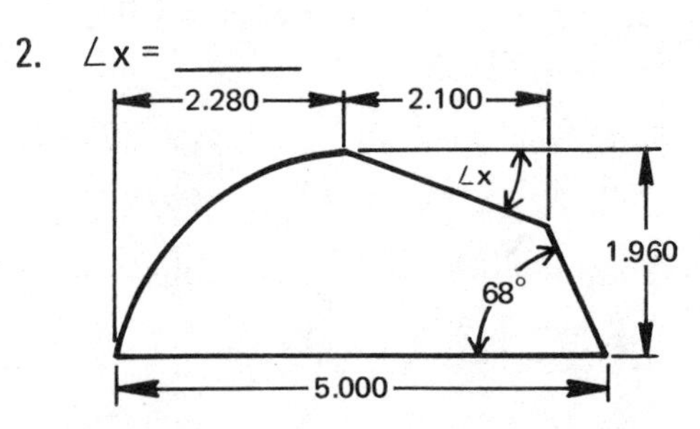

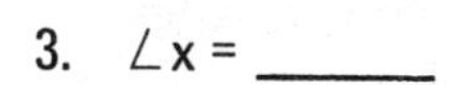

3. ∠x = ______

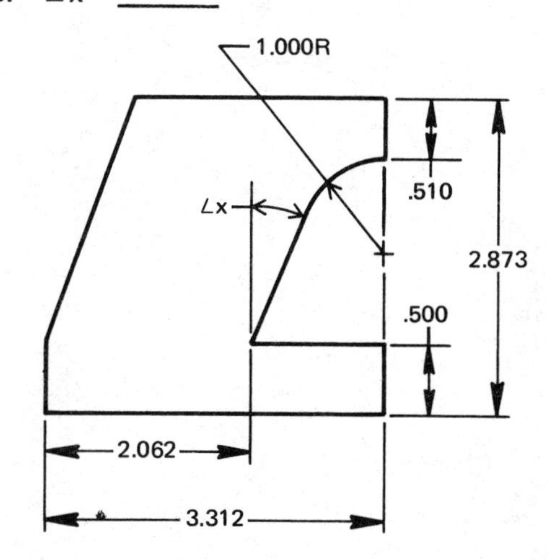

4. ∠y = ______

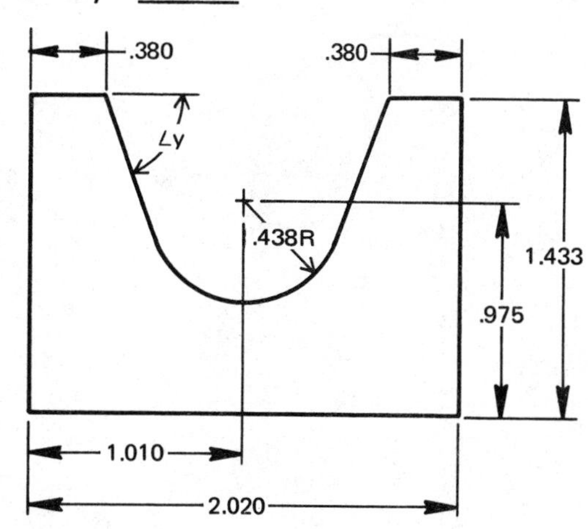

5. Gage dimension y = ______

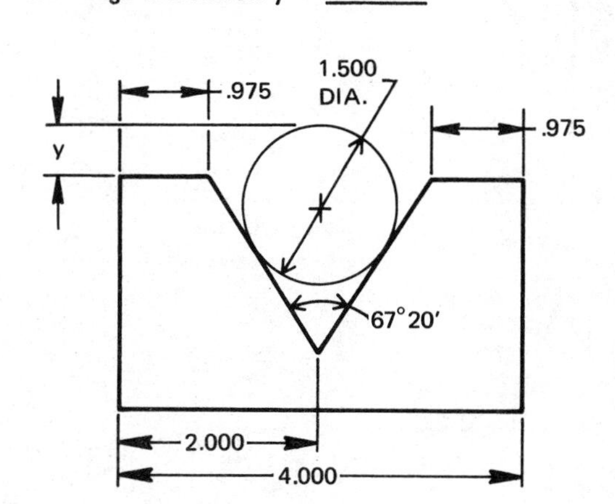

6. Dimension x = ______

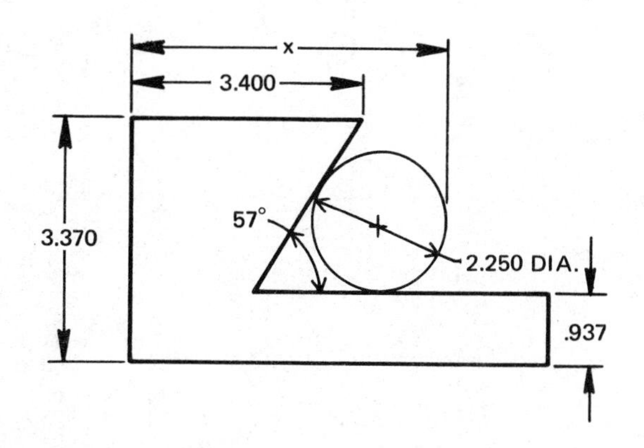

7. ∠x = ______

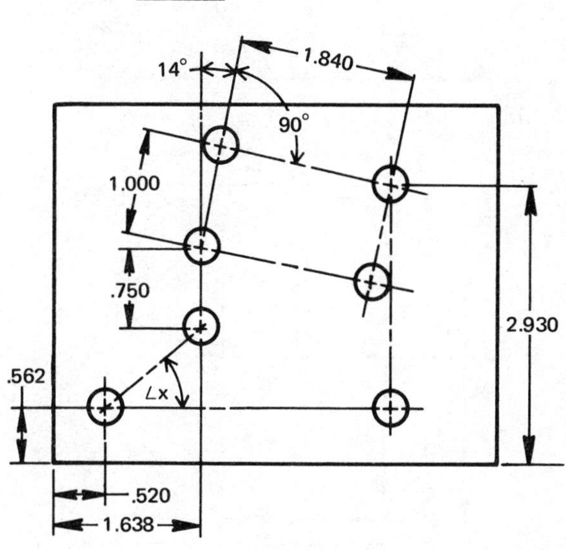

8. ∠y = ______

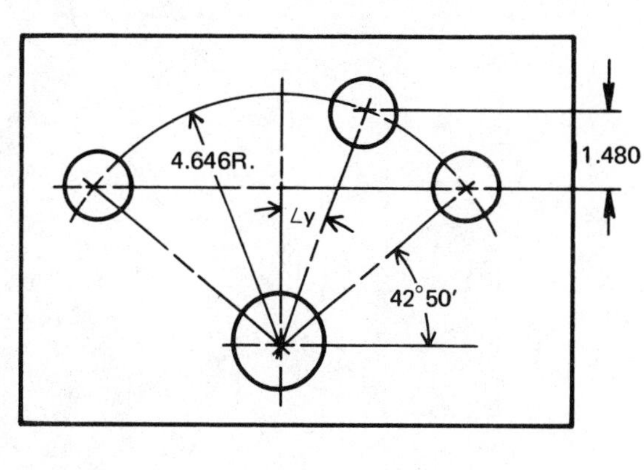

9. Length x = ______

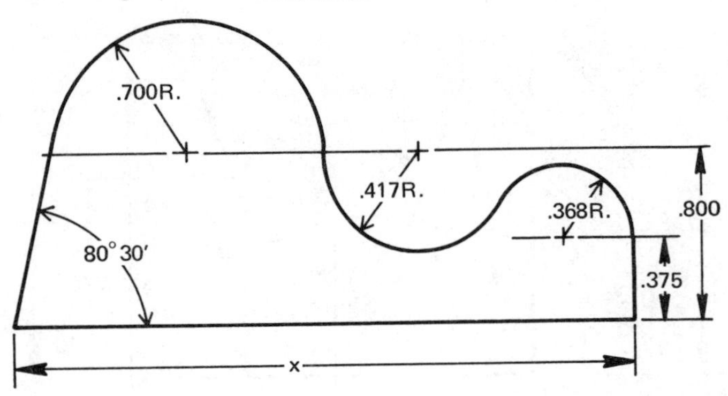

10. ∠y = ______

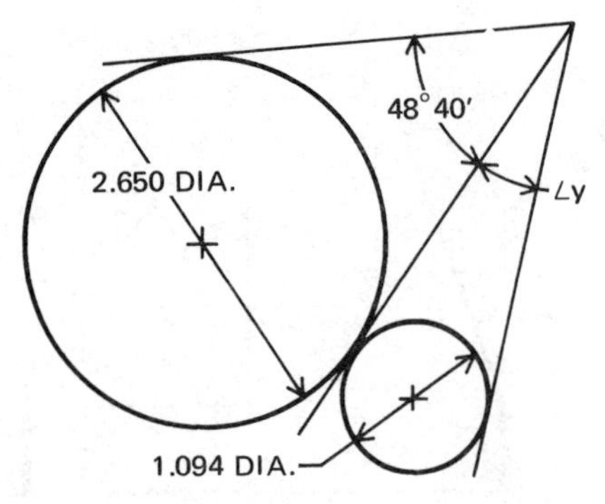

11. Dimension x = ______

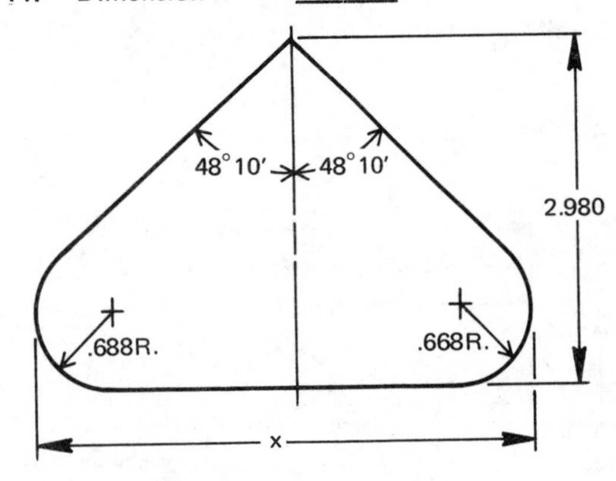

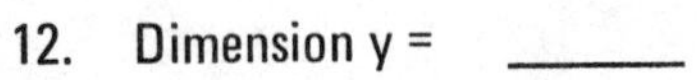
12. Dimension y = ______

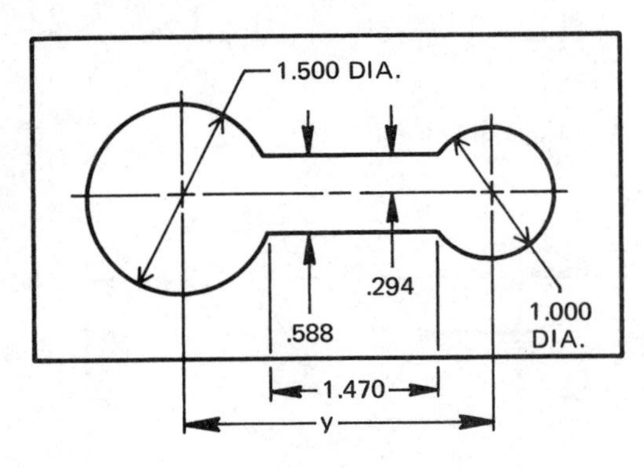

13. Dimension y = ______

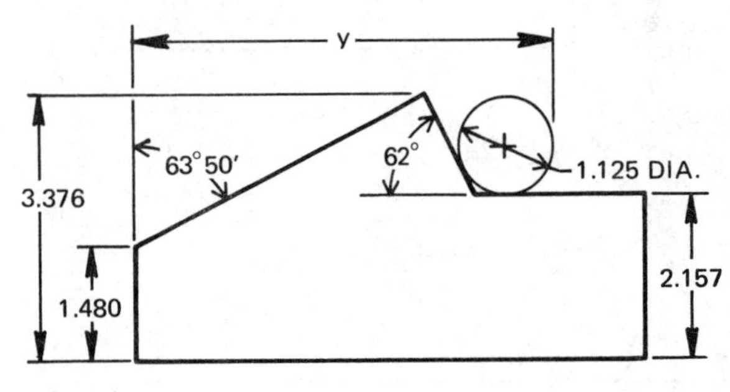

14. ∠x = ______

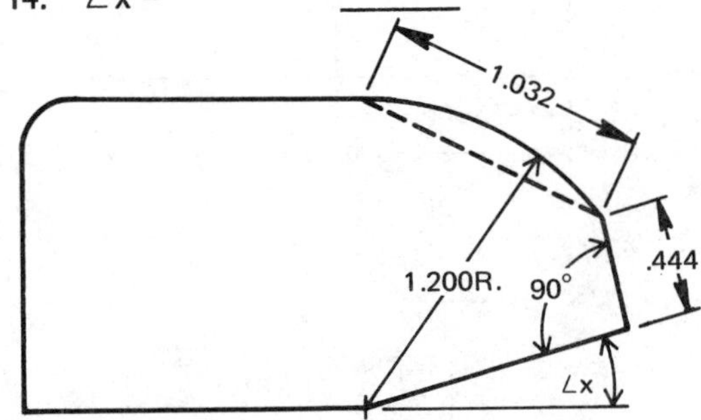

15. ∠x = ______

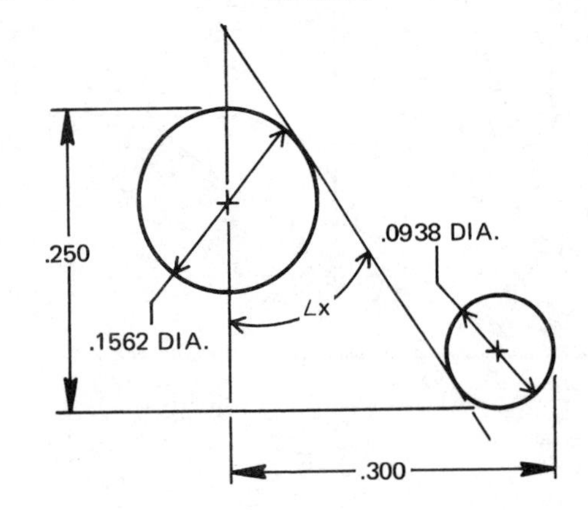

16. Dimension y = ______

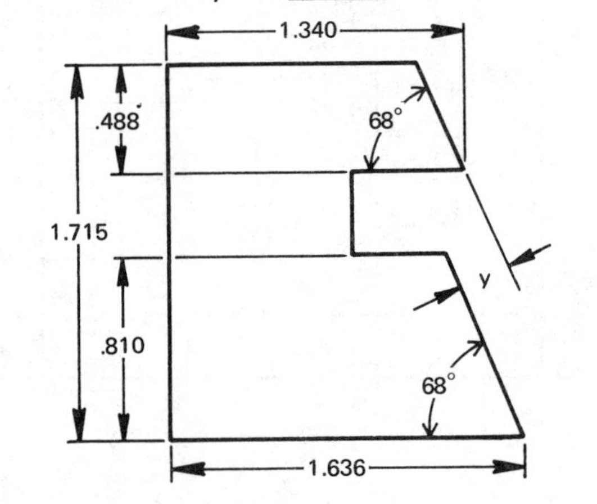

17. Dimension y = ______

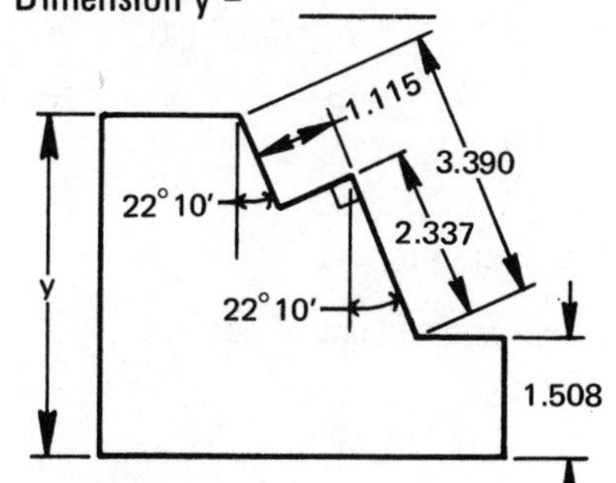

18. Dimension x = ______

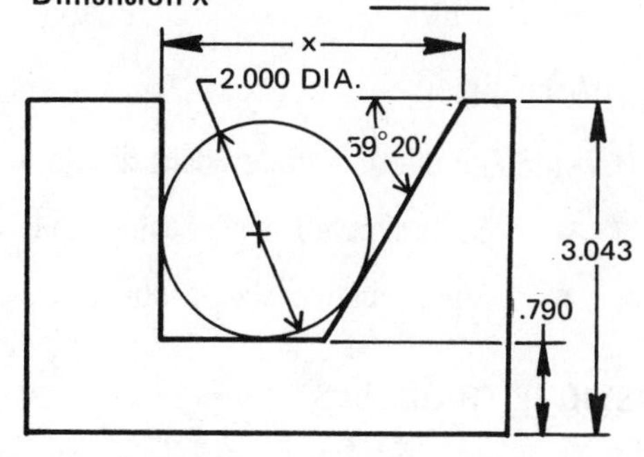

19. Dimension x = ______

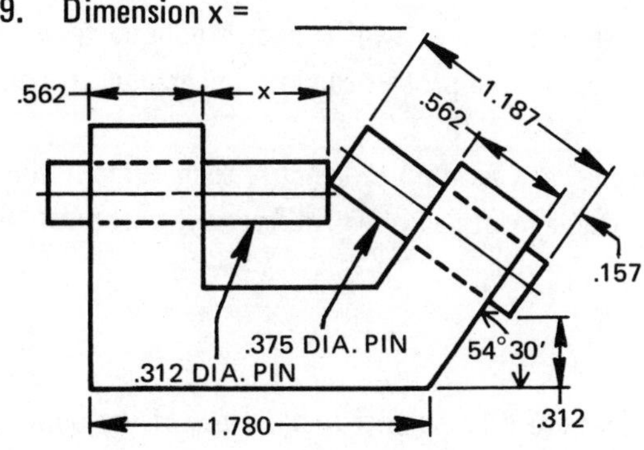

20. ∠x = ______

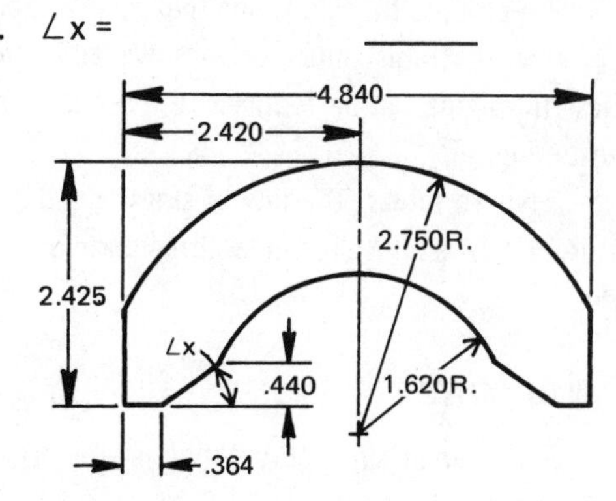

21. ∠x = ______

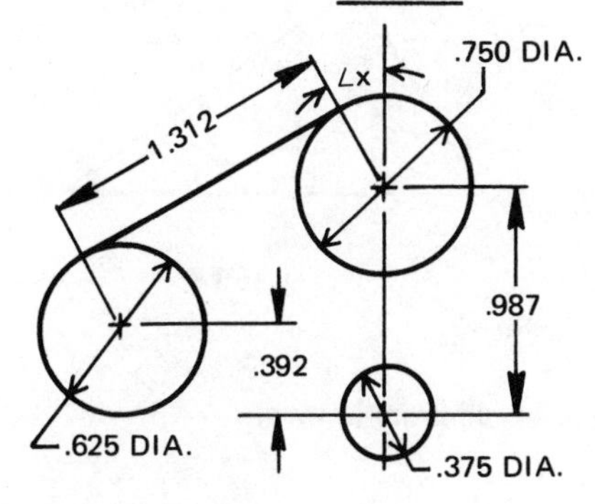

22. Check dimension y = ______

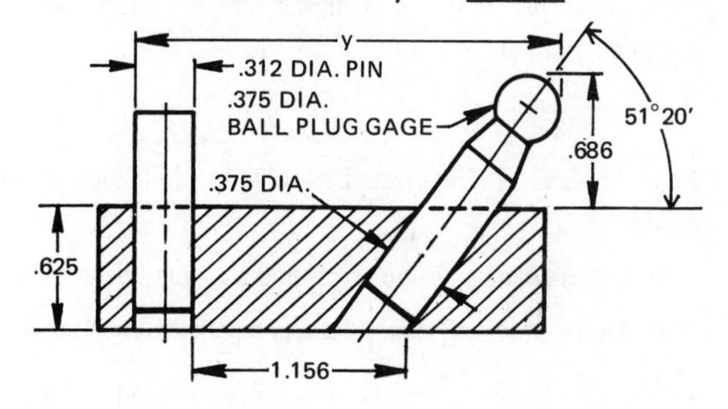

23. Dimension y = ______

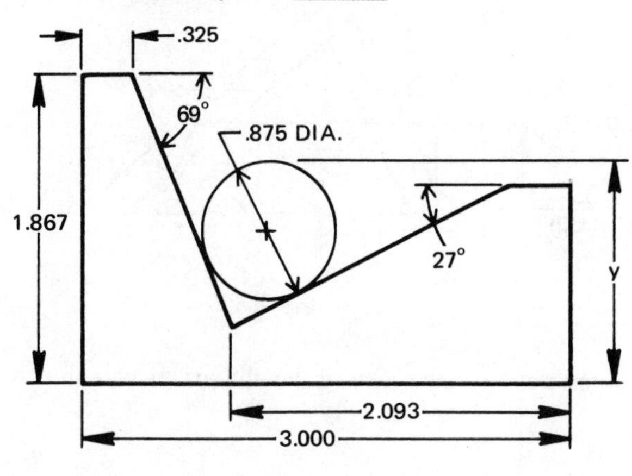

24. ∠x = ______

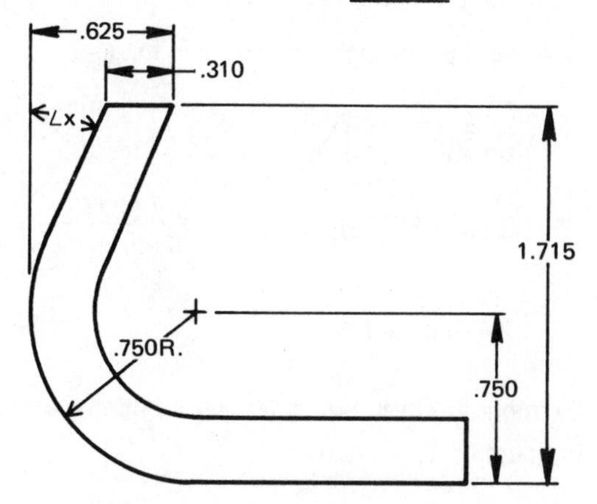

Unit 45 Oblique Triangles: Law of Sines, and Law of Cosines

OBJECTIVES

After studying this unit, the student should be able to

- Solve simple triangles using the law of sines and the law of cosines.
- Solve practical shop problems by applying the law of sines and the law of cosines.

OBLIQUE TRIANGLES

An oblique triangle is one that does not contain a right angle.

The machinist must often solve practical machine shop problems which involve oblique triangles. These problems can be reduced to a series of right triangles, but the process can be cumbersome and time consuming.

Two formulas, the law of sines and the law of cosines, can be used to simplify such computations. In order to use either formula, three parts of an oblique triangle must be known; at least one part must be a side.

LAW OF SINES

The law of sines states that in any triangle, the sides are proportional to the sines of the opposite angles.

In reference to the triangle shown in figure 45-1, the formula is stated:

$$\frac{a}{\sin A} = \frac{b}{\sin B} = \frac{c}{\sin C}$$

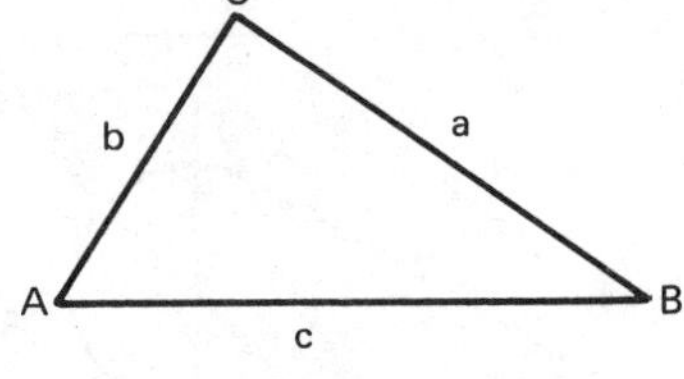

Fig. 45-1

Rule: The law of sines is used to solve the following kinds of problems:

- Those where any two angles and any one side are known.
- Those where any two sides and an angle opposite one of the given sides are known.

Example 1: Given two angles and a side, determine side x of the oblique triangle, figure 45-2.

▲ Set up a proportion and solve for x.

$$\frac{x}{\sin 36^\circ} = \frac{3.500''}{\sin 58^\circ};\ \frac{x}{.58778} = \frac{3.500}{.84805};\ .84805x =$$

$$3.500\ (.58778);\ x = \frac{3.500\ (.58778)}{.84805};\ x = \frac{2.05723}{.84805} =$$

2.426 inches

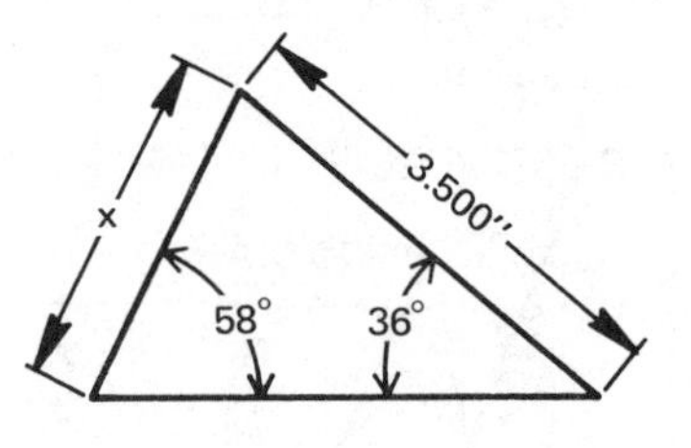

Fig. 45-2

Example 2: Given two sides and an opposite angle, determine $\angle x$ and side y of the oblique triangle shown in figure 45-3.

▲ Determine $\angle x$: $\frac{4.500}{\sin \angle x} = \frac{6.000}{\sin 63^\circ 50'};\ \frac{4.500}{\sin \angle x} = \frac{6.000}{.89751};$

$6.000 (\sin \angle x) = 4.500 (.89751);$

$\text{sine} \angle x = \dfrac{4.500 (.89751)}{6.000} = .67313;$

$\angle x = 42° 19'.$

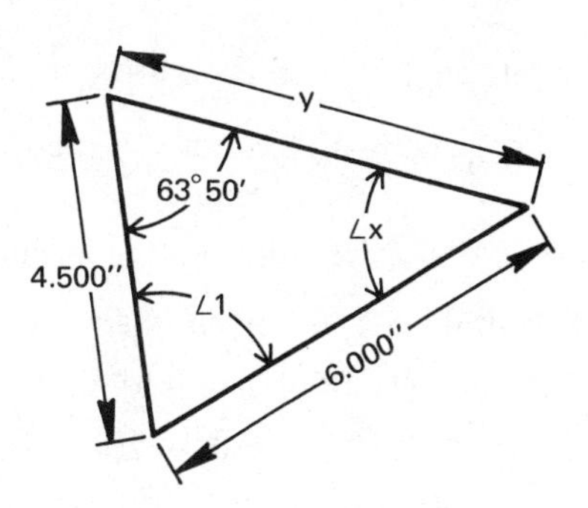

Fig. 45-3

▲ Determine side y:

Determine $\angle 1$: $\angle 1 = 180° - (63° 50' + \angle x) = 180° - 106° 9' = 73° 51';$

Set up a proportion: $\dfrac{6.000}{\sin 63° 50'} = \dfrac{y}{\sin 73° 51'};$

$\dfrac{6.000}{.89751} = \dfrac{y}{.96054}$; $.89751y = 6.000 (.96054)$; $y = \dfrac{6.000 (.96054)}{.89751}$; $y = 6.421$ inches.

Example 3: Solving for an angle greater than 90°: Determine $\angle x$ of the oblique triangle shown in figure 45-4.

Set up a proportion:

$\dfrac{1.400}{\sin 28° 10'} = \dfrac{2.750}{\sin \angle x}$; $\dfrac{1.400}{.47204} = \dfrac{2.750}{\sin \angle x}$;

$1.400 (\sin \angle x) = 2.750 (.47204);$

$\sin \angle x = \dfrac{2.750 (.47204)}{1.400}$; $\sin \angle x = .92722.$

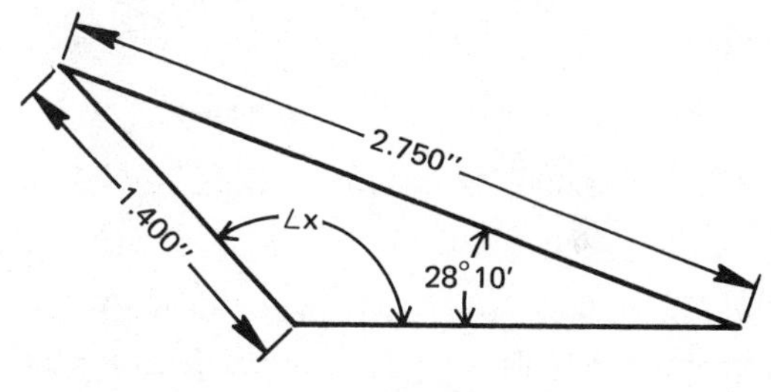

Fig. 45-4

The angle that corresponds to the sin function .92722 is 68°. Because $\angle x$ is greater than 90°, $\angle x$ = the supplement of 68°, Angle x = 180° - 68° = 112°.

Note: If necessary, refer to unit 41 to review the method of determining angles and functions in any quadrant.

LAW OF COSINES

There are two forms of the law of cosines:

- When two sides and an included angle are given, the law is stated: In any triangle, the square of any side is equal to the sum of the squares of the other two sides minus twice the product of these two sides multiplied by the cosine of their included angle.

  In reference to the triangle shown in figure 45-5, the formula is stated:

  $a^2 = b^2 + c^2 - 2\,bc \cos A$

  $b^2 = a^2 + c^2 - 2\,ac \cos B$

  $c^2 = a^2 + b^2 - 2\,ab \cos C$

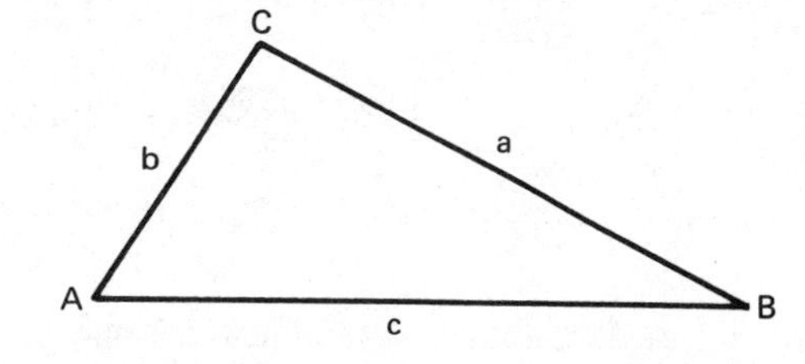

Fig. 45-5

- When three sides are given, the law is stated: In any triangle, the cosine of an angle is equal to the sum of the squares of the two adjacent sides minus the square of the opposite side, divided by twice the product of the two adjacent sides.

In reference to the triangle shown in figure 45-5, the formula is stated:

$$\cos A = \frac{b^2 + c^2 - a^2}{2\,bc}; \quad \cos B = \frac{a^2 + c^2 - b^2}{2\,ac}; \quad \cos C = \frac{a^2 + b^2 - c^2}{2\,ab}$$

Example 1: Given two sides and the included angle, determine side x of the triangle shown in figure 45-6.

▲ Substitute the values in their appropriate places in the formula and solve for x.

$x^2 = 5.600^2 + 6.200^2 - 2(5.600)(6.200)(\cos 36° \ 50')$

$x^2 = 31.360 + 38.440 - 2(5.600)(6.200)(.80038)$

$x^2 = 69.8000 - 55.5784$; $x = \sqrt{14.2216} =$ 3.771 inches

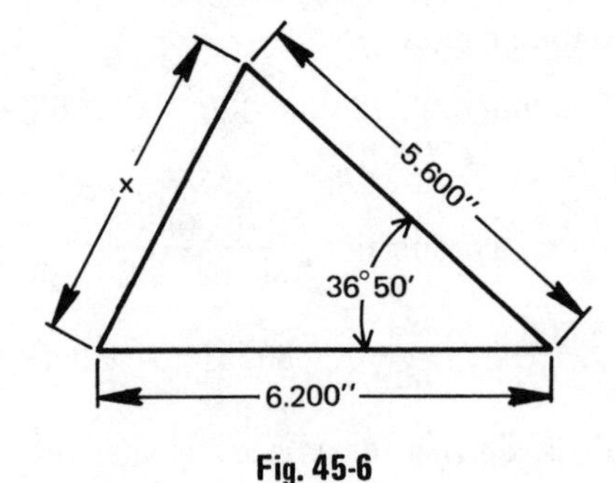

Fig. 45-6

Example 2: Given two sides and the included angle, determine side x and ∠ 1 of the triangle shown in figure 45-7.

▲ Substitute values to solve for side x:

▲ $x^2 = 3.800^2 + 4.100^2 - 2(3.800)(4.100)(\cos 123°)$

Note: Since the given angle is greater than 90°, the cosine of the angle is equal to the negative cosine of its supplement. Therefore, cos 123° = - cos (180° - 123°) = - cos 57° = - .54464. This negative value must be used in computing side x.

$x^2 = 3.800^2 + 4.100^2 - 2(3.800)(4.100)(-.54464)$.

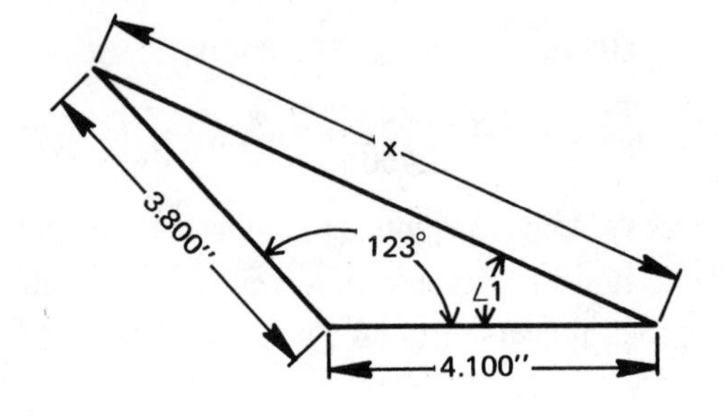

Fig. 45-7

The product of two negative values is a positive value; therefore, $-2(3.800)(4.100)(-.54464) = +16.971$.

$x^2 = 3.800^2 + 4.100^2 + 16.971 = 14.440 + 16.810 + 16.971 = 48.221$; $x = \sqrt{48.221} = 6.944$ inches.

▲ Solve for ∠ 1 using the Law of Sines: $\frac{3.800}{\sin \angle 1} = \frac{\text{side x}}{\sin 123°}$;

(Sin 123° = sin (180° - 123°) = sin 57° = .83867.)

$\frac{3.800}{\sin \angle 1} = \frac{6.944}{.83867}$; $6.944 (\sin \angle 1) = (3.800)(.83867)$; $\sin \angle 1 = \frac{3.800\ (.83867)}{6.944} = .45895$;

$\angle 1 = 27° \ 19'$.

Note: Since side x has been determined, ∠ 1 may be computed by using the Law of Cosines (given three sides), but it is simpler to use the Law of Sines.

Example 3: Given three sides, determine ∠ x of the triangle shown in figure 45-8.

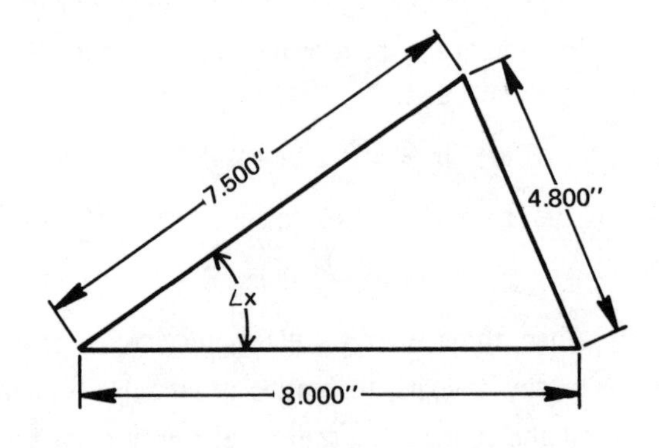

Fig. 45-8

▲ Substitute values in their appropriate places in the formula.

$$\cos\angle x = \frac{7.500^2 + 8.000^2 - 4.800^2}{2(7.500)(8.000)}, \quad \cos\angle x = \frac{56.250 + 64.000 - 23.040}{120} = \frac{97.210}{120} = .81008;$$

$x = 35° 54'$

Example 4: Given three sides, determine $\angle x$ of the triangle shown in figure 45-9.

▲ Substitute values in the formula.

$$\cos\angle x = \frac{3.000^2 + 3.700^2 - 5.300^2}{2(3.000)(3.700)}$$

$$\cos\angle x = \frac{9.000 + 13.690 - 28.090}{22.200}$$

$$\cos\angle x = \frac{22.690 - 28.090}{22.200}$$

$$\cos\angle x = \frac{-5.400}{22.200} = -.24324$$

Fig. 45-9

Since the $\cos\angle x$ is a negative value, $\angle x$ is equal to the supplement of the angle found in the function table. The angle corresponding to the cosine function .24324 is 75° 55'. Therefore, the angle whose cos function is - .24324 is equal to 180° - 75° 55' = 104° 5'. Angle x = 104° 5'.

APPLICATION

A. Solve the following problems using the law of sines.

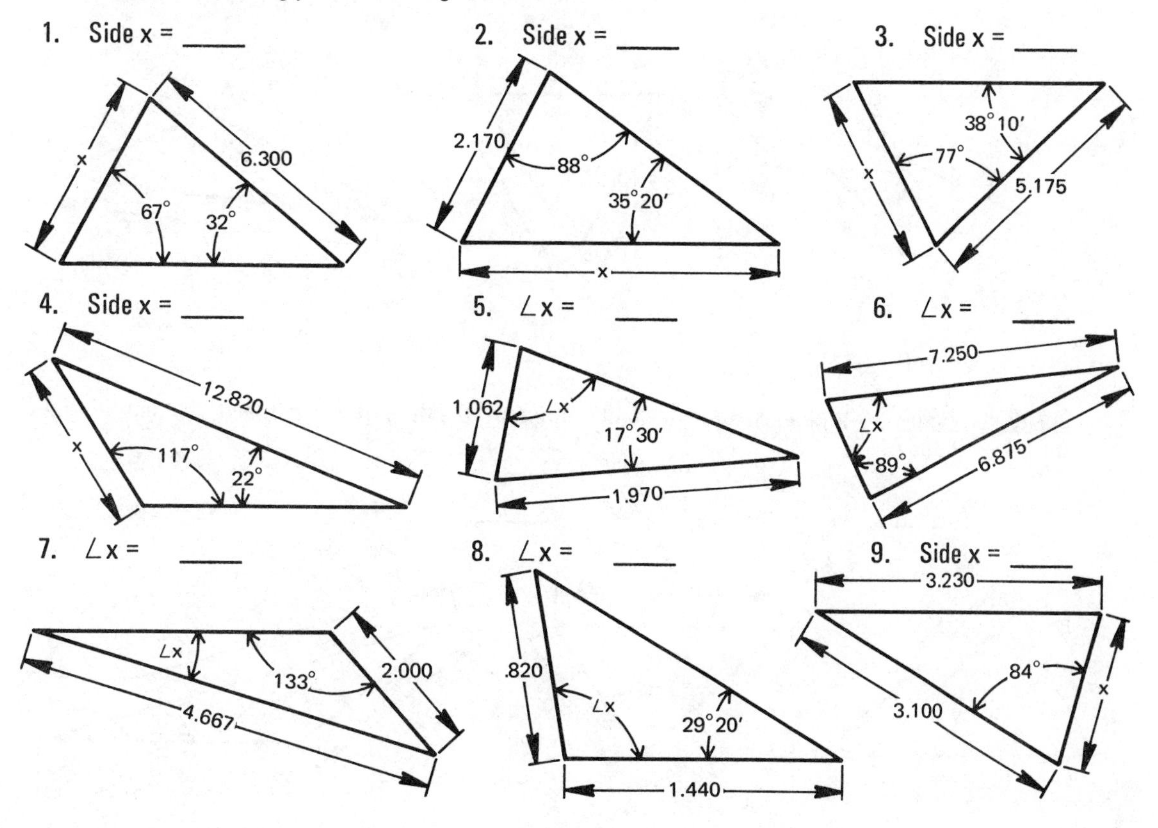

B. Solve the following problems using the law of cosines.

1. Side x = ______

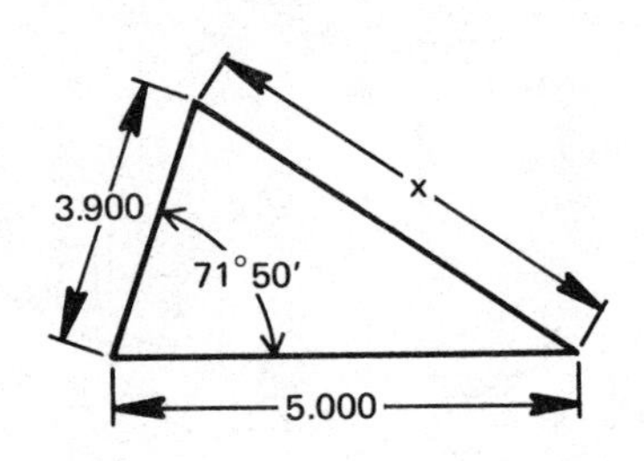

2. Side x = ______

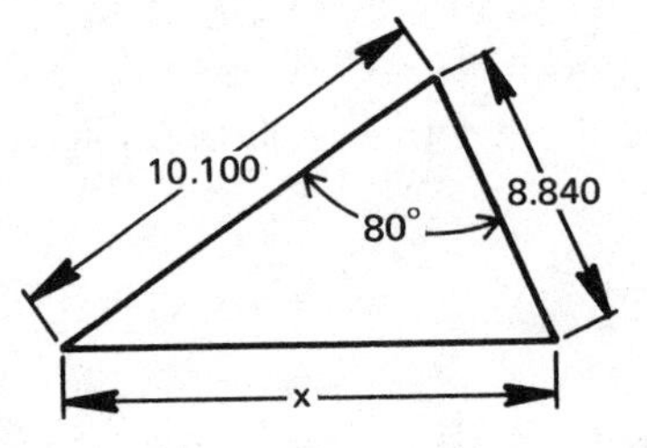

3. Side x = ______

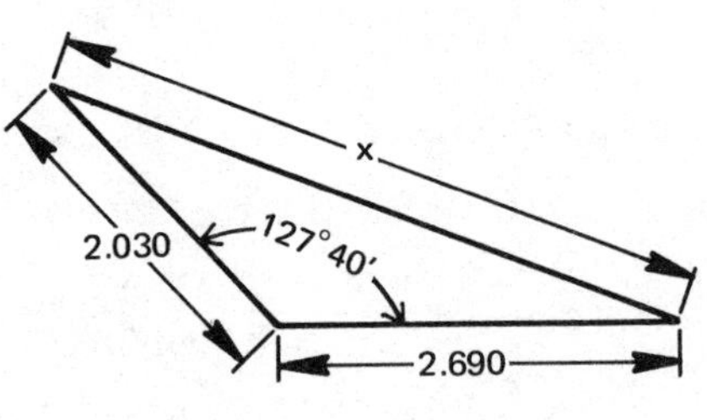

4. ∠x = ______

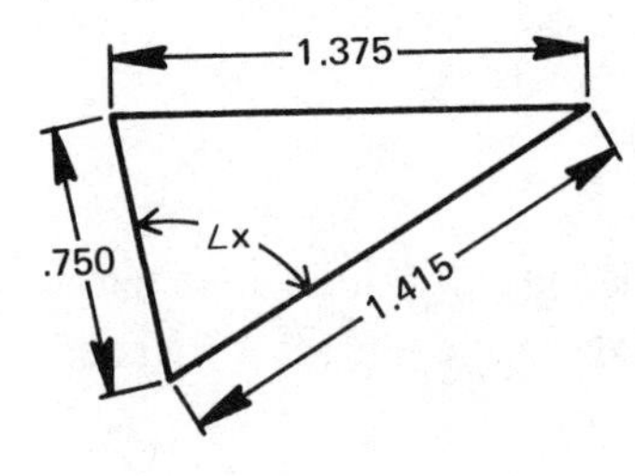

5. ∠x = ______

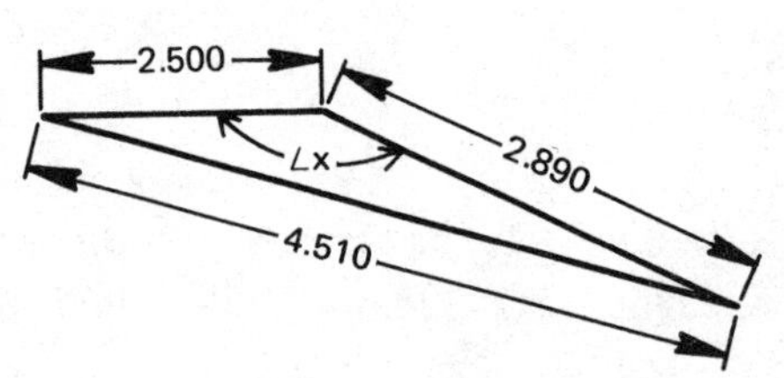

6. ∠x = ______

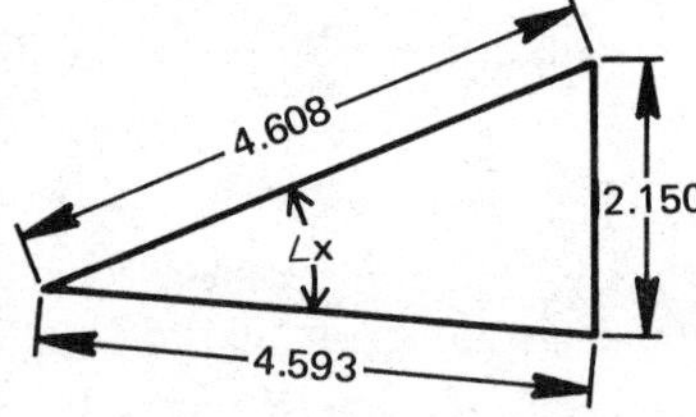

7. ∠x = ______

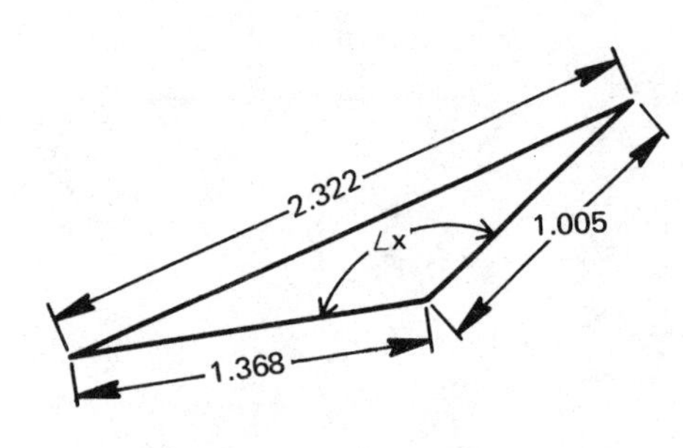

8. Side x = ______

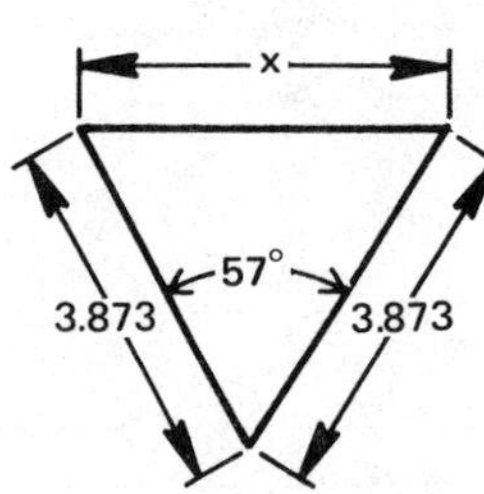

9. Side x = ______

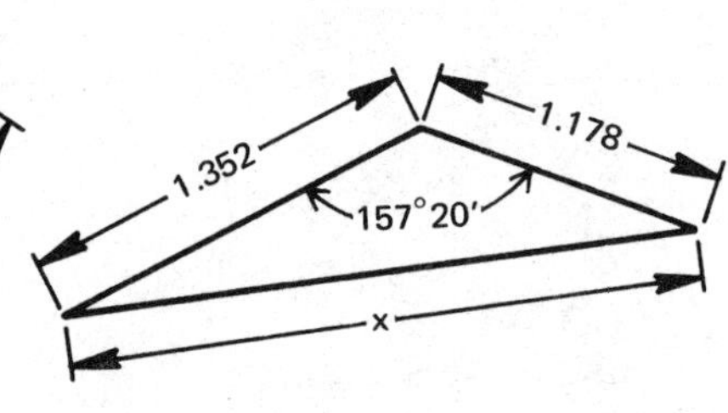

C. Solve the following problems using a combination of the law of cosines and the law of sines.

1. Side x = ______
   ∠y = ______

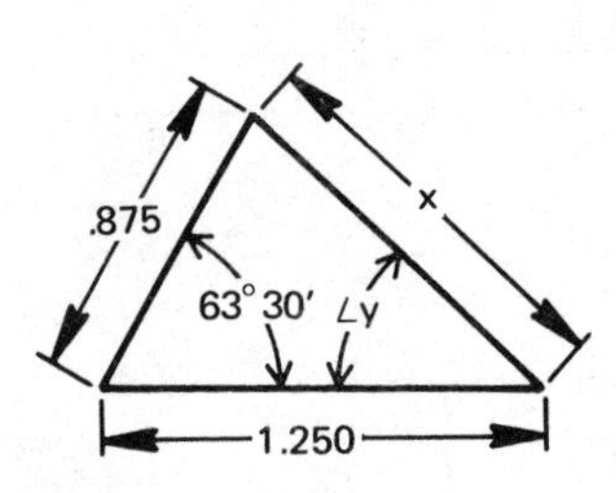

2. Side x = ______
   ∠y = ______

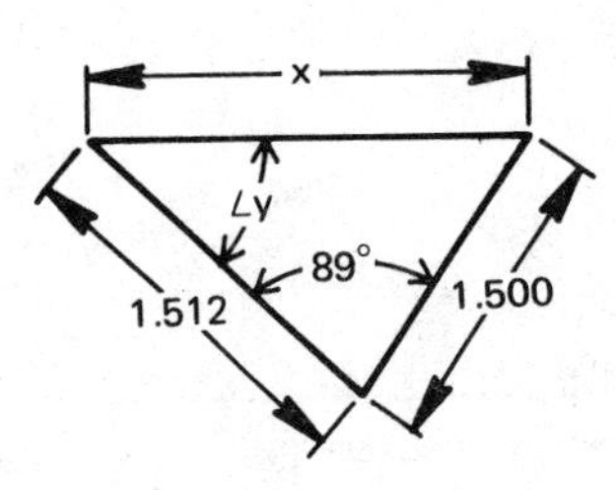

3. Side x = ______
   ∠y = ______

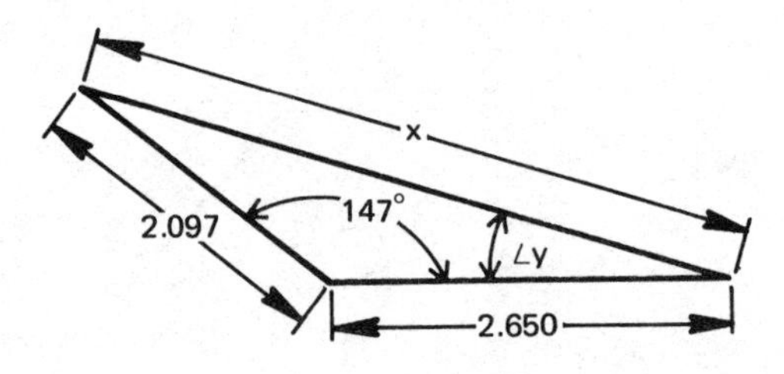

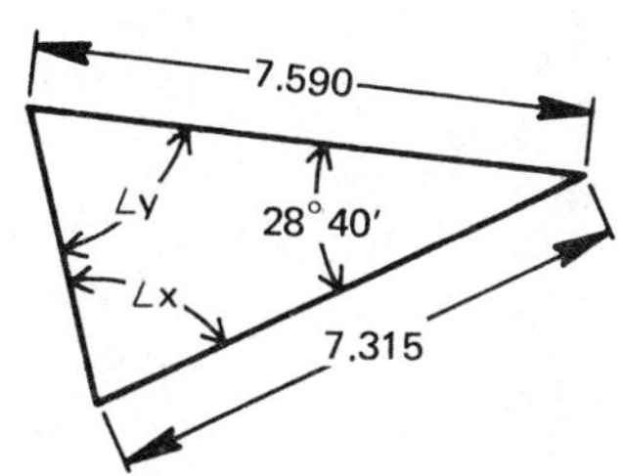

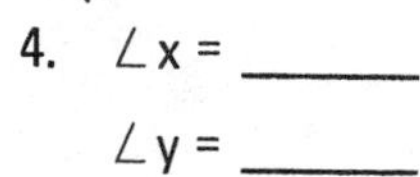

4. ∠x = ______

   ∠y = ______

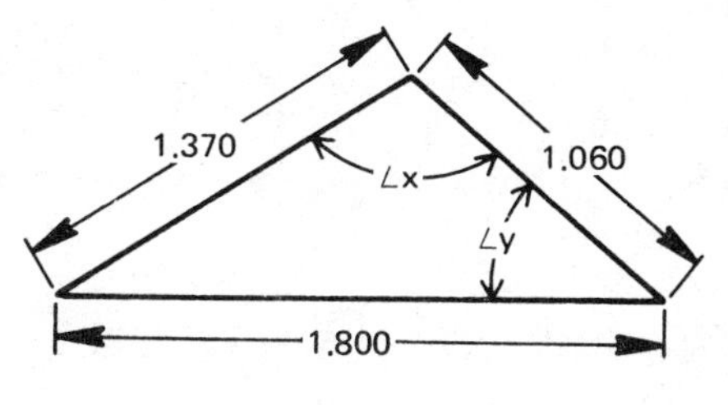

5. ∠x = ______

   ∠y = ______

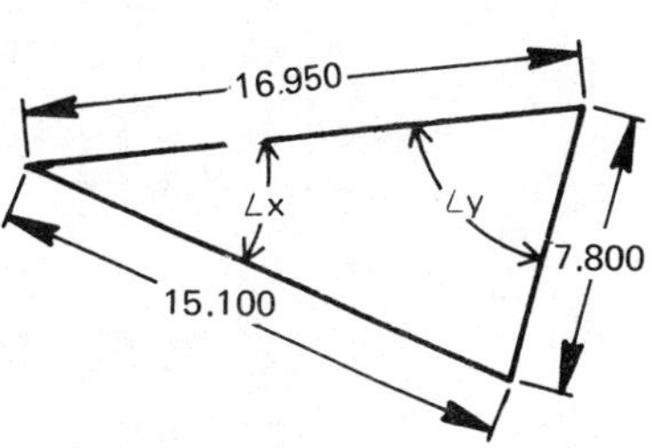

6. ∠x = ______

   ∠y = ______

D. Solve the following practical machine shop problems.

1. ∠x = ______

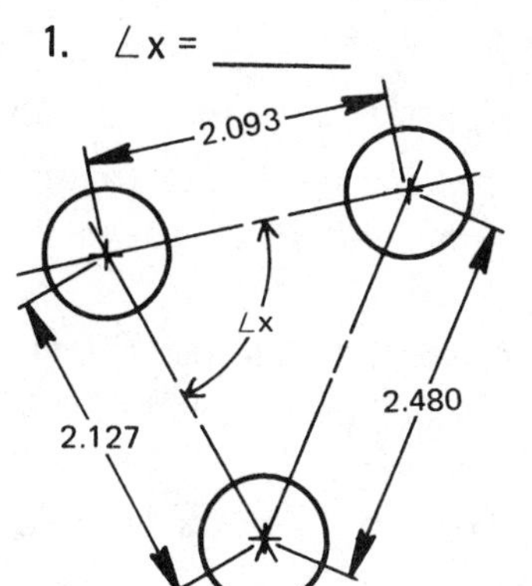

2. Distance y = ______

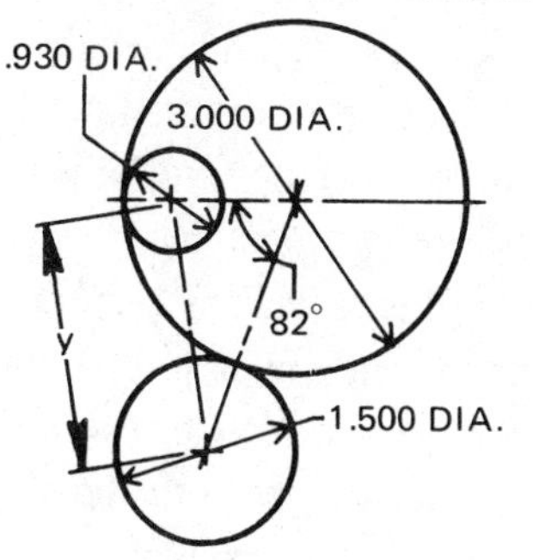

3. Distance x = ______

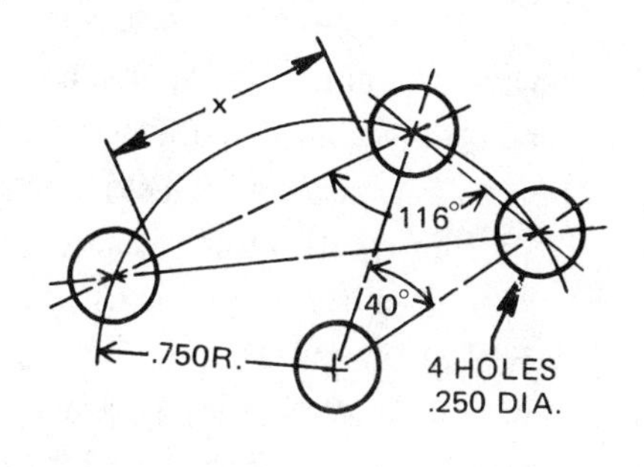

4. ∠x = ______

1.372
1.800
4.125R.
∠x

5. Dimension y = ______

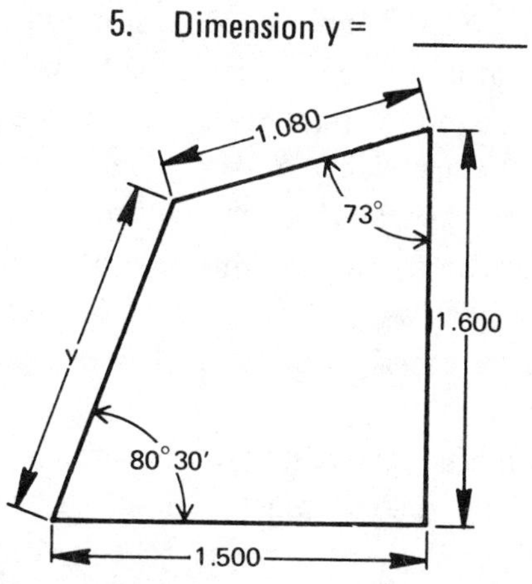

6. Dimension y = ______

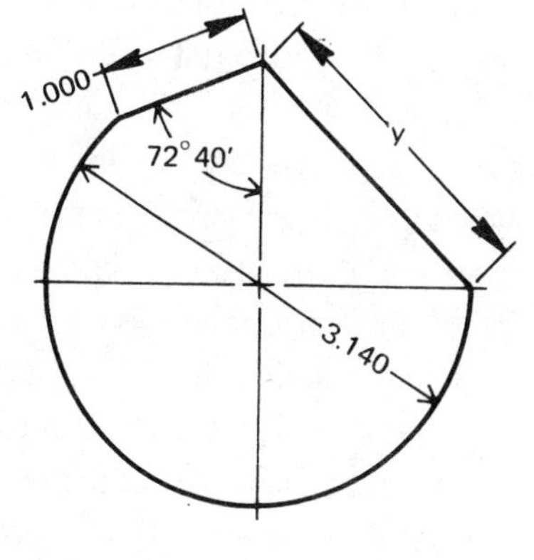

7. Dimension y = ______

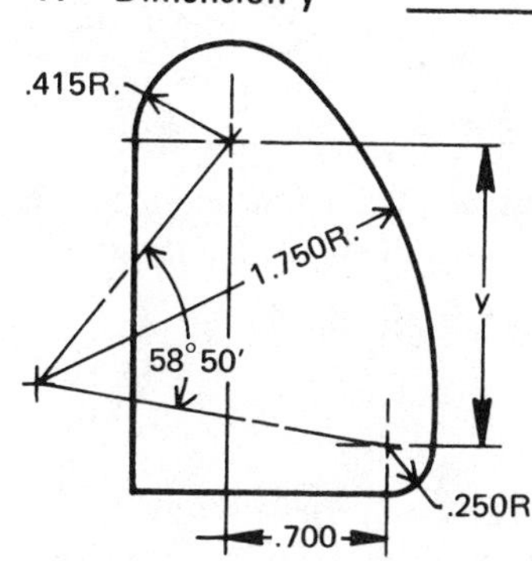

8. ∠x = ______

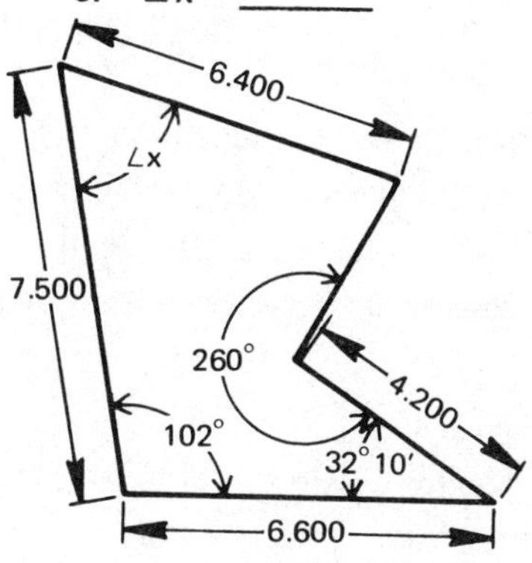

9. ∠y = ______

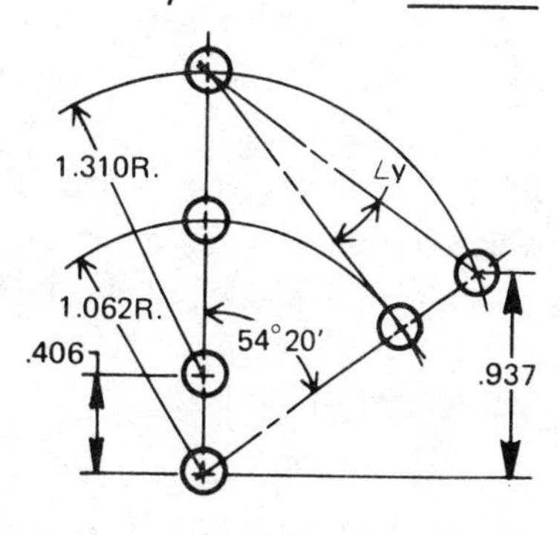

SECTION VI
Numerical Control

Unit 46 Introduction to Numerical Control

OBJECTIVES

After studying this unit, the student should be able to

- Locate points in a Cartesian coordinate system.
- Plot points in a Cartesian coordinate system.

Numerical control machines are widely used in the manufacture of machined parts with the trend toward greater application in the future. Therefore, at least a fundamental knowledge of numerical control should be acquired. Although the machinist does not usually write a program of operations, he should have a basic understanding of numerical control programming.

Numerical control machines are designed for a wide range of machining operations. Numerical control milling machines, engine lathes, turret lathes, and punch presses are widely used in industry, as are more specialized machines such as riveting, drafting, flame cutting, and inspection machines.

Although numerical control machines may vary as to the type and complexity of operations, the basic principles of operations are, in general, common to all machines.

The operations to be performed by the machines are first programmed. Programming consists of writing a program manuscript (order of instruction) from a conventional process sheet or blueprint. The data on the manuscript is then transferred to either punched or magnetic tape codes. The tape is fed through a tape reader which converts the tape codes into electrical signals. These signals are sent to a central control unit. The control unit then converts the signals into commands which are sent to the machine. By means of electrical circuits, switches, motors, and cams, the commands from central control determine the operation of the machine. As the numerical control machine operates, feedback signals are sent back to the central control unit. These signals are compared to the original tape signals and must agree before the machine proceeds to the next operation.

To write a program for numerical control machines, the programmer should have an understanding of proper machining practices such as feeds, speeds, properties of metals, and cutter characteristics. He must be trained in numerical control programming methods. He must know how the numerical control machine functions and how to write a program using proper programming coding. Coding determines where the signals are ultimately set in the machine. Some programs require only simple arithmetic, while others require a knowledge of advanced mathematics.

Because programming for numerical control machinery is a study within itself, no attempt is made in this text to discuss even a simple program thoroughly or to teach coding. However, two important concepts which are fundamental to numerical control machines can be discussed. They are numerical control programming with reference to the Cartesian coordinate system and the binary system of numeration.

LOCATION OF POINTS

Programming is based on locating points within the Cartesian coordinate system, which is discussed in unit 41. In a plane, a point can be located from a fixed point by two dimensions. For example, a point can be located by stating that it is three units up and five units to the right of a fixed point. The Cartesian coordinate system gives point location by using positive and negative values rather than location stated as being up or down and left or right from a fixed point.

A point is located in reference to the origin by giving the point an x and y value. The x value is always given first. The pair of x and y values is called the *coordinate* of the point.

Examples of point locations in the Cartesian coordinate system are given, figure 46-1.

Example 1: Locate point A which has a coordinate of (3,5). The x value is + 3 units and the y value is + 5 units. Therefore, point A is located in Quadrant I.

Example 2: Locate point B which has a coordinate of (- 6, 4). The x value is - 6 units and the y value is + 4 units. Therefore, point B is located in Quadrant II.

Example 3: Locate point C which has a coordinate of (- 7, - 3). The x value is - 7 units and the y value is - 3 units. Therefore, point C is located in Quadrant III.

Example 4: Locate point D which has a coordinate of (2, - 5). The x value is + 2 units and the y value is - 5 units. Therefore, point D is located in Quadrant IV.

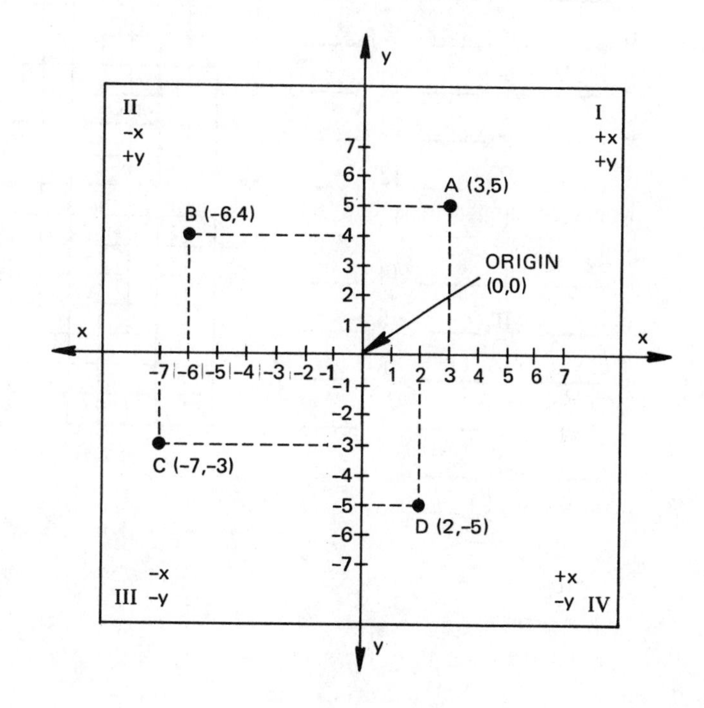

Fig. 46-1

APPLICATION

1. Using graph paper, draw an x- and a y-axis and plot the following coordinates.

   a. (- 2, 5), (3, 9), (- 7, - 2), (0, 3), (- 5, 0), (- 2, - 2)

   b. (5, - 7), (- 7, 5), (- 1, 0), (0, 0), (- 3, - 4), (8, - 6)

   c. (- 5, - 5), (- 3, - 3), (0, 0), (2, 2), (4, 4), (7, 7).

   Connect these points. What kind of a geometric figure is formed? What is the value of the angle formed in reference to the x and y axes?

   d. (- 9, - 7), (- 6, - 5.3), (- 3, - 3.5), (1, - 1), (4.5, 1), (7, 2.5), (6, 3), (4, 4), (2, 5), (0, 6), (- 2, 7), (- 3, 5), (- 5, 1), (- 6.5, - 2), (- 8, - 5)

   Connect these points in the order that they are given. What kind of a geometric figure is formed?

2. Refer to the points plotted on the illustrated Cartesian Coordinate plane. Give the x and y coordinates of the following points.

A ____ M ____ 7 ____

B ____ N ____ 8 ____

C ____ P ____ 9 ____

D ____ Q ____ 10 ____

E ____ R ____ 12 ____

F ____ S ____ 13 ____

G ____ T ____ 14 ____

H ____ U ____ 15 ____

J ____ V ____ 16 ____

K ____ 5 ____ 17 ____

L ____ 6 ____

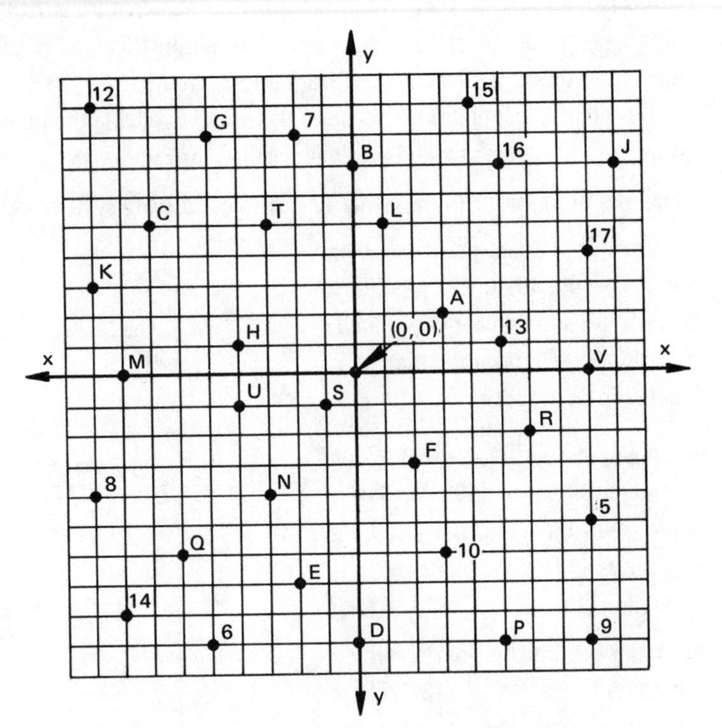

Unit 47 Point-to-point Programming on Two-axis Machines

OBJECTIVES

After studying this unit, the student should be able to

- Program dimension frum engineering drawings using a point-to-point control system with two-axis machines. Both absolute and incremental dimensioning are applied.

TYPES OF SYSTEMS

Some numerical control machines are designed so that they can be programmed for as many as six axes. Table movement left and right is the x-axis and table movement forward and back is the y-axis. Other axes are movement up and down by either the table or the spindle, the tilting of the table, the swiveling of the head, and the rotation of the table during the machining process.

A thorough study of the subject of numerical control programming is required in order to program for all axes. The discussion in this text is limited to the principles of programming for two-axis machines, the x- and y-axis. All other motions are controlled by the operator and not by tape commands.

Control systems are generally either point-to-point or continuous path (contouring) systems. The continuous path system is the more complex of the two systems, and programming for this system requires an analysis of the system which is beyond the scope of this text. This discussion of the principles of programming is limited to the fundamentals of programming for machining holes using the point-to-point system, although most point-to-point machines can also be used for straight cut milling (cuts parallel to the x- and y-axes.)

POINT-TO-POINT PROGRAMMING ON TWO-AXIS MACHINES

The movement of the machine as it performs the operations of machining holes in a part is similar to that of conventional machinery. The part to be machined is located for its various operations by either a movable table with a fixed spindle or a movable spindle with a fixed table.

The part to be machined must be positioned with respect to a zero reference point on the table. This zero reference point is either fixed or movable (floating zero) depending upon the design of the machine. Only the fixed zero is considered here.

The methods of dimensioning for the hole locations are either absolute or incremental.

ABSOLUTE DIMENSIONING

All dimensions are given from the fixed zero of the machine table. The workpiece must be located in reference to this fixed zero. For example, the x- and y-dimensions are established from two finished edges (setup point) of the workpiece to the fixed zero of the table. All other machining locations are then given from the fixed zero.

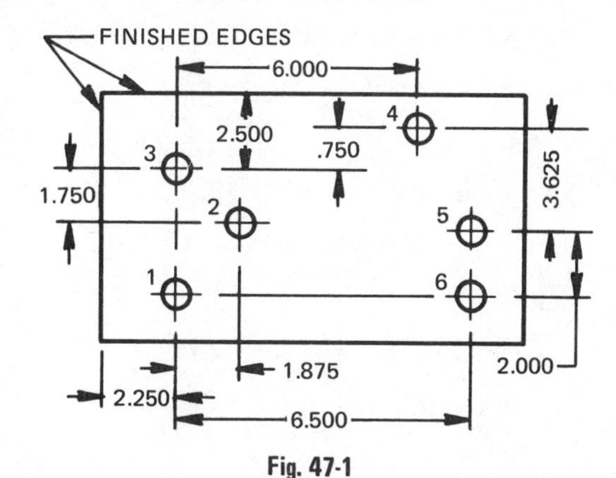

Fig. 47-1

Examples of Absolute Dimensioning

Example 1: Program dimension the hole locations in the part shown in figure 47-1. The figure shows the part as it is dimensioned on a blueprint before programming for numerical control.

▲ The workpiece is positioned at a convenient distance from the fixed zero of the machine table to the setup point (top left corner) of the workpiece, figure 47-2. In this example, x = 5″ and y = 4″.

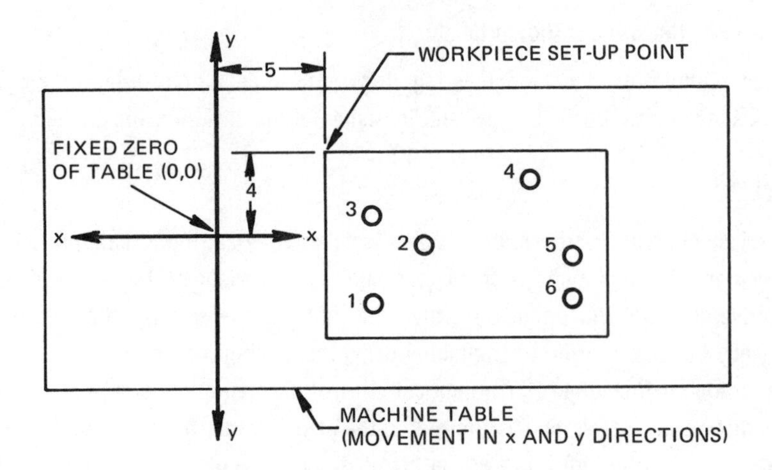

Fig. 47-2

▲ All the hole locations are programmed from the fixed zero of the table. Generally, the order in which the holes are machined is that which requires the least amount of machine movement. Note positive and negative x- and y-directions from (0, 0). The table in figure 47-3 lists the coordinates of the hole locations from (0, 0).

Hole	x	y
1	7.250	- 3.375
2	9.125	- .250
3	7.250	1.500
4	13.250	2.250
5	13.750	- 1.375
6	13.750	- 3.375

Fig. 47-3

Hole 1:

$x = 5 + 2.250 = 7.250$

$y = 4 - (2.500 - .750) - 3.625 - 2.000 =$
$4 - 1.750 - 3.625 - 2.000 = 4 - 7.375 = -3.375$

Hole 2:

$x = 5 + 2.250 + 1.875 = 9.125$

$y = 4 - (2.500 + 1.750) = 4 - 4.250 = -.250$

Hole 3:

x is the same as x of Hole 1 = 7.250

$y = 4 - 2.500 = 1.500$

Hole 4:

$x = 5 + 2.250 + 6.000 = 13.250$

$y = 4 - (2.500 - .750) = 4 - 1.750 = 2.250$

Hole 5:

$x = 5 + 2.250 + 6.500 = 13.750$

$y = 4 - (2.500 - .750) - 3.625 = -1.375$

Hole 6:

x is the same as x of Hole 5 = 13.750

y is the same as y of Hole 1 = - 3.375

Example 2: Figure 47-4 shows a part as it is dimensioned before programming for numerical control. Program dimension the hole locations.

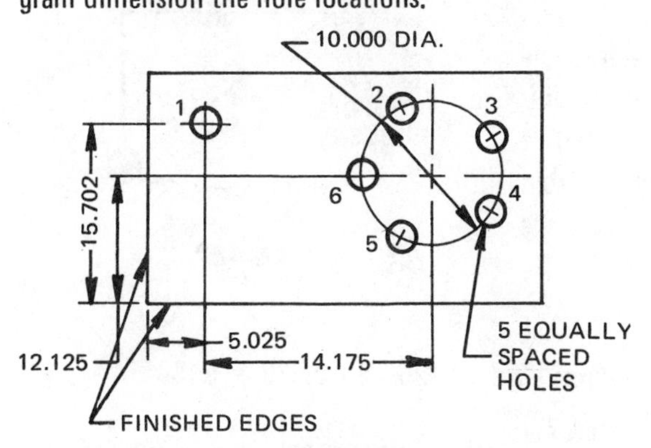

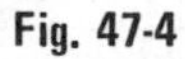
Fig. 47-4

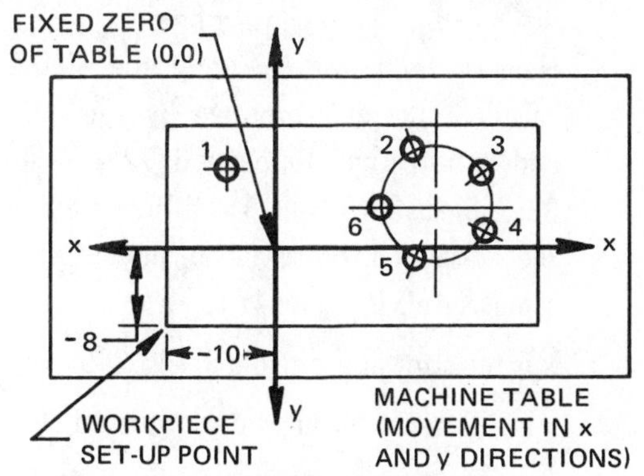

Fig. 47-5

▲ The workpiece is positioned at a convenient distance from the fixed zero of the machine table to the setup point (bottom left corner) of the workpiece, figure 47-5. In this example, x = - 10" and y = - 8".

▲ All hole locations are programmed from the fixed zero of the table. The table in figure 47-6 lists the coordinates of the hole locations from (0, 0).

Hole	x	y
1	- 4.975	7.702
2	7.655	8.880
3	13.245	7.064
4	13.245	1.186
5	7.655	-.630
6	4.200	4.125

Fig. 47-6

Hole 1:

$x = -10 + 5.025 = -4.975$

$y = 15.702 - 8 = 7.702$

Hole 2: Calculate the number of degrees between two consecutive holes on the 10.000 diameter circle: 360°/5 = 72°. From the center of the 10.000 diameter circle calculate a and b dimensions shown in figure 47-7.

Solve for a: $\sin 72^\circ = \frac{a}{5.000}$, $a = 5.000\ (.95106) = 4.755$

Solve for b: $\cos 72^\circ = \frac{b}{5.000}$, $b = 5.000\ (.30902) = 1.545$

HOLE 2 — 5.000 — a — 72° — b — CENTER OF 10.000 DIA. CIRCLE

Fig. 47-7

x = x distance to the center of the 10.000 dia. circle - b = -10 + 5.025 + 14.175 - 1.545 = 7.655

y = y distance to the center of the 10.000 dia. circle + a = 12.125 - 8 + 4.755 = 8.880

Hole 3: From the center of the 10.000 diameter circle calculate the angle formed by the horizontal centerline and Hole 3. Angle formed = 180° - 2(72°) = 180° - 144° = 36°.

Calculate a and b dimensions in figure 47-8.

Solve for a: $\sin 36^\circ = \frac{a}{5.000}$, $a = 5.000\ (.58778) = 2.939$

Solve for b: $\cos 36^\circ = \frac{b}{5.000}$, $b = 5.000\ (.80902) = 4.045$

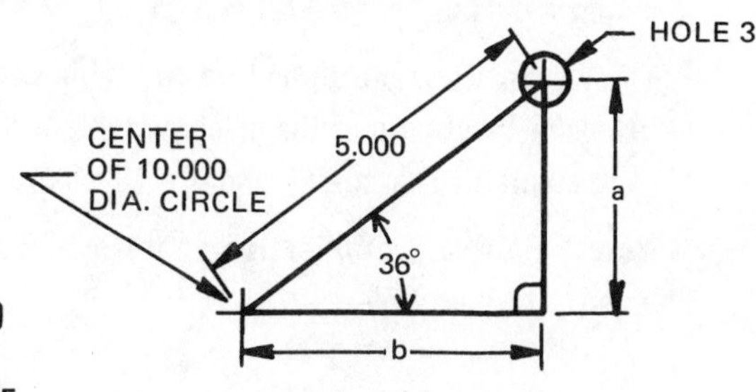

Fig. 47-8

x = x distance to the center of the 10.000 dia. circle + b = -10 + 5.025 + 14.175 + 4.045 = 13.245

y = y distance to the center of the 10.000 dia. circle + a =
12.125 - 8 + 2.939 = 7.064

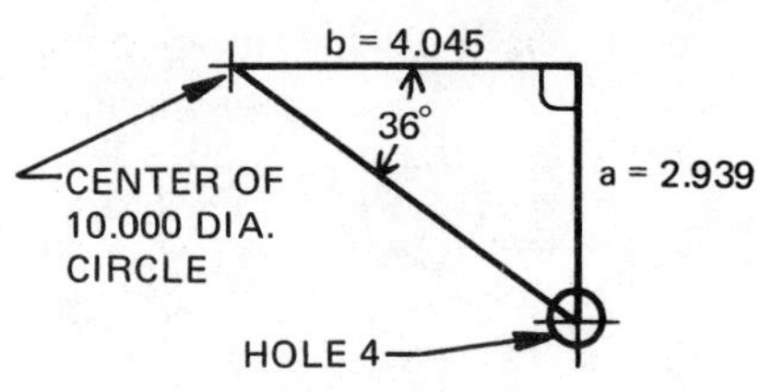

Fig. 47-9

Hole 4: From the center of the 10.000 diameter circle calculate the angle formed by the horizontal centerline and Hole 4. Angle formed = 3 (72°) - 180° = 216° - 180° = 36°. Since both Hole 4 and Hole 3 are projected 36° from the horizontal, the a and b dimensions of Hole 4 are the same as Hole 3, figure 47-9.

x is the same as x of Hole 3 = 13.245

y = y distance to the center of the 10.000 diameter circle - a = 12.125 - 8 - 2.939 = 1.186

Hole 5: Since both Hole 5 and Hole 2 are projected 72° from the horizontal, the a and b dimensions of Hole 5 are the same as Hole 2, figure 47-10.

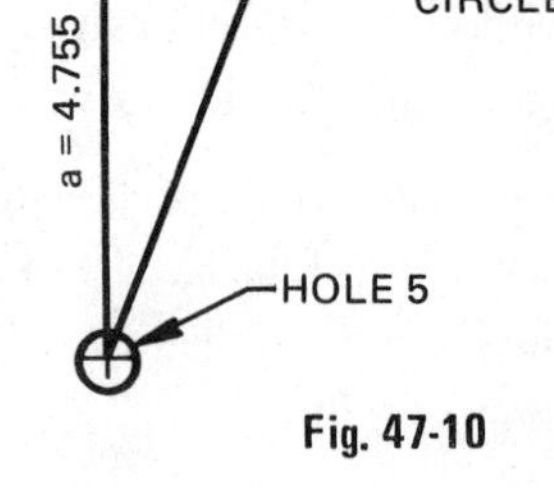

Fig. 47-10

x is the same as x of Hole 2 = 7.655

y = y distance to the center of the 10.000 dia. circle - a =
12.125 - 8 - 4.755 = - .630

Hole 6: x = x distance to the center of the 10.000 diameter circle - 5.000 = - 10 + 5.025 + 14.175 - 5.000 = 4.200

y = 12.125 - 8 = 4.125

INCREMENTAL DIMENSIONING

In incremental dimensioning, the dimensions to each location are given from the immediate previous location. The location of a hole is considered the origin (0, 0) of the x and y axis. From this origin, x and y dimensions are given to the next hole. Each new location in turn becomes the origin for the x and y dimensions to the next hole. The direction of travel, positive and negative, must be noted and is based upon the Cartesian coordinate system just as it was with absolute dimensioning. The first hole location is dimensioned with reference to the fixed zero of the table, while each subsequent hole is dimensioned from the hole directly preceding it. Each hole becomes the origin for the next hole to be machined.

Example of Incremental Dimensioning

- ▲ Dimension the part shown in figure 47-11, using incremental dimensioning. (This same part was used to illustrate absolute dimensioning in figures 47-1 and 47-2.)
- ▲ The workpiece is positioned on the table with the distance from the fixed zero of the machine table to the setup point of the workpiece as x = 5" and y = 4", figure 47-12.

Note: Be aware of direction of travel as + or - from one hole to the next hole.

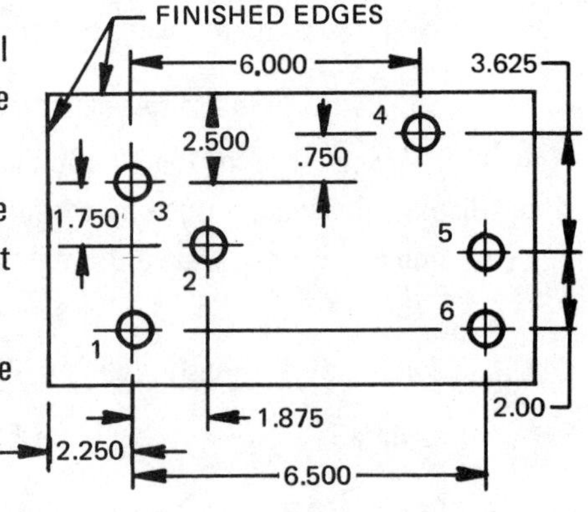

Fig. 47-11

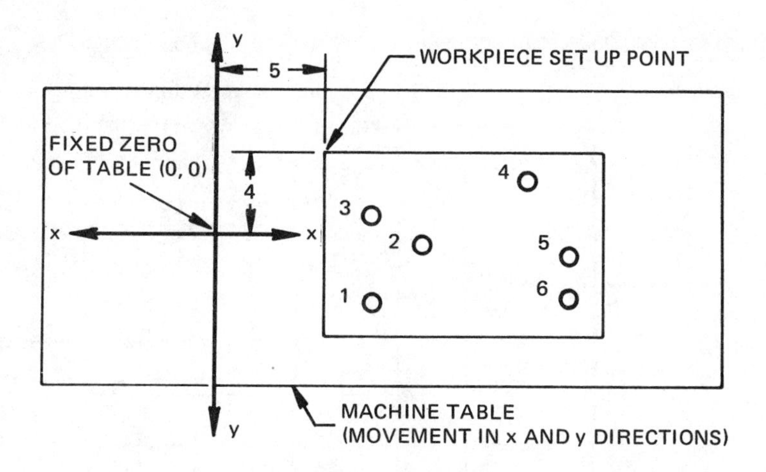

Fig. 47-12

The table in figure 47-13 lists the coordinates using incremental dimensioning.

Hole 1: x = 5 + 2.250 = 7.250

y = 4 - (2.500 - .750) - 3.625 - 2.000 = 4 - 1.750 - 3.625 - 2.000 = 4 - 7.375 = - 3.375

Note: It can be seen that the x and y dimensions for Hole 1 are identical to those using absolute dimensioning. This is true for the first hole only.

Hole	x	y
1	7.250	- 3.375
2	1.875	3.125
3	- 1.875	1.750
4	6.000	.750
5	.500	- 3.625
6	0	- 2.000

Fig. 47-13

- Each hole becomes the origin for the next hole

Hole 2: The center of Hole 1 is the origin for locating Hole 2.

x = 1.875

y = (2.500 - .750) + 3.625 + 2.000 - (2.500 + 1.750) = 3.125

Hole 3: The center of Hole 2 is the origin for locating Hole 3.

x = - 1.875 y = 1.750

Hole 4: The center of Hole 3 is the origin for locating Hole 4.

x = 6.000 y = .750

Hole 5: The center of hole 4 is the origin for locating Hole 5.

x = 6.500 - 6.000 = .500 y = - 3.625

Hole 6: The center of Hole 5 is the origin for locating Hole 6.

x = 0 y = - 2.000

APPLICATION

Program dimension the following parts. The dimensions given in the tables are taken from blueprints before programming for numerical control. Dimensions X′, and Y′, are the positioning dimensions from the fixed zero (0, 0) of the table to the setup point of the workpiece.

Use the hole location dimensions given in the tables. Write the program in table form similar to the examples shown in figures 47-3, 47-6 and 47-13. List the holes in sequence.

A. Refer to the drawing below and program dimension problems 1, 2, and 3 using

  a. absolute dimensioning b. incremental dimensioning

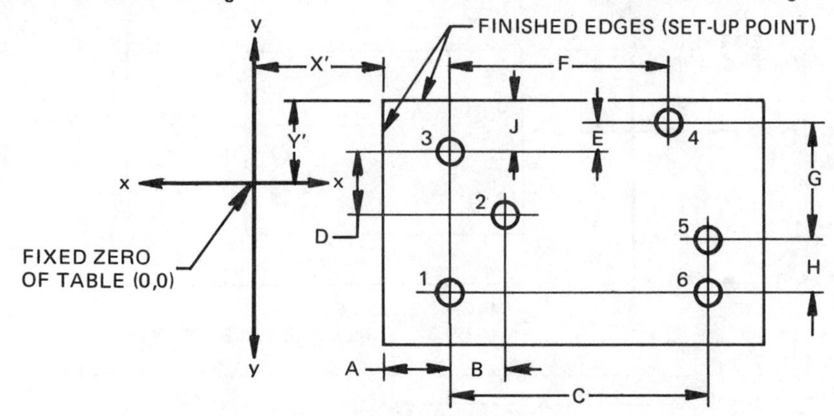

Problem	X'	Y'	Hole Location Dimensions									
			A	B	C	D	E	F	G	H	J
1	5	4	2.500	2.100	6.750	1.875	.825	6.125	3.875	2.125	2.375
2	6	5	2.710	2.615	7.010	2.070	.920	6.475	4.307	2.416	2.300
3	5	5	3.000	2.812	7.403	2.253	1.008	6.616	4.612	2.615	2.550

B. Refer to the drawing below and program dimension problems 4, 5, and 6 using

  a. absolute dimensioning b. incremental dimensioning

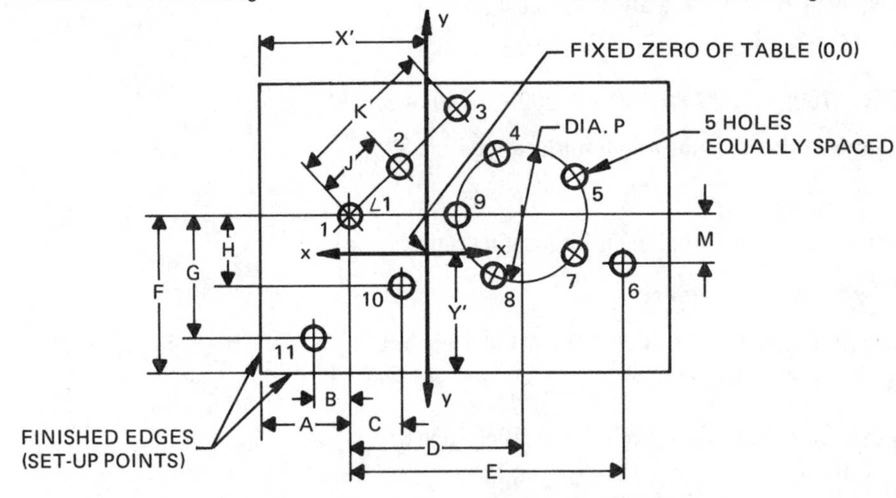

Prob-lem	X'	Y'	Hole Location Dimensions											Dia. P	∠1
			A	B	C	D	E	F	G	H	J	K	M		
4	- 10	- 8	5.175	1.300	3.250	14.250	22.100	12.500	9.150	5.150	5.500	11.250	4.625	10.000	42° 0'
5	- 10	- 8	5.250	1.412	3.562	14.400	22.250	12.750	9.375	5.270	5.600	11.300	4.850	10.200	43° 0'
6	- 9	- 7	6.120	1.500	3.600	15.125	22.300	12.870	9.512	5.620	5.200	11.400	4.910	9.800	41° 45'

C. Refer to the drawing below and program dimension problems 7, 8, and 9 using

a. absolute dimensioning

b. incremental dimensioning

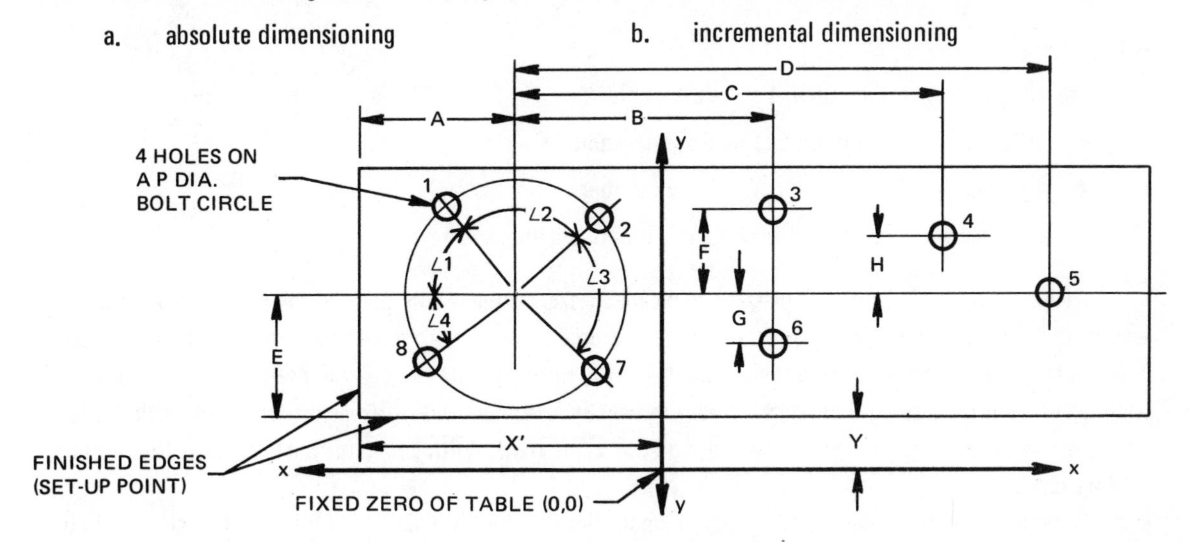

Problem	X'	Y'	Hole Location Dimensions A	B	C	D	E	F	G	H	P Dia.	∠1	∠2	∠3	∠4
7	- 18	5	10.185	13.700	19.215	26.750	7.500	5.750	3.170	4.250	12.200	75° 45'	55° 30'	95° 15'	20° 10'
8	- 20	5	10.520	13.975	19.504	27.315	7.615	5.942	2.623	4.514	12.400	77° 10'	57° 15'	93° 25'	15° 0'
9	- 20	4	9.815	12.750	18.917	26.070	7.125	5.340	2.975	3.984	11.200	72° 40'	61° 30'	98° 50'	18° 20'

Unit 48 Binary Numeration System and Binary Coded Tape

OBJECTIVES

After studying this unit, the student should be able to

- Convert binary numbers to decimal numbers.
- Convert decimal numbers to binary numbers.
- Code numerical control tape using a binary-decimal system.

After the numerical control programming manuscript is made, the data on the program manuscript is transferred to tape. The signals which control the operation of the machine are received from the tape. Numerical control tape is either punched or magnetic; both operate basically the same. Punched tapes are supplied in rolls of varying diameters which are usually one inch wide. Holes are punched in the tape in rows and columns (tracks). Normally the tape has eight tracks with one additional column for the tape-feeding sprockets.

A hole punched in the tape signals an open circuit; the absence of a hole signals a closed circuit. Because a circuit is either open or closed (ON or OFF), it can be represented by either of two digits: 0 to represent OFF and 1 to represent ON. By the use of a sufficient number of circuits, combinations of these two digits can be used to represent any number.

The mathematical system which uses only the digits 1 and 0 is called the *binary numeration system.* An understanding of the structure of the decimal system is helpful in discussing the binary system.

STRUCTURE OF THE DECIMAL SYSTEM

The elements of a mathematical system are the base of the system, the particular digits used, and the locations of the digits with respect to the decimal point (place value).

Examples of the Structure of Decimal Numbers:

16 = 1 ten + 6 units

216 = 2 hundreds + 1 ten + 6 units

4,216 = 4 thousands + 2 hundreds + 1 ten + 6 units

64,216 = 6 ten-thousands + 4 thousands + 2 hundreds + 1 ten + 6 units

All numbers are combinations of the digits 0 - 9. Each place value is ten times greater than the place value directly to its right.

The decimal system is built on powers of the base 10. Since any number with an exponent of 0 equals 1, 10^0 equals 1. An analysis of the number 64,216 shows this structure.

6	4	2	1	6	
10^4 = 10,000	10^3 = 1,000	10^2 = 100	10^1 = 10	10^0 = 1	60,000 4,000
6×10^4 = $6 \times 10{,}000$ = 60,000	4×10^3 = $4 \times 1{,}000$ = 4,000	2×10^2 = 2×100 = 200	1×10^1 = 1×10 = 10	6×10^0 = 6×1 = 6	200 10 + 6
60,000 +	4,000 +	200 +	10 +	6 =	64,216

The same principles of structure hold true for numbers that are less than one. A number with a negative exponent is equal to its positive reciprocal. When the number is inverted and the negative exponent changed to a positive exponent, the result is as follows.

$$10^{-1} = \frac{1}{10^1} = .1, \quad 10^{-2} = \frac{1}{10^2} = \frac{1}{100} = .01,$$

$$10^{-3} = \frac{1}{10^3} = \frac{1}{1000} = .001,$$

$$10^{-4} = \frac{1}{10^4} = \frac{1}{10,000} = .0001$$

An analysis of the number .8502 shows this structure again.

8	5	0	2	
10^{-1} = .1	10^{-2} = .01	10^{-3} = .001	10^{-4} = .0001	
8×10^{-1} = $8 \times .1$ = .8	5×10^{-2} = $5 \times .01$ = .05	0×10^{-3} = $0 \times .001$ = 0	2×10^{-4} = $2 \times .0001$ = .0002	.8 .05 .000 + .0002
.8 +	.05 +	0 +	.0002 =	.8502

BINARY SYSTEM

The same principles of structure apply to the binary system as to the decimal system. The binary system is built upon the base 2 and uses only the digits 0 and 1. As with the decimal system, the elements which must be considered are the base, the particular digits used, and the place value of the digits.

Since the binary system is built on the powers of the base 2, each place value is twice as large as the place value directly to its right.

Examples of the Structure of Binary Numbers

2^6	2^5	2^4	2^3	2^2	2^1	2^0	2^{-1}	2^{-2}	2^{-3}	2^{-4}
2 x 2 x	2 x 2 x	2 x 2 x	2 x 2 x	2 x 2 =	2 x 1 =	1	$\frac{1}{2}$ =	$\frac{1}{2^2}$ =	$\frac{1}{2^3}$ =	$\frac{1}{2^4}$ =
2 x 2 x	2 x 2 x	2 x 2 =	2 =	4	2		.5	$\frac{1}{2 \times 2}$ =	$\frac{1}{2 \times 2 \times 2}$ =	$\frac{1}{2 \times 2 \times 2 \times 2}$
2 x 2 =	2 =	16	8					$\frac{1}{4}$ =	$\frac{1}{8}$ =	= $\frac{1}{16}$ =
64	32							.25	.125	.0625

Numbers are shown as binary numbers by putting a 2 to the right and below the number (subscript) as shown; 11_2, 100_2, 1_2, 10001_2 are binary numbers.

Numbers in the decimal system are usually shown without a subscript. It is understood the number is in the decimal system. In certain instances, for clarity, decimal numbers are shown with the subscript 10.

CHANGING BINARY NUMBERS TO DECIMAL NUMBERS

The following examples show the method of converting binary numbers to equivalent decimal numbers. Remember that 0 and 1 are the only digits in the binary system.

▲ 11_2 means $1(2^1) + 1(2^0)$

$11_2 = 2 + 1 = 3_{10}$

▲ 111_2 means $1(2^2) + 1(2^1) + 1(2^0)$

$111_2 = 4 + 2 + 1 = 7_{10}$

▲ 11101_2 means $1(2^4) + 1(2^3) + 1(2^2) + 0(2^1) + 1(2^0)$

$11101_2 = 16 + 8 + 4 + 0 + 1 = 29_{10}$

▲ 101.11_2 means $1(2^2) + 0(2^1) + 1(2^0) + 1(2^{-1}) + 1(2^{-2})$

$101.11_2 = 4 + 0 + 1 + .5 + .25 = 5.75_{10}$

CHANGING DECIMAL NUMBERS TO BINARY NUMBERS

The following examples show the method of converting decimal numbers to equivalent binary numbers.

1. Change 25_{10} to the binary number equivalent:

▲ Determine the largest power of 2 in 25; $2^4 = 16$. There is one 2^4. Subtract 16 from 25; $25 - 16 = 9$.

▲ Determine the largest power of 2 in 9; $2^3 = 8$. There is one 2^3. Subtract 8 from 9; $9 - 8 = 1$.

▲ Determine the largest power of 2 in 1; $2^0 = 1$. There is one 2^0. Subtract 1 from 1; $1 - 1 = 0$. There is no remainder.

▲ Note that although there are no 2^2 and 2^1, the place positions for these values must be shown as zeros:

$25_{10} = 1(2^4) + 1(2^3) + 0(2^2) + 0(2^1) + 1(2^0)$

$25_{10} = 1 \quad 1 \quad 0 \quad 0 \quad 1$

$25_{10} = 11001_2$

2. Change 11.625_{10} to binary number equivalent:

▲ $2^3 = 8$; $11.625 - 8 = 3.625$

▲ $2^1 = 2$; $3.625 - 2 = 1.625$

▲ $2^0 = 1$; $1.625 - 1 = .625$

▲ $2^{-1} = 1/2 = .5$; $.625 - .5 = .125$

▲ $2^{-3} = 1/8 = .125$; $.125 - .125 = 0$

▲ Note that there are no 2^2 and 2^{-2}

$11.625_{10} = 1(2^3) + 0(2^2) + 1(2^1) + 1(2^0) + 1(2^{-1}) + 0(2^{-2}) + 1(2^{-3})$

$11.625_{10} = 1 \quad 0 \quad 1 \quad 1 \quad 1 \quad 0 \quad 1$

$11.625_{10} = 1011.101_2$

BINARY CODED TAPE

A hole in the numerical control tape indicates an open circuit and represents a 1 in the binary system. The absence of a hole indicates a closed circuit and represents a 0 in the binary system. The tracks of the tape determine place value.

Figure 48-1 shows a simplified binary coded tape and its relationship to equivalent decimal and binary numbers. In actual practice the coding on the tape is more involved with codings such as the sequence of operations, direction of machine table movement, spindle movement, and the feeds and speeds. This tape shows the coding for the decimal numbers 12, 7, 17, 23, 5, 31, 16, 15, and 1.

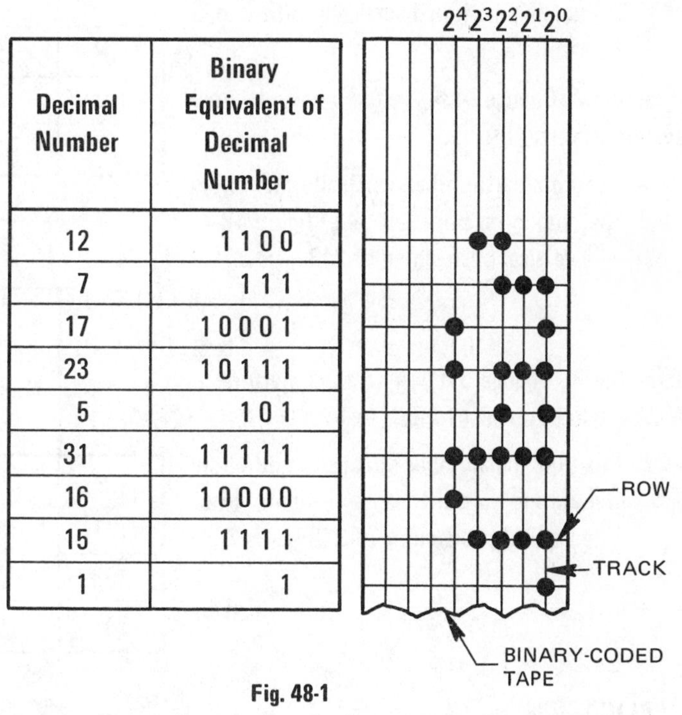

Decimal Number	Binary Equivalent of Decimal Number
12	1100
7	111
17	10001
23	10111
5	101
31	11111
16	10000
15	1111
1	1

Fig. 48-1

BINARY-DECIMAL SYSTEM

Most numerical control machines use a binary-decimal system rather than the pure binary system that has been discussed. In a pure binary system a large number of places are required to represent numbers.

For example, the decimal number 3173 when converted to the binary system consists of a 2^{11}, 2^{10}, 2^6, 2^5, 2^2, and a 2^0; in binary form it appears as 110001100101. To code this number on tape in pure binary form, 12 tracks are required as shown in figure 48-2.

	2^{11}	2^{10}	2^9	2^8	2^7	2^6	2^5	2^4	2^3	2^2	2^1	2^0
Binary Form	1	1	0	0	0	1	1	0	0	1	0	1
Coding on Punched Tape	•	•				•	•			•		•

Fig. 48-2

The binary-decimal system eliminates the need for a large number of tracks. It uses the decimal system for place locations, but converts each digit of the decimal number into a binary number.

Example 1: Change 243_{10} to binary-decimal system.

	10^2	10^1	10^0
Decimal System	2	4	3
Binary-Decimal System	10	100	11

Fig. 48-3

▲ Figure 48-3 shows the binary-decimal equivalent, 10 100 11, of 243_{10} in its horizontal form. In the actual coding of the numerical control tape, the number should be positioned vertically rather than horizontally.

Example 2: Change 243_{10} to the binary-decimal system in vertical form.

▲ Position the number vertically and show the coding on punched tape, figure 48-4. The vertical positioning of 243_{10} is 2
4
3

	Decimal Number	Vertical Binary-Decimal Number
10^2	2	10
10^1	4	100
10^0	3	11

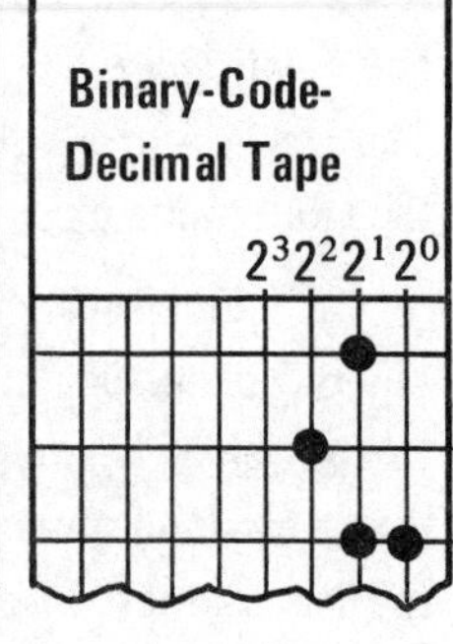

Fig. 48-4

Example 3: Change 3109_{10} to the binary-decimal system in vertical form.

▲ Position the number vertically and show the coding on punched tape, figure 48-5. The vertical positioning of 3109_{10} is 3
1
0
9

	Decimal Number	Vertical Binary-Decimal Number
10^3	3	11
10^2	1	1
10^1	0	0
10^0	9	1001

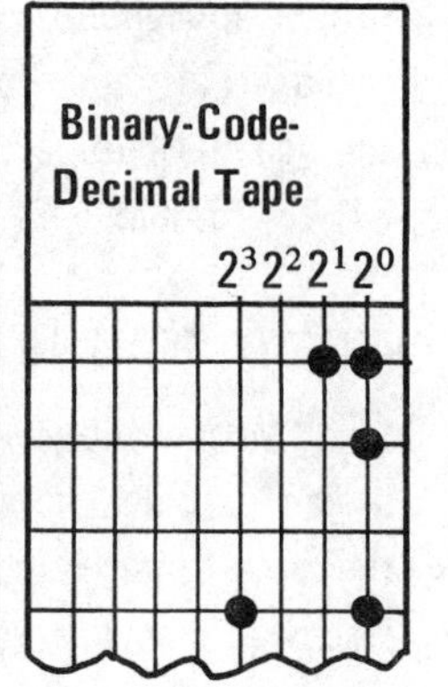

Fig. 48-5

APPLICATION

A. Analyze the following numbers using the method shown in the section on the structure of the decimal system.

1. 285
2. 2855
3. 90500
4. .704
5. 23.023
6. 105.009
7. 4751.107
8. 3006.0204
9. 175.0753

B. Change the following binary numbers to decimal numbers.

1. 10_2 ____
2. 1_2 ____
3. 100_2 ____
4. 1011_2 ____
5. 1101_2 ____
6. 1111_2 ____
7. 10100_2 ____
8. 1001_2 ____
9. 11000_2 ____
10. 10101_2 ____
11. 101010_2 ____
12. 110011_2 ____
13. 111010_2 ____
14. $.1_2$ ____
15. $.101_2$ ____
16. 11.11_2 ____
17. 11.01_2 ____
18. 10.001_2 ____
19. 1111.11_2 ____
20. 1001.0101_2 ____
21. 10011.1001_2 ____

C. Change the following decimal numbers to binary numbers.

1. 12 ____
2. 100 ____
3. 87 ____
4. 26 ____
5. 43 ____
6. 4 ____
7. 102 ____
8. 98 ____
9. 1 ____
10. 7 ____
11. 51 ____
12. 270 ____
13. .5 ____
14. .125 ____
15. .375 ____
16. 10.5 ____
17. 81.75 ____
18. 19.0625 ____
19. 111.25 ____
20. 1.125 ____
21. 163.875 ____

D. Convert the given decimal numbers to binary-decimal numbers in vertical form as shown in figures 48-4 and 48-5. Code the tape for each number. (The solution to the first problem is shown.)

		DECIMAL NUMBER	VERTICAL BINARY-DECIMAL NUMBER	BINARY-CODE-DECIMAL TAPE $2^3 2^2 2^1 2^0$
1. 2,403	10^3	2	10	
	10^2	4	100	
	10^1	0	0	
	10^0	3	11	
2. 3,173	10^3			
	10^2			
	10^1			
	10^0			
3. 9,157	10^3			
	10^2			
	10^1			
	10^0			
4. 803	10^2			
	10^1			
	10^0			
5. 736	10^2			
	10^1			
	10^0			
6. 74,932	10^4			
	10^3			
	10^2			
	10^1			
	10^0			
7. 87	10^1			
	10^0			
8. 1,029	10^3			
	10^2			
	10^1			
	10^0			

		DECIMAL NUMBER	VERTICAL BINARY-DECIMAL NUMBER	BINARY-CODE-DECIMAL TAPE $2^3 2^2 2^1 2^0$
9. 5,005	10^3			
	10^2			
	10^1			
	10^0			
10. 8,321	10^3			
	10^2			
	10^1			
	10^0			
11. 1,000	10^3			
	10^2			
	10^1			
	10^0			
12. 429	10^2			
	10^1			
	10^0			
13. 202	10^2			
	10^1			
	10^0			
14. 60,070	10^4			
	10^3			
	10^2			
	10^1			
	10^0			
15. 24	10^1			
	10^0			
16. 2,097	10^3			
	10^2			
	10^1			
	10^0			

APPENDIX

TABLE I DECIMAL EQUIVALENTS

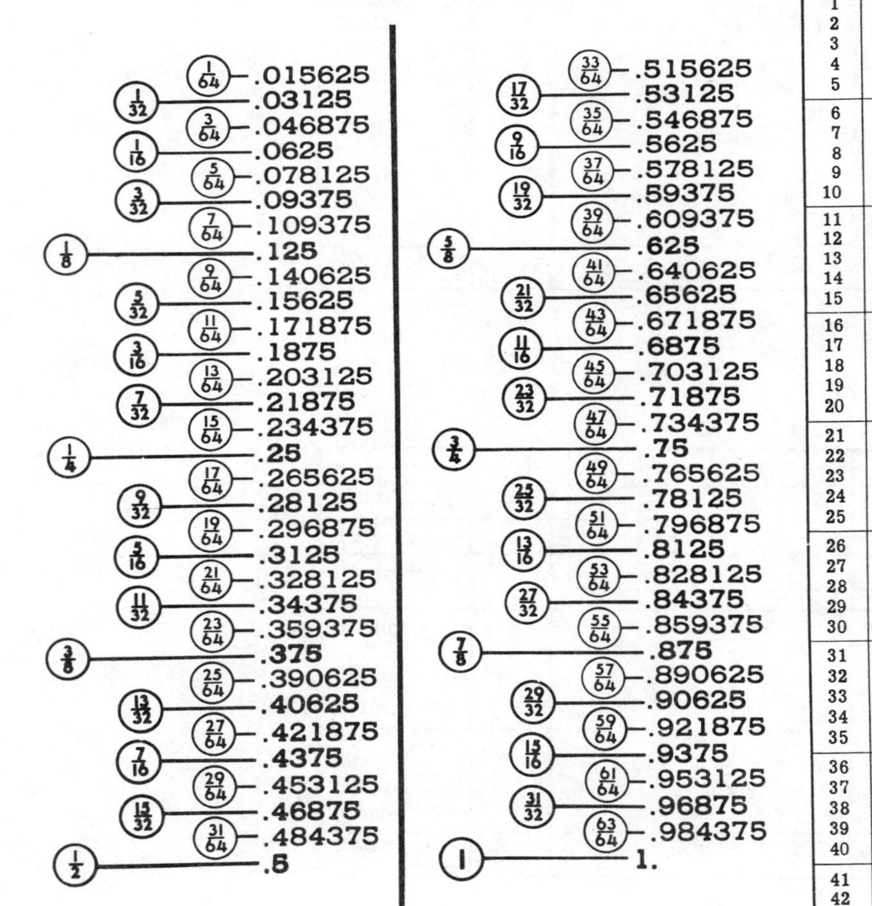

Fraction	Decimal	Fraction	Decimal
1/64	.015625	33/64	.515625
1/32	.03125	17/32	.53125
3/64	.046875	35/64	.546875
1/16	.0625	9/16	.5625
5/64	.078125	37/64	.578125
3/32	.09375	19/32	.59375
7/64	.109375	39/64	.609375
1/8	.125	5/8	.625
9/64	.140625	41/64	.640625
5/32	.15625	21/32	.65625
11/64	.171875	43/64	.671875
3/16	.1875	11/16	.6875
13/64	.203125	45/64	.703125
7/32	.21875	23/32	.71875
15/64	.234375	47/64	.734375
1/4	.25	3/4	.75
17/64	.265625	49/64	.765625
9/32	.28125	25/32	.78125
19/64	.296875	51/64	.796875
5/16	.3125	13/16	.8125
21/64	.328125	53/64	.828125
11/32	.34375	27/32	.84375
23/64	.359375	55/64	.859375
3/8	.375	7/8	.875
25/64	.390625	57/64	.890625
13/32	.40625	29/32	.90625
27/64	.421875	59/64	.921875
7/16	.4375	15/16	.9375
29/64	.453125	61/64	.953125
15/32	.46875	31/32	.96875
31/64	.484375	63/64	.984375
1/2	.5	1	1.

TABLE II POWERS AND ROOTS OF NUMBERS (1 through 100)

Number	Powers Square	Powers Cube	Roots Square	Roots Cube	Number	Powers Square	Powers Cube	Roots Square	Roots Cube
1	1	1	1.000	1.000	51	2,601	132,651	7.141	3.708
2	4	8	1.414	1.260	52	2,704	140,608	7.211	3.733
3	9	27	1.732	1.442	53	2,809	148,877	7.280	3.756
4	16	64	2.000	1.587	54	2,916	157,464	7.348	3.780
5	25	125	2.236	1.710	55	3,025	166,375	7.416	3.803
6	36	216	2.449	1.817	56	3,136	175,616	7.483	3.826
7	49	343	2.646	1.913	57	3,249	185,193	7.550	3.849
8	64	512	2.828	2.000	58	3,364	195,112	7.616	3.871
9	81	729	3.000	2.080	59	3,481	205,379	7.681	3.893
10	100	1,000	3.162	2.154	60	3,600	216,000	7.746	3.915
11	121	1,331	3.317	2.224	61	3,721	226,981	7.810	3.936
12	144	1,728	3.464	2.289	62	3,844	238,328	7.874	3.958
13	169	2,197	3.606	2.351	63	3,969	250,047	7.937	3.979
14	196	2,744	3.742	2.410	64	4,096	262,144	8.000	4.000
15	225	3,375	3.873	2.466	65	4,225	274,625	8.062	4.021
16	256	4,096	4.000	2.520	66	4,356	287,496	8.124	4.041
17	289	4,913	4.123	2.571	67	4,489	300,763	8.185	4.062
18	324	5,832	4.243	2.621	68	4,624	314,432	8.246	4.082
19	361	6,859	4.359	2.668	69	4,761	328,509	8.307	4.102
20	400	8,000	4.472	2.714	70	4,900	343,000	8.367	4.121
21	441	9,261	4.583	2.759	71	5,041	357,911	8.426	4.141
22	484	10,648	4.690	2.802	72	5,184	373,248	8.485	4.160
23	529	12,167	4.796	2.844	73	5,329	389,017	8.544	4.179
24	576	13,824	4.899	2.884	74	5,476	405,224	8.602	4.198
25	625	15,625	5.000	2.924	75	5,625	421,875	8.660	4.217
26	676	17,576	5.099	2.962	76	5,776	438,976	8.718	4.236
27	729	19,683	5.196	3.000	77	5,929	456,533	8.775	4.254
28	784	21,952	5.292	3.037	78	6,084	474,552	8.832	4.273
29	841	24,389	5.385	3.072	79	6,241	493,039	8.888	4.291
30	900	27,000	5.477	3.107	80	6,400	512,000	8.944	4.309
31	961	29,791	5.568	3.141	81	6,561	531,441	9.000	4.327
32	1,024	32,798	5.657	3.175	82	6,724	551,368	9.055	4.344
33	1,089	35,937	5.745	3.208	83	6,889	571,787	9.110	4.362
34	1,156	39,304	5.831	3.240	84	7,056	592,704	9.165	4.380
35	1,225	42,875	5.916	3.271	85	7,225	614,125	9.220	4.397
36	1,296	46,656	6.000	3.302	86	7,396	636,056	9.274	4.414
37	1,369	50,653	6.083	3.332	87	7,569	658,503	9.327	4.481
38	1,444	54,872	6.164	3.362	88	7,744	681,472	9.381	4.448
39	1,521	59,319	6.245	3.391	89	7,921	704,969	9.434	4.465
40	1,600	64,000	6.325	3.420	90	8,100	729,000	9.487	4.481
41	1,681	68,921	6.403	3.448	91	8,281	753,571	9.539	4.498
42	1,764	74,088	6.481	3.476	92	8,464	778,688	9.592	4.514
43	1,849	79,507	6.557	3.503	93	8,649	804,357	9.644	4.531
44	1,936	85,184	6.633	3.530	94	8,836	830,584	9.695	4.547
45	2,025	91,125	6.708	3.557	95	9,025	857,375	9.747	4.563
46	2,116	97,336	6.782	3.583	96	9,216	884,736	9.798	4.579
47	2,209	103,823	6.856	3.609	97	9,409	912,673	9.849	4.595
48	2,304	110,592	6.928	3.634	98	9,604	941,192	9.900	4.610
49	2,401	117,649	7.000	3.659	99	9,801	970,299	9.950	4.626
50	2,500	125,000	7.071	3.684	100	10,000	1,000,000	10.000	4.642

TABLE III TRIGONOMETRIC FUNCTIONS

Angles	Sin	Cos	Tan	Cot	Sec	Csc	Angles
0° 00′	.00000	1.0000	.00000	infinite	1.0000	infinite	90° 00′
10′	.00291	1.0000	.00291	343.77	1.0000	343.77	50′
20′	.00582	.99998	.00582	171.88	1.0000	171.89	40′
30′	.00873	.99996	.00873	114.59	1.0000	114.59	30′
40′	.01164	.99993	.01164	85.940	1.0001	85.946	20′
50′	.01454	.99989	.01455	68.750	1.0001	68.757	10′
1° 00′	.01745	.99985	.01746	57.290	1.0001	57.299	89° 00′
10′	.02036	.99979	.02036	49.104	1.0002	49.114	50′
20′	.02327	.99973	.02328	42.964	1.0003	42.976	40′
30′	.02618	.99966	.02619	38.189	1.0003	38.201	30′
40′	.02908	.99958	.02910	34.368	1.0004	34.382	20′
50′	.03199	.99949	.03201	31.242	1.0005	31.257	10′
2° 00′	.03490	.99939	.03492	28.636	1.0006	28.654	88° 00′
10′	.03781	.99929	.03783	26.432	1.0007	26.450	50′
20′	.04071	.99917	.04075	24.542	1.0008	24.562	40′
30′	.04362	.99905	.04366	22.904	1.0009	22.925	30′
40′	.04653	.99892	.04658	21.470	1.0011	21.494	20′
50′	.04943	.99878	.04949	20.206	1.0012	20.230	10′
3° 00′	.05234	.99863	.05241	19.081	1.0014	19.107	87° 00′
10′	.05524	.99847	.05533	18.075	1.0015	18.103	50′
20′	.05814	.99831	.05824	17.169	1.0017	17.198	40′
30′	.06105	.99813	.06116	16.350	1.0019	16.380	30′
40′	.06395	.99795	.06408	15.605	1.0020	15.637	20′
50′	.06685	.99776	.06700	14.924	1.0022	14.958	10′
4° 00′	.06976	.99756	.06993	14.301	1.0024	14.335	86° 00′
10′	.07266	.99736	.07285	13.727	1.0026	13.763	50′
20′	.07556	.99714	.07578	13.197	1.0029	13.235	40′
30′	.07846	.99692	.07870	12.706	1.0031	12.745	30′
40′	.08136	.99668	.08163	12.251	1.0033	12.291	20′
50′	.08426	.99644	.08456	11.826	1.0036	11.868	10′
Angles	Cos	Sin	Cot	Tan	Csc	Sec	Angles

Angles	Sin	Cos	Tan	Cot	Sec	Csc	Angles
5° 00′	.08716	.99619	.08749	11.430	1.0038	11.474	85° 00′
10′	.09005	.99594	.09042	11.059	1.0041	11.104	50′
20′	.09295	.99567	.09335	10.712	1.0043	10.758	40′
30′	.09585	.99540	.09629	10.385	1.0046	10.433	30′
40′	.09874	.99511	.09923	10.078	1.0049	10.127	20′
50′	.10164	.99482	.10216	9.7882	1.0052	9.8391	10′
6° 00′	.10453	.99452	.10510	9.5144	1.0055	9.5668	84° 00′
10′	.10742	.99421	.10805	9.2553	1.0058	9.3092	50′
20′	.11031	.99390	.11099	9.0098	1.0061	9.0651	40′
30′	.11320	.99357	.11394	8.7769	1.0065	8.8337	30′
40′	.11609	.99324	.11688	8.5556	1.0068	8.6138	20′
50′	.11898	.99290	.11983	8.3450	1.0071	8.4046	10′
7° 00′	.12187	.99255	.12278	8.1444	1.0075	8.2055	83° 00′
10′	.12476	.99219	.12574	7.9530	1.0079	8.0156	50′
20′	.12764	.99182	.12869	7.7704	1.0082	7.8344	40′
30′	.13053	.99144	.13165	7.5958	1.0086	7.6613	30′
40′	.13341	.99106	.13461	7.4287	1.0090	7.4957	20′
50′	.13629	.99067	.13758	7.2687	1.0094	7.3372	10′
8° 00′	.13917	.99027	.14054	7.1154	1.0098	7.1853	82° 00′
10′	.14205	.98986	.14351	6.9682	1.0102	7.0396	50′
20′	.14493	.98944	.14648	6.8269	1.0107	6.8998	40′
30′	.14781	.98902	.14945	6.6912	1.0111	6.7655	30′
40′	.15069	.98858	.15243	6.5606	1.0115	6.6363	20′
50′	.15356	.98814	.15540	6.4348	1.0120	6.5121	10′
9° 00′	.15643	.98769	.15838	6.3138	1.0125	6.3924	81° 00′
10′	.15931	.98723	.16137	6.1970	1.0129	6.2772	50′
20′	.16218	.98676	.16435	6.0844	1.0134	6.1661	40′
30′	.16505	.98629	.16734	5.9758	1.0139	6.0588	30′
40′	.16792	.98580	.17033	5.8708	1.0144	5.9554	20′
50′	.17078	.98531	.17333	5.7694	1.0149	5.8554	10′
Angles	Cos	Sin	Cot	Tan	Csc	Sec	Angles

TABLE III TRIGONOMETRIC FUNCTIONS (Cont'd)

Angles	Sin	Cos	Tan	Cot	Sec	Csc	Angles
10° 00′	.17365	.98481	.17633	5.6713	1.0154	5.7588	80° 00′
10′	.17651	.98430	.17933	5.5764	1.0159	5.6653	50′
20′	.17937	.98378	.18233	5.4845	1.0165	5.5749	40′
30′	.18224	.98325	.18534	5.3955	1.0170	5.4874	30′
40′	.18509	.98272	.18835	5.3093	1.0176	5.4026	20′
50′	.18795	.98218	.19136	5.2257	1.0181	5.3205	10′
11° 00′	.19081	.98163	.19438	5.1446	1.0187	5.2408	79° 00′
10′	.19366	.98107	.19740	5.0658	1.0193	5.1636	50′
20′	.19652	.98050	.20042	4.9894	1.0199	5.0886	40′
30′	.19937	.97992	.20345	4.9152	1.0205	5.0158	30′
40′	.20222	.97934	.20648	4.8430	1.0211	4.9452	20′
50′	.20507	.97875	.20952	4.7729	1.0217	4.8765	10′
12° 00′	.20791	.97815	.21256	4.7046	1.0223	4.8097	78° 00′
10′	.21076	.97754	.21560	4.6383	1.0230	4.7448	50′
20′	.21360	.97692	.21864	4.5736	1.0236	4.6817	40′
30′	.21644	.97630	.22169	4.5107	1.0243	4.6202	30′
40′	.21928	.97566	.22475	4.4494	1.0249	4.5604	20′
50′	.22212	.97502	.22781	4.3897	1.0256	4.5021	10′
13° 00′	.22495	.97437	.23087	4.3315	1.0263	4.4454	77° 00′
10′	.22778	.97371	.23393	4.2747	1.0270	4.3910	50′
20′	.23062	.97304	.23700	4.2193	1.0277	4.3362	40′
30′	.23345	.97237	.24008	4.1653	1.0284	4.2836	30′
40′	.23627	.97169	.24316	4.1126	1.0291	4.2324	20′
50′	.23910	.97100	.24624	4.0611	1.0299	4.1824	10′
14° 00′	.24192	.97030	.24933	4.0108	1.0306	4.1336	76° 00′
10′	.24474	.96959	.25242	3.9617	1.0314	4.0859	50′
20′	.24756	.96887	.25552	3.9136	1.0321	4.0394	40′
30′	.25038	.96815	.25862	3.8667	1.0329	3.9939	30′
40′	.25320	.96742	.26172	3.8208	1.0337	3.9495	20′
50′	.25601	.96667	.26483	3.7760	1.0345	3.9061	10′
Angles	Cos	Sin	Cot	Tan	Csc	Sec	Angles

Angles	Sin	Cos	Tan	Cot	Sec	Csc	Angles
15° 00′	.25882	.96593	.26795	3.7321	1.0353	3.8637	75° 00′
10′	.26163	.96517	.27107	3.6891	1.0361	3.8222	50′
20′	.26443	.96440	.27419	3.6471	1.0369	3.7816	40′
30′	.26724	.96363	.27732	3.6059	1.0377	3.7420	30′
40′	.27004	.96285	.28046	3.5656	1.0386	3.7031	20′
50′	.27284	.96206	.28360	3.5261	1.0394	3.6651	10′
16° 00′	.27564	.96126	.28675	3.4874	1.0403	3.6279	74° 00′
10′	.27843	.96046	.28990	3.4495	1.0412	3.5915	50′
20′	.28123	.95964	.29305	3.4124	1.0420	3.5559	40′
30′	.28402	.95882	.29621	3.3759	1.0429	3.5209	30′
40′	.28680	.95799	.29938	3.3402	1.0438	3.4867	20′
50′	.28959	.95715	.30255	3.3052	1.0448	3.4532	10′
17° 00′	.29237	.95630	.30573	3.2709	1.0457	3.4203	73° 00′
10′	.29515	.95545	.30891	3.2371	1.0466	3.3881	50′
20′	.29793	.95459	.31210	3.2041	1.0476	3.3565	40′
30′	.30071	.95372	.31530	3.1716	1.0485	3.3255	30′
40′	.30348	.95284	.31850	3.1397	1.0495	3.2951	20′
50′	.30625	.95195	.32171	3.1084	1.0505	3.2653	10′
18° 00′	.30902	.95106	.32492	3.0777	1.0515	3.2361	72° 00′
10′	.31178	.95015	.32814	3.0475	1.0525	3.2074	50′
20′	.31454	.94924	.33136	3.0178	1.0535	3.1792	40′
30′	.31730	.94832	.33460	2.9887	1.0545	3.1515	30′
40′	.32006	.94740	.33783	2.9600	1.0555	3.1244	20′
50′	.32282	.94646	.34108	2.9319	1.0566	3.0977	10′
19° 00′	.32557	.94552	.34433	2.9042	1.0576	3.0715	71° 00′
10′	.32832	.94457	.34758	2.8770	1.0587	3.0458	50′
20′	.33106	.94361	.35085	2.8502	1.0598	3.0206	40′
30′	.33381	.94264	.35412	2.8239	1.0608	2.9957	30′
40′	.33655	.94167	.35740	2.7980	1.0619	2.9713	20′
50′	.33929	.94068	.36068	2.7725	1.0630	2.9474	10′
Angles	Cos	Sin	Cot	Tan	Csc	Sec	Angles

TABLE III TRIGONOMETRIC FUNCTIONS (Cont'd)

Angles	Sin	Cos	Tan	Cot	Sec	Csc	Angles
20° 00′	.34202	.93969	.36397	2.7475	1.0642	2.9238	70° 00′
10′	.34475	.93869	.36727	2.7228	1.0653	2.9006	50′
20′	.34748	.93769	.37057	2.6985	1.0664	2.8778	40′
30′	.35021	.93667	.37388	2.6746	1.0676	2.8554	30′
40′	.35293	.93565	.37720	2.6511	1.0688	2.8334	20′
50′	.35565	.93462	.38053	2.6279	1.0699	2.8117	10′
21° 00′	.35837	.93358	.38386	2.6051	1.0711	2.7904	69° 00′
10′	.36108	.93253	.38721	2.5826	1.0723	2.7694	50′
20′	.36379	.93148	.39055	2.5605	1.0736	2.7488	40′
30′	.36650	.93042	.39391	2.5387	1.0748	2.7285	30′
40′	.36921	.92935	.39727	2.5172	1.0760	2.7085	20′
50′	.37191	.92827	.40065	2.4960	1.0773	2.6888	10′
22° 00′	.37461	.92718	.40403	2.4751	1.0785	2.6695	68° 00′
10′	.37730	.92609	.40741	2.4545	1.0798	2.6504	50′
20′	.37999	.92499	.41081	2.4342	1.0811	2.6316	40′
30′	.38268	.92388	.41421	2.4142	1.0824	2.6131	30′
40′	.38537	.92276	.41763	2.3945	1.0837	2.5949	20′
50′	.38805	.92164	.42105	2.3750	1.0850	2.5770	10′
23° 00′	.39073	.92050	.42447	2.3559	1.0864	2.5593	67° 00′
10′	.39341	.91936	.42791	2.3369	1.0877	2.5419	50′
20′	.39608	.91822	.43136	2.3183	1.0891	2.5247	40′
30′	.39875	.91706	.43481	2.2998	1.0904	2.5078	30′
40′	.40141	.91590	.43828	2.2817	1.0918	2.4912	20′
50′	.40408	.91472	.44175	2.2637	1.0932	2.4748	10′
24° 00′	.40674	.91355	.44523	2.2460	1.0946	2.4586	66° 00′
10′	.40939	.91236	.44872	2.2286	1.0961	2.4426	50′
20′	.41204	.91116	.45222	2.2113	1.0975	2.4269	40′
30′	.41469	.90996	.45573	2.1943	1.0989	2.4114	30′
40′	.41734	.90875	.45924	2.1775	1.1004	2.3961	20′
50′	.41998	.90753	.46277	2.1609	1.1019	2.3811	10′
Angles	Cos	Sin	Cot	Tan	Csc	Sec	Angles

Angles	Sin	Cos	Tan	Cot	Sec	Csc	Angles
25° 00′	.42262	.90631	.46631	2.1445	1.1034	2.3662	65° 00′
10′	.42525	.90507	.46985	2.1283	1.1049	2.3515	50′
20′	.42788	.90383	.47341	2.1123	1.1064	2.3371	40′
30′	.43051	.90259	.47698	2.0965	1.1079	2.3228	30′
40′	.43313	.90133	.48055	2.0809	1.1095	2.3087	20′
50′	.43575	.90007	.48414	2.0655	1.1110	2.2949	10′
26° 00′	.43837	.89879	.48773	2.0503	1.1126	2.2812	64° 00′
10′	.44098	.89752	.49134	2.0353	1.1142	2.2676	50′
20′	.44359	.89623	.49495	2.0204	1.1158	2.2543	40′
30′	.44620	.89493	.49858	2.0057	1.1174	2.2411	30′
40′	.44880	.89363	.50222	1.9912	1.1190	2.2282	20′
50′	.45140	.89232	.50587	1.9768	1.1207	2.2153	10′
27° 00′	.45399	.89101	.50953	1.9626	1.1223	2.2027	63° 00′
10′	.45658	.88968	.51319	1.9486	1.1240	2.1902	50′
20′	.45917	.88835	.51688	1.9347	1.1257	2.1778	40′
30′	.46175	.88701	.52057	1.9210	1.1274	2.1657	30′
40′	.46433	.88566	.52427	1.9074	1.1291	2.1536	20′
50′	.46690	.88431	.52798	1.8940	1.1308	2.1418	10′
28° 00′	.46947	.88295	.53171	1.8807	1.1326	2.1300	62° 00′
10′	.47204	.88158	.53545	1.8676	1.1343	2.1185	50′
20′	.47460	.88020	.53920	1.8546	1.1361	2.1070	40′
30′	.47716	.87882	.54296	1.8418	1.1379	2.0957	30′
40′	.47971	.87743	.54673	1.8291	1.1397	2.0846	20′
50′	.48226	.87603	.55051	1.8165	1.1415	2.0735	10′
29° 00′	.48481	.87462	.55431	1.8041	1.1433	2.0627	61° 00′
10′	.48735	.87321	.55812	1.7917	1.1452	2.0519	50′
20′	.48989	.87178	.56194	1.7796	1.1471	2.0413	40′
30′	.49242	.87036	.56577	1.7675	1.1489	2.0308	30′
40′	.49495	.86892	.56962	1.7556	1.1508	2.0204	20′
50′	.49748	.86748	.57348	1.7438	1.1528	2.0101	10′
Angles	Cos	Sin	Cot	Tan	Csc	Sec	Angles

TABLE III TRIGONOMETRIC FUNCTIONS (Cont'd)

Angles	Sin	Cos	Tan	Cot	Sec	Csc	Angles
30° 00′	.50000	.86603	.57735	1.7321	1.1547	2.0000	60° 00′
10′	.50252	.86457	.58124	1.7205	1.1566	1.9900	50′
20′	.50503	.86310	.58513	1.7090	1.1586	1.9801	40′
30′	.50754	.86163	.58904	1.6977	1.1606	1.9703	30′
40′	.51004	.86015	.59297	1.6864	1.1626	1.9606	20′
50′	.51254	.85866	.59691	1.6753	1.1646	1.9510	10′
31° 00′	.51504	.85717	.60086	1.6643	1.1666	1.9416	59° 00′
10′	.51753	.85567	.60483	1.6534	1.1687	1.9322	50′
20′	.52002	.85416	.60881	1.6426	1.1707	1.9230	40′
30′	.52250	.85264	.61280	1.6319	1.1728	1.9139	30′
40′	.52498	.85112	.61681	1.6213	1.1749	1.9048	20′
50′	.52745	.84959	.62083	1.6107	1.1770	1.8959	10′
32° 00′	.52992	.84805	.62487	1.6003	1.1792	1.8871	58° 00′
10′	.53238	.84650	.62892	1.5900	1.1813	1.8783	50′
20′	.53484	.84495	.63299	1.5798	1.1835	1.8697	40′
30′	.53730	.84339	.63707	1.5697	1.1857	1.8611	30′
40′	.53975	.84182	.64117	1.5597	1.1879	1.8527	20′
50′	.54220	.84025	.64528	1.5497	1.1901	1.8443	10′
33° 00′	.54464	.83867	.64941	1.5399	1.1924	1.8361	57° 00′
10′	.54708	.83708	.65355	1.5301	1.1946	1.8279	50′
20′	.54951	.83549	.65771	1.5204	1.1969	1.8198	40′
30′	.55194	.83389	.66189	1.5108	1.1992	1.8118	30′
40′	.55436	.83228	.66608	1.5013	1.2015	1.8039	20′
50′	.55678	.83066	.67028	1.4919	1.2039	1.7960	10′
34° 00′	.55919	.82904	.67451	1.4826	1.2062	1.7883	56° 00′
10′	.56160	.82741	.67875	1.4733	1.2086	1.7806	50′
20′	.56401	.82577	.68301	1.4641	1.2110	1.7730	40′
30′	.56641	.82413	.68728	1.4550	1.2134	1.7655	30′
40′	.56880	.82248	.69157	1.4460	1.2158	1.7581	20′
50′	.57119	.82082	.69588	1.4370	1.2183	1.7507	10′
Angles	Cos	Sin	Cot	Tan	Csc	Sec	Angles

Angles	Sin	Cos	Tan	Cot	Sec	Csc	Angles
35° 00′	.57358	.81915	.70021	1.4282	1.2208	1.7434	55° 00′
10′	.57596	.81748	.70455	1.4193	1.2233	1.7362	50′
20′	.57833	.81580	.70891	1.4106	1.2258	1.7291	40′
30′	.58070	.81412	.71329	1.4020	1.2283	1.7220	30′
40′	.58307	.81242	.71769	1.3934	1.2309	1.7151	20′
50′	.58543	.81072	.72211	1.3848	1.2335	1.7081	10′
36° 00′	.58779	.80902	.72654	1.3764	1.2361	1.7013	54° 00′
10′	.59014	.80730	.73100	1.3680	1.2387	1.6945	50′
20′	.59248	.80558	.73547	1.3597	1.2413	1.6878	40′
30′	.59482	.80386	.73996	1.3514	1.2440	1.6812	30′
40′	.59716	.80212	.74447	1.3432	1.2467	1.6746	20′
50′	.59949	.80038	.74900	1.3351	1.2494	1.6681	10′
37° 00′	.60182	.79864	.75355	1.3270	1.2521	1.6616	53° 00′
10′	.60414	.79688	.75812	1.3190	1.2549	1.6552	50′
20′	.60645	.79512	.76272	1.3111	1.2577	1.6489	40′
30′	.60876	.79335	.76733	1.3032	1.2605	1.6427	30′
40′	.61107	.79158	.77196	1.2954	1.2633	1.6365	20′
50′	.61337	.78980	.77661	1.2876	1.2661	1.6303	10′
38° 00′	.61566	.78801	.78129	1.2799	1.2690	1.6243	52° 00′
10′	.61795	.78622	.78598	1.2723	1.2719	1.6182	50′
20′	.62024	.78442	.79070	1.2647	1.2748	1.6123	40′
30′	.62251	.78261	.79544	1.2572	1.2778	1.6064	30′
40′	.62479	.78079	.80020	1.2497	1.2807	1.6005	20′
50′	.62706	.77897	.80498	1.2423	1.2837	1.5947	10′
39° 00′	.62932	.77715	.80978	1.2349	1.2867	1.5890	51° 00′
10′	.63158	.77531	.81461	1.2276	1.2898	1.5833	50′
20′	.63383	.77347	.81946	1.2203	1.2929	1.5777	40′
30′	.63608	.77162	.82434	1.2131	1.2960	1.5721	30′
40′	.63832	.76977	.82923	1.2059	1.2991	1.5666	20′
50′	.64056	.76791	.83415	1.1988	1.3022	1.5611	10′
Angles	Cos	Sin	Cot	Tan	Csc	Sec	Angles

TABLE III TRIGONOMETRIC FUNCTIONS (Cont'd)

Angles	Sin	Cos	Tan	Cot	Sec	Csc	Angles
40° 00′	.64279	.76604	.83910	1.1918	1.3054	1.5557	50° 00′
10′	.64501	.76417	.84407	1.1847	1.3086	1.5503	50′
20′	.64723	.76229	.84906	1.1778	1.3118	1.5450	40′
30′	.64945	.76041	.85408	1.1709	1.3151	1.5398	30′
40′	.65166	.75851	.85912	1.1640	1.3184	1.5345	20′
50′	.65386	.75661	.86419	1.1572	1.3217	1.5294	10′
41° 00′	.65606	.75471	.86929	1.1504	1.3250	1.5242	49° 00′
10′	.65825	.75280	.87441	1.1436	1.3284	1.5192	50′
20′	.66044	.75088	.87955	1.1369	1.3318	1.5141	40′
30′	.66262	.74896	.88473	1.1303	1.3352	1.5092	30′
40′	.66480	.74703	.88992	1.1237	1.3386	1.5042	20′
50′	.66697	.74509	.89515	1.1171	1.3421	1.4993	10′
42° 00′	.66913	.74314	.90040	1.1106	1.3456	1.4945	48° 00′
10′	.67129	.74120	.90569	1.1041	1.3492	1.4897	50′
20′	.67344	.73924	.91099	1.0977	1.3527	1.4849	40′
30′	.67559	.73728	.91633	1.0913	1.3563	1.4802	30′
40′	.67773	.73531	.92170	1.0850	1.3600	1.4755	20′
50′	.67987	.73333	.92709	1.0786	1.3636	1.4709	10′
43° 00′	.68200	.73135	.93252	1.0724	1.3673	1.4663	47° 00′
10′	.68412	.72937	.93797	1.0661	1.3710	1.4617	50′
20′	.68624	.72737	.94345	1.0599	1.3748	1.4572	40′
30′	.68835	.72537	.94896	1.0538	1.3786	1.4527	30′
40′	.69046	.72337	.95451	1.0477	1.3824	1.4483	20′
50′	.69256	.72136	.96008	1.0416	1.3863	1.4439	10′
44° 00′	.69466	.71934	.96569	1.0355	1.3902	1.4395	46° 00′
10′	.69675	.71732	.97132	1.0295	1.3941	1.4352	50′
20′	.69883	.71529	.97700	1.0236	1.3980	1.4310	40′
30′	.70091	.71325	.98270	1.0176	1.4020	1.4267	30′
40′	.70298	.71121	.98843	1.0117	1.4060	1.4225	20′
50′	.70505	.70916	.99420	1.0058	1.4101	1.4183	10′
45° 00′	.70711	.70711	1.0000	1.0000	1.4142	1.4142	45° 00′
Angles	Cos	Sin	Cot	Tan	Csc	Sec	Angles

Acknowledgments

Publications Director
Alan N. Knofla

Editor-in-Chief
Marjorie Bruce

Source Editor
Elinor Gunnerson

Director of Manufacturing and Production – Delmar
Frederick Sharer

Illustrators
Tony Canabush
Mike Kokernak
Sue Staucet

Production Specialists
Kathy Bottieri
Jean Le Morta
Betty Michelfelder

Contributors
Westinghouse Electric Corporation,
Gas Turbine Systems Division,
Philadelphia, Pa. – Title page photograph

Technical Reviewers
Gail Hendrix
Carolyn LaFevor
Students at State Area Vocational-Technical School, McMinnville, Tennessee

INDEX

A

Abscissa, 224
Absolute dimensions, 257-260
Absolute value, 106
Acute angle, 180
Addendum, 164
Addition
  algebraic operations, 112-113
  decimal fractions, 42
  equation, solution by, 129-130
  fractions, 12-14
  mixed numbers, 14
  signed numbers, 106
  subtractions, combing with, 28
Adjacent side, 217
Algebra
  See also Algebraic expressions; Algebraic operations
  algebraic expressions, 99-101
  arithmetic numbers, 99
  cutting speeds, formulas, 157
  cutting time, formulas, 159
  direct proportion, 152-153
  equations, 130
    addition, solution by, 129-130
    checking, the, 125
    combined operation solution, 141-142
    division solution, 130-131
    introduction to, 121-125
    multiplication solution, 135-136
    powers, 137
    rearranging formulas, 141-142
    roots solution, 136-137
    subtraction solution, 128-129
    transposition, 129
  formulas, 99
  fundamentals of, 99-168
  inverse proportion, 153-154
  literal numbers, 99
  proportion, 147
  ratio, 146-147
  Rpm, formulas, 157-158
  signed numbers, 105-109
  spur gears, application of formulas, 163-166
  symbolism, 99-101
Algebraic expression
  defined, 99
  evaluation of, 100-101
Algebraic operations
  additions, 112-113
  combined operations, 119
  division, 117
  factor, 112
  like terms, 112
  literal factors, 112
  multiplications, 114-115
  numerical coefficient, 112
  parentheses, removal of, 119
  powers, 118
  roots, 118-119
  subtraction, 113-114
  term, 112
  unlike terms, 112
Alternate interior angles, 181
Altitude, 192
Angle cuts, 235
Angle measurement
  complementary, 178
  supplementary, 178
Angles, 170
  acute, 180
  alternate interior, 181
  complementary, 224
  corresponding, 181
  formed by a transversal, 181
  formed inside a circle, 204
  formed on a circle, 205
  formed outside a circle, 205
  obtuse, 180
  right, 180
  straight, 180
Angular measure
  arithmetic operations, 171-172
  decimal degrees, conversion of, 173
  units of, 170
Arc, 198
Arithmetic numbers, 99
Axioms, 169

B

Base angles, 185

Basic dimension, 70
Bevel protractor, 177, 178
Bevels, 234
Bilateral tolerance, 70
Binary coded tape, 267
Binary-decimal system, 267-268
Binary numbers
  changing decimal numbers to, 206
  changing to decimal numbers, 266
Binary systems, 265
Bisect, 192
Bisecting an angle, 212

C

Caliper, 82-84
Cancellation, 20-22
Cartesian coordinate system, 224-225
Central angle, 198
Chord, 198
Circles, 198-201
  angles formed inside a, 204
  angles formed on a, 205
  angles formed outside a, 205
  defined, 198
  tangents to, 235
Circular pitch, 164
Circumference, 198
Clearance, 71-72, 164
Combined operations
  addition and subtraction, 28
  algebraic operations, 119
  equation solution, 141-142
  fractions, 28-30
  mixed numbers, 28-30
  multiplication and division, 29
  order of operations for, 28
Common denominator, 12
Common fractions
  changing decimal fractions, 38-39
  changing to decimal fractions, 37-38
Complementary angles, 178
  functions of, 224
Complex functions, 8, 30
Congruent triangles, 190
Continuous path system, 257
Corresponding angles, 181
Cosecant, 217
Cosine, 217
Cosines, law of, 249-251
Cotangent, 217
Cutting speed
  drill press, 158
  grinder, 158
  lathe, 157
  milling machine, 158
Cutting time
  drill press, 160
  formulas for, 159
  lathe, 159
  milling machine, 160

D

Decimal degrees, 173
Decimal equivalent table, 64
Decimal fractions
  adding, 42
  changing common fractions to, 37-38
  changing to common fractions, 38-39
  division of, 49-50
  explanation of, 33-35
  mixed, 35
  multiplication of, 46-47
  subtraction of, 43
Decimal numbers
  changing binary numbers to, 266
  changing to binary numbers, 266
Decimal system, 264-265
Decimals
  combined operations of, 65-66
  non-terminating, 38
  rounding-off, 37
  terminal, 38
Dedendum, 164
Denominator, 8
  changing to least common denominators, 12
Diameter, 198
Diameter, outside, 164
Diametral pitch, 164
Direct proportions, 152-153
Dividing a line into equal parts, 213
Division
  algebraic operations, 117

Division (continued)
decimal fractions, 49-50
equation, solution by, 130-131
fractions, 24
mixed numbers, 24
multiplication, combining with, 29
signed numbers, 108
Dovetails, 235
Drill press
cutting speed, 158
cutting time, 160
revolution per minute, 158-159
surface speed, 158

E

English system, 96
Equations
addition, solution by, 129-130
checking the, 125
combined operations solution, 141-142
division solution by, 130-131
equality, expression of, 121
from word questions to, 122-125
multiplication solution, 135-136
powers solution, 137
rearranging formulas, 142-143
roots solution, 136-137
subtraction, solution by, 128-129
transposition, 129
the unknown quantity, 122
Equilateral triangle, 185-192
Equivalent, 9
Equivalent fractions, 12-13
Exponent, 53
Extremes, 147

F

Factors, 53, 112
Forced fit *See* Interference fit
Formula, 53
rearranging, 142-143
Fractions
adding, 13-14
changing into equivalent fractions with the least common denominator, 12-13
complex, 30
decimal, 33-35
Fractions (continued)
definitions of, 8
dividing, 24
fractional parts, 7-8
inverting, 24-25
multiplying, 20
reduction of, 9
subtracting, 16

G

Gage blocks, 78-79
Gear calculations, 165-166
Gears, 163
Geometric constructions
bisecting an angle, 212
parallel to a line, 211
perpendicular bisecting of a line, 210
perpendicular to a line at a given point on a line, 211
tangents to a circle from an outside point, 212
Geometric principles, 181-182
Geometry *See* Plane Geometry
Grinder
cutting speed, 158
revolution per minute, 158-159
surface speed, 158

H

Height gage, 84
Hypotenuse, 185, 217

I

Improper fraction, 8
changing, 9
Incremental dimensioning, 260-261
Index, 58
Inscribed angle, 198-204
Instruments, measuring
gage blocks, 78-79
limitations of, 69
micrometers, 89-92
precision, degree of, 69
steel rules, 76-78
vernier caliper, 82-84
vernier height gage, 84
vernier micrometer, 91-92
Intercepted arc, 204
Interference fit, 72

Interpolation, 219-220
Inverse proportions, 153-154
Inverting, 24
Involute curve, 163
Isosceles triangle, 185, 192
  applications, 234-235

L

Lathe
  cutting speed, 157
  cutting time, 159
  revolution per minute, 157-158
Least common denominator, 12
Like terms, 112
Lines
  oblique, 170
  parallel, 170
  perpendicular, 170
Literal factors, 112
Literal numbers, 99
Lowest terms, 9

M

Machine applications
  angle cuts, 235
  bevels, 234
  complex practical, 240-244
  distance between holes, v-slots, 234-235
  dovetails, 235
  isosceles triangle, 234-235
  tangent to circle, 235
  tapers, 234
  thread wire checking dimensions, 235
  v-blocks, 235
Mean dimension, 71
Means, 147
Measurement
  clearance, 71-72
  English system, 96
  gage blocks, 78-79
  metric system, 95-96
  micrometer, 89-92
  precision, degree of, 69-70
  steel rules, 76-78
  tolerance, 70-72
  vernier caliper, 82-84
  vernier height gage, 84
Measurement (continued)
  vernier micrometer, 91-92
Measuring instruments *See* Instruments, measuring
Meter, 94
Metric system, 69, 95-96
  changing units from English system to, 96
Metric units, 94-95
Milling machine
  cutting speed, 158
  cutting time, 160
  revolution per minute, 158-159
  surface speed, 158
Micrometer
  description, 89
  proper procedure in the use of, 92
  reading and setting the, 90
  vernier, 91-92
Millimeter, 94
Mixed decimals, 35
Mixed number, 8
  changing, 9
  division of, 25
  subtracting, 17
Multiplication
  algebraic operation, 114-115
  decimal fractions, 46-47
  division, combining with, 29
  fractions, 20-22
  mixed numbers, 20-22
  signed numbers, 107
  solution of equations by, 135

N

Negative number, 105
Numbers
  arithmetic, 99
  literal, 99
  precision degree of, 70
Number scale, 105-106
Numerator, 8
Numerical coefficient, 112
Numerical control machines
  introduction, 254-255
  points, location of, 254-255
  programming, 257-261

O

Oblique lines, 170
Oblique triangles
cosines, law of, 249-251
defined, 248
sines, law of, 243-249
Obtuse angle, 180
Opposite side, 217
Ordinate, 224
Origin, 224

P

Parallel lines, 170
Parallelogram, 193
Parallel to a line, 211
Parentheses
removal of, 119
use of, 54-55
Perfect squares, 60
Perpendicular bisector of a line, 210
Perpendicular lines, 170
Perpendicular to a line at a given point on the line, 211
pi, (π), 55
Pinion, 163
Pitch *See* Diametral pitch
Pitch circles, 164
Pitch diameter, 164
Plane geometry
angles, 170, 180-182
angular measure
arithmetic operations, 171-172
angular measure, units of, 170
axioms, 169
bisecting an angle, 212
circles, 198-201, 204-206
decimal degrees, 173
defined, 169
dividing a line into equal parts, 213
fundamentals of, 169-215
geometric principles, 181-182
lines, 170
parallel to a line, 211
perpendicular bisector of a line, 210
perpendicular to a line at a given point on the line, 211
Plane geometry (continued)
polygons, 193-194
Pythagorean Theorem, 192-193
tangents to a circle from an outside point, 212
triangles, 185-187, 190-193
Point-to-point programming, 257-261
Polygons, 193-194
hexagon, regular, 193
parallelogram, 193
rectangle, 193
regular, 193
square, 193
Positive number, 105
Powers
algebraic operations, 118
description of, 53-54
equation solution, 137
parentheses, use of, 54-55
signed numbers, 108
Precision, degree of
measuring instruments, 69
numbers, 70
Proportion
direct, 152-153
extreme, 147
inverse, 153-154
means, 147
Protractors
bevel, 177
simple, 176-177
vernier, 177-178
Pythagorean Theorem, 192-193

Q

Quadrants
defined, 225
determining angles and functions in, 225-226

R

Radical sign, 58
Radius, 198
Ratio
description of, 146-147
order of terms and reduction of, 147
Rectangle, 193
Regular hexagon, 193
Regular polygon, 193

Revolutions per minute
drill press, 158-159
grinder, 158-159
lathe, 157-158
milling machine, 158-159
Right angle, 180
Right angle trigonometry, 216-220
interpolation, 219-220
Right triangle, 185, 192-193
calculation of angles and sides, 228-231
ratio of sides, 216
Root circle, 164
Root diameter, 164
Roots
algebraic operations, 118-119
description of, 58-59
equations, solution by, 136-137
index, 58
radical sign, 58
signed numbers, 109
Rounding-off a decimal, 37
Rpm *See* Revolutions per minute

S

Scalene triangle, 185
Secant, 198, 217
Sector, 198
Segment, 198
Sides
adjacent, 217
hypotenuse, 217
opposite, 217
right triangle, 217
Signed numbers
addition, 106
defined, 105
division, 108
multiplication, 107
negative numbers, 105
operations using, 106-109
positive numbers, 105
powers, 108
subtraction, 107
Similar triangles, 190-191
Simple protractor, 176,177
Sine, 217
Sine bar, 233
Sine plate, 233
Sines, law of, 248-249
Spur gears, 163-164
pitch circles, 164
standard formulas, 164-165
Square, 193
Square roots, 58-61
computing, method of, 60-61
Steel rule
correct procedure in use of, 77-78
description, 76-77
Straight angle, 180
Subtraction
algebraic operations, 113-114
combining with addition, 28
decimal fractions, 43
equation, solution by, 128-129
fractions, 16
mixed numbers, 17
signed numbers, 107
Supplementary angles, 178
Surface speed
drill press, 158
grinder, 158
milling machine, 158
Symbolism
algebraic expression, 99-101
arithmetic numbers, 99
literal numbers, 99

T

Tangent, 198, 217
Tangents to a circle from an outside point, 212
Tapers, 154-155, 234
Term, 112
Terminating decimals, 38
Thread wire checking dimensions, 235
Tolerance, 70-72
bilateral, 70
unilateral, 70
Tooth thickness, 164
Transposition
addition, 130

Transposition (continued)
subtraction, 129
Transversal, 181
Triangles
base, 185
congruent, 190
corresponding parts of, 186-187
defined, 185
equilateral, 185, 192
isosceles, 185, 192
oblique, 249-251
right, 185, 192-193, 228-231
scalene, 185
similar, 190
Trigonometric functions
analysis of, 223-224
cosecant, 217
cosine, 217
cotangent, 217
quadrants, 225-226
secant, 217
sine, 217
table, 218-219
tangent, 217
Trigonometry
bevels, 234
Cartesian coordinate system, 224-225
complementary angles, function of, 204
cosines, law of, 249-251
defined, 216
interpolation, 219-220
Trigonometry (continued)
machine applications
complex practical, 240-244
simple practical, 233
oblique triangle, 249-251
right angle, 216-220
right triangle, 228-231
right triangle sides, 126-217
sine bar, 233
sine plate, 233
sines, law of, 248-249
tapers, 234
Two-axis machines, 257-260

U

Unilateral tolerance, 70
Unlike terms, 112

V

V-blocks, 235
Vernier caliper
description, 82
reading and setting a measurement, 83-84
Vernier height gage, 84
Vernier micrometer
description, 91
reading and setting the, 91-92
Vernier protractor, 177-178

W

Whole depth, 164
Whole numbers, 14
Working depth, 164
Wringing, 78